Effects of Blending on Distribution System Water Quality

Awwa Research Foundation

About the Awwa Research Foundation

The Awwa Research Foundation (AwwaRF) is a member-supported, international, nonprofit organization that sponsors research to enable water utilities, public health agencies, and other professionals to provide safe and affordable drinking water to consumers.

The Foundation's mission is to advance the science of water to improve the quality of life. To achieve this mission, the Foundation sponsors studies on all aspects of drinking water, including supply and resources, treatment, monitoring and analysis, distribution, management, and health effects. Funding for research is provided primarily by subscription payments from approximately 1,000 utilities, consulting firms, and manufacturers in North America and abroad. Additional funding comes from collaborative partnerships with other national and international organizations, allowing for resources to be leveraged, expertise to be shared, and broad-based knowledge to be developed and disseminated. Government funding serves as a third source of research dollars.

From its headquarters in Denver, Colorado, the Foundation's staff directs and supports the efforts of more than 800 volunteers who serve on the board of trustees and various committees. These volunteers represent many facets of the water industry, and contribute their expertise to select and monitor research studies that benefit the entire drinking water community.

The results of research are disseminated through a number of channels, including reports, the Web site, conferences, and periodicals.

For subscribers, the Foundation serves as a cooperative program in which water suppliers unite to pool their resources. By applying Foundation research findings, these water suppliers can save substantial costs and stay on the leading edge of drinking water science and technology. Since its inception, AwwaRF has supplied the water community with more than $300 million in applied research.

More information about the Foundation and how to become a subscriber is available on the Web at **www.awwarf.org**.

Effects of Blending on Distribution System Water Quality

Prepared by:
J. S. Taylor, J. D. Dietz, A. A. Randall, S.K. Hong, C. D. Norris, L. A. Mulford, J. M. Arevalo, S. Imran, M. Le Puil, S. Liu, I. Mutoti, J. Tang, W. Xiao, C. Cullen, R. Heaviside, A. Mehta, M. Patel, F. Vasquez, D. Webb
University of Central Florida
4000 Central Florida Blvd., Orlando, FL 32816

Jointly sponsored by:
Awwa Research Foundation
6666 West Quincy Avenue, Denver, CO 80235-3098
and
Tampa Bay Water
2535 Landmark Drive, Clearwater, FL 33761

Published by:

Awwa Research Foundation
Advancing the Science of Water®

DISCLAIMER

This study was jointly funded by the Awwa Research Foundation (AwwaRF) and Tampa Bay Water (TBW). AwwaRF and TBW assume no responsibility for the content of the research study reported in this publication or for the opinions or statements of fact expressed in the report. The mention of trade names for commercial products does not represent or imply the approval or endorsement of AwwaRF or TBW. This report is presented solely for informational purposes.

TABLE OF CONTENTS

LIST OF TABLES

LIST OF FIGURES

FOREWORD

The Awwa Research Foundation is a nonprofit corporation that is dedicated to the implementation of a research effort to help utilities respond to regulatory requirements and traditional high-priority concerns of the industry. The research agenda is developed through a process of consultation with subscribers and drinking water professionals. Under the umbrella of a Strategic Research Plan, the Research Advisory Council prioritizes the suggested projects based upon current and future needs, applicability, and past work; the recommendations are forwarded to the Board of Trustees for final selection. The foundation also sponsors research projects through the unsolicited proposal process; the Collaborative Research, Research Applications, and Tailored Collaboration programs; and various joint research efforts with organizations such as the U.S. Environmental Protection Agency, the U.S. Bureau of Reclamation, and the Association of California Water Agencies.

This publication is a result of one of these sponsored studies, and it is hoped that its findings will be applied in communities throughout the world. The following report serves not only as a means of communicating the results of the water industry's centralized research program but also as a tool to enlist the further support of the nonmember utilities and individuals.

Projects are managed closely from their inception to the final report by the foundation's staff and large cadre of volunteers who willingly contribute their time and expertise. The foundation serves a planning and management function and awards contracts to other institutions such as water utilities, universities, and engineering firms. The funding for this research effort comes primarily from the Subscription Program, through which water utilities subscribe to the research program and make an annual payment proportionate to the volume of water they deliver and consultants and manufacturers subscribe based on their annual billings. The program offers a cost-effective and fair method for finding research in the public interest.

A broad spectrum of water supply issues is addressed by the foundation's research agenda; resources, treatment and operations, distribution and storage, water quality and analysis, toxicology, economics, and management. The ultimate purpose of the coordinated effort is to assist water suppliers to provide the highest possible quality of water economically and reliably. The true benefits are realized when the results are implemented at the utility level. The foundation's trustees are pleased to offer this publication as a contribution toward that end.

Walter J. Bishop
Chair, Board of Trustees
Awwa Research Foundation

James F. Manwaring, P. E.
Executive Director
Awwa Research Foundation

ACKNOWLEDGMENTS

The University of Central Florida (UCF) and UCF project team wish to express their sincere gratitude to Tampa Bay Water (TBW); Hillsborough County, Fla.; Pasco County, Fla.; Pinellas County, Fla.; City of New Port Richey, Fla.; City of St. Petersburg, Fla.; and City of Tampa, Fla. which are the Member Governments of TBW and the Awwa Research Foundation (AwwaRF) for their support and funding for this project.

Special recognition and gratitude is given to the following:

- Christine Owen – Water Quality Assurance Officer for TBW and also the TCP project coordinator for her endless support and coordination of all UCF project related activities.
- Robert M. Powell – Pinellas County, Fla. Director of Laboratory and Pick Tally – Pinellas County, Fla. Director of Utilities for their support during the proposal and conduct of the project.
- Roy Martinez – AwwaRF Project Officer for his continuous support and assistance in proposing, conducting and reporting and all other activities associated with the project.
- Jonathan Clement, PhD, P.E. –Black Veatch Engineers; Koby Cohen, VP Quality Assurance Suburban Water Systems; Bruce Johnson, Assistant Director of Utilities, Tucson AZ; Michael R. Schock, Research Chemist USEPA for their technical review and contributions as AwwaRF Project Advisory Committee Members
- Tampa Bay Water Staff: TBW staff at the Cypress Creek Water Treatment Facility continuously supported this project through design, construction, permitting, operation and all phases and are recognized for their continuous contributions.
- Dennis A. Marshall and Oz Wiesner – Pinellas County, Fla. Operation Support for their continuous service in project construction, field analysis, quality assurance, water supply, field maintenance and field management of the TCP.
- Luke A. Mulford, PhD, P.E. – Hillsborough County Water Quality Manager for his contribution in the design, construction, conduct and continual support of the TCP.
- The contributions of Shi Baoyou and Young Cheol Chang are acknowledged for their work as doctoral fellows in this project
- Kevin Young – This report is dedicated to Kevin Young; a UCF Student who helped extensively in the field and laboratory and passed away this year.

EXECUTIVE SUMMARY

INTRODUCTION

Drinking water utilities in Florida and similar environments have historically relied on groundwater, but over-reliance on the source has led to the pursuit of other, multiple sources. Tampa Bay Water and the University of Central Florida sought, through the Tailored Collaboration project with AwwaRF described below, understanding of the issues involved when multiple-source waters are blended, particularly the impacts to distribution system water quality.

Tampa Bay Water (TBW) manages drinking water resources for six member governments (MGs) on the west coast of central Florida: the cities of New Port Richey, St. Petersburg, and Tampa, and Hillsborough, Pasco, and Pinellas counties. This region, like most of Florida, has historically utilized groundwater for its drinking water supply; however, adverse environmental impacts have mandated that alternatives (non-groundwater sources such as desalinated seawater and treated surface water) be used for future supply. Recently, TBW has brought on line two alternative supplies: a 66-million gallon per day (mgd) surface water treatment facility, which utilizes enhanced coagulation, ozone, and biological filtration; and a 25-mgd desalination plant. Finished water from these facilities are blended with finished groundwater and distributed to the MGs through a large regional transmission system, which runs through the three-county area. Blends of the various resources depend on the location of delivery and supply availability. Although the regional water supply contract defines "acceptable water quality" (more stringently than SDWA requires), the delivered water quality was not constant, and, prior to this study, the effects of blended waters on the regional distribution system water quality were unknown.

IRON RELEASE

Surface Characterization and Thermodynamic Modeling

Identification of the chemical composition of the corrosion scales and relating their solubility to the total and dissolved iron concentrations in the pilot distribution system was investigated as a potential approach to predicting iron release in drinking water distribution systems. X-ray Photoelectron Spectroscopy (XPS) and X-ray Diffraction (XRD) surface techniques were used to identify the chemical composition of the upper most and deeper scales layers formed on the internal pipe surfaces, respectively.

Geothite (α-FeOOH), lepidocrocite (γ-FeOOH), ferric oxide, (Fe_2O_3,) magnetite (Fe_3O_4), and siderite ($FeCO_3$) were identified as major corrosion products in the unlined and galvanized iron pipe. Additionally, ZnO, and $Zn_5(CO_3)_2(OH)_6$ were identified on the galvanized iron pipe surfaces, while very small quantities of precipitated $CaCO_3$ and Fe_2O_3 were observed on the PVC and lined cast iron pipes. The following corrosion product maturation sequence was suggested; free metal ions to amorphous products ($Fe(OH)_2$), $Fe(OH)_3$) and finally to crystalline products.

In contact with air, siderite, $FeCO_{3(s)}$ decomposes to $Fe_2O_{3(s)}$. Dissolved iron was found to be highly correlated (R^2 = 0.88) to $Fe_2O_{3(s)}$ in the unlined cast iron pipe, suggesting that $FeCO_3$ due to its high solubility may be controlling dissolved/total iron concentrations in drinking water distribution systems. PDS receiving 100% conventionally treated ground water exhibited the lowest amounts (intensity peaks) of $FeCO_{3(s)}$, indicating that high alkalinity water may suppress iron release.

Thermodynamic solubility models were evaluated for $FeCO_3$, $Fe(OH)_2$ and $Fe(OH)_3$. The $FeCO_{3(s)}$ model gives the closest prediction of all three models, while the $Fe(OH)_{2(s)}$ model grossly over-predicts and the $Fe(OH)_{3(s)}$ grossly under-predicts. All three thermodynamic models, however fail to predict satisfactorily actual pilot distribution system iron concentrations. This suggests that a major portion of the iron release occurred in the form of particulate material. The major feature of the siderite model was that it predicted decreased iron release with increased alkalinity and pH.

Response Characterization

Strong positive correlations were found between total iron concentrations, apparent color and turbidity. Therefore apparent color (or turbidity), which are easily monitored in the field, are used as surrogate measures of total iron. This finding is consistent with a determination that the majority of iron (and color) is associated with particulates.

Empirical Model for Prediction of Elevation of Apparent Color

Empirical models were developed to predict color release in the hybrid PDS (ES 1) and galvanized iron (ES 2) using 12 months of operating data.

$$\Delta Apparent\ Color(CPU) = 0.04775\ (Alkalinity)^{-0.912}\ (Sulfate)^{0.118} (Chloride)^{0.485} \times (Sodium)^{0.561} (DO)^{0.967} (Temperature)^{0.813} (HRT)^{0.836} \quad \text{(ES 1)}$$

$$\Delta Apparent\ Color(CPU) = 10.71\ (Alkalinity)^{-1.63}\ (Silica)^{-0.55}\ (Sulfate)^{1.12} \times (Temperature)^{1.06} \quad \text{(ES 2)}$$

The empirical model for hybrid PDSs suggests that color release decreases with increasing alkalinity and increases with increase in chlorides, sulfate, sodium, dissolved oxygen, temperature and hydraulic retention time. Increasing alkalinity would be beneficial in reducing the iron release from the intermediate corrosion scales due to the thermodynamic considerations of siderite solubility. The effect of calcium and pH on color release could not be confirmed as all the waters were stabilized with respect to calcium carbonate.

Because these empirical models are specific for the pipe geometry and hydraulic conditions for the PDS, direct transfer of measured water quality in the PDS (total iron, color, or turbidity) to predict full-scale system response is not appropriate without consideration of actual site-specific features of pipe geometry and hydraulics. An iron flux model was developed that permits determination of the rate of release of iron from the pipe surface as a function of water quality, pipe material, and hydraulics (Reynolds Number). The iron flux value may be coupled with knowledge of the pipe geometry (length and diameter) and flow rate to define the impairment in water quality associated with passage through a pipe network. This impairment in water quality has been expressed as an increase in apparent color during contact with unlined cast iron and/or galvanized pipe materials. The other pipe materials (PVC and lined cast iron) were not implicated in release of iron corrosion products.

The relative color release can be assessed for different sources (or blends) based on their individual water chemistry. Both the surface and desalination sources present a potential concern for release of corrosion products. This adverse effect is attributed to altered water chemistry in comparison to the groundwater source. Specific differences include a reduction in alkalinity (both surface and desalination sources), an increase in chloride (desalination), an increase in sodium (both sources, especially desalination), and an increase in sulfate (both sources, especially surface water).

For systems that operate under laminar flow conditions, the iron flux (K_m) may be considered to be independent of Reynolds Number. The resulting increase in total iron concentration (or apparent color, or turbidity) would be proportional to the hydraulic residence time or pipe length and inversely proportional to the pipe diameter as shown in equation (ES 3). In the limiting case of zero velocity, the potential elevation of iron concentration would be extreme. This result is consistent with observations that adverse water quality is often associated with dead ends or dead zones that experience extended hydraulic residence times.

$$C_{EFFLUENT} - C_{FEED} = \Delta C = \frac{4\,K_m\,HRT}{D} \qquad \text{(ES 3)}$$

For systems that operate at fully turbulent conditions, the iron flux (K_m) may be considered to be proportional to the Reynolds Number. For large Reynolds Number, the resulting increase in total iron concentration would be independent of Reynolds Number or pipe diameter, but proportional to pipe length. Due to the limited lengths of unlined cast iron and/or galvanized pipe that remain in the Member Governments' distribution systems, the adverse water quality impact of blending source waters may not be significant for turbulent conditions. A robust analysis of the water quality effects is feasible with the iron flux model. Examples are provided in this report to illustrate application of the iron flux model for varying water quality, pipe material, pipe geometry, and hydraulic conditions.

Chemical Mitigation Measures – Alkalinity Supplementation

Control of release of iron corrosion products, and the associated increase in apparent color, may be mitigated by supplementation of alkalinity. Two alternatives are identified for the alkalinity source: sodium bicarbonate ($NaHCO_3$) and lime/carbon dioxide ($Ca(OH)_2 + CO_2$). Based on the empirical model for unlined cast iron pipe, alkalinity supplementation with a sodium salt may be less effective than use of calcium salts.

COPPER CORROSION

Surface Characterization and Theoretical Model Development

SEM, XRD, and XPS were used to characterize the corrosion scale formed on surface of copper coupons. XRD technique was unable to detect copper(II) species on copper coupons even after six months exposure to drinking waters. However, copper(II) was consistently observed in the scale using XPS. For all samples taken at one, two, three and six months the major corrosion scale layer on the copper coupons consisted of cuprite (Cu_2O).

The surface structure of the scale varied with water chemistry. A high alkalinity and low chloride groundwater produced a more compact top scale layer than a low alkalinity and high chloride desalinated water. Tenorite (CuO) and cupric hydroxide ($Cu(OH)_2$) were the major copper(II) species on the very top of the copper corrosion layer. Copper was also present in the very top layer, which indicated a continuous disproportionation reaction of cuprite (Cu_2O). Cuprite (Cu_2O) variation in the top layer was dependent on incubation period. Cu(II) species were not dependent on the incubation period.

Alkalinity and pH were the major factors affecting dissolved copper release. The trend in total copper release with pH and alkalinity was consistent with the trend predicted by the $Cu(OH)_2$ solid phase equilibrium model. The theoretical approach was useful qualitatively, but was limited quantitatively possibly due to the effect of particulate copper and kinetic limitation.

Total copper release in the corrosion loops depended on alkalinity, pH, sulfate, silica and temperature. The release of total copper from home plumbing systems can be affected by blending surface, ground and desalinated waters because of water quality changes. Adverse effects on total copper release can be mitigated by controlling water chemistry of blended finished waters (primarily by controlling alkalinity and pH). The concentrations of calcium in the blended PDS waters did not affect total copper release. Dissolved oxygen and chlorine residual were not demonstrated to have a significant correlation on copper release for this dataset. Total copper release decreased as silica increased indicating naturally occurring silica functioned as a corrosion inhibitor.

Empirical models as shown in equations (ES 4) and (ES 5) were developed and verified that described total copper release as a function of water quality at the 50th and 90th percentiles.

$$[Cu]_{T-50} = (Temperature)^{0.72}(Alkalinity)^{0.73}(pH)^{-2.86}(Sulfate)^{0.10}(Silica)^{-0.22} \quad \text{(ES 4)}$$

$$[Cu]_{T-90} = 0.28 + (Temperature)^{0.72}(Alkalinity)^{0.73}(pH)^{-2.86}(Sulfate)^{0.10}(Silica)^{-0.22} \quad \text{(ES 5)}$$

Simulation of Copper Release in Pipe Distribution System and Copper Rule Compliance

Sensitivity analysis indicated the copper action level would be violated for some blends. The predicted violation was primarily due to high alkalinity. Figure ES 1 represents the simulation of copper release that would be experienced at the 90th percentile as a result of blending treated ground, surface, and desalinated waters with the typical water chemistry listed in Table ES 1. Copper release models can be used to proactively manage copper release in drinking water systems. These models can predict copper release using feed water chemistry.

Table ES 1
Water quality range used to predict ninetieth percentile copper concentrations

Parameter	Unit	Ground Water	RO Water	Surface Water
pH		7.6	8.3	8.3
Alkalinity	mg/L as $CaCO_3$	225	50	50
SO_4^{-2}	mg/L	10	30	180
Silicate	mg/L as SiO_2	20	20	10

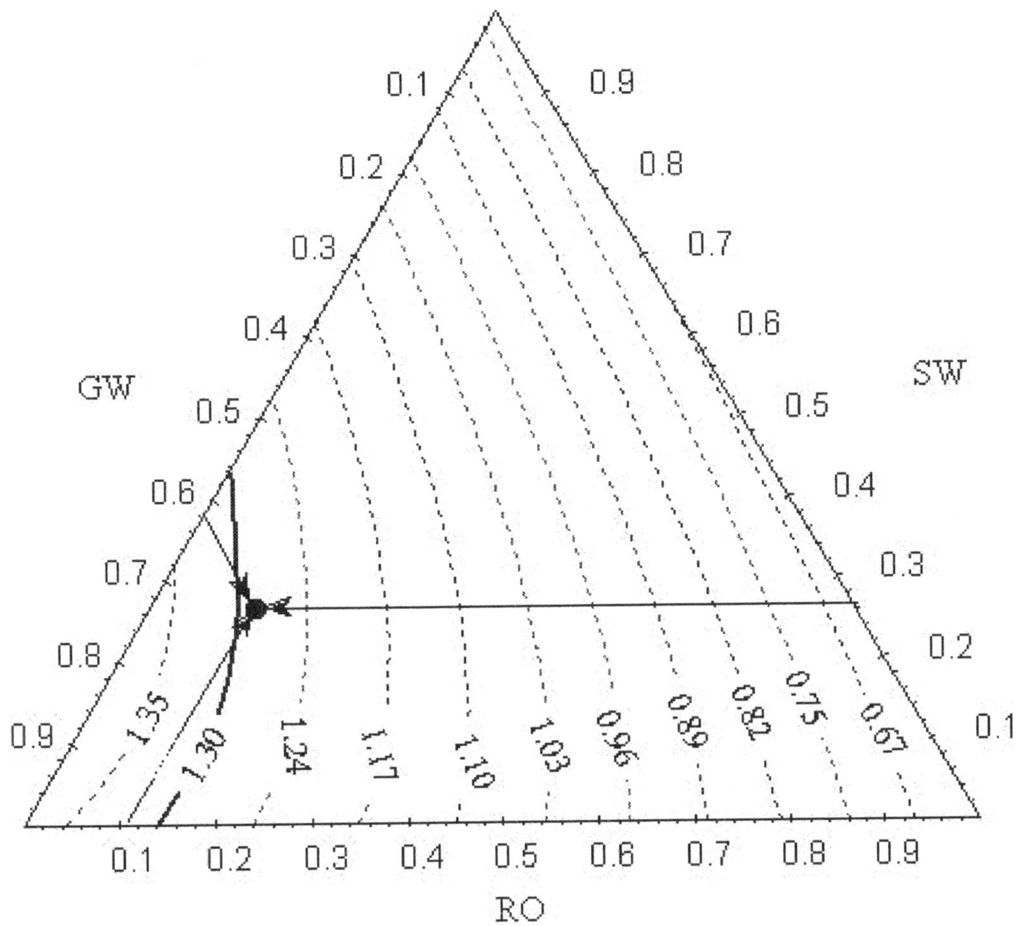

Figure ES 1 Ternary diagram showing predicted ninetieth percentile copper concentration for blends of finished ground, surface and desalinated water using non-linear model

LEAD CORROSION

The main crystalline compounds of corrosion products formed on the lead coupons consisted of lead dioxide (PbO_2), lead oxide (PbO), cerussite ($PbCO_3$), and hydrocerussite ($Pb_3(CO_3)_2(OH)_2$). Among these corrosion products, cerussite and hydrocerussite peaks were more pronounced. A large amount of amorphous phases of the above species was also observed. $Pb_3(CO_3)_2(OH)_2$ and $PbCO_3$ peak area percentage obtained from XPS analysis were directly correlated with lead release, which suggested $Pb_3(CO_3)_2(OH)_2$ may be the controlling solid phase for lead release.

Based on the results of surface characterization, theoretical hydrocerussite ($Pb_3(CO_3)_2(OH)_2$) model was developed to predict lead concentrations in water on the basis of the assumption that corrosion scales were in equilibrium with dissolved lead ions. The model showed that lead levels in water were primarily determined by pH and alkalinity. The complexes with sulfate and chloride also contribute to total lead concentration, but to a lesser degree.

The empirical models for total lead release at the 50th and 90th percentiles shown in equations (ES 6) and (ES 7) were developed and verified using project data. Total lead release increased as temperature, chloride, and alkalinity increased; and decreased as sulfate and pH for the conditions of this work. Calcium, dissolved oxygen, and chlorine residual did not significantly affect total lead release for the conditions of this work. Sensitivity analysis on the empirical model indicated that alkalinity, chloride, and sulfate were important (sensitive) parameters, with chloride being the most sensitive. Notably, higher temperature also increases lead release.

$$[Pb]_{T-50} = 1.027^{(T\text{-}25)}(Alkalinity)^{0.677}(pH)^{-2.726}(Chloride)^{1.462}(Sulfate)^{-0.228} \qquad \text{(ES 6)}$$

$$[Pb]_{T-50} = 8.96 + 1.027^{(T\text{-}25)}(Alkalinity)^{0.677}(pH)^{-2.726}(Chloride)^{1.46}(Sulfate)^{-0.228} \qquad \text{(ES 7)}$$

The 90th percentile of the empirical model for lead release over predicted lead in Member Government distribution systems probably

because the lead surface area in the pilot study was greater than in the member government households. The empirical model could be calibrated using a simple ratio of the actual and predicted lead as shown in the framed text.

$Pb_{90,Historical}$ = 5 μg/L
$Pb_{90,Historical,predicted}$ = 18.7 μg/L
Therefore correction factor (CF) = 5/18.7 = 0.27
Therefore,

$Pb_{90,actual} = 0.27 * Pb_{90,predicted}$

As for copper, a ternary diagram (Figure ES 2) was developed for the release of total lead as a function of source using the water quality presented in Table ES 1. The results presented indicate that low lead release occurs at higher percentage of surface water, while higher percentages of RO and groundwater had corresponding higher lead release.

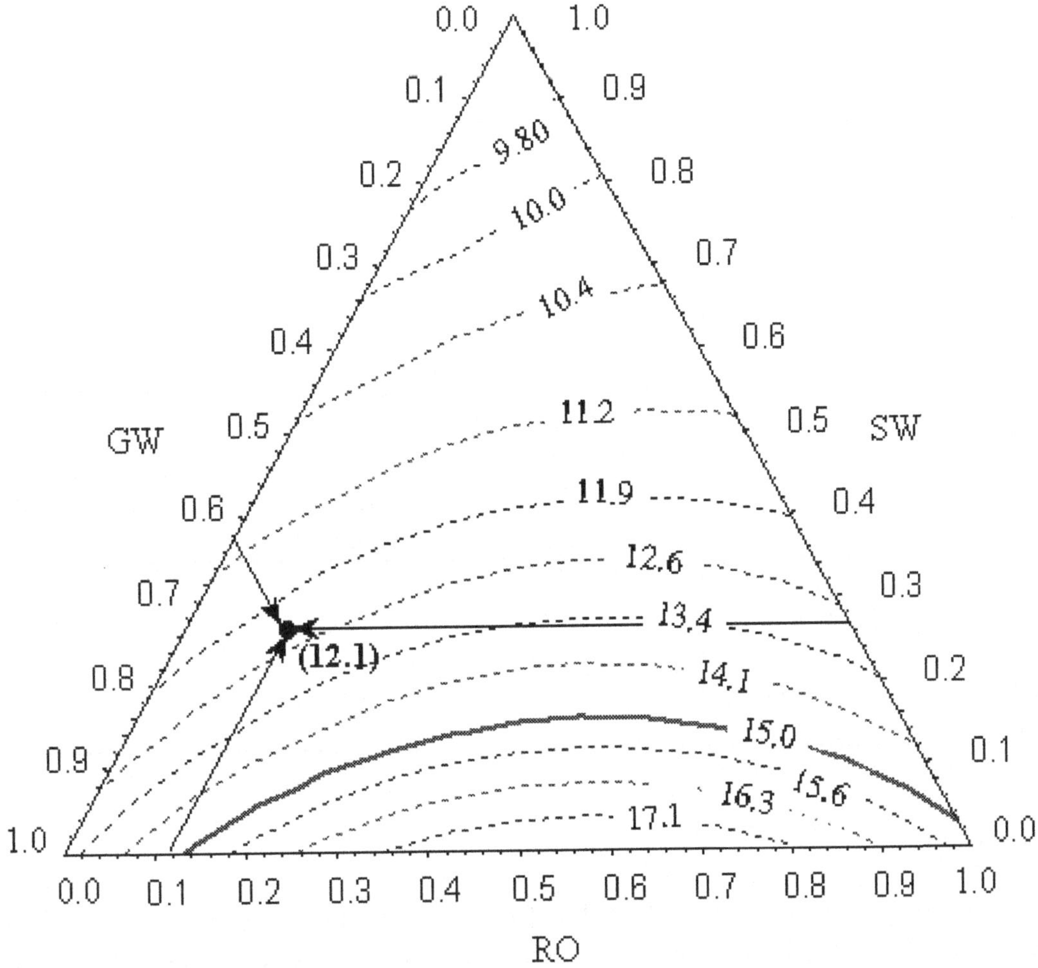

Figure ES 2 Ternary diagram showing predicted lead concentration for blends of finished ground, surface and desalinated water using non-linear model

AQUEOUS PHASE TRANSITION

Metal Transition in the Aqueous Phase

A model for transitional metal release following water quality change was developed and verified in the form shown in equation (ES 8).

$$M_t = M_{ss,1} + (M_{ss,2} - M_{ss,1})\left(1 - \exp\left(-k\theta^{(T-20)}t\right)\right) \quad \text{(ES 8)}$$

Where $M_{ss,2}$-$M_{ss,1}$ = driving force
$M_{ss,1}$ = steady-state predicted metal concentration before water quality change
$M_{ss,2}$ = steady-state predicted metal concentration after water quality change
M_t = transient lead concentration at 't' days after water quality change
T = bulk water temperature (°C)
θ = temperature correction factor
k = rate constant (day^{-1})
t = time (days) after water quality change

Nonlinear least-square estimates of the rate constant (k) and the temperature correction factors for the three different metals and four pipe materials are given in Table ES 2. The rate constant indicates the speed of response of the metal release.

Specifically, the time to reach a predicted steady state is shown in Table ES 2 to vary from 26 to 42 days, which means utilities have a significant period of time to take action to remediate any adverse effects of total metal release caused by water quality change. Examples of such actions are changing finished water quality, inhibitor addition or flushing.

Table ES 2
Nonlinear least squares estimate of the rate constant and temperature correction factor for all the pipe materials

Pipe material	Response Variable	Rate constant (k) (day^{-1})	Temperature correction factor (θ)	Time to reach predicted steady state release (days)
Copper pipe	Total copper mg/L	0.071	1.02	42
Lead coupon	Total lead μg/L	0.081	1.05	37
Cast iron pipe	Total iron mg/L	0.116	1.03	26
Galvanized pipe	Total iron mg/L	0.068	1.07	44

HPC Transition

An investigation was conducted on the effects of water quality change (phase change) on HPC transition in the presence of chloramine and free chlorine residuals. Conventionally treated ground water was switched with enhanced coagulated surface water and RO was switched with blended water so HPC transition in the presence of chloramines or free chlorine could be evaluated at identical PDSs and water qualities.

Log HPCs versus time are shown in Figure ES 3 and Figure ES 4 for free chlorine and chloramine residuals respectively. The transition date (water quality change) is shown by a vertical dotted line in Figure ES 3 and Figure ES 4. The sharp rise in HPCs after the transition data was irrespective of PDS residual concentration and form as the same magnitude of HPC change occurred in PDS 1-8 regardless of residual concentration and form. HPC trended downward at the end of the transition experiment in most of the PDSs.

Release of the HPCs was attributed to disruption of the biofilm, which could have been caused by physically disrupting the biofilm when changing the feed lines to the PDS or chemically disrupting the biofilm following the feed line switch.

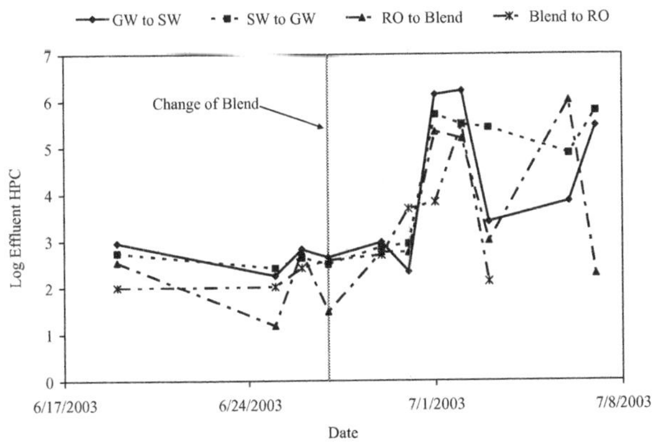

Figure ES 3 PDS effluent Log HPC by source switch as a function of time, for free chlorine residual

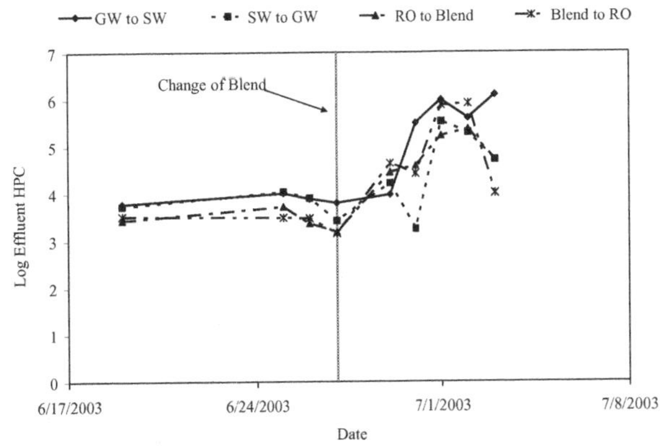

Figure ES 4 PDS effluent Log HPC by source switch as a function of time, for chloramine residual

HPC variations through PDS 1 through 8 were described as a function of temperature, time and residual for free chlorine and chloramines as shown in equations (ES9) and (ES10) using the data from the HPC transition experiments.

$$Log\left(HPC\right) = \frac{1.19^{(T-20)} \times 1.12^{t}}{1.51 \times Cl_2^{0.14}} \tag{ES9}$$

Where HPC = cfu/ml
T = temperature, °C
t = time, days
Cl_2 = free chlorine residual, mg/L as Cl_2

$$Log(HPC) = \frac{1.06^{(T-20)} \times 1.08^{t}}{0.42 \times NH_2Cl^{0.04}} \tag{ES10}$$

Where NH_2Cl = chloramine residual, mg/L as Cl_2

A three dimensional graph of actual and predicted Log HPCs normalized to 20 °C versus time and residual is shown in Figure ES 5. Log HPCs were normalized to a common temperature (20 °C) in order to remove the effects of temperature.

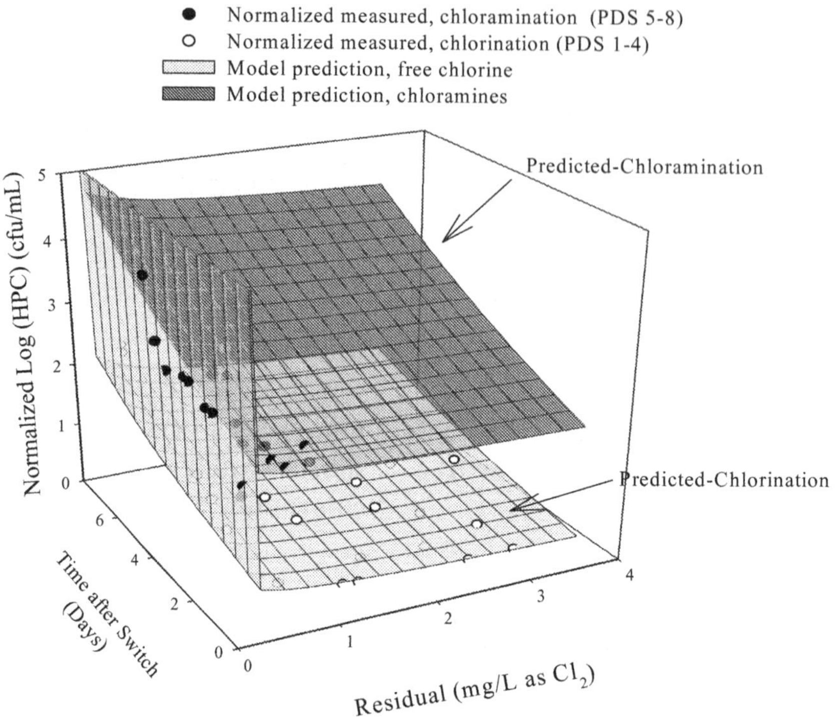

Figure ES 5 Actual and predicted HPCs versus residual and time following steady state water quality change

Both planes increase sharply at low residuals, which emphasize the importance of residual to microorganism growth in drinking water distribution systems. The chloramine plane is significantly above the free chlorine plane at all points except where both residuals approach zero. Free chlorine more effectively retarded HPC proliferation in the PDSs.

CHLORINE RESIDUAL MODELING

Chlorine residuals decayed at different rates in the pilot distribution system depending on pipe material, temperature, hydraulics and geometry. Lower rates of chlorine consumption occurred in PVC and lined cast iron pipes while chlorine concentration declined rapidly in the unlined cast iron and galvanized iron pipes. A general empirical model was developed to predict free chlorine and chloramine residual in distribution system as a function of time, water quality, pipe material and pipe geometry.

Free chlorine and chloramine decay was described using the general model shown in equation (ES 11). The coefficients and exponents for model use are shown in Table ES 3. The model describes residual decay using a first order model based on initial residual dose, wall reaction constants, bulk water reaction constant, pipe material, pipe diameters, temperature and time. Chlorine residual decreased in PDSs as time, UV-254 (organic content) and temperature increased for all the pipe materials. The models for residual were developed and verified using project data.

$$Cl_{2(t)} = Cl_{2(0)} \times \exp\left[-\left(K_b \times UV254 + \frac{K_W}{D_p}\right) \times A^{(Temp-20)}\right] \times time \quad \text{(ES 11)}$$

Where K_b = bulk decay constant (cm/hr)
K_W = wall decay constant (in/hr)
$Cl_{2(t)}$ = chlorine concentration at time t (mg/L as Cl_2)
$Cl_{2(0)}$ = initial chlorine concentration (mg/L as Cl_2)
A = temperature correction coefficient
T = temperature in degrees C
t = time expressed in hours
Dp = pipe diameter (inches)

Table ES 3
Free chlorine and chloramine dissipation model coefficients and R^2

Material	Free Chlorine			Chloramine		
	Parameter	Value	R^2	Parameter	Value	R^2
PVC	K_b	0.048	0.97	K_b	0.020	0.90
	K_W	0.007		K_W	0.007	
	A	1.15		A	1.16	
LCI	K_b	0.084	0.93	K_b	0.017	0.90
	K_W	0.007		K_W	0.011	
	A	1.18		A	1.16	
UCI	K_b	0.172	0.97	K_b	0.147	0.87
	K_W	0.063		K_W	0.084	
	A	1.13		A	1.11	
G	K_b	0.614	0.93	K_b	0.319	0.87
	K_W	0.054		K_W	0.073	
	A	1.04		A	1.12	

The rate of residual dissipation is expressed using two constants; one for the bulk liquid reactions, K_b, and the other for pipe wall reactions, K_W. The bulk rate constant is dependent on UV-254, temperature and time and the wall rate constant is dependent on temperature, pipe diameter and time. K_b was lower for lined cast iron and PVC was lower than unlined cast iron and galvanized iron. K_W was also lower for lined pipe materials (PVC and lined cast iron pipes) K_W for unlined cast iron and galvanized steel pipes was 7 to 10 times larger than PVC and lined cast iron pipes.

The rate of residual dissipation increases as temperature and organic content of water increased. The effect of temperature on residual dissipation can make residual maintenance difficult for any disinfectant during summer.

BIOLOGICAL STABILITY

Biostability is a relatively new concept to drinking water that involves interaction between the bulk water and biofilms in the distribution systems. Biostability is an assessment of overall distribution system quality with respect to microorganisms. Organic carbon is considered to be the limiting nutrient with respect to heterotrophic biostability in most drinking water distribution systems. The organic carbon that drives heterotrophic growth is referred to as biodegradable organic matter (BOM). BOM is typically quantified as assimilable organic carbon (AOC) or biodegradable organic carbon (BDOC). Biostability has been defined as AOCs of 10 μg/L as C or less without a residual and AOCs of 100 μg/L as C or less in the presence of 0.5 mg/L free chlorine residual or 1 mg/L as Cl_2 of a chloramine residual based on the occurrence of total coliforms in drinking water distributions systems.

Bulk Biostability

No total coliforms were ever found in the PDS in more than three hundred samples over 1.5 years of operation. Hence, bulk biostability was assessed by HPC population. A general assessment of biostability as determined by HPC populations with respect to residual is shown in Figure ES 6.

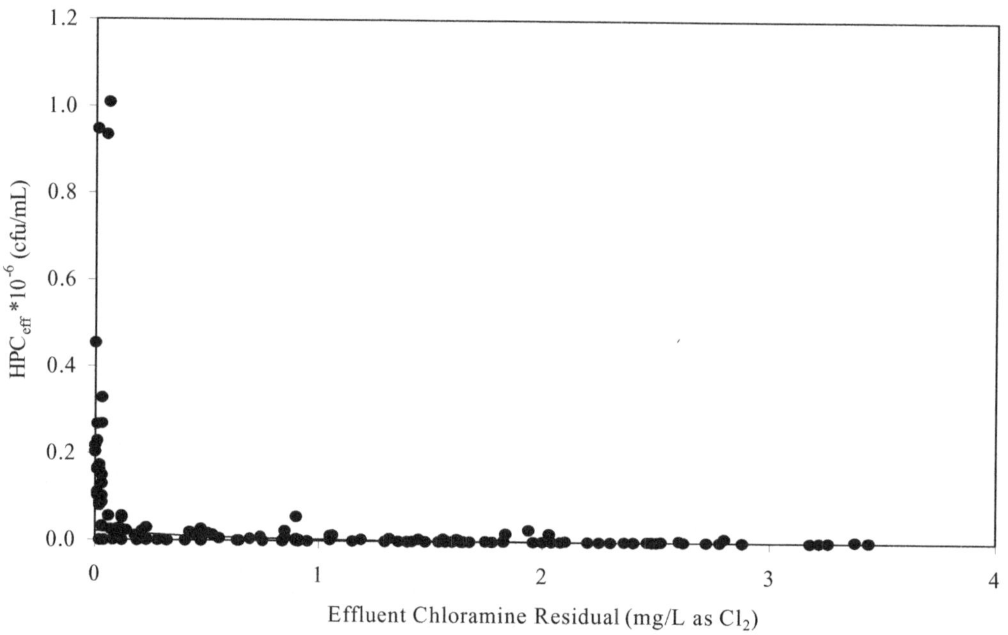

Figure ES 6 Effect of chloramine residual on effluent HPC

HPC populations dramatically increase when the PDS effluent residual is less than 0.1 mg/L as Cl_2, which marks the beginning of the zone of biological instability as defined by excessive HPC populations. Transition begins at chloramine residuals of 0.2 to 0.3 mg/L as Cl_2. The transition zone and the zone of biological instability were defined as beginning at 1,000 cfu/mL and 10,000 cfu/mL HPCs, respectively.

HPC Modeling

The predicted and actual Log HPCs versus residual and AOC are shown in Figure ES 7, which illustrate the impact of AOC and residual for this model. Residual impacts the predicted HPC population significantly more than AOC, which was practically and statistically insignificant in this work. The predicted difference in HPC populations at expected AOCs at the average residual is about 10% of the HPC population, which is not significant.

An empirical model was developed that described Log HPC as a function of temperature and effluent residual that was specific for each pipe material as shown in equation (ES 12). AOC, DOC or other inorganic parameters were not significant to Log HPC in this work.

$$LogHPC = \left(2.6407H + 2.5964UCI + 3.3322LCI + 2.9237PVC + 2.9111G\right) \times \frac{1.029^{(Temp-20)}}{(0.0310 + T\,Cl_{2-eff})^{0.11476}} \qquad \text{(ES 12)}$$

where: T = Temperature, °C

$T\,Cl_{2\text{-eff}}$ = Effluent chloramine residual, mg/L as Cl_2

A three dimensional plot of predicted and actual Log HPC versus chloramine residual and temperature using equation (ES 12) is shown in Figure ES 8. The dark curves within the net are curves for constant Log HPC 3 and 4; which defined the zones of biological stability.. Both residual and temperature must be considered simultaneously to determine biostability. Biostability increases as temperature increases and residual decreases, hence increased residual maintenance must be done in the distribution during warmer weather and at high demand locations.

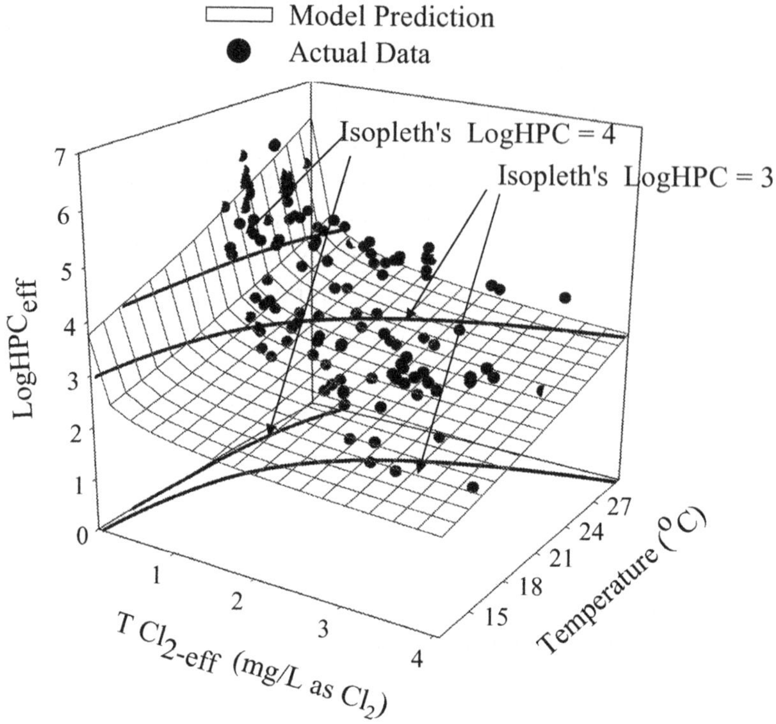

Figure ES 7 Predicted and actual Log HPCs versus chloramine residual and temperature for the hybrid PDSs

Given that utilities can do nothing to control temperature in a distribution system, residual is the only practically significant parameter on HPCs and requires more active management during high temperature periods. The Log HPC model shown in equation (ES 12) can be proactively utilized to anticipate biological instability by incorporating the chlorine residual model shown in equation (ES 11) into equation (ES 12).

A strong correlation was obtained between effluent DO and chloramine residual as shown in Figure ES 8. Unfortunately, the strong correlation between DO and residual kept both from being significant in all regression analyses as they were confounding. However as opposed to residual, DO depletion is a symptom and not a cause of biological instability, but can be used to track microbiological activity in the distribution system.

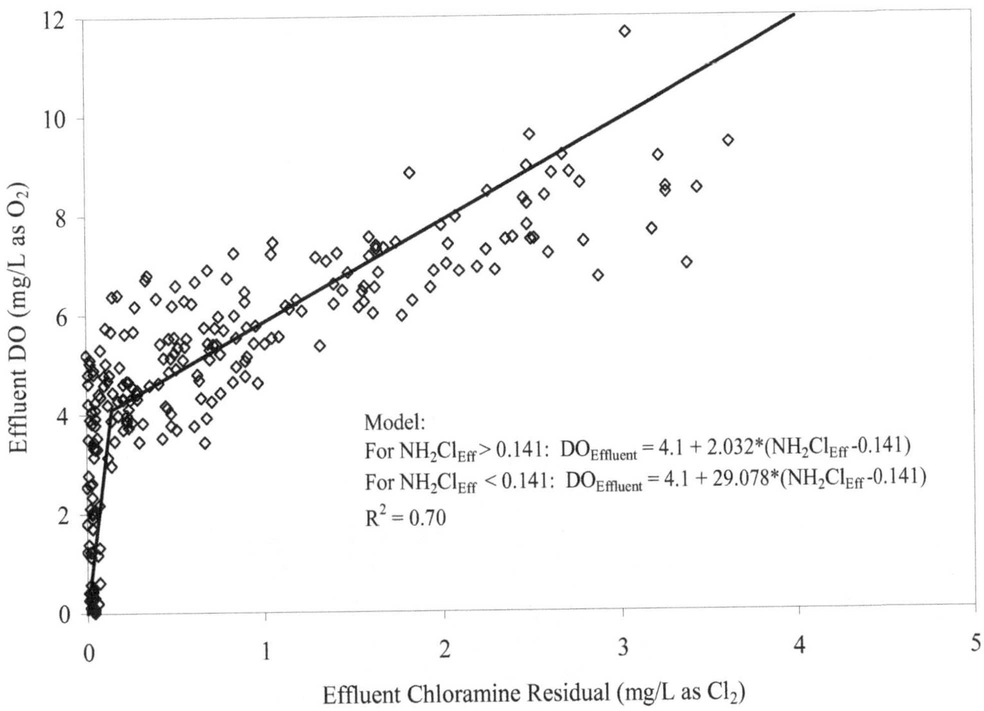

Figure ES 8 Correlation of effluent chloramine residual with effluent DO

Biofilm Biostability

Biofilms are always present, can never practically be removed and represent reservoirs for microorganisms in drinking water distribution systems. Hence, utilities have to successfully co-exist with biofilms. Distribution system biofilms should be kept at a steady-state in equilibrium with a strong residual. In such an environment, the bulk residual would be able to deactivate almost all of the microorganisms as they are released from the biofilm. In the opinion of the research team, there is little residual in a biofilm. Residuals enter biofilms by diffusion and dissipate via oxidation reactions with viable organisms and non-viable material that can function as an electron donor. Strong residuals can reduce "or beat back" biofilms somewhat but can not eliminate biofilms because the non-viable growth products of microorganism build a solid barrier that protects them from residuals. If a residual is lost, biofilm growth, like bulk microorganism growth, will increase and a non-desirable environment will develop.

Biofilm Modeling

An empirical model for biofilm as measured by Log PEPA was developed and is shown in equation (ES 13).

$$Log\ PEPA = (8.6966UCI + 8.3791LCI + 7.7651PVC + 8.3993G) \times \frac{1.0232^{(Temp-20)}}{T\ Cl_{2-eff}^{\ 0.0112}(LogHPC)^{0.0292}\ HRT^{0.0271}} \qquad \text{(ES 13)}$$

PEPA increased as temperature increases and decreased as residual increased, which is intuitively rational. However, PEPA increased as HPCs and HRT decreased, which may not be intuitively rational and it is hypothesized that this could be due to an equilibrium relationship between biofilms and bulk HPCs. As shown in Figure ES 9, chloramine residual has little impact on the magnitude of estimated log PEPA; however, temperature has a significant impact on log PEPA.

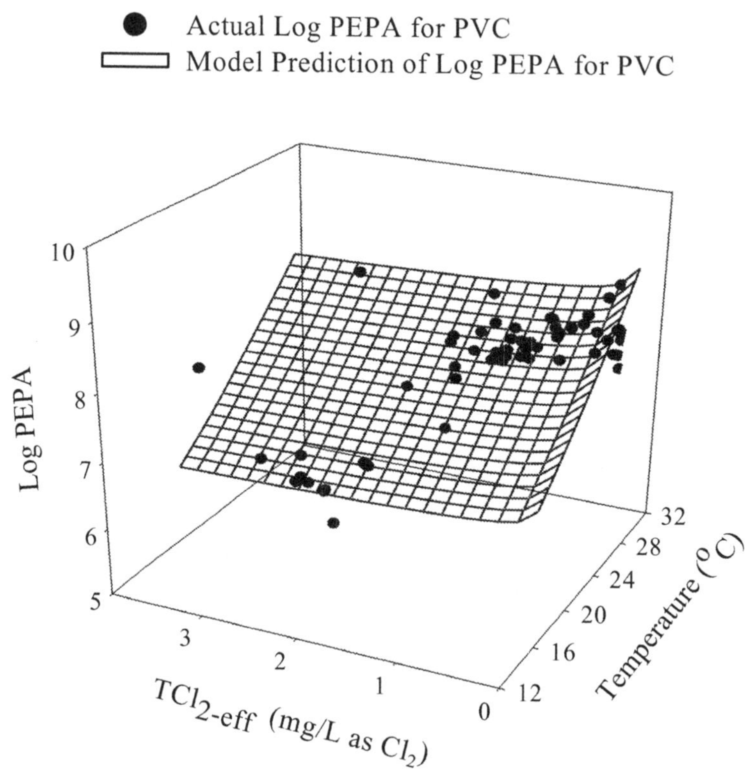

Figure ES 9 Predicted and normalized Log PEPA to 2 days HRT and 4 Log HPC versus temperature and T $Cl_{2\text{-eff}}$

The biofilm density hardly changes from 7 logs beginning at near zero residual to a chloramine residual of 4 mg/L as Cl_2; whereas, the biofilms density increases from approximately 7 logs to more than 9 logs from 15 °C to 30 °C. The impacts of HRT and log HPC_{eff} are also slight relative to temperature.

Process Biostability

As shown in Table ES 4, the assessment of process biostability was done by comparing finished water bulk stream measures of AOCs, BDOCs and HPCs and finished water biofilm measures of PEPA relative to process and pipe material.

Table ES 4
Process biostability parameters

Parameter	Units	G1	G2	G3	G4	S1	S2	RO
Log HPC	Log cfu/mL	4.2	3.7	3.3	3.8	3.7	3.4	2.7
AOC	μg/L C	110	89	113	57	96	55	43
BDOC	mg/L C	1.0	0.6	0.3	0.3	0.4	0.5	0.3
Log PEPA	log cells/cm^2	9.1	9.0	8.9	8.7	8.8	8.7	9.1

Although arbitrary, some quantitative assessment of the water treatment processes can be made in regard to biological quality of the finished water. As shown in Table ES 5, the processes were ranked by relative position among processes based on lowest biological parameter. For example, RO received 1 point for have the lowest average Log HPC, 1 point for having the lowest average AOC, 1 point for having the lowest average BDOC and 6 points for having the second highest average BDOC, which summed to 9 as shown in the weight column of Table ES 5. The data in Table ES 5 indicates that biological water quality in the finished water increases as degree of treatment increases.

Biofilm cell density (PEPA) varied by material, was greatest in the unlined metal galvanized steel and unlined cast iron coupons and least in the PVC and lined cast iron coupons. The order and average cell densities by material were PVC ($10^{8.37}$) < LCI ($10^{8.83}$) < UCI ($10^{9.01}$) < G ($10^{9.16}$). Consequently, the biofilm data gather in this work indicates replacing unlined metal pipes with lined pipes will lower the biofilm cell density in the drinking water distribution system.

Table ES 5
Process ranking by relative position based on lowest biological parameter

Points for Position	HPC	AOC	BDOC	PEPA	Rank Order	Weight
1	RO	RO	RO	G4	RO	9
2	G3	S2	G4	S2		
3	S2	G4	G3	S1	G4, S2, S1	12, 12, 12
4	G2	G2	S1	G3		
5	S1	S1	S2	G2	G3	16
6	G4	G1	G2	RO	G2	19
7	G1	G3	G1	G1	G1	27

NITRIFICATION

Episode

An unplanned nitrification episode occurred during this work. Nitrification was observed to proceed systematically as indicated by the variation of nitrites, nitrates, pH, DO and ammonia-N with residence time in the pilot distribution systems. Biochemical nitrification was associated primarily with loss of residual, which was followed by the stoichiometrically expected variation in water quality and is most likely to occur at temperatures common to spring and summer when residual dissipation is greatest. The average nitrogen profiles in the hybrid PDSs are shown in Figure ES 10, which exemplify nitrification. Nitrification was not specific to any source water in the PDSs nor was any source water protected from nitrification.

Controlled Investigation

The reported NO_X-N concentrations include the actual nitrite and nitrate concentration. The NO_X-N concentrations were about 137% of nitrite concentrations and so the nitrate concentrations were about 37% of the nitrite concentrations. Higher ammonia in the influent resulted in higher nitrite in the effluent, because there was more available food for AOBs that could be converted to nitrite and nitrate (Figure ES 11). Higher nitrite was associated with lower Cl_2:NH_3 ratio (Figure ES 12). Lower Cl_2:NH_3 ratio means higher free ammonia or lower chlorine residual or both. The former contributes more food to the system and the latter means less capacity to inhibit the nitrifier growth. In support of this, the nitrification potential was found to be negatively related to Cl_2:NH_3 ratio.

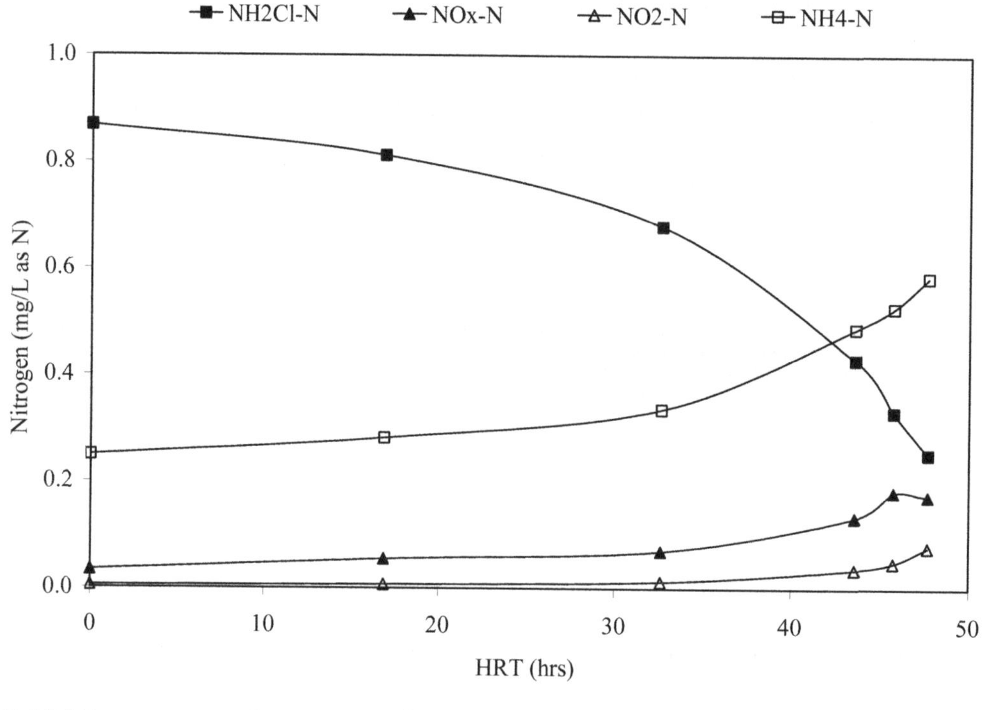

Figure ES 10 The average nitrogen profiles in the hybrid PDSs.

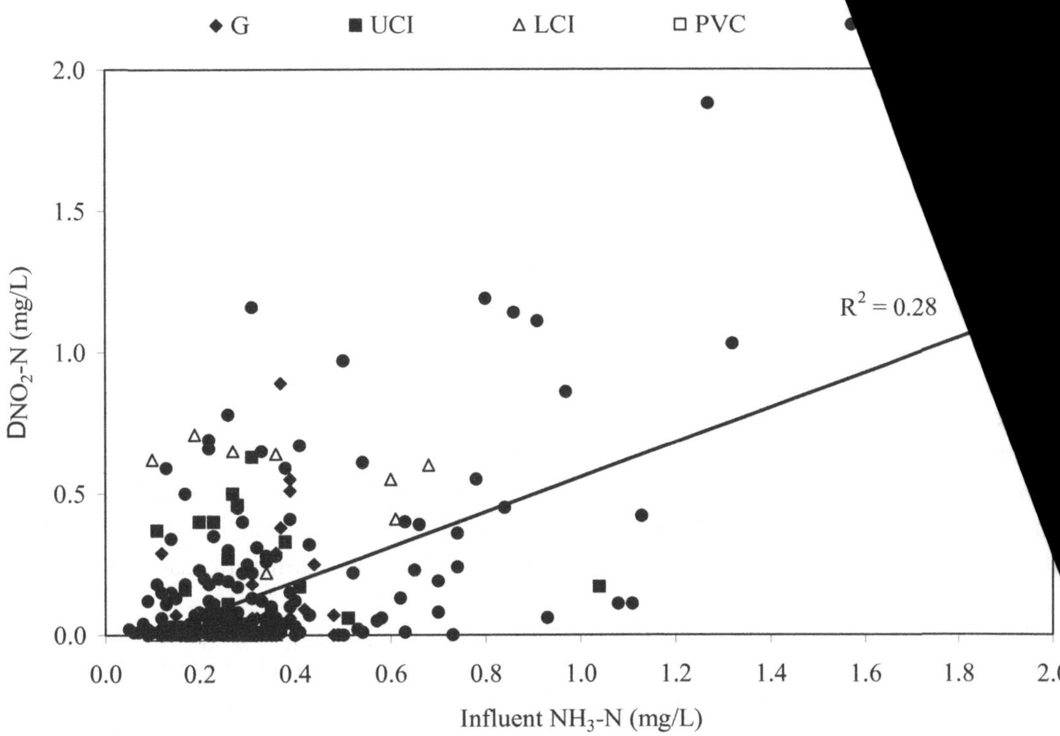

Figure ES 11 Relationship of influent ammonia and NO_2-N production

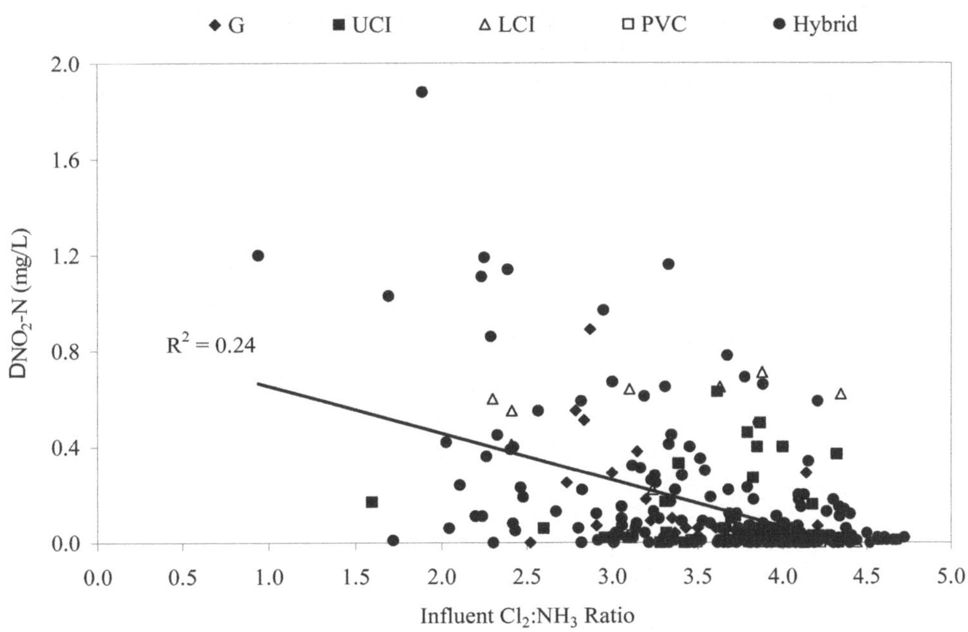

Figure ES 12 Relationship of nitrite and influent Cl_2:NH_3 ratio

ımmonia loss, nitrite production and nitrate production 14 are shown in Figure ES 13 for flush frequency and ual loss during flush is shown in the upper right hand ss ranges from 2.9 to 3.3 mg/L as Cl_2 isopleths and show oramine concentration in the flush decreased and as flush g more frequently and with a higher strength chloramine during normal operation, which indicates the PDS demand cted by flush frequency and strength. The residual loss rength (chloramine concentration), and increased with flush

e burn was instigated from 12/15/02 to 12/23/02 with 5.0 mg/L ation was restored on 12/23/02. As shown in Figure ES 14, nated by free chlorine. Low nitrite concentrations were found in after May 1, 2003 with the highest concentration of nitrites at 0.2 e 27, 2003. However, nitrite did not appear in the PDS effluent until h corresponds to a 15 to 30 °C temperature increase and three to four trification.

dual threshold for nitrification was about 1.0 mg/L. Effluent nitrite ations and there were no signs of nitrification when the effluent residual (Figure ES 15). The residual of chloramine acted as a biocide, and as ure ES 16, AOBs were negatively correlated to residual. The AOB ncreased exponentially as chloramines approached zero. When residuals were /L as Cl_2, the AOB growth was significantly inhibited. This is very similar to the relationship between HPCs and chloramine residual shown in the biostability Higher residuals certainly reduced if not eliminated nitrification. As shown in Figure , the highest AOBs occurred in sources that had the least residuals.

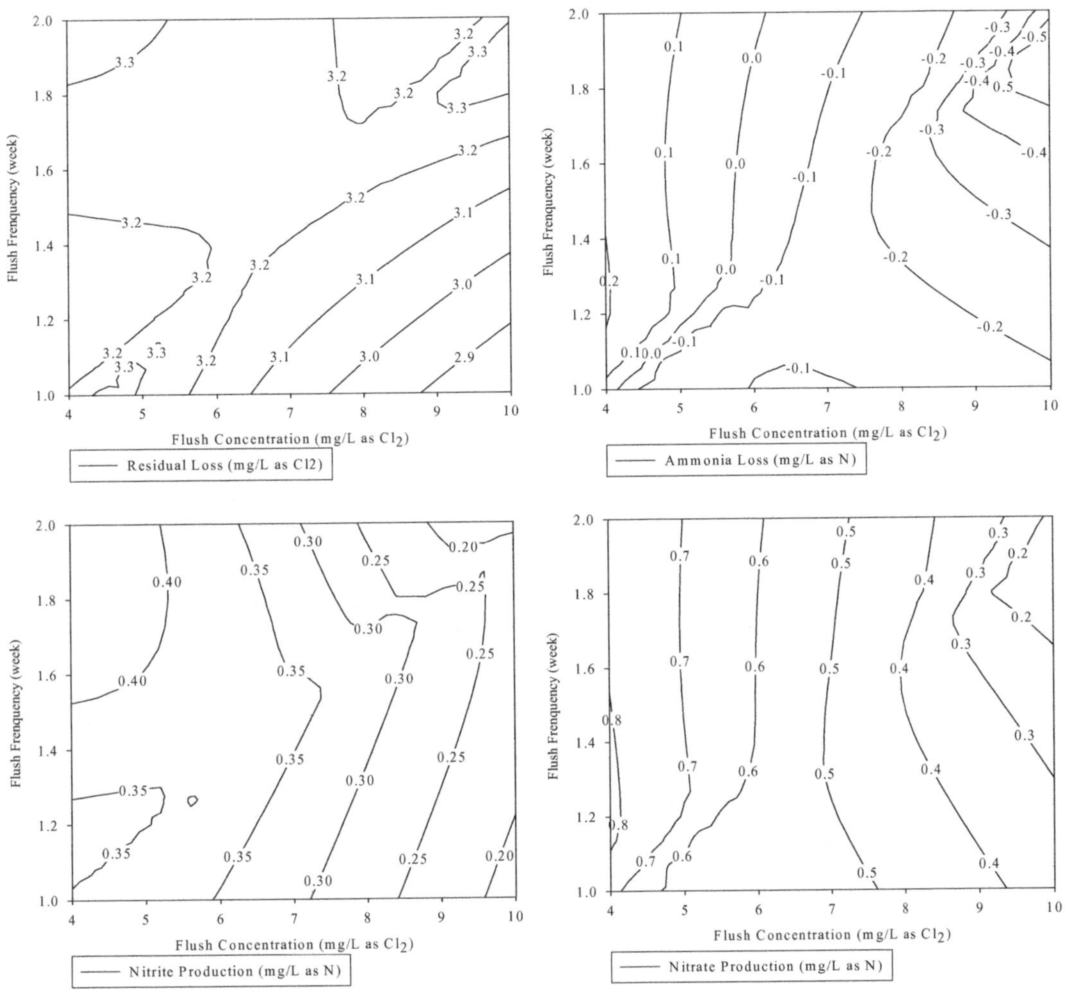

Figure ES 13 The effect of flush residual concentration and frequency on residual loss, ammonia loss, nitrite production, and nitrate production

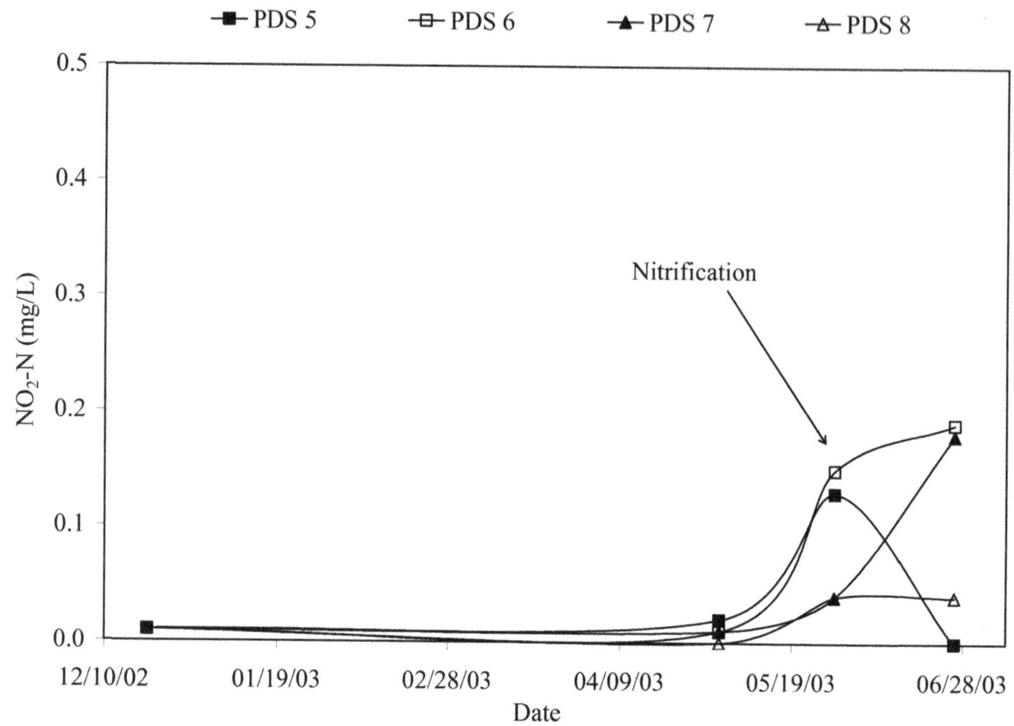

Figure ES 14 NO_2-N trend after free chlorination (PDS 5-8)

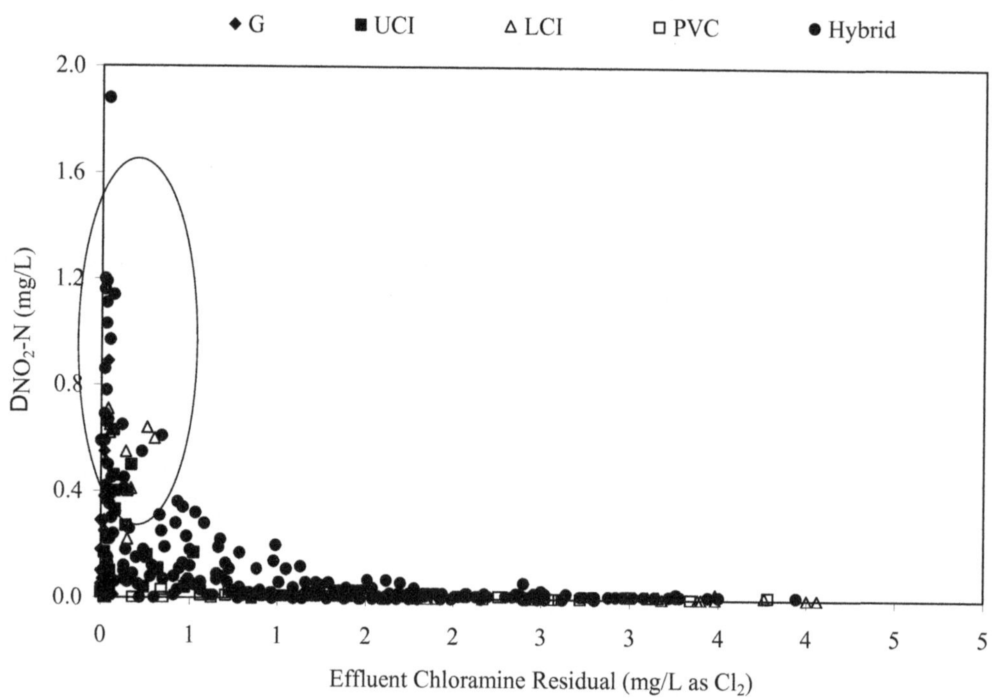

Figure ES 15 Relationship of nitrite production and effluent chloramine residual

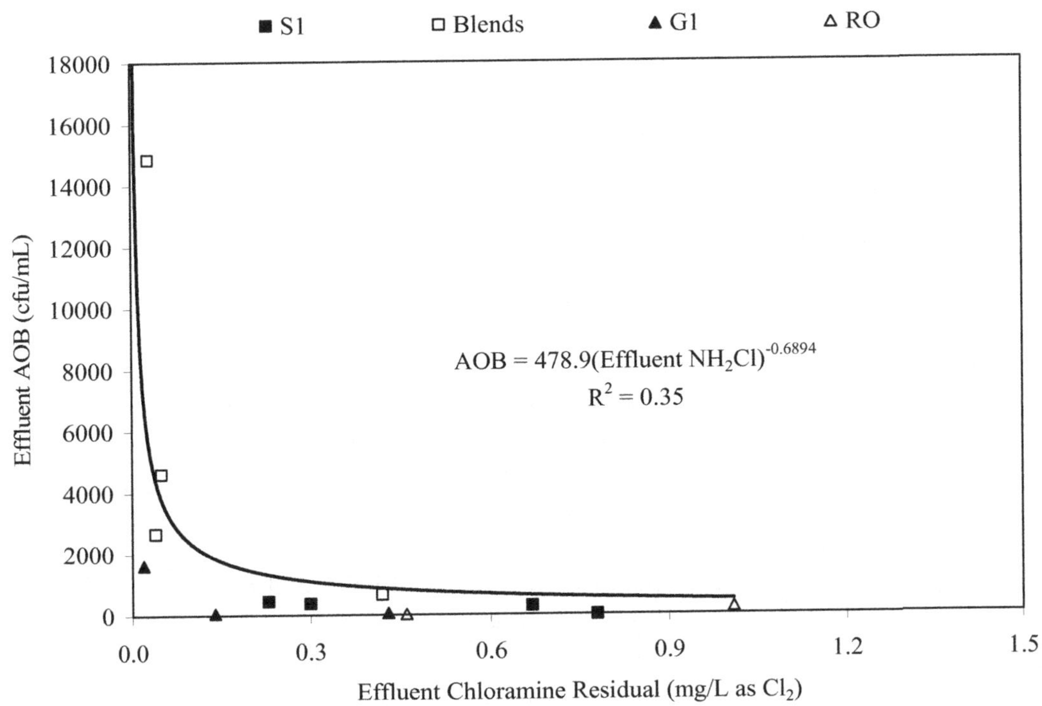

Figure ES 16 Relationship of effluent chloramine residual and AOB in the bulk water

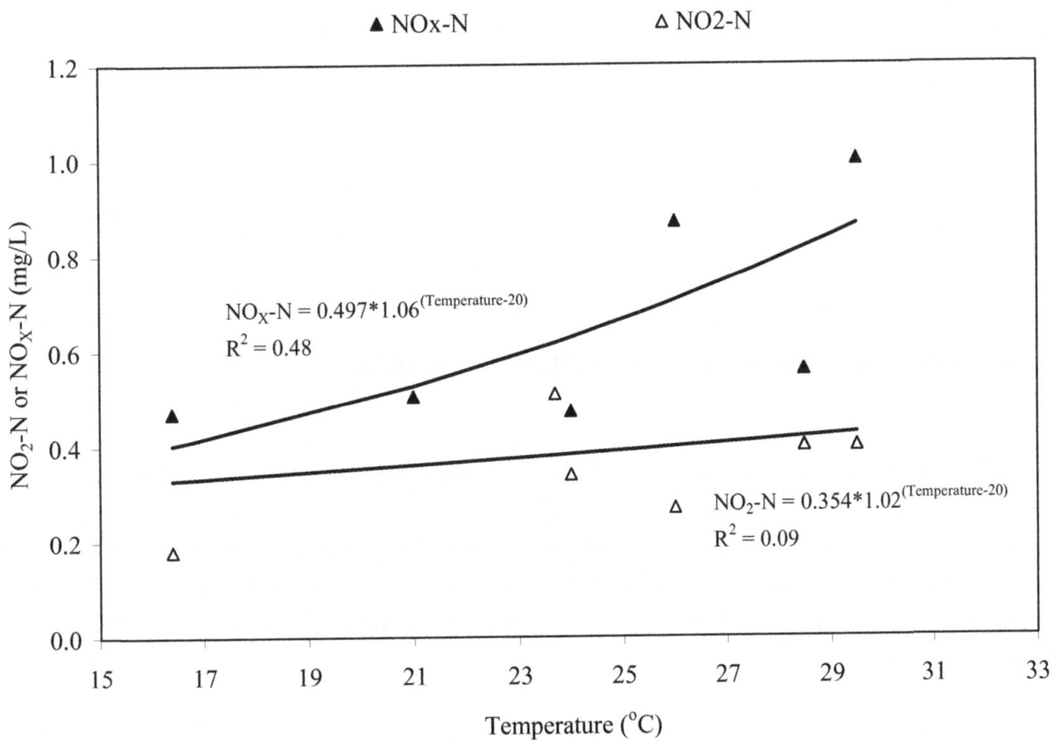

Figure ES 17 Relationship of effluent nitrite and NO_X-N with temperature in the bulk water

Figure ES 17 illustrates the influence of temperature on nitrite. The data utilized was collected from 09/16/02 to 12/14/02. During this period, water chemistry did not vary and provided an unbiased observation of temperature effects on nitrification. Nitrite correlated well with temperature. Hence, nitrification occurred at high temperatures and increased with temperature.

The relationship of HPC and AOB proliferation in the biofilm is presented in Figure ES 18. Similar to the bulk water, the correlation was significant. The highest AOB growth occurred on the G material, which was up to 3.8×10^5 cfu/cm^2, while the AOB density on the other materials was less than 1×10^5 cfu/cm^2. The heterotrophic growth on the G material was also the highest.

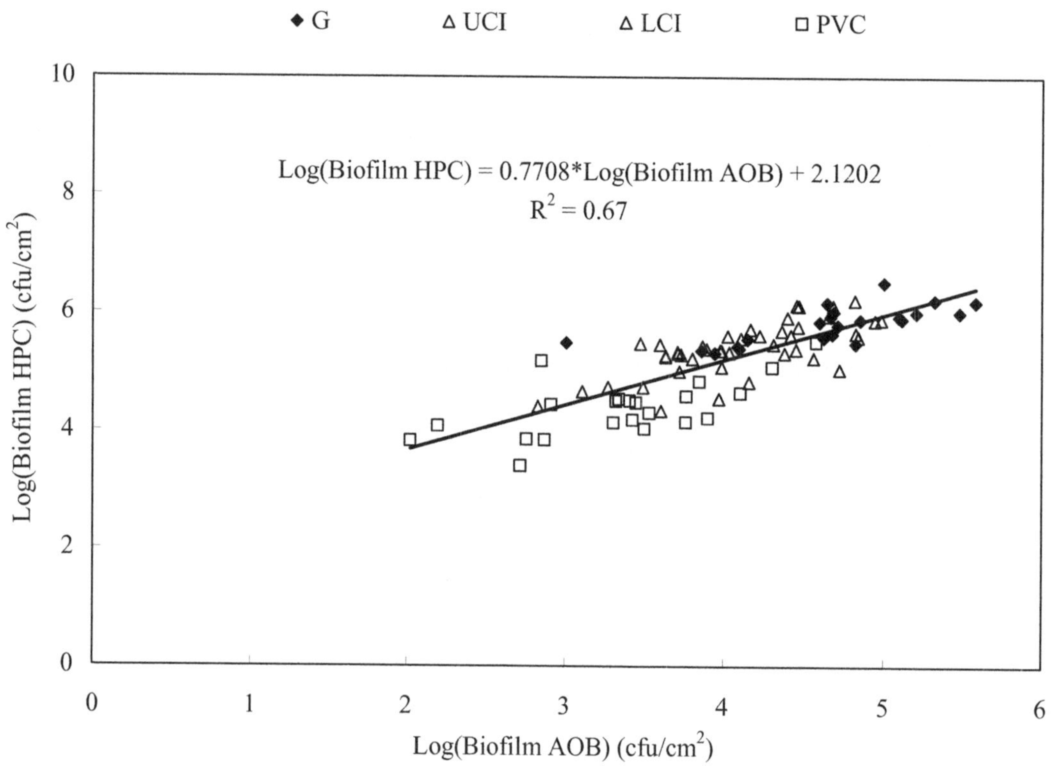

Figure ES 18 Relationship of HPC and AOB in the biofilm

A statistical model was developed to determine the significance of independent water quality variables to nitrite formation in the PDSs. The data set from the nitrification episode and the controlled nitrification study was used for model development. The statistical assessment provides insight into mechanisms involving nitrification, which were used in development of the theoretical models and provided a means of predicting nitrification. The models are shown in equations (ES 14) and (ES 15).

$$NO_2\text{ - }N = \frac{1.06^{(T\text{-}20)}(0.160 \times H + 0.186 \times UCI + 0.169 \times LCI + 0.037 \times G)}{\left(NH_3\text{-}N\right)^{0.310}(DO)^{0.315}\left(\left(NH_2Cl\right)^{0.722} + 0.188\right)} \qquad \text{(ES 14)}$$

where: NO_2-N = Nitrite N, mg/L as N
T = Temperature, °C
H = Hybrid PDS, (0,1)
UCI = Unlined Cast Iron PDS, (0,1)
LCI = Lined Cast Iron PDS, (0,1)
G = Galvanized Steel PDS, (0,1)
NH_3-N = Influent Ammonia N, mg/L as N
DO = Effluent DO, mg/L as O_2
NH_2Cl = Effluent Chloramine Residual, mg/L as Cl_2

$$NO_X\text{ - }N = \frac{1.05^{(T\text{-}20)}(0.430 \times H + 0.776 \times UCI + 0.400 \times LCI + 0.103 \times G)}{\left(NH_3\text{-}N\right)^{-0.312}(\mathrm{DO})^{0.242}\left(\left(NH_2\mathrm{Cl}\right)^{0.585} + 0.295\right)} \qquad \text{(ES 15)}$$

where: NO_x-N = Nitrate + Nitrite N, mg/L as N

COMPARISON OF FREE CHLORINE AND CHLORAMINES

The effects of free chlorine and chloramines on distribution system water quality as measured by residual dissipation, total iron release, apparent color release, turbidity release, total copper release, total lead release, residual maintenance, biological stability and nitrification were compared in a three month study using identical water sources and identical hybrid PDSs. The same PDS operating conditions were used in the comparison study as were used for the last year of operation. The PDS HRT was two days, which corresponded to a 10.8 mile equivalent pipe length in an actual distribution system flowing at 0.33 ft/sec. All comparisons of free chlorine and chloramine distribution water quality were made using paired t tests to determine statistically significant differences at a 95 % CL.

Chloramine residuals were lower than free chlorine residuals in all PDS effluents due to the reaction of chloramines with the pipe wall in the UCI and G sections of the PDS, which was predicted by the residual model. Chloramine wall decay constants were higher and bulk decay constants were lower than the same for free chlorine for all the pipe materials studied Simulation of chloramine and free chlorine residuals for equal conditions shows free chlorine residuals are slightly higher than chloramine residuals in G pipes with less than 1-ft diameters, and in UCI pipes with less than 0.5-ft diameters. Simulated chloramine residuals exceed free chlorine residuals in all PVC and LCI pipe regardless of diameter.

The release of total iron, turbidity and apparent color was higher in the presence of chloramines relative to free chlorine. Enhanced coagulated surface water and RO waters using chloramine had total iron concentration around 0.6 mg/L., in contrast the same water had total iron concentration of around 0.25 mg/L when using free chlorine. PDSs fed with RO and enhanced coagulated surface water had higher average total iron release than conventionally treated ground water and the blend regardless of the disinfectant utilized. Similar trends were observed for turbidity and apparent color due to the particulate nature of iron release in the PDS.

The release of total copper and total lead was higher in the presence of chloramines than in the presence of free chlorine for PDSs receiving GW, SW and a GW, SW, RO blend. There was no significant difference of released total copper in RO for either free chlorine or chloramines. Total lead release was higher in RO and the GW, SW, RO blend when using chloramine compared with the same waters using free chlorine. There was no significant difference in total lead release in the presence of either free chlorine or chloramines for GW or SW.

Biological nitrification was not observed in the presence of free chlorine and was practically insignificant in the presence of chloramines. Average NO_2 and NO_3 in the chloramine PDSs were 0.07 mg/L as N and 0.08 mg/L as N, respectively. The comparison demonstrated that nitrification could be practically non-existent when chloramines are used for residual maintenance when a residual is maintained and dead zones are avoided.

Biological Stability

In general, biological indicators did not show differences high enough (greater than one order of magnitude) to make conclusive statements concerning the relative biostability of free chlorine and chloramines. The data show that there was a higher average residual, less ammonia, less nitrite and less nitrate in the free chlorine PDSs than in the chloramine PDSs. However, there was very little nitrification occurring in the chloramine PDSs. Bulk HPCs were lower in the presence of free chlorine than chloramine. However, the lower HPCs might be due to the higher influent free chlorine residual. AOBs were present in the bulk in all PDSs regardless of disinfectant. Free chlorine did not appreciably inhibit AOB growth, the concentrations of AOB within the biofilm were similar for chlorinated and chloraminated waters. The presence of AOBs in the free chlorinated PDS even after three months of free chlorine use is significant.

HPC growth in bulk and in the biofilm was greater in the presence of monochloramines relative to free chlorine. Hence, DO consumption was higher in the presence of chloramines than in the presence of free chlorine. There were more biofilm HPCs and AOBs in 7 of 8 observations of the chloraminated PDSs and free chlorine PDSs. These differences were statistically significant for biofilm HPCs. Biofilm densities as indicated by PEPA, biofilm HPCs and biofilm FISH AOBs were higher for the chloraminated GW and blended sources than PEPA, biofilm HPC and biofilm FISH AOB for the chlorinated GW and blend. There was not a significant difference among biofilm densities associated with finished RO and SW.

The AOB and HPC biofilm data shows that AOB and HPC biofilm density increased as the surface roughness of the supporting material increased. Biofilm HPC and biofilm AOB were lower for PVC coupons than for LCI, UCI, and G coupons regardless of water source and disinfectant. The highest PEPA, biofilm HPC and biofilm AOB were found on UCI and G coupons.

AOBs were present in the free chlorine biofilms after three months of exposure to free chlorine. A three month free chlorine exposure reduced the AOBs in the biofilm by no more than one order of magnitude. Hence, three-month and probably longer free chlorine burns will not remove AOBs from biofilms, which indicates free chlorine burns are a temporary control for nitrification. However, as shown in the residual study, nitrification can be successfully controlled if a strong chloramine residual is maintained in a near dead-zone environment.

CHAPTER 1
INTRODUCTION

This document is the final report for a tailored collaboration project (TCP) funded by Tampa Bay Water (TBW) and AwwaRF. The TCP was conducted by the University of Central Florida (UCF) and had the objective of determining and minimizing the adverse water quality effects in the Member Governments (MGs) distribution systems that potentially could be caused by blending finished waters produced from ground, surface and saline sources. TBW manages drinking water supply for six MGs, which are Hillsborough, Pasco and Pinellas Counties and the Cities of New Port Richey, St. Petersburg, and Tampa.

PURPOSE

Drinking water utilities in Florida and similar environments have historically relied on groundwater, but over-reliance on the source has led to the pursuit of other sources. Tampa Bay Water and the University of Central Florida sought, through the Tailored Collaboration project with AwwaRF described below, to develop an understanding of the distribution system water quality issues involved when multiple-source waters are blended.

Tampa Bay Water (TBW) manages drinking water resources for six member governments (MGs) on the west coast of central Florida: the cities of New Port Richey, St. Petersburg, and Tampa, and Hillsborough, Pasco, and Pinellas counties. This region, like most of Florida, has historically utilized groundwater for its drinking water supply; however, adverse environmental impacts have mandated that alternatives (non-groundwater sources such as desalinated seawater and treated surface water) be used for future supply. Recently, TBW has brought on line two alternative supplies: a 66-million gallon per day (mgd) surface water treatment facility, which utilizes enhanced coagulation, ozone, and biological filtration; and a 25-mgd desalination plant. Finished water from these facilities are blended with finished groundwater and distributed to the MGs through a large regional transmission system, which runs through the three-county area. Blends of the various resources depend on the location of delivery and supply availability. Although the regional water-supply contract defines "acceptable water quality" (more stringently than SDWA requires), the delivered water quality was not constant, and, prior to this study, the effects of these blended waters on distribution system water quality were unknown.

As noted, the TCP was proposed to determine the effects of the pending wholesale system impacts on Member Government distribution system water quality. MGs distribution systems differed in age, size, pipe, quality and treatment systems and receive different blends of finished water dependent on where water is taken from the wholesale system. Presently, five of the MGs receive finished water that is a blend of desalinated sea water, treated ground water and treated surface water. Significant water quality variation from the aggregate blend occurs depending on season and location, which can adversely affect distribution system water quality. The effects of expected combinations of finished water quality on distribution water quality and methods of optimizing distribution water quality were determined.

The research objectives were (1) Determination of seasonal changes in aesthetic and regulated water quality from variations of blending finished waters from significantly different sources; and (2) Determination of methods to mitigate adverse seasonal changes in aesthetic and regulated water quality from variations of blending finished waters from significantly different sources. The technical approach was field plant evaluation and modeling of water quality changes in pilot distribution systems (PDSs) receiving finished waters, which were produced entirely or from blends of desalinated sea water, treated surface and ground waters using advanced and conventional treatment processes that simulated the MGs water treatment facilities of interest.

LITERATURE

Existing literature on blending different waters in distribution systems is presented in Chapter 2 and was limited but showed extensive turbidity and color water quality problems can possibly result from blending when different disinfectants are combined and pipe films are dissolved. Existing information on blending effects on MG distribution system water quality consisted of a corrosion semi-batch study and a theoretical model to predict solute mixing. This field investigation dramatically expanded on existing information by (a) investigating biological stability, disinfection by-products and taste & odor, (b) correlating existing information to distribution system water quality and (c) interrelating corrosion, biological stability and disinfection by-products to MG distribution water quality and relating those consequences to source, treatment, blend, frequency of blending change, disinfectant, season and distribution system residence time.

METHODS

As described in Chapter 3, eighteen different PDSs and seven different pilot treatment systems were designed and built by UCF and TBW-MG personnel. Four UCF faculty, two post doctoral fellows, twelve UCF graduate students and two MG field personnel were involved with design and construction of the field facility. The resulting facility has 8500 ft^2 under roof and 4400 ft^2 of 6 inch reinforced cement pad for tank support. This facility was manned and operated seven days a week twenty-four hours a day by nine UCF graduate students and two MG field personnel. Twenty thousand gallons a week of finished water was produced from the seven pilot processes. Combinations of different disinfected and stabilized finished waters were discharged to pilot distribution systems as singular finished waters or blends of two or three finished waters. Physical, chemical and biological changes resulting from the steady and non-steady state blending of different sources were monitored in the raw waters, finished waters and through the PDSs. The seven different finished waters were (G1) a conventionally treated groundwater, (G2) a softened groundwater, (G3) a softened blend of desalinated groundwater, coagulated surface water and finished groundwater, (G4) a nanofiltered blend of desalinated groundwater, coagulated surface water and finished groundwater, (S1) a coagulated surface water, (S2) a nanofiltered surface water and (RO) a desalinated groundwater. The eighteen PDSs consisted of PVC, galvanized steel (G), unlined cast iron (UCI) or lined cast iron (LCI) pipe taken from the existing MG distribution systems. Four of the distribution systems consisted of single pipe materials. Fourteen distribution systems consisted of the four different pipe materials. The eighteen distribution systems received blends from the various treatment systems simulating the actual blends in the MG distribution systems.

Quality assurance plans (QAP) were developed for the laboratory and field, and followed through the course of the study. The QAP is also described in Chapter 3. Field and laboratory procedures for operation and analysis were developed and approved by the project advisory committee before any operation and analyses were conducted. Extensive analyses were done in the field and the laboratory. Separate field operation logs were kept for each different process, the PDSs and the corrosion shed. Separate field and laboratory logs were kept for each analysis and duly recorded by work order number to maintain a paper trail for all field and laboratory activities. Physical and aesthetic water parameters measured were temperature, turbidity, taste & odor and color. Chemical water quality parameters included DBPs, NPDOC, chlorine demand & residual, alkalinity, pH, total & calcium hardness, LSI, sulfates, TDS, iron, sodium, chlorides, aluminum, silica, lead and copper. Nitrogen and phosphorous forms were also measured to enable nutrient balances across the PDSs. Biological parameters included AOC, BDOC, HPCs, SPCs, TC, FC and attached cell growth.

INFORMATION GAINED

Data was collected from the field facility for one and one-half years in three month phases. The first year of data collection was directed at the investigation of changing distribution system water quality as affected by changing water quality, different pipe materials and season, regarding (a) the release of total iron, apparent color and turbidity, (b) residual dissipation, (c) lead and copper corrosion and (d) biological stability as defined by residual, HPCs, AOC change and biofilm change. (e) Nitrification, (f) the comparison of chlorine and chloramines and (g) confounding factors on total iron, apparent color and turbidity release were investigated in the last two phases of the project. Before any finished water was produced in the field, laboratory studies were done for softening, coagulation, ozonation, membrane and disinfection processes to support pilot and correlate to full scale plant operation. Laboratory studies were also done to determine DBP and disinfectant change with time to support field studies.

CHAPTER 2
APPLICABLE LITERATURE

IRON RELEASE

The basic requirement for drinking water distribution systems is to be able to convey water from the source to the consumer without significant deterioration in water quality. Water quality changes in drinking water distribution systems occur as a result of complex and often interrelated physico-chemical and biological processes (Hedberg and Johansson 1987). The mineral composition of the source water usually determines whether adverse conditions (conditions that will lead to corrosion and subsequent deterioration of quality) exist within the distribution systems (Pisigan and Singley 1987). Biological conditions are also inextricably linked to the mineral content of the source water (Carter et al. 2000; Bade et al. 2000; Melo and Vieira 1999; Fransolet et al. 1988).

Red water problems are mostly due to the release of corrosion byproducts from old cast iron and galvanized iron pipes. The iron released may be due to either direct corrosion of the pipe material or due to the dissolution and/or scour of corrosion scales in the interior. The release of iron is mainly in the form of particulates and flakey material from the corrosion scales formed on these pipes (Clement et al. 2002; Lin, Ellaway and Adrien 2001).

The problems associated with red water conditions are mainly aesthetic (Smith et al. 1999). However, a more serious consequence could be the loss of disinfectant residual and subsequent microbial proliferation in the distribution systems (Hallam et al. 2002; Vikesland and Valentine 2002a, 2002b, 2000; Frateur et al. 1999). Corrosion products attached to pipe surfaces or accumulated as sediments in the distribution systems can shield microorganisms from disinfectants (Schock 1999). Release of corrosion byproducts in the form of turbidity may provide safe havens for pathogens.

Increase in anionic species like sulfates and chlorides cause a corresponding increase in corrosion. This phenomenon is hypothesized to be due to the disruption of protective corrosion scales (Schock 1999).

The problem of red water release is often confused with the corrosion of iron pipes. In old iron pipes red water problems are mainly due to dissolution and scour of internal corrosion scales. This disruption may be due to changes in water chemistry and/or flow patterns within the distribution systems (Clement et al. 2002; Sarin et al. 2001).

Corrosion indices have been developed based on laboratory studies and field observations. The commonly used corrosion indices are based on calcite solubility. Other indices that incorporate other water quality parameters have also been developed (Schock 1999; Rossum and Merrill 1983; Singley 1981).

Statistical models provide a valuable tool in predicting corrosion related problems like red water (Pisigan and Singley 1984). Some researchers have also suggested the use of a group of water quality parameters and/or indices to evaluate the corrosive behavior of water (Dye 1964; Pisigan and Singley 1984; Giljanovic 1999).

Corrosion of Iron in Drinking Water Distribution Systems

Iron corrodes, albeit in varying degrees, under any given water quality. The most stable thermodynamic forms of iron are the ferric compounds. Hence, corrosion is also called an extractive metallurgy in reverse where metallic iron is re-converted to its oxide. Iron corrosion has been studied in detail under varying conditions. Control of iron release in drinking water distribution systems is complicated by the many potential iron sources (raw water, ferric coagulation for surface waters, iron pipe corrosion) and release mechanisms (corrosion product solubility and film release). There are conflicting views on the exact mechanism of iron release, with some authors (Smith et al. 1999; Kuch 1988; Pisigan and Singley 1987) suggesting direct release from corrosion processes. Sarin et al. (2000) state that iron release bears no simple relation to the rate of iron corrosion, while Sander et al. (1997) emphasizes the need to distinguish between red water problems and corrosion rate and they suggest that corrosion-related water quality problems are mainly a function of the precipitation and dissolution properties of corrosion products formed. McNeill and Edwards (2001) concur with Sander et al. (1996, 1997) that red water quality problems do not depend primarily on the corrosion process, but on a large reservoir of corrosion product layers making it difficult to model scale behavior. It is however important to note that corrosion, an electrochemical thermodynamic process, ultimately supplies the initial source of corrosion related iron within the water supply system. The solubility of passive layers and film release processes predominantly contribute to the total iron concentration causing red water problems.

Several studies (Kuch 1988; Sander et al. 1997; Sarin et al. 2001, and Lin, Ellaway and Adrien 2001) have identified the following species as commonly found within the passive layers, goethite (α-FeOOH), lepidocrocite (γ-FeOOH), Magnetite (Fe_3O_4), Siderite ($FeCO_3$), ferric hydroxide ($Fe(OH)_3$), ferrous hydroxide ($Fe(OH)_2$), and calcium carbonate ($CaCO_3$). The contribution of the dissolved iron from the passive layers is controlled by the solubility of the solid species, available pipe surface area, flow velocity and the concentration gradient of ions between the pipe water and bulk liquid. Clement et al. (2002), suggest that siderite controls the concentration of dissolved ferrous ion in water due to its higher solubility product. However the siderite model does not adequately predict total iron concentrations leading to speculations that mechanisms other than dissolution govern iron released in distribution systems. Recent research on iron corrosion in chloride and sulfate environments has revealed the existence of green rusts that are unstable $Fe(II)$ - $Fe(III)$ hydroxyl salts that oxidize in the presence of oxygen. The crystal structure of green rusts is characterized as composed of a stacking of $Fe(OH)_2$-like layers carrying a positive charge due to the presence of $Fe(III)$ and interlayers constituted of anions and water molecules. They are transient compounds between metallic iron and final corrosion products, and therefore may govern the mechanisms and kinetics of corrosion and passivation of iron in aqueous media. A number of green rusts like GR(Cl^-), GR(CO_3^{2-}), GR(SO_3^{2-}) and GR(SO_4^{2-}) have been identified (Refait et al. 1998, 2003).

$$2FeOOH_{(s)} + 4Fe_{aq}^{2+} + 2H_2O + SO_4^{2-} + 6OH^- \rightarrow Fe_4^{II} Fe_2^{III} (OH)_{12} SO_4 \tag{2.1}$$

The importance of green rusts in the corrosion and scale release mechanisms is that green rusts differ significantly in their structure and composition. If green rusts are the controlling solid phase for iron solubility, they complicate the evaluation of corrosion byproduct release by thermodynamic considerations. At present, not enough literature exists to delineate the role of green rusts in corrosion prevention and scale formation.

The impact of various water quality parameters, notably chlorides, sulfates and dissolved carbon on corrosion products are well documented in literature (Sazou and Pagitsas 2002; Sarin et al. 2001; Rushing, McNeill and Edwards 2002 and Jones 1996). Savoye et al. (2001) suggest that the general corrosion to passivity transition of iron is mainly affected by pH as well as some anion species such as chloride, sulfate and carbonate. McNeill and Edwards (2001) discussed the effects of pH, alkalinity, buffer intensity and dissolved oxygen on iron release and contend that most of the studies are not reflective of conditions within the distribution system. However there is a general agreement that increased pH, buffer intensity, alkalinity and dissolved oxygen all lead to decreased red water complaints. In contrast weight loss experiments using coupons show that elevated dissolved oxygen result in increased corrosion rate. Flow velocity has been shown to increase corrosion rate at high dissolved oxygen levels, while excessively high velocities may scour away the protective layer (McNeill and Edwards 2001). Stagnation has been associated with high turbidity spikes especially at night (Sarin et al. 2001; Beckett et al. 1998; Van Rijsbergen et al. 1998 and McNeill and Edwards 2001). These complex physical–chemical interactions within the pipe system lead to the release of iron from the passive corrosion product layer in particle form. The particulate release rate is independent of the concentration gradient; hence a different approach to prediction of the total iron concentration is required.

Effect of Water Quality

A number of corrosion indices have been developed that relate the water quality to the corrosion potential of iron pipes. The Langelier Index has a prominent historical role among these indices as a measure of the stability of water quality being introduced into iron distribution systems. A positive LI indicates that the water is saturated with respect to $CaCO_3$. The predominant mechanism expected here is the deposition of a protective $CaCO_{3(s)}$ scale that will prevent corrosion. A negative LI indicates that the water is unsaturated with respect to $CaCO_3$. Such waters tend to remove the protective $CaCO_{3(s)}$ scale and expose the metal surface of the pipes to further corrosion. The general practice is to increase pH by 0.2 – 1 units above pH_s to prevent corrosion in distribution systems (Marangou and Savvides 2001).

A simplified form of the Langelier Index has traditionally been used as shown in equation (2.2).

$$LI = pH - (pK_2 + pK_{so} + pCa^{2+} + pHCO_3^-) \tag{2.2}$$

where pK_2 = Negative log of equilibrium constant for bicarbonate dissociation
pK_{so} = Negative log of equilibrium constant for $CaCO_3$ dissociation
pCa = Negative log of molar calcium concentration
$pHCO_3^-$ = Negative log of molar bicarbonate concentration

A number of other related corrosion indices have been developed by various researchers that incorporate the calcium carbonate solubility criteria to evaluate the stability of the source waters. The calcium carbonate precipitation potential (CCPP) is a the amount of calcium carbonate that will precipitate or dissolve from solution as it comes to equilibrium with solid calcium carbonate (Merrill and Sanks 1978).

McCauley (1960) used a rearranged form of the calcium carbonate solubility expression, termed the Driving Force Index (DFI), to calculate the tendency to deposit calcium carbonate. The DFI was modified by Rossum and Merrill (1983) to include the effect of alkalinity and pH. A similar qualitative attempt at predicting calcium carbonate precipitation was proposed by Dye (1964) and termed the momentary excess (ME).

Other researchers have proposed empirical correlations of the pH_s to the corrosion potential in iron pipes. Ryznar developed a curve based on the actual field results showing the encrustation or corrosion of steel mains as a function of the stability index (RI) (Singley 1981).

Rossum and Merrill (1983) conducted a review of some common calcium carbonate based indices. Based on the correlation among the indices the authors conclude that the Langelier Index (LI), Ryznar index (RI) and Aggressiveness Index (AI) are linear functions of pH and bear no clear relationship to calcium carbonate precipitation potential (CCPP). The driving force index (DFI) and momentary excess (ME) display a hump-backed curve characteristic of CCPP. The authors conclude that that though LI, DFI and ME are satisfactory indicators of whether the water is saturated with $CaCO_{3(s)}$. They should not be used for any other purpose. The authors further observed a redundancy between certain indices like DFI, AI and LI. They recommend the CCPP as the index best suited to describe $CaCO_{3(s)}$ saturation state.

However, other researchers (Larson and Skold 1958; Rossum 1987) observed that formation of calcium carbonate films do not correlate well to the LI. Based on a study in which the authors found that the waters with a positive Langelier Index (LI) was at least as corrosive as those with a negative LI, Pisigan and Singley (1985b) recommend that LI should be used with caution. They observed that though LI gives an indication of calcium carbonate solubility, it is not an infallible measure of corrosion. Trussel and Thomas (1971) noted that individually stable waters could have a blend that is not stable with regards to calcium carbonate solubility and vice versa. They caution against the misconception that individually stable waters would have a stable blend.

The drawbacks of the Langelier Index for predicting the stability of waters for corrosion control fueled the search for other water quality parameters that affected corrosion and iron release in distribution systems. One such measure is the Larson Ratio. Larson ratio is developed from the relative corrosive behavior of chlorides and sulfates to the protective properties of bicarbonate. The authors deduced that the corrosivity of air-saturated waters is dependent on the proportion of corrosive agents to the inhibitive agents and their concentrations (Larson and Skold 1958). Larson and Skold (1957) observed that calcium in the presence of alkalinity, regardless of pH or saturation index, is an effective inhibitor of corrosion. The authors concluded that tubercles are formed when inhibitors to corrosion are present in insufficient concentration. However, unsaturated water exhibited generalized corrosion.

The Larson Ratio (LR) is defined as shown in equation (2.3). A LR above 0.5 is considered corrosive.

$$LR = \frac{[Cl^-]+[SO_4]}{[HCO_3]} \qquad (2.3)$$

where [ ] = concentration in meq/L

Chlorides and sulfates were identified as major factors that increased the corrosion rate through interference with the formation of the protective calcite and siderite layers. Sulfates in particular may be involved in biological reactions with the indigenous microbial populations within the distribution system.

In a slightly alkaline solution a simplified model for corrosion of iron may be given as shown in equation (2.4).

$$Fe^{2+} + 2H_2O + 2Cl^- \Leftrightarrow Fe(OH)_2 + 2HCl \tag{2.4}$$

The resultant decrease in pH is localized in pits and can lead to further breakdown of the passive film on the metal surface. Other anions of strong acids, such as sulfates and nitrate, will also hydrate to acidic pH, but chloride is far more mobile in solution and more aggressive in both pitting and crevice corrosion.

Sulfate reduction within corrosion products can be demonstrated by tests for the presence of sulfate-reducing bacteria. In the Stadtwerke-Hannover laboratory, tests for sulfate reducers in corrosion products were carried out for many years. Almost every test was positive. The few exceptions were attributed to sampling conditions and not the absence of sulfate reducing bacteria. Similar findings of Tuovinen et al. (1980) and Lee, O'Connor and Banerji (1980) were reported by AWWARF (1985) for U.S. pipe. AWWARF (1985) reported that Miller and King (1975) proposed the following reaction for sulfate reduction under microbial action as shown in equation (2.5).

$$Fe^{2+} + 4Fe + SO_4^{2-} + 4H_2O \Leftrightarrow 4Fe(OH)_2 + FeS \tag{2.5}$$

Norton and LeChevallier (1997) noted that Larson Ratio values > 1 (associated with corrosive conditions) also interfere with the inactivation of biofilm bacteria.

Miyata and Asakura (2002) studied the oxygen reduction reaction at rust free iron surface in neutral unbuffered chloride solutions. Based on their observations the authors concluded that the concentration of Cl^- and pH did not affect the cathodic current. This implies that oxygen reduction is the principal corrosion mechanism on rust free iron surface. This study lends support to the hypothesis that the major role of chlorides in corrosion is destabilization and dissolution of passive films.

Simard, Menard and Brossard (1998) conducted studies on the passive film breakdown of 1024 mild steel induced by 0.05 M chloride ions. Pre-passivated mild steel anodes were used in this study. The authors observed that the breakdown potential increased with increasing bicarbonate and pH. The breakdown of passive films is attributed to its dissolution by chloride ions. Localized corrosion was then demonstrated due to the interstitial (crevice) corrosion at the mild steel.

Vatankhah, Drogowska and Menard (1998) evaluated the electrochemical behavior of iron electrodes in the presence of bicarbonate solutions at different sulfate levels. The authors observed that in the presence of sulfate ions only, iron electrodes experience uniform corrosion and sulfate ions were aggressive. However, in the presence of bicarbonate the sulfate ions became less aggressive and pitting-passivation behavior was observed. The authors recommended that a molar ratio of 10:1 of bicarbonate to sulfate results in a complete suppression of any localized attacks induced in the presence of sulfate ions. The authors explained that the main difference between the pitting behavior exhibited by chlorides and sulfates could be related to the ratio of their ionic radii (1:8). It was deduced that chloride ions are more mobile than sulfate ions. Another explanation is based on the nature of the corrosion products formed. The chloride corrosion products are highly soluble and therefore not protective. Sulfates react with iron forming relatively stable products and may compete with carbonates species. However, since the transient compounds have higher selectivity for carbonates, passivation occurs with the accumulation of carbonate species on the electrode.

The bicarbonate and carbonate species affect many important reactions in corrosion chemistry, including the ability of water to form protective metallic carbonate scale or passivating film such as $CaCO_{3(s)}$ or $FeCO_{3(s)}$. Sander et al. (1996) concluded that in waters with moderate or high total carbonate concentrations the surface complex is formed from free carbon dioxide whereas waters with low total carbonate content involve calcium complexation. The authors observed that at low total carbonate concentrations the corrosion rate is dependent on the calcium content while at high total carbonate concentrations no dependence is observed.

Based on laboratory loop studies and field investigations, Pisigan and Singley (1987) observed that an increase in buffer capacity at constant alkalinity (100 mg/L as $CaCO_3$) in the pH range of 6.0-9.0 decreased the corrosion rate of mild steel. However, raising the buffer capacity by raising alkalinity was counterproductive due to a corresponding increase in ionic strength and conductivity. Increasing flow rates was observed to increase the corrosion rates due to increased supply of oxygen. The authors recommend buffer capacity as a better indicator of corrosion potential than alkalinity. They related corrosion rates to the ratio defined in equation (2.6). The authors report that increasing the ratio causes a decrease in corrosion rate. Based on observations from field studies, the authors note that an elevated Total Dissolved Solids (TDS) or chloride content can lead to higher corrosion rates.

$$\beta / Cond^{3/2} \qquad (2.6)$$

Where β = Buffer capacity in M/L

Cond = Conductivity in μΩ/cm

Increased total dissolved solids (TDS) cause an increase in the ionic strength of the waters. This can cause a change in the activities of the ionic species in the water. Another effect of increased TDS is increase in conductivity of the water which increases the corrosive current and facilitates electrochemical reactions. TDS may also effect the formation of protective films depending upon its nature. If TDS is present as chlorides and sulfates, it will cause increased corrosivity of iron. If the TDS is composed of bicarbonate and hardness ions the water tends to be non-corrosive towards iron (Schock 1999).

Singley (1981) conducted a survey of existing corrosion indices and concluded that all indices developed were based on certain simplifying assumptions that were applicable to the specific cases for which they were developed. Therefore, no single corrosion index is applicable universally. The author observes that based on the review of literature and experience; there are a number of water quality parameters that need to be considered in addition to calcium carbonate solubility, including calcium, magnesium, alkalinity, carbonate, carbon dioxide, pH, chlorides, sulfates, ionic strength, conductivity, total dissolved solids, color, hydrogen sulfide, buffer capacity, phosphate, silica, dissolved oxygen, chlorine and temperature. The author advises that the most useful form of corrosion index would consider all relevant factors for the water, but must be designed in such a way that the protective mechanisms for a given water would predominate in the calculation of the index.

Dissolved oxygen significantly influences corrosion of metals in water. Oxygen accepts electrons from corroding metal, allowing corrosion reactions to continue. Oxygen reacts with ferrous iron and oxidizes it to the more stable ferric form. Ferric iron accumulates at the point of corrosion causing tubercles or settles out at dead-ends and interfering with flow (Schock 1999). However, dissolved oxygen has a converse effect on the protective nature of the iron corrosion scales. High dissolved oxygen content of water causes the oxidation of ferrous iron formed during corrosion and fixes it within the scales. These ferric hydroxides later dehydrate or further oxidize to form more protective oxide scales like goethite and lepidocrocite (Lin, Ellaway and Adrien 2001; Clement et al. 2002). Schock (1999) has reported research conducted by Legrand and Leroy (1990, 1995) in which the authors indicate that dissolved oxygen plays a beneficial role in the creation of passivating films on unlined iron piping. Ferric hydroxide layer on pipes has been observed to be more compact and stronger than when formed from chlorine in the absence of oxygen. Similar findings have been reported by other researchers (Clement et al. 2002). Dissolved oxygen is reported to facilitate the formation of favorable passivating film layers consisting of a mixture of calcium carbonate, ferrous carbonate and ferric oxide passivating layers that are referred to as the ferricalcic deposit.

Dissolved oxygen content influences the speciation of the microbial population within the distribution system. Low dissolved oxygen content of the source waters can cause anoxic zones and proliferation of sulfate reducing bacteria (SRB) (McNeill and Edwards 2002). SRBs have been implicated in the production sulfides and the consequent taste and odor problems.

In the absence of dissolved oxygen, previously deposited ferric scales (lepidocrocite, $\gamma - FeOOH$) can act as electron acceptors as shown in equation (2.7). This Kuch reaction produces ferrous ions and allows the corrosion reaction to continue even after DO is depleted (McNeill and Edwards 2001).

$$Fe_{(Metal)} + 22FeOOH_{(scale)} + 2H^{+} \Leftrightarrow 3Fe^{2+} + 4OH^{-} \tag{2.7}$$

McNeill and Edwards (2002) conducted a literature review of the effect of temperature in assessing iron pipe corrosion in water distribution systems. The authors state that, quantifying the effect of temperature on distribution systems is limited by the fact that most are buried underground and therefore temperature variations are minimized. The authors illustrate that at higher temperatures an increase in biological activity could be expected. The solubility of oxygen is reduced which could lead to the depletion of electron acceptors and predominance of Kuch type reactions. Lowered viscosity at higher temperatures could increase the diffusion rates in scales and bulk liquid and affect diffusion limited corrosion reactions. Thermodynamic changes could affect solid solubility or stability. An increased rate of ferrous oxidation could lead to changes in the solid phase. Thermal stresses induced due to the different expansion coefficients of the metal and scales could lead to disruption of passivating layers. Schock (1999) notes that the temperature effects on corrosion are complex and multiple phenomena often occur simultaneously. For instance, temperature changes cause changes in solubility of solids, changes in formation of complex ions, changes in diffusion rates of dissolved gases and solution species, changes in the physical properties of the metals and changes in water properties. The authors conclude that generalizations on the effect of temperature are rarely accurate due to the multitude of often confounding and conflicting effects.

Low hydraulic residence time (HRT) sometimes aid the formation of protective oxide coatings on the metal by transporting the protective material to the pipe surface at a higher rate. Conversely, high distribution system water velocity can also transport the corroding constituents in water at a faster rate and increase corrosion. A change from lower to higher velocities can cause a scouring effect that transports sediments to the tap. Schock (1999) notes that a water quality which may be satisfactory at high to medium velocities may cause colored water at lower velocities due to the reduced diffusion of protective ingredients to the metal surface. A general rule based on the work of many researchers would be to avoid dead-ends, low velocities and fluctuations in flow patterns (Clement et al. 2002, Schock 1999).

Iron Corrosion and Iron Release

Iron-based pipe materials may corrode to form iron corrosion products when exposed to an aqueous environment. Important chemical reactions include oxidation-reduction reactions and precipitation-dissolution reactions. The oxidation reactions represent iron corrosion. The precipitation reactions represent the formation of corrosion layers on iron pipe surface causing passivation. The dissolution reactions represent the dissolution of corrosion products (iron release). Water quality and many physical factors affect these reactions.

Red water can occur if iron corrosion scales are released from the pipe wall. There are several proposed mechanisms to explain the red water phenomena. Kuch (1988) proposed that the red-water problem might be the result of cyclic periods of flow and stagnation in a water pipe. When the dissolved oxygen concentration is low, ferric oxides serve as the alternate oxidants and ferrous ions are generated by both oxidation and reduction reactions. The ferrous ions may either precipitate or remain soluble. The fraction that remains soluble migrates towards the bulk solution, where it can be oxidized, precipitated, and carried to the consumer's tap, forming red-water. Other mechanisms include dissolution of corrosion scale components, and the hydraulic scouring by flowing water or sloughing of crystalline iron oxides or iron-oxidizing bacteria, which often include amorphous iron oxide particles. In most cases, iron release is a cumulative effect of all the above (AWWA 1996).

Surface Characterization

Surface analytical techniques have been widely used in corrosion studies and have provided information on surface characteristics of corrosion layers, which has aided the elucidation of corrosion and metal release mechanisms.

Among the surface techniques, EDS, XRD, XPS and RMS are the general choices of studying iron corrosion products and corrosion layers qualitatively and quantitatively. By using these techniques, α-FeOOH, Fe_2O_3, Fe_3O_4, and γ-FeOOH were identified as the major corrosion products formed from iron corrosion processes (Lin, Ellaway and Adrien 2001; Sarin et al. 2001, Oh, Cook, and Townsend 1999; Marco et al. 2000; Jeong and Kim 2001; Ruiz, Seal, and Reinhart 2000). $FeCO_3$ may be formed in an aqueous environment, but it is unstable and quickly decomposes to Fe_2O_3 when exposed to air. Therefore, it is rarely found in XPS scan results (Heuer and Stubbins 1999).

Factors Affecting Iron Release

The processes of iron corrosion and iron release are affected by many factors (AWWA 1996; Hem, Vik and Langen 2001). Alkalinity/pH can affect the rate of iron corrosion, the formation of protective film, and the solubility of iron (Hem, Vik and Langen 2001; Price and Jefferson 1997; Shull 1980). Increasing pH and/or alkalinity can help control corrosion and minimize color release, especially pH, which favors the formation of calcium carbonate scale and affects iron uptake and corrosion. A corrosion inhibitor, such as orthophosphate, might assist in controlling iron uptake if coupled with pH adjustment. The formation of a protective film on pipe surface, such as $CaCO_3$ film, can reduce iron corrosion rate and lessen iron release from the surface (Larson and Skold 1957; Shams EL Din 1986; Facey and Smith 1995). Higher temperature was reported to increase iron corrosion rate (Volk et al. 2000). Surface complexes were found to favor iron corrosion by Sander et al. (1996, 1997).

LEAD RELEASE

Lead Corrosion

Lead in drinking water results mostly from corrosion of lead materials in the distribution system, and is generally not present in source waters. Significant sources of lead and copper in drinking water are lead solder pipe joints, lead service connections, brass fixtures, and copper pipes found in existing plumbing systems of homes. Lead is anodic in distribution systems and produces Pb (II) and Pb (IV) ions, which form stable lead complexes in water. The most common oxidant for lead in drinking water is DO. Primary lead corrosion products are Pb (II) species in most cases (AWWA 1996; Pankow 1991). The oxidized solid phases are primarily PbO, PbO_2, PbCO3 and $Pb_3(CO_3)_2(OH)_2$. Other lead species that occur in drinking water are lead carbonate, hydroxyl, chloride or sulfate complexes, which affect the solubility of lead in water. The major Pb (II) complex in drinking water is $PbCO_3^0$. $PbCO_{3(S)}$ and $Pb_3(CO_3)_2(OH)_{2(S)}$ are the primary solid phases in drinking water, but do not adequately passivate lead corrosion because of discontinuous formation (Evans 1960). The discontinuous character of lead deposits is a probable reason for the persistent corrosion of new lead materials.

The leaching of lead and copper from plumbing materials is a source of metals in drinking water and is a public health concern. To protect public health, the US Environmental Protection Agency (EPA) has set the maximum action level for leads as 15 ug/L at 90 percentile (AWWA 1996; Birden, Calabrese and Stoddard 1985; Scottish Executive 2002; United States Army Environmental Hygiene Agency 2003).

Surface Characterization

Surface analytical techniques (XPS, XRD, SEM-EDS, ICP-AES, etc.) have been widely used to study the corrosion products of lead. Among them, XPS and XRD were used by many researchers (Chen et al. 2002; Elzenga and Graveland 1980; Paparazzo and Moretto 1995; Young and McCaslin 1985) to identify the lead corrosion compounds and investigate surface corrosion layers. Researches showed $Pb_3(CO_3)_2(OH)_2$, $PbCO_3$ and PbO as the major lead corrosion products formed in general aqueous conditions.

Factors Affecting Lead Release

Water quality and other factors can influence the lead concentration in drinking water by altering corrosion rate, affecting the formation and physical structure of lead corrosion species, changing the solubility of lead, forming protective films, and scouring lead corrosion products from pipe walls or loosen deposits in distribution systems and home plumbing.

pH is the most important factor controlling lead release. Normally, pH is associated with alkalinity, but lead solubility is relatively more sensitive to pH than to alkalinity. Lead solubility decreases with increasing pH in alkalinity range common to drinking water. Optimum alkalinity is associated with pH, and the best combinations of alkalinity and pH have to be determined experimentally in each case. pH and alkalinity adjustment has been used in many cases to reduce lead level in drinking water. It should be noted that increasing pH may negatively affect disinfection and increase formation of trihalomethanes (THM) (Boffardi 1988, 1990 and 1995; Chen, McAnally and Kumaraswamy 1994; Deb, Sabir and Garlow 1990; Frey 1989; Johnson et al. 1993; Karalekas, Ryan and Taylor 1983; Lee, Becker and Collins 1989; Patterson and O'Brien 1979; Risser 1997; Schock 1980, 1989; Schock and Gardels 1983; Vinci 1991).

High DO can increase lead concentration in water (Boffardi 1988). Lead concentration increases with temperature as lead solubility increase with temperature (Boffardi 1988; Lee, Becker and Collins 1989). Sulfate and silicate inhibited lead corrosion while decreasing pH and increasing chlorides stimulated corrosion (Oliphant 1983). Beccaria et al. (1985) found that some lead sulfate solids formed when a high sulfate concentration was present. Chloride increased the corrosion rate, and may complex lead, which would increase lead solubility. Silicate can react with lead to form a protective lead silicate film that inhibits lead corrosion (AWWA 1996; Schock and Gardels 1983; Anon 1994). Nitrate does not have a significant effect on lead release (Sheiham and Jackson 1981).

Modeling of Lead Release

Several approaches have been used by researchers to predict lead release in water. Thermodynamic models determine the solubility of the controlling solid phase and associated reactions. Research (Hem, Vik and Langen 2001; Colling, Whincup and Hayes 1987; Marani, Macchi and Pagano 1995) showed that either $Pb_3(OH)_2(CO_3)_2$ or $PbCO_3$ was the controlling solid phase for lead in drinking water systems. A $Pb_3(OH)_2(CO_3)_2$ thermodynamic model was developed by Schock (1980, 1989) and showed the response of the theoretical solubility curves for lead as a function of dissolved inorganic carbonate concentration and pH. The model agreed very well with the experimental data. Experiential models were developed from statistical analysis of actual data Haring (AWWA 1996). Hydraulic models were also developed based on the hydraulic properties of pipe and flow (Kuch and Wagner 1983). However, accurate prediction of lead release in water is difficult because of the variability of lead release, which is caused by the release of particulate lead and to a lesser extent by the precision and accuracy associated with the low concentrations in relation to the EPA "Action Level".

COPPER RELEASE

Copper Chemistry

Copper and copper alloys have been employed extensively in home plumbing systems because of their corrosion resistance and ease of installation. Without DO (Dissolved Oxygen), residual chlorine, or other electron acceptors, copper is very corrosion-resistant because of its noble electronic configuration (Reiber 1989; Eilbeck 1984; Stone et al. 1987).

In the presence of oxidizers, Cu(0) is oxidized to Cu(I) (cuprous) and Cu(II) (cupric) following a sequential one-electron transfer mechanism (Hamann, Hammett and Vielstich 1998; Lowe 1993; Wieckowski 1999). As newly formed free cuprous and cupric ions, as well as soluble complexes accumulate, copper precipitates form. Copper concentration increases until the precipitation reaction reaches equilibrium and stabilizes the final copper concentration (Pankow 1991). Equilibrium is lost as water flows through the copper pipes and the corrosion process is repeated until the copper pipe fails or the copper precipitate completely passivates the elemental copper.

Important species affecting copper corrosion in drinking water are hydroxide (i.e. pH), carbonate (i.e. Alkalinity or Dissolved Inorganic Carbon (DIC)), sulfate, chloride, ammonia, and orthophosphate. Among these, hydroxide, carbonate, sulfate and chloride species are most important due to their relative higher activity.

For Cu (I), the predominant solid phase is Cu_2O (cuprite). Important Cu (II) solid phases formed include tenorite ($CuO_{(S)}$) and cupric hydroxide ($Cu(OH)_{2(S)}$). $Cu(OH)_{2(S)}$ is a dominant copper solid phase in drinking water stagnation conditions under twenty-four hours. The dominant solid phase in aged copper pipes at neural ,pH in drinking water is malachite ($Cu_2CO_3(OH)_{2(S)}$) (Ives and Rawson 1962; Merkel et al. 2002). However, in new copper pipes, $Cu(OH)_{2(S)}$ is the dominant solid phase (Edwards and Sprague 2001). The detailed mechanisms controlling the transformation between the new and aged corrosion by-product, unfortunately, are not well understood.

Substantial research has been done to elucidate the primary water quality factors controlling copper release in drinking water systems, assuming copper release is controlled by precipitation/dissolution mechanism in drinking water. It is now widely accepted that alkalinity and pH are the most important factors. Other inorganic anions such as chloride, phosphate, dissolved oxygen, chlorine residual, natural organic matter as well as temperature, and microorganisms influence copper release to a lesser extent.

A long-term stagnant experiment conducted by Merkel et al. (2002) indicated that the copper release maximized after several hours following immersion into synthetic drinking water, and then decreased until equilibrium was achieved (Merkel et al. 2002). Under stagnant conditions, oxygen (or other electron sink species) concentration decreases at the interface continuously, while released copper increases within the short-term and then decreases until equilibrium. However, copper release over time did not follow simple first or second order kinetics. Mass transfer coefficients for released copper decreased during the experiment in an erratic fashion. Accurate descriptions of copper release in actual environments are further complicated by hydraulic shearing of copper scales during water transmission.

Numerous studies and models exist which can predict steady state copper concentration for varying water quality. However, relatively few or no studies have been conducted to understand copper transition from one steady state water quality environment to another. From coupon tests Broo, Berghult and Hedberg (1998) found that copper concentrations were kinetically controlled and that copper concentrations depended on water quality. Copper concentrations either reached very high values and then decreased to steady state or slowly increased steady state over time of exposure following a change in the aqueous environment. No models that quantitatively describe copper corrosion under actual drinking water conditions over time of exposure were found in the literature.

Although the effects of single factors have been exhaustively investigated, accurate prediction of copper release at a known water quality is difficult. The literature review for copper found five reasons for this: (i) kinetic limitations; (ii) controlling solid phase limitations; (iii) inconsistent equilibrium constants; (iv) confounding effects; and (v) particulate interference (Schock, Lytle and Clement 1995; Ugorets, Buketov and Akhmetov 1968; Chalyi 1963). All of these facts make the actual prediction based on theoretical consideration more difficult, and in most cases, only wide range and general trends could be estimated. As with many water quality issues facing the water community, workable solutions are best found through field studies that simulate actual conditions.

Except for the inorganic ligands, organic and biological mechanisms for copper release impact cuprosolvency. The effect of microbes on copper release results from microbiological attack on copper by exopolymers, which deteriorates the passivating film. NOM (Natural Organic Matter) is an aggregate parameters used to characterize organic materials in drinking water. As an aggregate parameter, NOM contains humic or fulvic substances, which are composed of many organic functional groups, which in turn can chemically bind Cu (II) either in the bulk or on the surface of the copper scale. If surface scale binding (adsorption) occurs, the adsorbed organic can form a diffusion barrier or as a less soluble corrosion film and reduce copper release (Edwards and Sprague 2001). If the bound species is a copper complex that thermodynamically favors the bulk water, then copper concentration and copper corrosion are increased. Presently, the effect of NOM on copper corrosion can only be determined by laboratory or field investigation for a specific aqueous environment.

Surface Characterization

Surface characterization techniques can be used to identify possible coexisting solid phases on copper scales following exposure to an aqueous environment,. X-ray diffraction (XRD) and X-ray Photoelectron Spectroscopy (XPS) are primarily utilized for surface characterization. Other surface characterization techniques such as Scanning Electron Microscopy (SEM) and SEM Energy Dispersive Spectroscopy (SEM/EDS) can be used to enhance visual identification of the surface and cross section of surface scales. XPS spectras identifies the top chemical components on the surface of the corrosion layer, and XRD identifies the chemical components in the bulk sample. SEM magnifies the physical structure of the surface in order to provide meaningful visual examination. The EDS spectra identify elements in the sample. (Patterson, Richard and Marani 1991; Edwards, Schock and Meyer 1996; Stumm 1990; Townsend, Simpson and Johnson 1995).

TRANSIENT METAL RELEASE

Transient metal release can be problematic in the drinking water community. The City of Tucson, Arizona switched from 100 % groundwater to 100 % surface water in order to comply with the 1980 Groundwater Management Act by constructing a $125,000,000 surface water treatment facility that could supply 100 MGD. After placing the new facility online in 1995, the City of Tucson Water Department received approximately 12,000 customer complaints (McGuire 1997) in a six month period. The consequences were severe and included law suits, personnel replacement, no utilization of the surface water facility, return to the ground water supply and the cost of investigations to enable successful surface water utilization.

In 1941, the Southern Metropolitan Water District of Southern California first delivered treated Colorado River water to its Member Governments through its old distribution system which had traditionally carried treated groundwater. The chlorine residual, TDS, DO and hardness of the ground water were different than that of the surface water. The release of rust deposits or 'red water' and biofilms causing numerous taste and odor complaints were attributed to disruption of the calcium carbonate scale. These conditions persisted during the transition period until new chemical scales and biofilms that were compatible with the new environment were formed in the drinking water distribution system (Streicher 1956).

In general, changes in water quality can cause extreme disturbances in existing distribution systems. Due to the large population growth and industrialization, most of the distribution systems lay beneath huge structures. Removal of these systems is difficult and extremely costly. American Water Works Association (AWWA) estimates that US water utilities will pay $325 billion over the next 20 years to upgrade water distribution systems (AWWA 1999). The promulgation of the "Lead and Copper Rule" by the United States Environmental Protection Agency (USEPA) in 1991 created 1.3 mg/L copper and 15 µg/L lead action levels at the tap. Eighty thousand water utilities were forced to monitor for copper and lead and several instigated changes in water treatment to meet these action levels (Schock, Lyte and Clement 1995). It is not easy to unravel the changes in water quality occurring during the transport of water through the distribution system.

Iron Release

Historically distribution systems have dead-ends where residuals are lost, taste and odors are generated and other adverse water quality issues arise. These problems increase when UCI or G pipes are used in drinking water distribution systems because the pipe material is not inert and reacts with water. The use of LCI, PVC or other pipes with inert interiors reduces but does not eliminate water quality issues in a drinking water distribution system. Urbanization, capital costs, huge structures are some of the factors that prevent the removal and replacement of older distribution system pipes.

After decades of transporting treated water, corrosion byproducts and precipitates form on the interior of distribution system pipes and eventually reached steady-state with the finished water in the distribution system. Changes in water chemistry can however disrupt this equilibrium and generate reactions with pipe materials and deposits in distribution systems, which cause discoloration and other water quality concerns (McIntyre and Mercer 1993). This deterioration in water quality reaching the consumer is related to destabilization of the surface films due to changes in source water characteristics or hydraulic conditions. Trussell and Thomas (1971) showed that blending two stable waters can produce unstable water with a potential for disruption in distribution system water quality.

Color and turbidity are significant parameters for the description of distribution system water quality. Color or turbidity changes in the distribution system are indicative of physiochemical or biological changes in the distribution system. Iron in the particulate form contributes to both apparent color and turbidity. Changes in color, turbidity or total iron can be correlated in iron pipes and fixtures exposed to water in a distribution system.

Iron Release and Related Parameters

Alkalinity, chlorides, sulfates and calcium are the primary water quality parameters believed to influence corrosivity. Intra-water corrosion variability and secondary water quality factors (e.g. microbial, flow patterns, flow velocity, etc.) could also play a vital role in mitigating corrosion by-product release (Elzenga, Graveland and Smeenk 1987).

Several researchers have shown that increased chlorides and sulfates can cause instantaneous spikes in corrosion rates of iron irrespective of the saturation state with respect to calcium carbonate of water (Elzenga, Graveland and Smeenk 1987; Larson and Skold 1957). In an actual distribution system study Elzenga, Graveland and Smeenk (1987) found an increase in chlorides and sulfates of about 2.9 meq/L, increased total iron and turbidity release three-fold within 24 hours of the exposure in a 6 inches UCI pipe in a drinking water distribution system.

Chloride generally has a much greater tendency to penetrate oxide films than sulfates, on the other hand sulfate results in the formation of a less protective oxide coating compared to chloride (Hatch and Rice 1959). The increased corrosion rate due to the increase in chlorides and sulfates can be explained by their interference in the formation of protective layers (Sander et al. 1996) and the increased ionic conductivity of water (Pisigan and Singley 1985a).

Several studies have been performed to study the short-term effects of increasing alkalinity on iron corrosion (Pisigan and Singley 1987; Kashinkunti et al. 1999). The studies documented that increased alkalinity can cause increased corrosion rates but decreased iron concentration in water. The increased corrosion rates were attributed to increased conductivity due to addition of HCO_3^- and the decreased dissolved iron concentration was attributed to carbonate precipitation of some iron (Pisigan and Singley 1987).

In both laboratory (Larson and Skold 1957) and full scale studies (Schimpff and Pearson 1980) calcium has been found as an effective inhibitor to iron corrosion. The reactions involving calcium are slow and a certain length of time is required for the effectiveness of calcium to become apparent.

Both free chlorine and chloramines are capable of redox reaction with soluble metallic and organic solutes in water. Many authors have documented that the reactions involving both free chlorine (Boffardi 1992; Tuthill et al. 1998; Stone et al. 1987) and chloramines (Vikesland and Valentine 2000; Treweek et al. 1985) are spontaneous and can cause sudden increase in corrosion rates.

Stagnation time of water is an important parameter for many reactions in distribution system. Increased turbidity has been observed in cast iron distribution system during nighttime and which decreased in early morning as demand increased and stagnation time decreased (Van Rijsbergen et al. 1998). Supporting this, Elzenga, Graveland and Smeenk (1987) also found that an increase in the retention time of 24 hours in UCI pipes (6 inch) increased the turbidity and iron concentration by approximately 2.5 times.

Many parameters that influence corrosion can vary with temperature. These parameters include DO solubility, solution properties (e.g. viscosity and ion mobility), ferrous ion oxidation rate and biological activity. Several studies have shown that the rate of any particular corrosion reaction will increase significantly with temperature within a very short period of time (Butler and Ison 1996; Stone et al. 1987).

Water quality change in drinking water distribution systems is influenced by the pipe material. Inert pipes, such as PVC and LCI, cause minimal or no changes in the water quality when the bulk liquid water chemistry changes. UCI and G pipes are prone to iron corrosion and adversely affect distribution system water quality, when the bulk distribution system water chemistry is changed.

Transitional Iron Release

Several models exists today that can predict steady state iron concentrations. Little or no research has been done to understand the transition of iron or any metal in drinking water distribution systems from one steady state to another.

Transition models do not exist but could be developed using simple reversible differential equations to describe transition to steady state. Kinetic constants can be utilized to define rates of chemical reactions or mixing in the absence of chemical releases. Iron release in drinking water distribution systems however is a dynamic process that could be described by either a plug flow or a complete mix reaction with or without a reaction. Although, neither a plug flow nor complete mix environment is ideally suited to distribution systems or home plumbing systems, arguments can be made for the description of stagnant or dead-end environments using plug flow models. Such models would require development of a global rate constant, which would be independent of water quality environments but dependent on pipe material and temperature. A general rule which applies to most chemical reactions in solution is "the rate constant doubles for every 10 ^{o}C increase in temperature" (Frost and Pearson 1961). Hence iron or the release of any metal in a drinking water distribution system is likely to increase as temperature increases.

A twelve month distribution system study carried at Florida Water Utilities showed that corrosion rate of mild steel exponentially dropped with time in distribution systems carrying highly corrosive water (TDS = 431mg/L) (Pisigan and Singley 1987). Another twelve month laboratory study on iron and galvanized nipples carrying varying source waters showed that corrosion rates decreased significantly with time for the first six months, whereas varied relatively little in the last six months of the study (Streicher 1956). Sixteen day laboratory studies on cast iron specimens showed that corrosion rates significantly increased with time in waters within the pH range of 7.5 to 8.5 (Larson and Skold 1957). Coupon tests targeting iron corrosion control by three different corrosion control strategies over a 150 day period showed that iron levels exponentially dropped within the first 20 days of the study and thereafter remained relatively stable during the rest of the experimental period (Kashinkunti et al. 1999).

Copper Release

Copper is an essential trace element found in food and drinking water, but acute toxicity can occur with excess ingestion (Chuttani et al. 1965). In past studies, levels as low as 0.3 mg/L Cu have been linked to gastrointestinal upset (Buchanan et al. 1994). The rate limiting process controlling the corrosion rate of copper is the diffusion of copper ions through the oxide film given the different water quality conditions under which corrosion film development can occur in real distribution systems (Feng et al. 1996). Studies performed on the short-term effects of various water quality parameters have revealed that chlorides, sulfates, alkalinity, disinfectant residual, pH and temperature are the major parameters affecting the release of copper over time.

The order of decreasing rate of corrosion by sodium salt solutions of equal conductivity to isolate anions was Cl^-, SO_4^{2-} and HCO_3^- (Stone et al. 1987). In an effort to evaluate the short-term response (60 minutes) of chlorides and sulfates on aged copper specimens, Stone et al. (1987) found that, both increased copper corrosion and the effect of chlorides was greater than that of sulfates. However, Broo, Berghult and Hedberg (1998) reported that chloride ions greatly affect the kinetics of corrosion process by retarding the oxidation of Cu (I) in natural waters (lowering corrosion rates) and can be helpful on a long-term basis. Pisigan and Singley (1987) stated that "For waters with sufficient buffering capacity the addition of chemicals for boosting alkalinity may reduce corrosion as enhanced corrosion of copper may result due to the increased ionic strength and conductivity". However, two studies performed by Pisigan and Singley (1987), and Obrecht and Quill (1961) reported bicarbonate had little or no effect on initial rates of copper corrosion.

Very few studies have documented the short-term effects of pH change on copper corrosion. One laboratory study showed that within 60 minutes the corrosion rate of aged copper specimens decreased by approximately 55% as pH increased from 6 to 8 (Stone et al. 1987). Several laboratory studies have revealed that the rate of copper leaching from pipes is greatly accelerated in waters with high free chlorine (Pisigan and Singley 1987, Stone et al. 1987). Pisigan and Singley (1987) showed that in the presence of oxygen the corrosive attacks led by chlorine residual (1-1.5 mg/L) can be more severe. Laboratory studies performed by Hong et al. (1998) showed that in the absence of free chlorine; DO alone had a little effect on the leaching of copper. Temperature was found to have a profound effect on copper corrosion. In a laboratory study Stone et al. (1987) showed that within 60 minutes the corrosion rate of aged copper specimens increased by 50% with the increase in temperature from 10°C to 25°C. The transitional responses of NOM and microorganisms on copper are not well understood, however the reactions for total copper release appear to be relatively slow.

Transitional Copper Release

To date numerous studies and models exist which can predict the steady state copper concentrations with changing water quality. However, relatively few or no studies have been conducted for copper to understand transition from one steady state to another steady state. Using coupon tests Broo, Berghult and Hedberg (1998) found that total copper release in a drinking water environment was kinetically controlled and that the total copper released depended on water quality. Total copper released either reached very high concentrations before falling back to a steady state concentration or slowly increased with time to a steady state concentration depending on reaction kinetics (Broo, Berghult and Hedberg 1998). There is no empirical equation describing kinetics or dynamics of copper corrosion under actual drinking water conditions.

Lead Release

Lead is an active and cumulative toxicant and has been associated with hyperactivity in children (David, Clark and Voeller 1972), neuro-psychological dysfunction in children (Landrigan et al. 1975), high blood pressure and cardiovascular diseases (Beevers et al. 1976), and renal insufficiency (Campbell 1977). Previous studies performed on the effects of various parameters have revealed that pH, alkalinity, temperature and plumbing age were dominant independent parameters that influenced the rate and stead state concentration of total lead released.

Studies performed by Kashinkunti et al. (1999) and Lee, Berker and Collins (1989) found lead release was more sensitive to pH than alkalinity. Both studies suggested that pHs above 8 provided superior lead control compared with lower pH. Short-term lead pipe studies performed by Patterson and O'Brien (1979) found that increasing pH from 7 to 8.5 reduced lead concentrations by approximately 50% within the first 5 days but consistent release of significant amounts of particulate lead began after 2 weeks of the corrective action. They found that pH adjustment from 7 to 8.5 coupled with addition of 20 mg/L HCO_3^- as $CaCO_3$ reduced lead concentrations by 85% and achieved steady state within three days. They also found the release of total lead was reduced by 75% within the first 10 days following alkalinity addition with no pH adjustment.

Studies performed by Wong and Berrang (1976) and, Lee, Berker and Collins (1989) showed that lead concentrations increased as the stagnation period of water in lead soldered or lead based pipes increased. Research performed by Lee, Berker and Collins (1989) and Murrell (1988) on the effect of aging of pipe on lead release showed that pipes older than 5 years had significantly low lead concentrations when compared with newer pipes.

Lead solubility has been shown to increase with temperature. In an actual distribution study performed in Boston, a direct trend was seen between water temperature and lead levels. In the study Karalekas, Ryan and Taylor (1983) found that the lead concentrations also increased with the seasonal fluctuations between winter and summer.

Several potential effects of chlorine, which could increase lead solubility or accelerate lead dissolution have been discussed, but no data is available to support these hypotheses. However, Lee, Berker and Collins (1989) reported that free chlorine had no apparent effect on the amount of lead contributed from plumbing materials.

A study found that increased sulfate and silicate inhibited lead corrosion while decreased pH and increased chloride stimulated lead corrosion (Oliphant 1983). Beccaria et al. (1985) found that some lead sulfate solids formed when a high sulfate concentration was present. They found that chlorides increased lead corrosion rate and postulated that it may complex with lead and increase dissolved lead.

Transitional Lead Release

Several researchers have successfully attempted to model lead release as a function of water quality. As a result of this, many models exist to predict steady state lead concentration for varying water quality. However, little information is available on lead transition for changing water quality. Patterson and O'Brien (1979) performed lead mitigation laboratory studies which resulted in a unsteady state change due to changing water quality. In two of the three corrosion control techniques (bicarbonate addition alone and pH adjustment accompanied with bicarbonate addition) they found that lead corrosion rate dropped by approximately 35-40% in the first 5 days and by another 40-45% within the following 5-10 days.

Corrosion rates remained nearly stable during the rest of the experimental period (60 days). Coupon tests targeting lead corrosion control using these three control strategies over a 150 day period showed that total lead release was not instantaneous but resulted from a steep drop during the first 15 days of the study and remained nearly constant during the rest of the experimental period (Kashinkunti et al. 1999). There is no model that described release of total lead within the transitional period between the steady states driven by changing water quality.

RESIDUAL

Free chlorine and chloramines are the most widely used disinfectants in the world. Historically, free chlorine has been used more commonly than chloramines because it is a stronger oxidant and provides multiple benefits such as rapid microbial inactivation as well as color and odor oxidation. The major disadvantage of using free chlorine is a reaction with NOM to form disinfection byproducts which are potential carcinogens and regulated by USEPA (Vikesland, Ozekin and Valentine 2001). The need to reduce the potential for carcinogen formation in drinking water has forced many utilities to change to other methods of disinfection and residual maintenance than using free chlorine.

The concentration of free chlorine and chloramines in water is reduced by their reactions with organic and inorganic matter. Monochloramines are significantly more stable than free chlorine but are still relatively unstable and decay in the presence of humic materials by auto decomposition and direct reactions with NOM. A comprehensive mechanistic model was developed by Duirk et al. (2002) to predict the loss of monochloramines due to reactions with NOM. According to the model, the auto decomposition of monochloramines in absence of NOM started with hydrolysis reactions that produced free chlorine and ammonia. The molecules of free chlorine then reacted to form dichloramines, which rapidly dissociated into chlorides and ammonia.

In addition to organic matter, chlorine reacts with inorganic compounds such as iron, manganese, nitrites and sulfates. Iron is usually present in solution in its reduced form: ferrous iron Fe (II). Free chlorine reacts very rapidly with iron as shown in Equation (2.8).

$$2Fe^{2+} + HOCl + H^{+} \rightarrow 2Fe^{3+} + Cl^{-} + H_2O \tag{2.8}$$

The reaction between chloramine and ferrous ion Fe (II) is as shown in Equation (2.9) (Vikesland and Valentine 2002a).

$$2Fe^{2+} + NH_2Cl + H^{+} \rightarrow 2Fe^{3+} + Cl^{-} + NH_3 \tag{2.9}$$

From a distribution system viewpoint the reactions between chlorine/chloramines and iron are also important. Kiéné, Lu and Lévi (1998) found that chlorine consumption in old cast iron and steel pipes without a protective coating was higher than in synthetic material pipes. The rate of chlorine decay was shown to be directly dependent on iron corrosion in field scale investigations. Frateur et al. (1999) found that the effect of corrosion on free chlorine decay is mostly due to the chemical reactions between chlorine and ferrous ions released by iron dissolution and that little or no free chlorine was electrochemically consumed at the metal surface.

Chlorine Decay Modeling

Disinfectant loss (also referred to as decay or dissipation) in drinking water distribution systems is a major concern for water utilities. The rate of chlorine dissipation in a distribution system is influenced by a number of factors. It is an accepted fact that the decay of chlorine residual in the distribution systems is caused by chlorine reactions within the bulk liquid phase and with materials associated with the pipe wall such as pipe surface, corrosion products and biofilms (Biswas, Lu and Clark 1993; Rossman, Clark and Grayman 1994). Vasconcelos et al. (1997) postulated that chlorine decay models could be based on first-order kinetics for bulk reactions, and either zero-order or first-order kinetics for pipe wall reactions. The overall chlorine decay expression for water flowing through a distribution system is shown in Equation (2.10).

$$\left(\frac{dC}{dt}\right)_{Total} = \left(\frac{dC}{dt}\right)_{Bulk} + \left(\frac{dC}{dt}\right)_{Wall} \tag{2.10}$$

Bulk decay

Usually chlorine decay in the bulk liquid has been modeled using first-order kinetics with respect to chlorine concentration, but this approach is site specific and fails to explain the discrepancies between kinetic constants for different waters (Powell et al. 2000). The kinetic constants for reactions in the bulk liquid have been found to be dependent on the water quality, specifically on temperature and organic content of the water (Kiéné, Lu and Lévi 1998, Hua et al. 1999). Kiéné, Lu and Lévi (1998) described bulk chlorine decay using a first-order kinetic rate with respect to chlorine where the constant K_b is related with TOC and temperature as shown in Equation (2.11) and Equation (2.12).

$$\left(\frac{dC}{dt}\right)_{Bulk} = -K_b \times C \tag{2.11}$$

$$K_b = a \times [TOC] \times e^{\left(\frac{-b}{T}\right)} \tag{2.12}$$

where K_b = bulk decay constant
C = chlorine concentration in the bulk
a = 1.8 x 106 L/mg-h
b = 6050 °K
T = temperature in °K

This expression is similar to the Arrhenius equation, with b equal to Ea/R.

Wall decay

Disinfection loss due to reactions at the wall must take into account the reaction constant, mass transfer limitations and the available surface area to volume geometry (Frateur et al. 1999;

Vikesland, Ozekin and Valentine 2001; Vikesland and Valentine 2002b; Hallam et al. 2002). The expression for the wall reaction rate stated by Rossman (2000) and Walski, Chase and Savic (2001) is shown in Equation (2.13).

$$\left(\frac{dC}{dt}\right)_{Wall} = -\left(\frac{A}{V}\right)K_{W1}C_W \tag{2.13}$$

where K_{W1} = wall reaction rate constant (L/T)
A = is the surface area (L^2)
V = is the pipe volume (L^3)

For a pipe running full, the hydraulic radius reduces to the pipe diameter divided by 4. Equation (2.13) can be written as Equation (2.14):

$$\left(\frac{dC}{dt}\right)_{Wall} = -\frac{4}{D_p}K_{W1}C_W \tag{2.14}$$

K_{W1} must be adjusted for mass transfer limitations that are dependant on the flow regime within the pipe segment. A film resistance model of mass transfer with coefficients dependent on the flow regime and proportional to the difference of chlorine concentration between the bulk liquid and the pipe wall can be expressed as Equation (2.15) (Rossman, Clark and Grayman 1994; Walski, Chase and Savic 2001).

$$\left(\frac{dC}{dt}\right)_{Mass\ transfer} = -\frac{4}{D_p}K_F(C - C_W) \tag{2.15}$$

where K_F = mass transfer coefficient
Dp = pipe diameter
C = chlorine concentration in the bulk
C_W = chlorine concentration at the pipe wall

The rate at which chlorine moves towards the pipe wall is dependent on the flow conditions. Rossman (2000) and Walski, Chase and Savic (2001) provided a relationship between the mass transfer constant, K_F and the dimensionless Sherwood Number, S_h, as shown in Equation (2.16).

$$K_F = S_h\frac{d}{D_p} \tag{2.16}$$

where d = molecular diffusivity of the species being transported (L^2/T)
Dp = pipe diameter (L)

Equation (2.17) and Equation (2.18) are differential and integrated equations based on the assumption that chlorine reacts as it reaches the pipe wall and there is no accumulation at the wall.

$$\left(\frac{dC}{dt}\right)_{Wall} = \left(\frac{dC}{dt}\right)_{Mass\ transfer} \qquad (2.17)$$

$$K_{W1}C_W = K_F(C - C_W) \qquad (2.18)$$

Solving for C_W and substituting in Equation (2.13), the first-order wall reaction rate is expressed as shown in Equation (2.19). K_W is an overall wall decay constant combining the mass transfer constant, K_F and the wall reaction constant, K_{W1}.

$$\left(\frac{dC}{dt}\right)_{Wall} = -\frac{4}{D_p}\frac{K_{W1}K_F}{(K_{W1} - K_F)} \times C = -\frac{K_W}{D_p} \times C \qquad (2.19)$$

Temperature effects: The rate constants can be adjusted for temperature variations using the van't Hoff – Arrhenius equation shown in Equation (2.20). Using this equation, the rate constant at operation temperature, K_{Temp}, can be calculated from the rate constant defined at a base temperature (20 °C).

$$K_{Temp} = K_{20} \times A^{(Temp-20)} \qquad (2.20)$$

where K_{Temp} = dissipation constant at temperature T
K_{20} = dissipation constant at 20 °C

BIOSTABILITY

General

The presence of bacteria in soils and natural waters is related to bacterial growth in water distribution systems. Most bacteria participate in one of the following elementary cycles: 1) The carbon cycle, 2) the nitrogen cycle or 3) the sulfur cycle. Iron oxidizing and reducing bacteria also play a significant role in the environment. The bacteria involved in the carbon cycle are heterotrophic organisms, and have very diverse metabolic pathways. They can use several electron acceptors (i.e. oxygen, nitrates, nitrites, sulfates and ferric ions). The autotrophs (such as nitrifiers) can initiate the colonization of the inner surface of the distribution systems.

By converting inorganic carbon to organic carbon, the autotrophs produce a nutrient source for the heterotrophs. Thus the autotrophs are the primary initiators of a more complex food chain and support the colonization by heterotrophs even under low carbon level conditions. For this reason, multi-nutrient monitoring is necessary to determine which nutrients are significant for microorganism growth in drinking water distribution systems. Limiting carbon only in the distribution system does not lead to biostability if other nutrients are present in large concentrations.

Biological Instability of Potable Waters

Potable water in distribution systems transports an extremely diversified microbial flora (Block et al. 1997; Doggett 2000; Percival et al. 2000) and complex organic matter (Croué et al. 2000). A relatively significant fraction of the organic matter is biodegradable (Block et al. 1992). Even when the organic fraction is considered to be small: so called oligotroph ground waters (<0.5 mg DOC/L) or nanofiltered waters (<0.2 mg DOC/L), bacterial growth is observed (Crissot-Laruade et al. 1999; Devender 1995; Levi et al. 1992; Sibille et al. 1997). Moreover, in most potable water distribution systems, the water/material interface is a favored site for cells and organic matter accumulation and for multiplication of bacteria (Block et al. 1993; Servais et al. 1992, 1995; Van der Wende et al. 1989). This proliferation may be followed by the bacterial detachment (Ascon-Cabrera, Thomas and Lebeault 1995) or by the displacement from the pipe surface to the bulk liquid (Rittman 1989) and transportion in the bulk liquid.

Biofilm Activity in Drinking Water Distribution Systems

The density of fixed microorganisms has been reported to be up to $10^6 - 10^8$ cells per cm^2 in most distribution systems. However technical limitations make it difficult to obtain a precise idea of the quantity, diversity, and activity of the biofilm (Byrd, Xu and Colwell 1991; Colwell and Grimes 2000; Roszak and Colwell 1987). New non-destructive techniques are an option which circumvents some of the problems associated with traditional techniques in which biofilm is disrupted, resuspended, and then cultured on agar plates. These techniques significantly reduce the variability associated with destructive techniques in which the biofilm is scraped off the material surface. In addition the bias associated with selective media is reduced and the non-destructive techniques make it possible to quantify non-culturable organisms which are believed to be the majority of bacteria as a whole. One of these non-destructive techniques is the Protein ExoProteolytic Activity (PEPA) assay (Laurent and Servais 1995). The technique quantifies the rate of hydrolysis for an intact biofilm, and this rate can be directly related to biomass density. The technique quantifies hydrolysis by using an organic molecule which fluoresces when it is hydrolyzed. Thus rate of hydrolysis can be measured by the rate of change in fluorescence, and this can be directly correlated to the amount of fixed biomass in the distribution system sample. Thus these techniques identify both culturable and non-culturable bacteria.

Factors Controlling Biofilm Accumulation in Distribution Systems

The accumulation and proliferation of fixed bacteria or biofilms is usually controlled by a large number of parameters which have been well studied at bench scale and rarely in full-scale systems to account for hydraulics, temperature, nature and concentration of nutrients, bacterial density and species introduced in the system, nature of pipe materials and predators. These interactions are shown graphically in Figure 2.1.

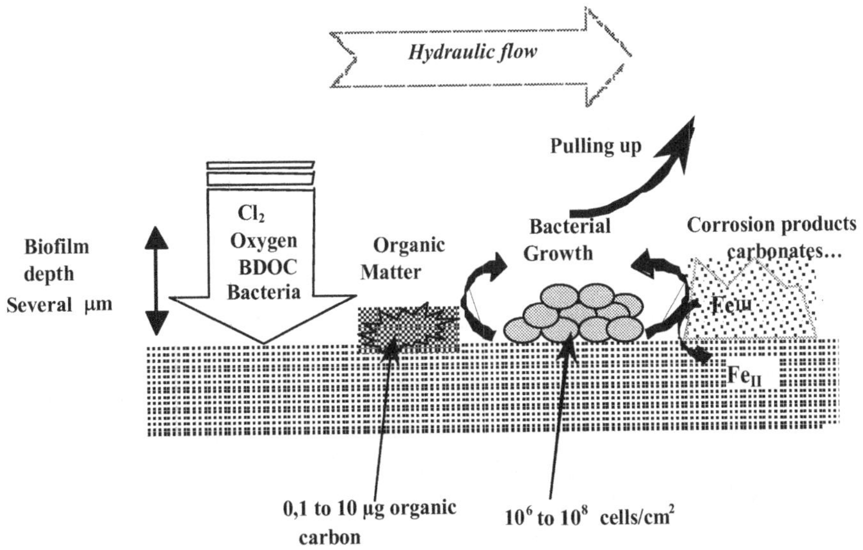

Figure 2.1 Scheme of several mechanisms taking place in the accumulation of biofilms on a surface in contact with potable water

The biofilm appears to be a metastable system, in one hand "fed" by the entrance of cells from the bulk (the velocity of deposit/fixation of cells on the surface of materials is correlated to the cell density of moving waters) and by the multiplication of bacteria, and in the other hand reduced by the constant sloughing and release of cells from the biofilm to the bulk. Under these conditions, the organization of the biofilm and the structure of the settlements mainly depend on several factors including:

- Hydraulic regime of the system (Bryers 1987, Characklis and Marshall 1990, Characklis and Wilderer 1989) and its variations (Stoodley, de Beer, and Lewandowski 1994).
- Bacterial species introduced in the system from the treated or non-treated source.
- Competition and advantage given to certain species due to their nutritional requirements (autotrotrophic nitrifiers vs heterotrophic populations) (Lipponen et al. 1998).
- Nature of the pipe materials used for potable water distribution systems. All of them are largely colonized by microorganisms (Niquette, Servais and Savoir 2000; Zacheus et al. 2000) but the supporting materials play a significant role in the selection of biomass and its organization (Kerr et al. 1999; Kielemoes, Hammes and Verstraete 2000; Pedersen 1990; Rogers et al. 1994; Van der Kooij et al. 1995). Indeed material determines the adsorption efficiency of the "pioneers" and can be a source of nutrients or growth factors. Finally, in the case of materials susceptible to corrosion, it has been clearly demonstrated that the presence of iron corrosion products enhances the activity and the production of heterotrophic biomass (Appenzeller et al. 2001).
- Resistance of fixed biomass to oxidants (Morin et al. 1999) can be partially explained by the reducing ability of the biofilm, fixed organic matter and sometimes of the

supporting material, as well as poor diffusion of the oxidant (De Beer, Srinivasan and Stewart 1994, Stewart and Raquepas 1995). The exposure of the biofilm bacteria to sub lethal oxidative stresses (Storz and Zheng 2000) leads to a remarkable cellular defense (surproduction of intracellular glutathion) (Saby 1999) and an increased resistance of the bacteria to oxidants (Saby et al. 2001).

NITRIFICATION

Chloramines were first used as a disinfectant in the drinking water system in 1916 in Ottawa, Canada (Race 1918, 1931). The first use of chloramines in the United States was in 1917 in Denver, Colorado (DeBerard 1917). By 1938, 16 percent of 2,541 utilities surveyed by AWWA were using chloramines, but the usage was dramatically reduced due to the shortage of ammonia during World War II (Gilcreas 1941, Means et al. 1986).

The EPA accepted chloramines as a secondary disinfectant in 1978 to limit the concentration of THMs, which was mainly responsible for the switch to chloramination at many utilities. In 1990, a survey revealed that approximately 22 percent of 438 surveyed utilities were utilizing chloramines as a secondary disinfectant (Wilczak et al. 1996). USEPA accepted chloramines as a primary disinfectant in 1983 (White 1999). Currently Stage 1 Disinfectants/Disinfection By-products (D/DBP) Rule enforces a THM4 limit of 80 μg/L and 60 μg/L for the sum of five HAA5 (USEPA 1998). The rule required large surface water systems to comply with the D/DBP Rule by January 2002, and ground water systems and small surface water systems to comply by January 2004.

The stepwise reactions by which chlorine and ammonia react to form chloramines are shown in Equation (2.21) to Equation (2.23) (Haas 1999).

$$NH_4^+ + HOCl \rightarrow NH_2Cl + H_2O + H^+ \tag{2.21}$$

$$NH_2Cl + HOCl \rightarrow NHCl_2 + H_2O \tag{2.22}$$

$$NHCl_2 + HOCl \rightarrow NCl_3 + H_2O \tag{2.23}$$

As shown in Equation (2.24) adding more chlorine will lead to breakpoint reaction:

$$2NH_4^+ + 3HOCl \rightarrow N_2 + 3Cl^- + 3H_2O + 5H^+ \tag{2.24}$$

The products of the reactions between chlorine and ammonia are chloramines. All three of these forms are called combined chlorine. The sum of their individual values is equal to the combined chlorine content (Haas 2000).

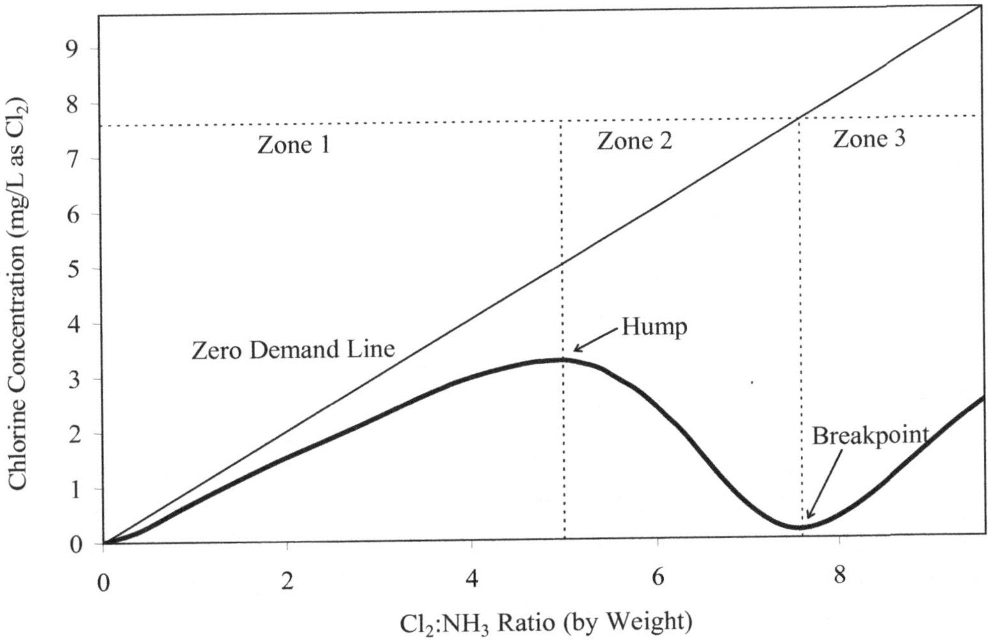

Figure 2.2 Chlorine-ammonia breakpoint curve.

The graphical illustration of these chemical reactions is called breakpoint curve. The breakpoint curve exists as varying amounts of chlorine is added to water containing a certain amount of ammonia nitrogen, as shown in Figure 2.2 (Taylor et al. 2001).

The principal reaction in zone 1 is the reaction between chlorine and the ammonium ion to form monochloramine (Equation (2.21)). All the way to the hump in the curve this reaction results in a chlorine residual containing only monochloramine. The hump occurs, theoretically, at chlorine to ammonia nitrogen weight ratio of 5:1 (molar ratio of 1:1). However, this is not always observed in practice. This ratio indicates the point where the reacting chlorine and ammonia nitrogen molecules are present in solution in equal amount (Haas 2000).

In zone 2, as the weight ratio begins to exceed 5:1 but less than 7.6:1 (molar ratio 1.5:1), some of the monochloramine reacts to form dichloramine (Equation (2.22)) and trichloramine (Equation (2.23)). At the breakpoint, all the ammonia nitrogen is oxidized to nitrogen with a chlorine to nitrogen weight ratio of 7.6:1 (Equation (2.24)). To the right of the breakpoint (zone 3), all ammonia compounds have been consumed, and additional chlorine dose in excess of 7.6:1 chlorine to nitrogen weight ratio results in the buildup of free chlorine residual (HOCl) (Haas 2000).

A survey revealed that 63% of the utilities have experienced evidence of nitrification episodes in the US (Wilczak et al. 1996; Odell et al. 1996). Another analytical survey of five chloraminated distribution systems in South Australia showed that 64% of samples tested positive for nitrifying bacteria (Cunliffe 1991; Regan, Harrington and Noguera 2002). High chloramine residual, up to 5-8 mg/L may not be a sufficient factor to control nitrification (Cunliffe 1991; Skadsen 1993; Wilczak et al. 1996). Adverse water quality effects caused by the nitrification episodes included a rapid decline in the chloramine and total ammonia-nitrogen residuals, elevated levels of nitrite and HPC, and decrease in alkalinity, pH and DO concentration (Wolfe et al. 1988; Valentine 1984; Wilczak et al. 1996).

The occurrence of nitrification was more frequent during summer or when the temperature was greater than 15 °C (Srinath, Loehr and Prakasam 1976, Wong-Chong and Loehr 1975; Lieu, Wolfe and Means 1993); nevertheless, there was also evidence of nitrification in cold water conditions (below 10°C) (Lieu, Wolfe and Means 1993).

Many systems have reported that 1-2 mg/L total residual is enough to prevent onset of nitrification (Odell et al. 1996). The pH range for nitrification is from 6.5 to 10 (Odell et al. 1996). Longer detention times increase the potential for nitrification to occur (Ike, Wolfe and Means 1988; Cunliffe 1991). The presence of sediment or tubercles in the distribution system can enhance the potential for nitrification by reducing residuals and providing favorable conditions for biological activity. Nitrification associated with longer storage times, or increased drawdown elevations has been observed and thus storage time is an important factor in storage tanks and covered reservoirs (Odell et al. 1996).

A survey performed by Wilczak et al. (1996), detected no meaningful trend between plant effluent ammonia concentration and Cl_2:N ratio, except for a general qualitative relationship of a decrease total ammonia concentrations with an increase in nitrite and nitrate-nitrogen. Skadsen (1993) documented that increasing Cl_2:N ratio did not prevent nitrification in the distribution system. However, a bench-scale experiment by Lieu, Wolfe and Means (1993) showed that at 25 °C the microbicidal activity of chloramines increased when the Cl_2:N ratio was increased from 3:1 to $\leq$ 4:1. However, the study showed that Cl_2:N ratio was not a factor at temperatures of 10 °C and 15 °C.

Wolfe et al. (1990) stated that AOB were not only capable of surviving in the presence of 1.2-1.5 mg/L of monochloramine but were also capable of growing in the presence of these disinfectant levels. Feben (1935) found that AOB could survive at chloramine residuals of 2 mg/L for as long as 60 minutes. The resistance of nitrifying bacteria to disinfection may result from their tendency of attaching to the surface of sediment and other debris (Watson et al. 1989). High levels of AOB have been detected in sediment and on the floors and walls of reservoirs (Ike, Wolfe and Means 1988; Wilczak et al. 1996).

Potential short-term control methods for nitrification include: reducing detention times (Wolfe et al. 1988; Ike, Wolfe and Means 1988; Negrin, Heyer and Cheng 1990; Lieu, Wolfe and Means 1993), increasing the ratio of Cl_2:N in order to minimizing excess ammonia in the finished water (Wolfe et al. 1988; Negrin, Heyer and Cheng 1990; Lieu, Wolfe and Means 1993), periodically chlorinating the water to breakpoint (Wolfe et al. 1988; Negrin, Heyer and Cheng 1990; Lieu, Wolfe and Means 1993), draining and refilling the storage tank (Burlingame and Brock 1985; Lieu, Wolfe and Means 1993), and regularly flushing (Wolfe et al. 1988).

MONOCHLORAMINES AND FREE CHLORINE

The use of monochloramine as a secondary disinfectant is increasing throughout drinking water utilities in the United States due to more stringent regulations. The D/DBP rule (USEPA, 1998) states that running annual average (RAA) levels of THMs shall not exceed 80 μg/L and the RAA levels of HAAs shall not exceed 60 μg/L and chloramines indeed reduce the potential for DBPs formation with respect to free chlorine (Brodtmann and Russo 1979; Norman, Harms and Looyenga 1980; Mitcham, Shelley and Wheadon 1983). In 1990, 22% of the 438 surveyed utilities were using chloramines as secondary disinfectant (Wilczak et al. 1996). Chloramines are generally accepted to be less reactive in the bulk aqueous stream than free chlorine. Some advantages of the use of chloramines are a more stable residual concentration, less odor and taste, lower HPC (Neden et al. 1992) and less production of disinfection by-products.

Even though chloramines have several advantages in light of the new regulations; however there are some disadvantages relative to chlorine. Whereas chloramines do not form regulated DBPs, and are the principle means of controlling DBPs, the D/DBP Rule also limits average distribution system concentrations to 4 mg Cl_2/L or less (USEPA 1998). Chloramines require prohibitively high CTs (concentration times contact time) to comply with regulations for pathogen inactivation, which severely limits use as primary disinfectant (USEPA 1990). Consequently, chloramines are not approved or used as a primary disinfectant in the state of Florida. Monochloramines decompose to ammonia, which increases ammonia in the distribution system. The presence of free ammonia in a distribution system promotes biological instability due to growth of ammonia-oxidizing bacteria which in turn deplete the chloramines residual. Two thirds of medium and large US distribution systems that use chloramines have experienced nitrification with 25% reporting moderate to severe nitrification problems (Wilczak et al. 1996).

Once an episode of nitrification has begun, chloramine concentration is quickly reduced. Therefore, to maintain good water quality, both AOB inactivation and AOB regrowth prevention strategies must be considered. Odell et al. (1996) showed a marked difference in AOB regrowth potential for different chloramines dosage in Metropolitan Water District of Southern California (MWDSC). At 1.7 mg/L as Cl_2, AOB regrowth occurred in 77% of the samples tested; at 2.0 mg/L as Cl_2, AOB regrowth appeared in 67 % of the samples tested; and with a combined residual of 2.5 mg/L as Cl_2, AOB regrowth appeared in 26% of test conditions.

Survival of AOB under disinfectant stress has been reported at different levels. A study conducted in California (Wolfe et al. 1990) indicated that nitrifiers were 13 times more resistant to monochloramine than to free chlorine and chloramine residuals at or below 1.5 mg/L may not control the growth of these organisms in the distribution system. Odell et al. (1996) showed that AOB survive and grow at 1.2-1.5 mg/L as Cl_2 in MWDSC waters, while Skadsen (1993) reported nitrification occurrence in the distribution system at 5-6 mg/L as Cl_2. AOB seem to have a significant resistance to disinfectant. CT with free chlorine for AOB is indeed far greater than CT for *E. coli.* This resistance may be due to the tendency of AOB to attach to surface and sediments. Nitrifying bacteria can grow in tight clusters, forming cyst-like structures, which are likely to be more resistant to disinfectant.

Lipponen, Suutari and Martikainen (2002) compared the occurrence of nitrification and the number of nitrifying bacteria in different drinking water distribution systems in Finland. The study found that the numbers of ammonia-oxidizing bacteria, nitrite oxidizing bacteria and their oxidation potentials were higher in distribution systems delivering chloraminated water when compared with networks carrying chlorinated waters. An observed increase in turbidity was associated with detachment of sediment with high concentration of nitrifiers from the pipe surface. A positive correlation was found between the occurrence of nitrifying bacteria and the numbers of heterotrophic bacteria and turbidity of the water.

Several parameters such as nitrite, nitrate, chloramines dose and residual, ammonia, pH, HPC, and DO can be used as indicators of nitrifying bacteria activity in a distribution system (Wilczak et al. 1996; Odell et al. 1996). Also, accelerated rates of monochloramine decay have been associated with elevated numbers of nitrifiers (Cunliffe 1991).

CHAPTER 3
METHODS AND PROCEDURES

PILOT DISTRIBUTION SYSTEM (PDS) DESIGN

The pilot distribution system was designed to mimic water quality changes resulting from single sources and blends of seven significantly different source waters in old distribution pipe systems. The feed water sources represented both the extreme and typical water chemistry likely to be experienced by utilities. The existing distribution systems traditionally received conventionally treated groundwater. Care was taken to maintain the internal film physical and chemical structures during pipe excavation, transportation and construction of the PDSs. Both the physical systems and pipe geometries selected represent typical scenarios experienced in a real distribution system.

The PDSs are composed of 18 different distribution lines. Lines 1 to 14 are hybrid lines that have four different materials: PVC, unlined iron, lined iron and galvanized iron pipes. The PDS was constructed of aged pipes that were obtained from existing utility distribution systems to represent the pipe materials used in the TBW Member Government's distribution systems. The pipes were removed from the Member Government's distribution networks, wetted, capped and transported to the pilot site at CCW (Cypress Creek Wellfield). Once onsite, the pipes were assembled and allowed to equilibrate with TBW (Tampa Bay Water) groundwater over a period of 4 months. After equilibrium was established, different blends were introduced into the PDS. The project was divided into 6 phases; each of three months duration. Similar blends were used in alternate phases to evaluate the effect of seasonal conditions on the PDS and related water quality.

Pilot Distribution System Components

Relevant characteristics of the individual lines include pipe identification, material of construction, pipe length and diameter and feed rates. The pilot distribution systems were identified sequentially (PDS01 to PDS18). The pilot distribution lines were operated to maintain either a five-day or two-day hydraulic residence time (HRT). Standpipes were located at the beginning and end of each PDS. The standpipes were made from translucent plastic pipe that were 48 inches long and had a 4 inches diameter. The retention time in the PDS feed standpipe was 3.1 and 7.8 hours for the two and five-day HRTs because of the low velocities associated with the two and five-day HRTs. These low velocities and high standpipe HRTs created a long holding time in the standpipes where HPCs grew. Microorganism mass transfer by diffusion is very possible in stagnant environments like the feed stream standpipe and PDS. Thus HPC back diffiusion from the PDS to the standpipe does occur and could account for elevated HPCs in influent samples taken from the standpipes, which were used to represent PDS influent.

Pilot distribution systems 01 to 14 were composed of four materials, laid out sequentially as:

- Approximately 20 feet (6.1 m)of 6-inch (0.15 m) diameter polyvinylchloride (PVC) pipe,
- Approximately 20 feet (6.1 m) of 6-inch (0.15 m) diameter lined cast iron (LCI) pipe,

- Approximately 12 feet (3.7 m) of 6-inch (0.15 m) diameter unlined cast iron (UCI) pipe,
- Approximately 40 feet (12.2 m) of 2-inch (0.05 m) diameter galvanized iron (G) pipe

PDS 15 to 18 were composed of a single material each as follows:

- PDS15: Eight reaches of approximately 12 feet (6.1 m) of 6-inch (0.15 m) diameter cast iron each,
- PDS16: Four pipe reaches of approximately 20 feet (6.1 m) of 6-inch (0.15 m) diameter lined cast iron plus 10 feet of 6-inch lined cast iron,
- PDS17: Five pipe reaches of approximately 20 feet (6.1 m) of 6-inch (0.15 m) diameter PVC each,
- PDS18: Six pipe reaches totaling 135 feet (41.1 m) of 2-inch (0.05 m) diameter galvanized iron pipe.

All pilot distribution systems were provided with a sampling port after each pipe reach to allow an assessment of water quality changes associated with each pipe reach. Pictures of the structures build for the field study are shown in Figure 3.1 to Figure 3.12. As shown in Figure 3.1, a truck and a stainless steel food grade trailer was used to haul surface water. The surface water was collected as shown in Figure 3.2 from an intake built by TBW specifically for this project. The raw surface water was stored in two 7000 gallon storage tanks as shown in Figure 3.3 before treatment. The large process area used to prepare the finished waters is shown in Figure 3.4 and was covered by a 4400 ft^2 of 6" cement pad and hurricane rated roof. The five trailers shown in Figure 3.5 from front to back contained an ozone pilot plant, a reverse osmosis pilot plant, a nanofiltration pilot plant, a storage facility and a field laboratory.

The annular reactors shown in Figure 3.6 were used for biofilm studies on the various finished waters produced for this investigation. The peristaltic pumps used to control the blends of the various finished waters are shown in Figure 3.7 with the feed stream standpipes used to regulate flow through the pilot distribution systems (PDSs), which can also be seen in Figure 3.8.

The PDSs were followed by coupon cradles as shown in Figure 3.9 and Figure 3.10. The cradles were six inch PVC pipes that housed four inch PVC pipes, which had been cut in half and supported pipe coupons for surface characterization studies. Finally the PDS effluent was directed to a corrosion shed as shown in Figure 3.11 that contained eighteen looped copper pipes and lead coupons as shown in Figure 3.12 for the copper and lead corrosion study.

Figure 3.1 Truck and stainless steel trailer used to haul raw surface water

Figure 3.2 Collection of raw surface water using TBW project intake

Figure 3.3 Raw surface water storage

Figure 3.4 Covered tanks for process treatment

Figure 3.5 Field trailers

Figure 3.6 Biological annular reactors

Figure 3.7 Peristaltic pumps and feed standpipes

Figure 3.8 Pilot distribution systems

Figure 3.9 Cradles for housing coupons

Figure 3.10 Mounted coupons

Figure 3.11 Corrosion shed

Figure 3.12 Copper lines containing lead coupons

PILOT PLANT OPERATION

Finished Waters

The following is a description of the finished waters produced by pilot water treatment systems (PWTS) located at the TBW Cypress Creek water treatment facility (WTF) project site. All finished waters discharged to the pilot distribution systems (PDSs) were produced at the project site. The seven PWTS and the finished water goals for the project are identified in Table 3.1 and Table 3.2.

Table 3.1
PDS source water descriptions

Symbol	System Description
G1	Ground water source. Treatment by aeration, disinfection by free chlorine with a residual of 4 mg/L after a 5 minute contact time, 4.0 mg/L chloramine residual.
G2	Ground water source. Treatment by lime softening to total hardness of 120 mg/L $CaCO_3$ disinfection by free chlorine with a residual of 4.0 mg/L after a 5 minute contact time, , 4.0 mg/L chloramine residual.
G3	Blend of finished G1, S1 and RO water source. Treatment by lime softening to total hardness of 120 mg/L $CaCO_3$ or alkalinity of not less than 50 mg/, 4.0 mg/L chloramine residual
G4	Blend of finished G1, S1 and RO water source. Treatment by membrane nanofiltration aeration, 4.0 mg/L chloramine residual.
S1	Surface water source. Treatment by ferric sulfate coagulation flocculation settling filtration disinfection by ozonation biologically activated carbon filtration, 4.0 mg/L chloramine residual.
S2	Surface water pretreated by ferric sulfate coagulation, flocculation and sedimentation, followed by nanofiltration, aeration and disinfection by free chlorine with a residual of 4 mg/L after a 5 minute contact time, 4.0 mg/L chloramine residual.
RO	Ground water source. Treatment by membrane reverse osmosis aeration disinfection by free chlorine with a residual of 4 mg/L after a 5 minute contact time , 4.0 mg/L chloramine residual.

Table 3.2
Pilot plant finished water treatment goals

Parameter	Standard	Target
PH units	7.4 min.	0.2 above pHs
Alkalinity (mg/L as CaCO3)	40 min	50
Calcium (mg/L as CaCO3)	50 min, 250 max	60 mg/L
Chloramine Residual (mg/L Cl_2)		4.5 mg/L
Na (mg/L)	80 max	<80
Cl (mg/L)	100 max	<100
Sulfate & Chloride Sum.	3.8 meq/L max	<3.8 meq/L
TDS (mg/L)	500 max	<500
Fe (mg/L)	0.15 max	<0.15
Color (CPU)	15 max	<15
TOC (mg/L)	3.6 max avg./6.5 max	<3.6
Ammonia (mg/L as N)	1 max	<0.5
Turbidity (NTU)	At filter 0.3 max/ 0.1 (95%), 0.25(100%)	<0.2
	Finished 1 max avg.	<0.3

Groundwater Pilot Plant

The groundwater, G1, pilot unit was designed to simulate the finished water of member governments that utilize conventional treatment of ground water. The G1 unit used the Cypress Creek well field as a raw water source. The main unit processes were aeration, disinfection and pH stabilization. Before disinfection, G1 was aerated to reduce hydrogen sulfide, produce an aerobic water supply and stabilize the water with respect to calcium carbonate. Aeration was achieved in G1 by pumping the raw water to the top of the finished water tank through a nozzle that sprayed the water inside the tank. This same aeration technique was used to stabilize G2 and G3.

Sodium hypochlorite was used as the source of free chlorine for primary disinfection and was dosed to provide a 4 mg/L residual after a 5 minute contact time. Afterwards, ammonium chloride was added to produce a 4 to 4.5 mg/L monochloramine residual. Ammonia was added in the form of NH_4Cl at a 4:1 to 5:1 ratio. The NH_3:Cl_2 ratio was initially 4:1 to protect against DBP formation. This ratio was increased to 5:1 in Phase III to reduce free ammonia. Chloramines were formed and boosted in the finished water storage tanks before discharge to the PDSs. During the first six months of the project, preformed monochloramines were used for boosting residual, which sometimes resulted in an 3:1 Cl_2:NH_3 ratio during the first six months of the project. Free chlorine was used for boosting residual after the first six months, which maintained a 5:1 Cl_2:NH_3 ratio during the last year of the project. Five thousand gallons of G1 were produced each week.

Lime Softening Pilot Plant

The lime softening groundwater treatment pilot system that produced G2 used raw groundwater from the Cypress Creek well field. Three to four meq/L of CaO were added to the raw groundwater to achieve a settled total hardness of 120 mg/L as $CaCO_3$. G2 Alkalinity following groundwater softening was always greater than 50 mg/L as $CaCO_3$ and did not require alkalinity addition to meet the alkalinity for finished water.

$Ca(OH)_2$ slurry at 111,000 mg/L was used for lime feed. The slurry fed to the raw water during rapid mix. The water overflowed into a tank where flocculation was achieved using a submersible pump that suspended and recirculated the floc until softening was achieved. After the tank was full, the pump operated for 1 hour and then the softened water settled for 24 hours. The settled water was pumped out of the settling tank and filtered using 1 micron bag filtration followed by filtration with 0.35-micron cartridge filtration. Two thousand gallons of G2 were produced weekly.

Blended Lime Softened Pilot Plant

The three main finished waters (S1, G1 and RO) were blended at two different ratios and then further treated by softening (G3) and nanofiltration (G4). The percentages of finished waters used in this blend for each study period is presented in Table 3.3.

Table 3.3
Percentage of finished waters blended to make G3 and G4 feed source

Study Period	Blend Percentage (%)		
	G1	S1	RO
Phase I and III	23	45	32
Phase II and IV	60	30	10

G3 water was produced by lime softening of blended S1, RO and G1 finished waters to achieve less than 120 mg/L as $CaCO_3$ total hardness, or greater than 50 mg/L as $CaCO_3$ alkalinity. The process was the same as for G2. Softening was always required to meet the total hardness criteria of the G1, S1 and RO blends that were used in the field study. However, it is very possible that a blend of these waters would have a total hardness less than 120 mg/L as $CaCO_3$ and an alkalinity greater than 50 mg/L as $CaCO_3$.

The softened water (G3) and membrane treated water (G4) process did not require a free chlorine contact time since G1, S1 and RO had been previously disinfected. The chloramine residual after the softening and filtration needed to be increased to 4.0 mg/L as Cl_2 using preformed chloramines. Sodium hydroxide was used for stabilization. One thousand gallons of G3 were produced every week.

Blended Nanofiltration Pilot Plant

G4 water was produced by an integrated membrane system (IMS) treating the same blend of G1, S1 and RO that was used to produce G3. A conventional Osmonics nanofilter was used to produce G4. Hypersperse MDC700TM was dosed at 2.7 mg/L to the NF feed stream to control scaling. The NF membrane unit was operated at 75 % recovery, which produced 1.3 gpm of permeate. The G4 permeate stream was aerated using the aeration towers shown in Figure 3.4.

Calcium chloride and sodium bicarbonate addition were added when needed to achieve the hardness and alkalinity goals of 1 meq/L as $CaCO_3$. The pH of the G4 finished water was maintained at 0.1 to 0.3 units above pH_s. A chloramine residual was maintained in the G4 feed stream at 4 mg/L as Cl_2.

Surface Water Pilot Plant

S1 was produced by enhanced coagulation, ozonation, biologically activated carbon (BAC) filtration, aeration and chloramination. The S1 production processes mirrored the TBW Regional Surface Water Treatment Facility's treatment process as closely as possible. The raw surface water was taken from the Hillsborough River and hauled weekly to the field facility for treatment. The raw SW was initially treated by coagulation, sedimentation and filtration (CSF) using ferric sulfate coagulation. The dose of ferric sulfate was determined based on water quality and treatment. Total organic carbon (TOC) concentration of the finished SW had to be less than 3.6 mg/L, which was the maximum allowable TOC concentration in the finished SW. The combined sulfate and chloride concentration had to be less than 3.8 meq/L, which controlled the maximum coagulant dose. Inline addition of 0.5 mg/L of Cytec N-1986 coagulant aid was used. The flocculated water was settled for 24 hours, and filtered through a 1 micron bag filter and followed by a 0.35 micron cartridge filter in series before transfer to the CSF storage tank.

Ozone and monochloramines were used for primary disinfection and residual maintenance respectively. A 1.4 mg-min/L CT as specified for 99.9 percent reduction of Giardia Lamblia at 10 ^{o}C was maintained during ozonation. The required contact time for this CT was determined by a tracer study on the 100 gallon contact tank, which determined that 10 % of the water exited the tank in 2.5 minutes (T10). A minimum ozone residual of 0.6 mg/L in the effluent stream was maintained to achieve the 1.4 mg/L-min CT. The ozonated water was filtered through 2 feet of BAC with a 5 minute EBCCT and then again through a 0.35 micron bags filter. The ozone pilot plant was housed in the front trailer as shown in Figure 3.5. Chloramine addition to S1 was different than for other processes because there was no need to use free chlorine as a primary disinfectant. Preformed chloramines were added from a stock that was made on the same day S1 was made until a 4 mg/L chloramine residual as Cl_2 was achieved in the S1 storage tank. Maintenance of chloramine residual in the storage tank was done as previously described.

Integrated Membrane System Pilot Plant

S2 pilot process was an integrated membrane system which consisted of nanofiltration (NF) of CSF surface water, which was housed in the second trailer from the front as shown in Figure 3.5. A conventional Osmonics nanofilter was used to produce S2. Hypersperse MDC700TM was dosed at 2.7 mg/L to the NF feed stream to control scaling. The NF membrane unit was operated at 75 % recovery, which produced 1.3 gpm of permeate. The S2 permeate stream was aerated using the aeration towers as previously described.

Calcium and alkalinity were added to the S2 permeate stream to meet project specifications. Calcium chloride and sodium bicarbonate were added to reach 1 meq/L as $CaCO_3$ goal for Ca hardness and carbonate alkalinity. Sodium hydroxide was added to achieve a finished pH that was 0.1-0.3 pH units above pH_s. Primary disinfection was achieved using a solution of sodium hypochlorite (free chlorine) with a minimum 4 mg/L residual and allowing a

2-minute contact time. A 4 mg/L chloramine as Cl_2 residual for secondary disinfection was produced in the S2 storage tank by adding ammonium chloride. Chloramine residual in the S2 storage tank was maintained as previously described.

Reverse Osmosis Pilot Plant

Reverse osmosis (Desalination) treatment pilot system produced a finished water that was described as RO. The RO pilot plant was housed in the third trailer from the front as shown in Figure 3.5 and utilized raw groundwater for the feed stream. The RO PWTS required the addition of TDS, calcium and alkalinity to the permeate to represent the finished water produced by the TBW Regional Desalination Facility. RO pretreatment consisted of 2.7 mg/L antiscalent addition (Hypersperse MDC700TM) followed by 5-micron cartridge filtration.

The Reverse osmosis (high rejection) membrane filtration unit was operated at 75% recovery, producing 10 gpm permeate flow, which was aerated by a 10-inch diameter aeration tower filled with tripack plastic packing as shown in Figure 3.4 previously. After aeration, 50 mg/L of sea salt was added to the aerated permeate stream to simulate the TBW desalination process. Calcium chloride and sodium bicarbonate were also added to meet the calcium and alkalinity specifications. The finished was stabilized with sodium hydroxide to 0.1 to 0.3 pH units above pH_s.

A diagram showing the water flow route through the PWTS is presented in Figure 3.13. The finished water qualities from these pilot processes were set to match the existing or future sources of water proposed by TBW. For all the waters, a new batch was prepared at the beginning of every week. The water quality analysis schedule required in the field for monitoring finished water storage and process operation is presented in Table 3.4. Each of the seven finished water storage tanks was measured for water stability parameters, disinfection residual, color and UV-254 using field methods. Additionally, each of the storage tanks was measured daily for disinfection residual maintenance and pH. The feed, permeate and concentrate from the membrane process were measured prior to post treatment to evaluate membrane performance and productivity. Additional water quality monitoring was conducted to evaluate specific unit processes within a treatment system. The water quality sampling schedule for parameters measured at the UCF Laboratory is presented in Table 3.5. The average water quality for the different source waters are enumerated in Table 3.9.

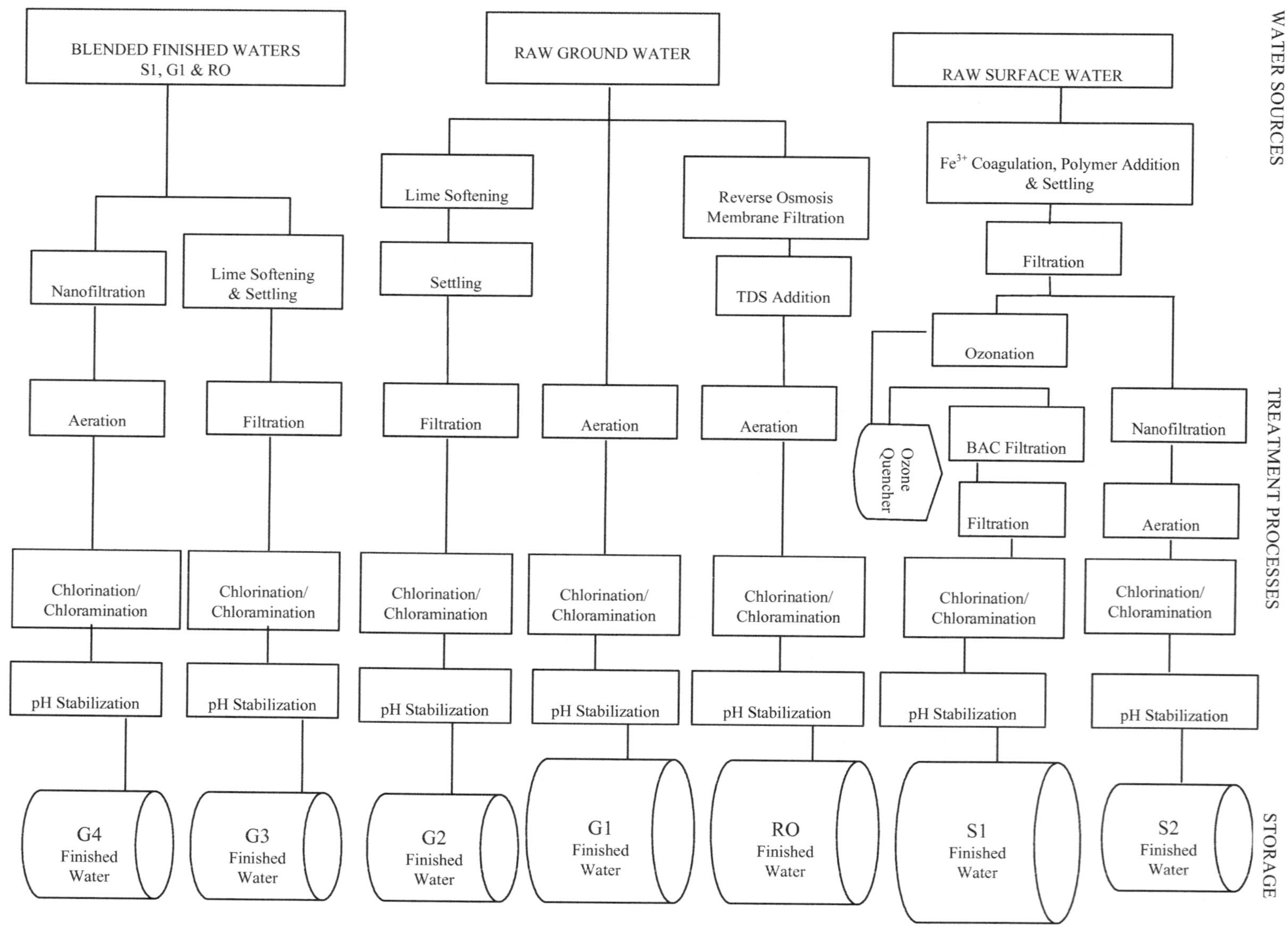

Figure 3.13 Process train layout

Table 3.4
UCF laboratory analysis schedule for water treatment processes

Parameter	G1 Raw	S1 Raw	G1 Fin	G2 Fin	G3 Fin	G4 Fin	S1 Fin	S2 Fin	RO Fin
Ammonia	Biwk	Biwk	Biwk	Biwk	Biwk	Biwk	Biwk	Biwk	Biwk
PH	Biwk	Biwk	Biwk	Biwk	Biwk	Biwk	Biwk	Biwk	Biwk
Alkalinity	Biwk	Biwk	Biwk	Biwk	Biwk	Biwk	Biwk	Biwk	Biwk
Ca	Biwk	Biwk	Biwk	Biwk	Biwk	Biwk	Biwk	Biwk	Biwk
Mg	Biwk	Biwk	Biwk	Biwk	Biwk	Biwk	Biwk	Biwk	Biwk
UV-254	Biwk	Biwk	Biwk	Biwk	Biwk	Biwk	Biwk	Biwk	Biwk
True Color	Biwk	Biwk	Biwk	Biwk	Biwk	Biwk	Biwk	Biwk	Biwk
Conductivity	Biwk	Biwk	Biwk	Biwk	Biwk	Biwk	Biwk	Biwk	Biwk
TDS	Biwk	Biwk	Biwk	Biwk	Biwk	Biwk	Biwk	Biwk	Biwk
Sodium	Biwk	Biwk	Biwk	Biwk	Biwk	Biwk	Biwk	Biwk	Biwk
Chloride	Biwk	Biwk	Biwk	Biwk	Biwk	Biwk	Biwk	Biwk	Biwk
Sulfate	Biwk	Biwk	Biwk	Biwk	Biwk	Biwk	Biwk	Biwk	Biwk
Iron	Biwk	Biwk	Biwk	Biwk	Biwk	Biwk	Biwk	Biwk	Biwk
TOC	Biwk	Biwk	Biwk	Biwk	Biwk	Biwk	Biwk	Biwk	Biwk
Turbidity	Biwk	Biwk	Biwk	Biwk	Biwk	Biwk	Biwk	Biwk	Biwk

Note: Biwk = Biweekly Sample collection (every two weeks)

Table 3.5
Field analysis schedule for water treatment processes

	G1 Raw	S1 Raw	G1 Fin	G2 Fin	G3 Fin	G4 Fin	S1 Fin	S2 Fin	RO Fin	G3/G4, S1/S2 feed	RO, S2, G4 Permeate
Free Cl_2	n/a	n/a	D	D	D	D	D	D	D		
Total Cl_2	n/a	n/a	D	D	D	D	D	D	D		
Ammonia	B	B	D	D	D	D	D	D	D		
Temperature	B	B	D	D	D	D	D	D	D	B	B
PH	B	B	D	D	D	D	D	D	D	B	
Alkalinity	B	B	B	B	B	B	B	B	B	B	
CaH	B	B	B	B	B	B	B	B	B	B	B
T. Hardness	B	B	B	B	B	B	B	B	B	B	
MgH	B	B	B	B	B	B	B	B	B	B	
UV-254	B	B	B	B	B	B	B	B	B	B	
Color	B	B	B	B	B	B	B	B	B	B	
Turbidity	B	B	B	B	B	B	B	B	B	B	
D.O.	B	B	B	B	B	B	B	B	B	B	
Conductivity	B	B	B	B	B	B	B	B	B	B	B
TDS	B	B	B	B	B	B	B	B	B	B	B
Chlorides		B					B		B		
Sulfates		B					B				
Iron							B				

Note: D = Daily, B = Batch = after each batch is finished.
Color and UV-254 are filtered but pH is not adjusted. Mg is determined by difference between CaH and TH. pHs is determined after all chemicals are added to adjust water quality to targeted goals. RO, S2 and G4 concentrate was measured for TDS by conductivity probe.

Process Finished Water Quality

Water quality monitoring during production was conducted to evaluate specific unit processes within a given treatment system. Raw surface water experienced the most seasonal water quality variations as shown in Table 3.6. Average raw and finished water quality parameters for the duration of the pilot study are presented in Table 3.7, Table 3.8, and Table 3.9.

The following variations in average water quality for raw surface water were observed between the dry season (December-May) and the wet season (May-November), alkalinity dropped from of 143 mg/L as $CaCO_3$ to 87 mg/L as $CaCO_3$, total hardness dropped from 238 mg/L as $CaCO_3$ to 156 mg/L as $CaCO_3$, color increased from 36 CPU to 192 CPU, TDS dropped from 314 mg/L to 200 mg/L. The low alkalinity during the wet season necessitated the supplementation of raw surface water alkalinity using sodium hydroxide in order to ensure proper coagulation and achieve the target water quality goals. Raw water coagulation was conducted at pH 4.5. During the wet season, alkalinity was seldom less than 1 meq/L, and during the dry season, alkalinity was seldom higher than 4 meq/L. NaOH addition was not required during the dry season due to high alkalinity, low color and a relatively low ferric sulfate dose. During the wet season as much as 4 meq/L of ferric sulfate were required to meet color and TOC removal goals, hence 3 meq/L of NaOH were required to offset the acid strength of ferric sulfate, (0.7 meq H^+/meq $Fe_2(SO_4)_3$). The range of sodium addition to S1 was from zero to approximately 70 mg/L as Na.

The finished waters for all processes were produced to meet target water quality objectives designed to mimic actual and expected TBW blended water quality shown in Table 3.7.

Table 3.6
Seasonal water quality fluctuations for raw surface water

Season	Water quality parameters								
	pH	Alk	TH	Color	UV-254	Turbidity	DO	TDS	SO_4^{2-}
Dry	7.9	143	238	36	0.10	2	6.3	314	120
Wet	7.2	87	156	192	0.80	1.6	3.0	200	79

Table 3.7
Water quality objectives for finished process waters

Parameters	Finished water type						
	G1	G2	G3	G4	S1	S2	RO
pH	pHs + 0.3	>pHs	>pHs	>pHs	>pHs	>pHs	>pHs
T. Hardness (mg/L as $CaCO_3$)	240	120	120	60 min	60 min	60 min	60 min
Alkalinity (meq/L)	4	1 min	1 min	1 min	1 min	1 min	1 min
Total Cl2 (mg/L)	4.0	4.0	4.0	4.0	4.0	4.0	4.0
Na (mg/L)	80 max	80 max	80 max	80 max	80 max	80 max	80 max
SO_4 & Cl^- (meq/L)	3.8 max	3.8 max	3.8 max	3.8 max	3.8 max	3.8 max	3.8 max
TDS (mg/L)	500 max	500 max	500 max	500 max	500 max	500 max	500 max
Color (CPU)	15 max	15 max	15 max	15 max	15 max	15 max	15 max
TOC (mg/L)	3.6 max	3.6 max	3.6 max	3.6 max	3.6 max	3.6 max	3.6 max

Despite the reported fluctuations in raw surface water quality, the surface water PWTS consistently met the set finished water quality targets (Table 3.2 and Table 3.7), reducing turbidity to 0.3 NTU, color to 1 CPU and UV-254 to 0.038 cm^{-1}. The finished surface water experienced elevated sulfates (180 mg/L) due to the addition of ferric sulfate as a coagulant. Conductivity and TDS also increased to 591 µS/cm, and 394 mg/L respectively.

Raw groundwater water quality did not change appreciably for the duration of the project. The average conductivity was 517 µS/cm with a standard deviation of 30 µS/cm. The average TDS was 344 mg/L with a standard deviation of 20 mg/L. Hence, 95 % of the time the range of TDS was from 304 tp 384 mg/L. G1 finished water had similar characteristics to the raw ground water except for DO which increased from an average of 0.2 mg/L to 7.5 mg/L due to the aeration process.

Minimal variations were experienced in most finished water quality parameters measured as shown by the standard deviation values presented in Table 3.9. The largest fluctuations were observed in G1, for which color had an average of 5 CPU with a standard deviation of 4 CPU, and turbidity with an average of 0.5 NTU and standard deviation of 0.5 NTU. Groundwater softening process (G2 finished water) significantly reduced calcium hardness (212 to 88 mg/L), alkalinity (211 to 93 mg/L as $CaCO_3$), TDS (314 to 198 mg/L). Similarly, softening of the blended S1/G1/RO water resulted in reduction in calcium hardness from 187 mg/L as $CaCO_3$ to 99 mg/L as $CaCO_3$ for G3. As expected, this calcium concentration was higher than that for RO at 64 mg/L as $CaCO_3$. G4, produced by nanofiltration of the blended S1/G1/RO, had significantly reduced average calcium (77 mg/L), TDS (207 mg/L) and color (0 CPU) with respect to the G3/G4 feed as shown in Table 3.9.

All the processes performed as expected during the project, consistently meeting water quality constraints set for the project (Table 3.2 and Table 3.7). Average finished water quality for selected parameters are provided in Table 3.8 and Table 3.9.

RO water process produced water with the lowest concentration of organics. The TOC values for RO and NF water were the lowest followed by surface water, softened water and groundwater as shown in Table 3.8. The complete order for the waters produced is RO<S2<G4<SW<G3<G4<GW. Membrane treated waters had the lowest turbidity, sulfates, and silica. Chlorides, sodium, calcium, magnesium and alkalinity were high for RO water due the addition sea salt to mimic desalination of sea water.

All the waters were maintained above the alkalinity limit of 1 meq/L (50 mg/L as CaCO3) as required. The total hardness target of 120 mg/L as $CaCO_3$ for the softened waters (G2 and G3) was obtained consistently during the project. All finished waters were stabilized with respect to the Langelier index, maintaining a LSI of between +0.1 and +0.3.

Table 3.8
UCF laboratory process waters selected average water quality characteristics

Parameter	Units	G1	G2	G3	G4	S1	S2	RO
Turbidity	NTU	0.5	0.2	0.3	0.1	0.3	0.2	0.2
Sodium	mg/L	16	13	38	28	81	45	52
Chlorides	mg/L	28	22	48	41	32	67	79
Calcium	mg/L as $CaCO_3$	212	88	99	77	134	62	64
Alkalinity	mg/L as $CaCO_3$	208	93	64	88	59	68	68
SO_4	mg/L	27	26	76	5	184	12	6
TOC	mg/L	3.3	2.2	1.5	0.8	1.2	0.4	0.3
Si	mg/L	14	14	10	10	9	9	1
HPC	cfu/mL	96	9	62	109	29	19	66

Table 3.9
Water quality averages for the water pilot treatment processes

Process	Free Cl_2	T Cl_2	NH_3	pH	Alkalinity	Ca H	TH	HPC	UV 254	Color	Turbidity	D.O.	Cond.	TDS	Temp	Cl^-	SO_4	Fe
Statistic	mg/L	mg/L	mg/L		mg/L $CaCO_3$	mg/L $CaCO_3$	mg/L $CaCO_3$	cfu/ml	cm^{-1}	CPU	NTU	mg/L	µS/cm	mg/L	°C	mg/L	mg/L	mg/L
G1																		
Average	0.1	4.7	0.2	7.9	208	212	243	96	0.070	5	0.5	7.5	515	346	23	25	28	0.06
Std Dev	0.1	0.6	0.1	0.2	20	10	12	10	0.005	4	0.5	0.6	31	39	3	6	8	0.04
G2																		
Average	0.1	4.8	0.2	8.0	93	88	121	9	0.065	1	0.2	7.8	292	194	23	0.1	4.8	0.2
Std Dev	0.1	0.4	0.1	0.2	12	10	9	6	0.004	1	0.2	0.7	27	18	4	0.1	0.4	0.1
G3																		
Average	0.1	4.6	0.3	8.1	64	99	123	62	0.055	0	0.3	8.6	429	285	23			
Std Dev	0.1	0.7	0.1	0.2	17	11	10	9	0.048	1	0.5	0.9	37	25	4			
G4																		
Average	0.1	4.6	0.2	8.2	88	77	89	109	0.030	0	0.1	8.4	312	207	23			
Std Dev	0.1	0.7	0.1	0.1	19	16	18	2	0.007	0	0.1	0.8	33	21	5			
RO																		
Average	0.2	4.8	0.1	8.3	68	64	70	66	0.023	0	0.2	8.2	411	274	23	79	3	0.01
Std Dev	0.1	0.6	0.1	0.2	4	4	5	4	0.005	0	0.1	0.6	23	18	3	8	5	0.01
S1																		
Average	0.1	4.5	0.2	8.2	59	134	157	29	0.038	1	0.3	9.5	591	394	22	30	180	0.04
Std Dev	0.0	0.4	0.1	0.2	5	54	65	2	0.049	1	0.2	1.8	87	58	4	6	46	0.03
S2																		
Average	0.2	5.0	0.1	8.3	68	62	69	19	0.030	1	0.2	8.2	357	238	23	67	11	0.02
Std Dev	0.1	0.7	0.1	0.2	6	5	6	3	0.007	1	0.1	0.9	47	32	5	7	6	0.02
G3/G4 Feed																		
Average	0.1	3.7		8.1	126	163	187	-	0.046	3	0.4	8.7	517	344	24			
Std Dev	0.1	0.4		0.1	25	26	30	-	0.009	2	0.3	1.0	30	20	3			
S1/S2 Feed																		
Average				5.4	16	150	179	-	0.067	5	4.3	6.8	465	312	23	25	269	
Std Dev				0.8	13	52	58	-	0.123	10	4.6	1.3	90	56	4	4	77	
Raw - GW																		
Average				7.5	211	212	241	-	0.063	6	0.3	0.2	475	314	23			
Std Dev				0.1	7	10	12	-	0.009	4	0.1	0.4	27	24	2			
Raw - SW																		
Average				7.4	102	143	173	-	0.6	135	1.6	4.7	346	228	22	19	75	
Std Dev				0.4	33	55	65	-	0.5	101	1.3	2.0	125	83	5	4	75	

The chloramine concentration for all finished water tanks was maintained at 4.5 mg/L, which dropped to 4.0mg/L prior to entering the pilot distribution system. HPC in all the finished water tanks were at or below 100 cfu/mL, indicating little or no microbiological activity in the storage tanks. However, HPCs were over 10^4 cfu/ml on some occasions in the initial PDS samples, which were taken from the PDS standpipes. As previously noted, the standpipe HRT was 3 to 8 hours which could have accounted for some HPC growth, and the HPCs taken in the standpipe samples were affected to some extent by back diffusion from the water in the PDS. The exact cause of high HPCs in the standpipe samples is unknown.

Overall, all parameters were maintained within the set limits. Warning limits for the different parameters were set at two standard deviations from water quality goals, while action limits were set at three standard deviations from the target limit. No corrective action on finished water was ever required indicating consistent and good control in the production of all the finished waters.

PILOT DISTRIBUTION SYSTEM OPERATION

After construction, the PDSs were equilibrated for a period of 144 days with a conventionally treated groundwater. A requirement to provide a substantial equilibration period (approximately three months) had been reported in the literature to allow stabilization of water quality parameters after construction of pilot testing facilities. It is believed that mechanical disturbance of the pipe interior films (chemical and/or biological) could result from harvesting, transport, and reassembly of the aged pipe materials. During the first part of the equilibration period, finished water from the Tampa Bay Water Cypress Creek Facility was used. This source was characterized by a low dissolved oxygen concentration and free chlorine residual. In the latter stage of equilibration, groundwater from the pilot treatment facilities (Source G1) was employed. This water was produced by aeration of raw groundwater from the Cypress Creek Well Field, followed by stabilization and chloramination. The influent and effluent from the PDS lines were monitored during the equilibration period to verify that any transient effects associated with PDS construction had been eliminated. Attainment of equilibration was defined by an absence of increase of apparent color between the PDS feed and effluent.

In November 2001, it was determined that the PDS lines were equilibrated and a decision was made to implement the blending study. The pilot distribution system was operated at a 5 day hydraulic retention time (HRT) initially (Phase I to III, 12/8/01 to 8/30/02) to simulate dead end conditions. The HRT was changed to 2 days during the final two weeks of Phase III through Phase VI (8/31/02 to 6/28/03) to permit maintenance of chlorine residual. Different source waters and their blends were introduced into the PDS by dosing pumps feeding individual influent standpipes for each PDS. These blends were selected to model the anticipated operations by TBW. The source waters for the PDSs were changed during phase II and IV with reassignment of individual blends to different pilot distribution lines. In most cases, the blend composition was not altered, relative to the composition in Phase I and III. In these instances, the blend was redirected to feed a different hybrid line.

Preparation of G3 and G4 was accomplished with a revised ratio of the three principal sources as presented previously in Table 3.3. The input to the single material lines was also modified with the same ratio of G1, S1, and RO used to prepare G3 and G4. The source waters for phases I to IV for each PDS by phase are summarized in Table 3.10. Phases V and VI were used for specific studies and the PDS operation is presented in the section called "additional studies".

The high frequency line (PDS14) operation includes twice-daily changes in the blend ratio. These adjustments are selected to maintain long-term average flows in line with ultimate capacities of the three major sources in the TBW service area: G1 (62%), S1 (27%), and RO (11%). Selection of actual blends was defined by Monte Carlo simulation methods. The three scenarios represent blends resulting from removal of one principal source (G1, S1, or RO) from production. Actual alteration of the blend was performed at the beginning and end of the production day (approximately 8 AM and 5 PM, respectively).

Table 3.10
Blends used for different pilot distribution systems (PDS) over the period of the study.

PDS – Material	Phase I (12/8/01 to 3/15/02) & Phase III (6/15/02 to 9/13/02)	Phase II (3/16/02 to 6/14/02) & Phase IV (9/14/02 to 12/13/02)	Comments
01 – Hybrid	GW (100)	G2 (100)	switch with PDS02
02 – Hybrid	G2 (100)	GW (100)	switch with PDS01
03 – Hybrid	SW (100)	S2 (100)	switch with PDS13
04 – Hybrid	G4 (100)	G3 (100)	switch with PDS12
05 – Hybrid	RO (100)	SW (100)	switch with PDS03
06 – Hybrid	GW (55) SW (45)	GW (68) RO (32)	switch with PDS07
07 – Hybrid	GW (68) RO (32)	GW (55) SW (45)	switch with PDS06
08 – Hybrid	GW (23) SW (45) RO (32)	GW (60) S2 (30) RO (10)	switch with PDS09
09 – Hybrid	GW (60) S2 (30) RO (10)	GW (23) SW (45) RO (32)	switch with PDS08
10 – Hybrid	G2 (50) SW (50)	G2 (62) SW (24) RO (14)	switch with PDS11
11 – Hybrid	G2 (62) SW (24) RO (14)	G2 (50) SW (50)	switch with PDS10
12 – Hybrid	G3 (100)	G4 (100)	switch with PDS04
13 – Hybrid	S2 (100)	RO (100)	switch with PDS05
14 – Hybrid	GW (62) SW (27) RO (11)	GW (62) SW (27) RO (11)	high frequency line
15 – Unlined Cast Iron	GW (23) SW (45) RO (32)	GW (60) SW (30) RO (10)	change blend pcts
16 – Lined Cast Iron	GW (23) SW (45) RO (32)	GW (60) SW (30) RO (10)	change blend pcts
17 – PVC	GW (23) SW (45) RO (32)	GW (60) SW (30) RO (10)	change blend pcts
18 – Galvanized Iron	GW (23) SW (45) RO (32)	GW (60) SW (30) RO (10)	change blend pcts

Note: Values in () indicate the percentage of the source water in blend

PDS Water Quality Schedule

Samples were collected and analyzed in the field and at the UCF laboratory. Monitoring and analyses of many physical, chemical and biological water quality parameters was carried out on the influent and effluent to the PDS. Table 3.11 shows the different parameters that were monitored during the study. Weekly analyses were completed in the field laboratory and the bi-weekly samples were brought back to UCF laboratory for analysis.

Table 3.11
Pilot distribution system monitoring program phase I through IV

Parameter	Field Analysis		UCF Laboratory	
	PDS Influent	PDS Effluent	PDS Influent	PDS Effluent
pH	Weekly	Weekly	Biweekly	Biweekly
Alkalinity	Weekly	Weekly	Biweekly	Biweekly
Turbidity	Weekly	Weekly	Biweekly	Biweekly
Free chlorine	Weekly	Weekly		
Chloramine residual	Weekly	Weekly		
Nitrite-N	As needed	As needed	Biweekly	Biweekly
Nitrate-N	As needed	As needed	Biweekly	Biweekly
Ammonia-N	Weekly	Weekly		
Dissolved oxygen	Weekly	Weekly		
Temperature	Weekly	Weekly		
Conductivity	Weekly	Weekly	Biweekly	Biweekly
Apparent Color	Weekly	Weekly		
True color			Biweekly	Biweekly
UV-254	Weekly	Weekly	Biweekly	Biweekly
TDS			Biweekly	Biweekly
Aluminum			Biweekly	Biweekly
Silica			Biweekly	Biweekly
Bromide			Biweekly	Biweekly
Chloride			Biweekly	Biweekly
Sulfate			Biweekly	Biweekly
Fluoride			Biweekly	Biweekly
Sodium			Biweekly	Biweekly
NPDOC			Biweekly	Biweekly
Calcium	Weekly	Weekly	Biweekly	Biweekly
Magnesium	Weekly	Weekly	Biweekly	Biweekly
Total hardness	Weekly	Weekly		
Total iron			Biweekly	Biweekly
Dissolved iron			Biweekly	Biweekly
HAA			Start/end Phase	Start/end Phase
THM			Start/end Phase	Start/end Phase
Pb (corr. loops)			Biweekly	Biweekly
Cu (corr. loops)			Biweekly	Biweekly
Manganese			Start	Start
Ortho-Phosphate			Biweekly	Biweekly
HPC			Start/end Phase	Start/end Phase
TC/E. Coli			Start/end Phase	Start/end Phase
AOC			Start/end Phase	Start/end Phase
BDOC			Start/end Phase	Start/end Phase
Biofilm PEPA			End Phase	End Phase
BFHPC			End Phase	End Phase

Quality assurance and quality control of both the laboratory and field determinations of water quality parameters was established by duplicating analyses of at least 10% of the samples. Where appropriate standards were available, 10% of the samples were spiked with known concentrations of the parameter being analyzed and the recovery measured. Blind duplicates and spikes were also used to determine the accuracy of measurements. Dynamic control charts were used to determine whether the results were acceptable.

To prevent non-representative biofilm thicknesses and sediment build-up caused by the slow velocity of flow through the PDSs the lines were flushed regularly. The objective of the flushing was to briefly attain the shear forces expected in full scale distribution systems. The PDS lines were flushed once a week during the 5-day HRT period and once every two weeks for the 2-day HRT period. The flush velocity was 1 ft/s (0.3 m/s) for at least 3 pipe volumes.

CORROSION LOOPS

A corrosion shed was built that housed eighteen corrosion loops that are shown in Figure 3.11 and Figure 3.12 respectively. A portion of the flow from each PDS discharge to a designated corrosion loop to simulate a kitchen environment and tap monitoring for LCR (lead and copper rule) compliance. The corrosion loop consisted of 30 feet copper tubing with a diameter of 1.6 cm (5/8 inch), which could hold approximately 1.8 L of water. One pure lead coupon was placed within the copper tubing between two standard copper tube fittings (brass) for each loop system. All other fittings and materials were PVC or other plastic polymers. The coupons had a surface area of 3.38 square inches. Assuming that 1/8th inch bead on the inside of each 0.5 inch diameter joint was the surface area exposed to water using solder, the coupon surface area was equivalent to 17 joint-ends or 7 to 8 fittings and was reasonable for a kitchen sink. A picture of the copper loops is presented in Figure 3.8.

To simulate the water that is delivered to customers, the corrosion loop units received water after passing through the associated pipe distribution system. The corrosion loops were flushed with 2 gallons of water every morning. The effluent of the copper loops was monitored regularly for copper and lead release. Samples were usually taken biweekly as shown in Table 3.11 with extra samples taken when needed for specific purposes. To sample the copper loops, a 1 liter sample was collected after six hours of water stagnation, shaken to mix completely, and then 125 ml of it was used for metal measurements. Various water quality parameters like pH, alkalinity, chloride, sulfate, silicate, calcium, sodium, DO etc. were also monitored. During phases V and VI specific dates of sampling were selected for groups of PDS lines as shown in Table 3.20, Table 3.21 and Table 3.26.

PIPE COUPON STUDY

Biofilm and surface characterization investigations were conducted using coupons made from the same MG pipes that were used to construct the PDSs. The coupons were housed in cradles following every PDS as shown in Figure 3.9 and Figure 3.10. Each cradle consisted of 4-inch diameter PVC pipe housing and 3-inch diameter PVC pipe cut length-wise as sliding pipe coupon holders. The length of the cradles was approximately 12 feet. The pipe coupon holders were easily removed from the cradles and replaced after each experimental phase for bio-film and chemical deposit analysis. The holders were transported in a large plastic bin with the coupons still in place. The humidity in the bin was elevated by placing a wet sponge inside. Upon arrival at the UCF labs the coupons were harvested for surface or biofilm studies.

The coupons for biofilm studies were circular while the surface characterization coupons were rectangular. The surface characterization coupon holders (containing much larger coupons) were placed downstream of the biological cradles to minimize transport of corrosion products onto the biofilm coupons.

Surface Scale Characterization

Pipe surface characterizations were conducted at the MCF (Material Characterization Facility) of the University of Central Florida and MAIC (Major Analytical Instrumentation Center) of University of Florida. To protect the corrosion layers formed in the pilot pipe distribution system, the coupon samples were stored in clean plastic boxes containing the same water in which the coupons had been incubated. These boxes were covered with Clingwrap (polyethylene film) to avoid reacting with the atmosphere after the coupon samples were harvested and then the coupons were transported back to UCF.

The two techniques employed for surface characterization analysis were X-ray diffraction (XRD) and X-ray Photoelectron Spectroscopy (XPS). Scanning Electron Microscopy/ Energy Dispersive Spectroscopy (SEM/EDS) was also used for identification of the corrosion products and the distribution of corrosion products in the scales formed in the aqueous environments. The XPS spectra patterns identify chemical components of the top surface of the corrosion layer, while XRD determines crystalline chemicals in the scale bulk. SEM visually magnifies the physical structure of the scale surface. The EDS spectra determine elements in the scale.

For the iron corrosion study, four different aged pipe materials namely: UCI, LCI, PVC and G were used. The aged pipes were obtained from actual TBW distribution systems located in Pinellas County, Pasco County, and the City of St. Petersburg, in the state of Florida. The four different pipe materials (UCI, LCI, PVC and G) were first cut into bigger rectangular coupons (4 in ×6 in) and then each bigger coupon was cut into two smaller rectangular coupons (4 in ×3 in): the initial and the final sample. The initial sample was analyzed to determine initial conditions of the coupon and the final sample was incubated inside the coupon cradle. The back of each pipe coupon was wrapped by Teflon tape to prevent possible contamination from exterior corrosion. For the lead and copper study pure rectangular (0.5 in ×3 in) lead and copper coupons were used. Table 3.12 shows the analysis matrix for surface characterization of the coupons.

Biological Film Characterization

Pilot distribution system pipe coupons for biological sampling were approximately 3 cm in diameter with a small PVC peg made from PVC welding rod attached to what was the outer surface of the pipe. The material is drilled from aged or pristine pipe and then de-burred to give smooth edges. In each integrated pilot distribution system cradle, there were duplicate aged coupons of each of the four materials. The sequence of the coupons in these cradles from upstream to downstream was PVC, LCI, UCI and finally G coupons.

The coupons were placed in this order to avoid transport of corrosion materials downstream to contaminate the less easily corroded materials such as PVC and CICL. Thus the two more easily corroded materials (i.e. CI and G) were placed in the downstream locations. The aged pipe coupons were obtained from used pipe segments from actual member government networks. However the cradles containing the coupons were placed in series with the hybrid PDSs during the first year of the project, so there were opportunities for transport of corrosive materials to the cradles. Factors that offset this transport, are (1) sedimentation of corrosive material at the very low (0.00023 or 0.00058 ft/sec) velocities in these lines, and (2) sedimentation in the 20-ft long ½-inch connection line from the PDS to the cradles.

Table 3.12
Analysis matrix for steady state surface characterization of coupons

Pipe Materials	PDS	Phase	XPS	XRD	EDS	SEM
Unlined Cast Iron	G1 (100%)	III Initial	√	√√	√	√
Unlined Cast Iron	G1 (100%)	III Final	√	√√	√	√
Unlined Cast Iron	RO (100%)	III Final	√	√√	√	√
Unlined Cast Iron	S1 (100%)	III Final	√	√√	√	√
Unlined Cast Iron	G1/S1 (55/45%)	III Final	√	√√	√	√
Unlined Cast Iron	G1/RO (68/32%)	III Final	√	√√	√	√
Unlined Cast Iron	G1/S1/RO (62/27/11%)	III Final	√	√√	√	√
Copper	G1 (100%)	II Final	√√	√√		
Copper	RO (100%)	II Final	√√	√√		
Copper	S1 (100%)	II Final	√√	√√		
Lead	G1 (100%)	III Final	√√	√√		
Lead	RO (100%)	III Final	√√	√√		
Lead	S1 (100%)	III Final	√√	√√		

In each cradle for single material pilot distribution systems, there were duplicated aged and pristine coupons of the same material as the pilot distribution system material. The pristine coupons allowed separate quantification of the bioassay variability and were purely for QA/QC. They were only being used during the first year of the study. They were placed upstream of aged pipe coupons to avoid transport of corrosion products from the aged material onto the pristine coupons. Each coupon/material pair was separated by 4 inches and duplicates were within 1 inch of each other.

Since there was not enough of each member government's coupons to include origin as an experimental variable, coupon pairs were randomly distributed among the PDSs so that any bias resulting from origin would not cause a systematic bias in the data. Each coupon/material pair was separated by 4 inches and duplicates were within 1 inch of each other. The biofilm coupons were usually harvested near the end of each Phase to quantify the amount of biological growth that has occurred during the Phase. Table 3.13 shows the scheduled biological sampling events for each phase of the pilot study.

Table 3.13
Sampling schedule for biological indicators

Phase	Coupon Type	PDS #	BFHPC	PEPA
I-IV	Pristine	1-14		End Phase
	Aged	15-18	End Phase	End Phase
		1-14		End Phase
		15-18	End Phase	End Phase
V	Aged	11-14	End Phase	End Phase
VI	Aged	1-8	End Phase	End Phase
		9-10	3 weeks†	3 weeks†
		11-14	3 weeks*	3 weeks*

† Sampling dates: 5/22/03, 6/12/03, 7/03/03
* Sampling dates: 4/24/03, 5/15/03, 6/05/03, and 6/26/03

BIOLOGICAL ANNULAR REACTORS (BAR)

The BAR is shown in Figure 3.6 and is a device designed to allow measuring and monitoring accumulation of film, whether microbial, organic or inorganic. A BAR consists of two concentric cylinders; a stationary outer cylinder and a rotating inner cylinder. Process fluids moves in a region between the cylinders with biofilms accumulating on the coupons (cylinder walls). Rotational speed is controlled by a variable speed DC motor. Twenty removable coupons are flush mounted on the internal, rotating cylinder.

The BAR reactor was fed the same way as the PDSs using an electronic metering pump (max. flow rate: 6gpd) connected to the influent manifold. This way the BAR was connected directly to any matrix in use during an operational phase. The BAR was used independently and in parallel with the operation of the PDS units.

Two BARs were used and allowed parallel comparison of finished waters. Two cycles of studies were performed using the BARs as shown in Table 3.14 The BARs were operated at a flow of 3 gpm to obtain a 2-hour HRT. Samples were taken and analyzed weekly for different bulk liquid water quality parameters as well as biofilm growth as shown in Table 3.14.

Table 3.14
Sampling schedule for biological annular reactors

BAR #	Exper. #	Water type	Parameters Analyzed									
			pH	DO	Cl_2	NO_2	NO_3	Alkalinity	HPC	BFHPC	PEPA	AOC
1	1	G4	x	x	x	x	x	x	x	x	x	x
1	2	G1	x	x	x	x	x	x	x	x	x	x
2	1	S1	x	x	x	x	x	x	x	x	x	x
2	2	S1	x	x	x	x	x	x	x	x	x	x

Note: All parameters were analyzed weekly for the influent and effluent of the BARs.

The main advantage of the BAR was its sample and operation flexibility, allowing sampling of coupons/bulk liquid more frequently for parameters of interests (PEPA, HPC, AOC, nitrate, nitrite, residual) than it was with the PDS, in which coupons were harvested only at the end of an operational phase. Therefore, it was possible to analyze trends occurring in the PDS and perform kinetic studies on biofilm accumulation.

Two separate experiments were conducted using the BARs. The source water used for each BAR unit during the experiments is shown in Table 3.14. The duration of each experiment with the BAR was as follows:

- Experiment #1 from 9/19/2002 to 11/21/2002
- Experiment #2 from 1/30/2003 to 3/13/2003

ADDITIONAL STUDIES

Confounding Effects

The operation of the pilot distribution system during phase I through IV included confounding effects between factors that may be relevant to release of corrosion products. Specifically, the water source(s) with the highest alkalinity (GW and blends enriched in GW) also contain the lowest chlorides. The water source(s) with the greatest sulfate (SW and blends enriched in SW), exhibited low alkalinity. Similarly, the water source(s) with the highest chlorides (RO and blends enriched in RO), also report low alkalinity. Additional studies were needed to isolate the effect of individual parameters on effluent water quality.

Phase five (12/14/02 to 4/4/03), was conducted to identify the individual effects of alkalinity, chlorides and sulfates on iron release and associated distribution water quality parameters and models. The range of alkalinity, chlorides and sulfates used in phase V was wider than in Phases I through IV (12/8/01 to 12/13/02). Table 3.15 presents the variable matrix used for PDS 1 through 6 by month during phase V.

Source waters were prepared by supplementing alkalinity in the form of sodium bicarbonate, chlorides from sea salt, and sulfates from calcium sulfate to existing finished waters. The purpose of these additions was to simulate a particular water chemsitry rather than a specific process. PDS I through VI were used for determining the effects of various anions. For instance, in the first month of Phase V, alkalinity was varied by PDS (80 – 300 mg/L as $CaCO_3$) while sulfates (13 ± 2 mg/L) and chlorides (34 ± 6 mg/L) were kept constant. The second month evaluated the impact of sulfates (12 – 150 mg/L) at fixed alkalinity (80 ± 5 mg/L as $CaCO_3$) and chlorides (32 ± 3 mg/L). In the third month chlorides were varied (30 – 94 mg/L) while alkalinity (76 ± 7 mg/L as $CaCO_3$) and sulfates (23 ± 7 mg/L) were kept constant. An additional set of verification experiments were conducted from in Phase V (12/14/02 to 4/4/03) using PDS 8 through 10. Blends were changed every two weeks during this investigation and alkalinity was varied from 60 to 250 mg/L as $CaCO_3$. This data is presented in Chapter 4, Empirical model verification (Table 4.5, Table 4.6, and Table 4.7).

Table 3.15

Phase V study to evaluate singular parameter effects (12/24/02 to 4/4/03)

Parameter	Month 01	Month 02	Month 03
Alkalinity (mg/L as $CaCO_3$)	80 – 300	80	76
Chlorides (mg/L)	34	32	30 - 94
Sulfates (mg/L)	13	12 - 150	23
DO (mg/L)	9	9	8
Temperature (^{o}C)	18	22	25

Variation of Hydraulic Conditions for Iron Flux (K_m) Experiment

Normal operation of the pilot distribution systems was limited to extremely small velocities in order to achieve the desired hydraulic residence times (two or five days). Consequently, the database developed from normal PDS operation would not be suitable for evaluation of the effect of variant hydraulic conditions (for example, Reynolds Number) on the iron flux values. A series of experiments were conducted to investigate the effect of Reynolds Number on the flux values. These experiments were conducted using PDS 15 (unlined cast iron) and PDS 18 (galvanized) to define values for each pipe material.

The flux experiments were carried out by feeding a blend of treated surface water – SW (30%), conventionally-treated groundwater – GW (60%) and desalinated water - RO (10%) sources at selected Reynolds Numbers in the unlined cast iron and galvanized pilot scale distribution system pipes as shown in Table 3.16. Water quality data for the blend used in the flux experiment is provided in Table 3.17. The feed and outlet water quality were measured to determine the color and turbidity at various time intervals.

Table 3.16
Mass transfer studies using cast iron and galvanized steel lines (PDS15 & 18)

Exp. #	Reynolds No.	Velocity (ft/sec)	Volume (gal.)	Duration (min.)	Flow Rate (gpm)	HRT (min.)
1	100	0.002	1200	6316	0.19	685
2	500	0.01	1200	1290	0.93	137
3	1000	0.02	1200	631	1.9	68.5
4	5000	0.10	1200	129	9.3	13.7
5	10000	0.21	1200	63	19	6.8
6	20000	0.42	1200	32	37	3.4
7	30000	0.63	1200	21	56	2.3

Table 3.17
Average feed water quality

	Turbidity (NTU)	Total Fe (mg/L)	Ca^{2+} (mg/L)	Mg^{2+} (mg/L)	Na^{+} (mg/L)	pH	Alkalinity (mg/L as $CaCO_3$)	Conductivity (μS/cm)	TDS (mg/L)	Cl^{-} (mg/L)	SO_4^{2-} (mg/L)
Feed	0.306	0.045	68	6.4	31	7.96	148	557	377	38	66

Nitrification Study

A planned nitrification study was conducted to investigate how to control nitrification in a system receiving varied water sources and blends of those waters. Several factors were evaluated to determine their impact on the beginning, continuation and decline of nitrification. These factors were NH_3 concentration, residual, temperature, pH, Cl_2/NH_3 ratio, AOC, AOB, and time.

The planned nitrification investigation was conducted during Phases V and VI. The waters utilized for the nitrification study were 100 % RO, 100 % S1, 100 % G1 and a blend. The blend B1 (40% G1, 30% S1, and 30% RO) was used from 12/24/2002 to 04/04/2003, and the blend B2 (62% G1, 27% S1, and 11% RO) from 04/05/2003 to 07/04/2003. The feed streams to PDSs 11- 14 were changed on 6/13/03. The different changes in water sources for PDSs 9 through 18 are presented in Table 3.18.

Initially, steps were taken to establish a nitrifying environment in PDS 11 to 14 from 12/24/2002 to 04/04/2003. The influent stream to these lines was fed with a 4 mg/L chloramine residual as Cl_2 and 1 mg/L of excess ammonia. Sodium thiosulfate was dosed to quench influent residual and create a microbiological nitrifying environment. The pilot plant pumping and piping scheme could not accommodate separate chloramination facilities, so chloramines quenching was necessary for PDS 11 through 14.

Table 3.18
Feed matrices for PDSs under nitrification study.

	01/30/03- 04/04/03	04/05/03- 06/12/03	06/13/03- 07/04/03
PDS 9	-	Blend 2	Blend 2
PDS 10	-	Blend 2	Blend 2
PDS 11	RO	RO	G1
PDS 12	S1	S1	Blend 2
PDS 13	G1	G1	RO
PDS 14	Blend 1	Blend 2	S1
PDS 15	-	Blend 2	Blend 2
PDS 16	-	Blend 2	Blend 2
PDS 17	-	Blend 2	Blend 2
PDS 18	-	Blend 2	Blend 2
Blend Ratio	G1 (40%), S1 (30%) RO (30%)	G1 (62%), S1 (27%) RO (11%)	G1 (62%) ,S1 (27%) RO (11%)

Beginning on 04/05/2003, PDSs 9, 10, 15, 16, 17 and 18 were incubated for nitrifiers for three weeks with 1.0 mg/L excess ammonia-N and sodium thiosulfate quenching of chloramine residuals, which provided another 0.8 mg/L of ammonia N. The HRT for the first three weeks was set at 5 days to accelerate the occurrence of nitrification, thereafter a 2-day HRT was used to prevent anaerobic conditions. Table 3.19 shows the different operating conditions for PDS lines 9 to 18 during the nitrification study.

Different combinations of excess ammonia-N, Cl_2:N ratio, flushing frequency and flushing strength were studied in the controlled nitrification investigations as shown in Table 3.19. During the nitrifier incubation period an excess NH_3 concentration of 1 mg/L was added to the feed of each PDS. After incubation, three different Cl_2:N ratios were investigated. Excess NH_3 concentration of 0, 0.2, 0.5 mg/L were used in the feed stream of the PDSs to reach a Cl_2:N ratio of 5:1, 4:1, and 3:1 respectively.

Since PDSs 11 through 14 were seeded during phase V, they were available for four cycles of study while PDSs 9,10 and 15 through 18 were only available for three study cycles as noted in Table 3.19.

From 04/05/2003 to 07/04/2003, PDSs 9 and 10 were used for investigating the effects of flushing the system using high residual finished water. The monochloramine concentrations used for flushing PDSs 9 and 10 were 10 mg/L as Cl_2 and 7 mg/L as Cl_2, respectively. Before flushing, the finished waters were pumped to a storage water tank and the chloramine concentration was adjusted to the target concentration for flushing. The effect of flushing frequency in nitrification was also studied, from 05/02/03 to 06/06/03, the PDSs were flushed every one and half weeks totaling three flushes, and from 06/06/03 to 07/04/03 the PDSs were flushed on a weekly basis, totaling four flushes.

The influent and effluent to each PDS were analyzed for several parameters weekly as shown in Table 3.20 and Table 3.21. Several parameters were also analyzed for the internal ports located between the different types of pipe material, the sampling dates for each PDS line are shown in Table 3.22.

Table 3.19
Nitrification operational and sampling schedule

Nitrification Study from 12/24/2202 to 07/04/2003

		PDS 9	PDS 10	PDS 11	PDS 12	PDS 13	PDS 14	PDS 15	PDS 16	PDS 17	PDS 18
		Phase V									
Source				RO	SW	GW	B1				
NH_2Cl	mg/L Cl_2			0 mg/L	0 mg/L	0 mg/L	0 mg/L				
NH_3-N	mg/L NH_3			1 mg/L	1 mg/L	1 mg/L	1 mg/L				
HRT	days			2 days	2 days	2 days	2 days				
Flush Freq.	weeks			no flush	no flush	no flush	no flush				
Flush Disinf.	NH_2Cl			no flush	no flush	no flush	no flush				
Flush Dis. Conc.	mg/L Cl_2			no flush	no flush	no flush	no flush				
		Phase VI Cycle 1									
Source		B2	B2	RO	SW	GW	B1	B2	B2	B2	B2
NH_2Cl	mg/L Cl_2	0	0	4 mg/L	4 mg/L	4 mg/L	4 mg/L	0	0	0	0
NH_3-N	mg/L NH_3	1 mg/L	1 mg/L	0.5 mg/L	0.5 mg/L	0.5 mg/L	0.5 mg/L	1 mg/L	1 mg/L	1 mg/L	1 mg/L
Cl_2/NH_3	mg/mg			3/1	3/1	3/1	3/1				
HRT	days	5 days	5 days	5 days	5 days	5 days	5 days	5 days	5 days	5 days	5 days
Flush Freq.	weeks			3 weeks	3 weeks	3 weeks	3 weeks				
Flush Time	minutes			5 min	5 min	5 min	5 min				
Flush Disinf.	NH_2Cl			NH_2Cl	NH_2Cl	NH_2Cl	NH_2Cl				
Flush Dis. Conc.	mg/L Cl_2			4 mg/L	4 mg/L	4 mg/L	4 mg/L				
		Phase VI Cycle 2									
Source		B2	B2	RO	SW	GW	B2	B2	B2	B2	B2
NH_2Cl	mg/L Cl_2	4 mg/L	4 mg/L	4 mg/L	4 mg/L	4 mg/L	4 mg/L	4 mg/L	4 mg/L	4 mg/L	4 mg/L
NH_3-N	mg/L NH_3	0.2 mg/L	0.2 mg/L	0.2 mg/L	0.2 mg/L	0.2 mg/L	0.2 mg/L	0.2 mg/L	0.2 mg/L	0.2 mg/L	0.2 mg/L
Cl_2/NH_3	mg/mg	4/1	4/1	4/1	4/1	4/1	4/1	4/1	4/1	4/1	4/1
HRT	days	2 days	2 days	2 days	2 days	2 days	2 days	2 days	2 days	2 days	2 days
Flush Freq.	weeks	1.5 weeks	1.5 weeks	3 weeks	3 weeks	3 weeks	3 weeks	3 weeks	3 weeks	3 weeks	3 weeks
Flush Time	minutes	5 min	5 min	5 min	5 min	5 min	5 min	5 min	5 min	5 min	5 min
Flush Disinf.	NH_2Cl	NH_2Cl	NH_2Cl	NH_2Cl	NH_2Cl	NH_2Cl	NH_2Cl	NH_2Cl	NH_2Cl	NH_2Cl	NH_2Cl
Flush Dis. Conc.	mg/L Cl_2	10 mg/L	7 mg/L	4 mg/L	4 mg/L	4 mg/L	4 mg/L	4 mg/L	4 mg/L	4 mg/L	4 mg/L
		Phase VI Cycle 3									
Source		B2	B2	RO	SW	GW	B2	B2	B2	B2	B2
NH_2Cl	mg/L Cl_2	4 mg/L	4 mg/L	4 mg/L	4 mg/L	4 mg/L	4 mg/L	4 mg/L	4 mg/L	4 mg/L	4 mg/L
NH_3-N	mg/L NH_3	0.16 mg/L	0.16 mg/L	0.16 mg/L	0.16 mg/L	0.16 mg/L	0.16 mg/L	0.16 mg/L	0.16 mg/L	0.16 mg/L	0.16 mg/L
Cl_2/NH_3	mg/mg	5/1	5/1	5/1	5/1	5/1	5/1	5/1	5/1	5/1	5/1
HRT	days	2 days	2 days	2 days	2 days	2 days	2 days	2 days	2 days	2 days	2 days
Flush Freq.	weeks	1.5 weeks	1.5 weeks	3 weeks	3 weeks	3 weeks	3 weeks	3 weeks	3 weeks	3 weeks	3 weeks
Flush Time	minutes	5 min	5 min	5 min	5 min	5 min	5 min	5 min	5 min	5 min	5 min
Flush Disinf.	NH_2Cl	NH_2Cl	NH_2Cl	NH_2Cl	NH_2Cl	NH_2Cl	NH_2Cl	NH_2Cl	NH_2Cl	NH_2Cl	NH_2Cl
Flush Dis. Conc.	mg/L Cl_2	10 mg/L	7 mg/L	4 mg/L	4 mg/L	4 mg/L	4 mg/L	4 mg/L	4 mg/L	4 mg/L	4 mg/L
		Phase VI Cycle 4									
Source		B2	B2	RO	B2	GW	SW	B2	B2	B2	B2
NH_2Cl	mg/L Cl_2	4 mg/L	4 mg/L	4 mg/L	4 mg/L	4 mg/L	4 mg/L	4 mg/L	4 mg/L	4 mg/L	4 mg/L
NH_3-N	mg/L NH_3	0.5 mg/L	0.5 mg/L	0.5 mg/L	0.5 mg/L	0.5 mg/L	0.5 mg/L	0.5 mg/L	0.5 mg/L	0.5 mg/L	0.5 mg/L
Cl_2/NH_3	mg/mg	3/1	3/1	3/1	3/1	3/1	3/1	3/1	3/1	3/1	3/1
HRT	days	2 days	2 days	2 days	2 days	2 days	2 days	2 days	2 days	2 days	2 days
Flush Freq.	weeks	1 weeks	1 weeks	3 weeks	3 weeks	3 weeks	3 weeks	3 weeks	3 weeks	3 weeks	3 weeks
Flush Time	minutes	5 min	5 min	5 min	5 min	5 min	5 min	5 min	5 min	5 min	5 min
Flush Disinf.	NH_2Cl	NH_2Cl	NH_2Cl	NH_2Cl	NH_2Cl	NH_2Cl	NH_2Cl	NH_2Cl	NH_2Cl	NH_2Cl	NH_2Cl
Flush Dis. Conc.	mg/L Cl_2	10 mg/L	7 mg/L	4 mg/L	4 mg/L	4 mg/L	4 mg/L	4 mg/L	4 mg/L	4 mg/L	4 mg/L

Table 3.20
Analytical plan for PDSs 11-14 from 04/05/2003 to 07/04/2003

Influent, Cradles, Effluent
PDS 11-14 sampling begins at Cycle 1

	Start 4/5/03		Cycle 1-Ports Sampled			Cycle 2-Ports Sampled			Cycle 3-Ports Sampled				Cycle 4-Ports Sampled		
	Parameters	Poss Ports	4/11/2003 week 1	4/18/2003 week 2	4/25/2003 week 3	5/2/2003 week 1	5/9/2003 week 2	5/16/2003 week 3	5/23/2003 week 1	5/30/2003 week 2	6/6/2003 week 3	6/13/2003 week 4	6/20/2003 week 1	6/27/2003 week 2	7/4/2003 week 3
Field	Temperature	0, 1, 3	0, 1, 3	0, 1, 3	0, 1, 3	0, 1, 3	0, 1, 3	0, 1, 3	0, 1, 3	0, 1, 3	0, 1, 3	0, 1, 3	0, 1, 3	0, 1, 3	0, 1, 3
	pH	0, 1, 3	0, 1, 3	0, 1, 3	0, 1, 3	0, 1, 3	0, 1, 3	0, 1, 3	0, 1, 3	0, 1, 3	0, 1, 3	0, 1, 3	0, 1, 3	0, 1, 3	0, 1, 3
	Total Cl_2	0, 1, 3	0, 1, 3	0, 1, 3	0, 1, 3	0, 1, 3	0, 1, 3	0, 1, 3	0, 1, 3	0, 1, 3	0, 1, 3	0, 1, 3	0, 1, 3	0, 1, 3	0, 1, 3
	Free Cl_2	0, 1, 3													
	DO	0, 1, 3	0, 1, 3	0, 1, 3	0, 1, 3	0, 1, 3	0, 1, 3	0, 1, 3	0, 1, 3	0, 1, 3	0, 1, 3	0, 1, 3	0, 1, 3	0, 1, 3	0, 1, 3
	NH_3-N	0, 1, 3	0, 1, 3	0, 1, 3	0, 1, 3	0, 1, 3	0, 1, 3	0, 1, 3	0, 1, 3	0, 1, 3	0, 1, 3	0, 1, 3	0, 1, 3	0, 1, 3	0, 1, 3
	NO_2-N	0, 1, 3	0, 1, 3	0, 1, 3	0, 1, 3	0, 1, 3	0, 1, 3	0, 1, 3	0, 1, 3	0, 1, 3	0, 1, 3	0, 1, 3	0, 1, 3	0, 1, 3	0, 1, 3
	NO_3-N	0, 1, 3	0, 1, 3	0, 1, 3	0, 1, 3	0, 1, 3	0, 1, 3	0, 1, 3	0, 1, 3	0, 1, 3	0, 1, 3	0, 1, 3	0, 1, 3	0, 1, 3	0, 1, 3
	Alkalinity	0, 1, 3	0, 1, 3	0, 1, 3	0, 1, 3	0, 1, 3	0, 1, 3	0, 1, 3	0, 1, 3	0, 1, 3	0, 1, 3	0, 1, 3	0, 1, 3	0, 1, 3	0, 1, 3
	UV-254	0, 1, 3													
	Apparent Color	0, 1, 3	0, 1	0, 1	0, 1	0, 1	0, 1	0, 1	0, 1	0, 1	0, 1	0, 1	0, 1	0, 1	0, 1
	Conductivity	0, 1, 3	0, 1	0, 1	0, 1	0, 1	0, 1	0, 1	0, 1	0, 1	0, 1	0, 1	0, 1	0, 1	0, 1
	Turbidity	0, 1, 3	0, 1	0, 1	0, 1	0, 1	0, 1	0, 1	0, 1	0, 1	0, 1	0, 1	0, 1	0, 1	0, 1
Laboratory	TKN	0, 1, 3	0, 1, 3	0, 1, 3	0, 1, 3	0, 1, 3	0, 1, 3	0, 1, 3	0, 1, 3	0, 1, 3	0, 1, 3	0, 1, 3	0, 1, 3	0, 1, 3	0, 1, 3
	NPDOC	0, 1, 3	0, 1, 3	0, 1, 3	0, 1, 3	0, 1, 3	0, 1, 3	0, 1, 3	0, 1, 3	0, 1, 3	0, 1, 3	0, 1, 3	0, 1, 3	0, 1, 3	0, 1, 3
	Ca	0, 1, 3	0, 1	0, 1	0, 1	0, 1	0, 1	0, 1	0, 1	0, 1	0, 1	0, 1	0, 1	0, 1	0, 1
	Mg	0, 1, 3	0, 1	0, 1	0, 1	0, 1	0, 1	0, 1	0, 1	0, 1	0, 1	0, 1	0, 1	0, 1	0, 1
	Cu	2													
	Pb	2													
	Fe(Total)	0, 1, 3			0, 1			0, 1				0, 1			0, 1
	Fe(dissolved)	0, 1, 3			0, 1			0, 1				0, 1			0, 1
	Si	0-1-3													
	Na	0-1-3													
	Cl	0-1-3													
	SO_4	0-1-3													
	Br	0-1													
	THMs	0-1													
	HAAs	0-1													
Biological	HPC	0-1-3	0, 1, 3	0, 1, 3	0, 1, 3	0, 1, 3	0, 1, 3	0, 1, 3	0, 1, 3	0, 1, 3	0, 1, 3	0, 1, 3	0, 1, 3	0, 1, 3	0, 1, 3
	AOC	0-1-3			0-1-3			0, 1				0-1-3			0, 1
	AOB	0-1-3			0-1-3			1				0-1-3			1
	TC/EC	0-1-3			1			1				1			1
	Aerobic Spores	0-1-3			0,1										0,1
	Biofilm-Pepa	3			3			3				3			3
	Biofilm-AOB	3			3			3				3			3
	Biofilm-HPC	3			3			3				3			3

Note: 0 = influent
1 = effluent
2 = copper loops
3 = coupon cradle

Table 3.21
Analytical plan for PDSs 9, 10, 15-18 from 04/05/2003 to 07/04/2003

Influent, Cradles, Effluent
PDS 9, 10, 15-18 sampling started May 8(Biological)/9(WQ)

	Start 4/5/03 Parameters	Poss Ports	Cycle 1-4 wk 4/11/2003 week 1	4/18/2003 week 2	4/25/2003 week 3	5/2/2003 week 1	Cycle 2-Ports Sampled 5/9/2003 week 1	5/16/2003 week 2	5/23/2003 week 3	Period 3-Ports Sampled 5/30/2003 week 1	6/6/2003 week 2	6/13/2003 week 3	Period 4-Ports Sampled 6/20/2003 week 1	6/27/2003 week 2	7/4/2003 week 3
Field	Temperature	0, 1, 3	1	1	1	1	0, 1, 3	0, 1, 3	0, 1, 3	0, 1, 3	0, 1, 3	0, 1, 3	0, 1, 3	0, 1, 3	0, 1, 3
	pH	0, 1, 3	1	1	1	1	0, 1, 3	0, 1, 3	0, 1, 3	0, 1, 3	0, 1, 3	0, 1, 3	0, 1, 3	0, 1, 3	0, 1, 3
	Total Cl_2	0, 1, 3	1	1	1	1	0, 1, 3	0, 1, 3	0, 1, 3	0, 1, 3	0, 1, 3	0, 1, 3	0, 1, 3	0, 1, 3	0, 1, 3
	Free Cl_2	0, 1, 3													
	DO	0, 1, 3	1	1	1	1	0, 1, 3	0, 1, 3	0, 1, 3	0, 1, 3	0, 1, 3	0, 1, 3	0, 1, 3	0, 1, 3	0, 1, 3
	NH_3-N	0, 1, 3					0, 1, 3	0, 1, 3	0, 1, 3	0, 1, 3	0, 1, 3	0, 1, 3	0, 1, 3	0, 1, 3	0, 1, 3
	NO_2-N	0, 1, 3	1	1	1	1	0, 1, 3	0, 1, 3	0, 1, 3	0, 1, 3	0, 1, 3	0, 1, 3	0, 1, 3	0, 1, 3	0, 1, 3
	NO_3-N	0, 1, 3	1	1	1	1	0, 1, 3	0, 1, 3	0, 1, 3	0, 1, 3	0, 1, 3	0, 1, 3	0, 1, 3	0, 1, 3	0, 1, 3
	Alkalinity	0, 1, 3					0, 1, 3	0, 1, 3	0, 1, 3	0, 1, 3	0, 1, 3	0, 1, 3	0, 1, 3	0, 1, 3	0, 1, 3
	UV-254	0, 1, 3													
	Apparent Color	0, 1, 3					0, 1	0, 1	0, 1	0, 1	0, 1	0, 1	0, 1	0, 1	0, 1
	Conductivity	0, 1, 3					0, 1	0, 1	0, 1	0, 1	0, 1	0, 1	0, 1	0, 1	0, 1
	Turbidity	0, 1, 3					0, 1	0, 1	0, 1	0, 1	0, 1	0, 1	0, 1	0, 1	0, 1
Laboratory	TKN	0, 1, 3					0, 1, 3	0, 1, 3	0, 1, 3	0, 1, 3	0, 1, 3	0, 1, 3	0, 1, 3	0, 1, 3	0, 1, 3
	NPDOC	0, 1, 3					0, 1, 3	0, 1, 3	0, 1, 3	0, 1, 3	0, 1, 3	0, 1, 3	0, 1, 3	0, 1, 3	0, 1, 3
	Ca	0, 1, 3					0, 1	0, 1	0, 1	0, 1	0, 1	0, 1	0, 1	0, 1	0, 1
	Mg	0, 1, 3					0, 1	0, 1	0, 1	0, 1	0, 1	0, 1	0, 1	0, 1	0, 1
	Cu	2													
	Pb	2													
	Fe(Total)	0, 1, 3					0, 1		0, 1			0, 1			0, 1
	Fe(dissolved)	0, 1, 3					0, 1		0, 1			0, 1			0, 1
	Si	0-1-3													
	Na	0-1-3													
	Cl	0-1-3													
	SO_4	0-1-3													
	Br	0-1													
	THMs	0-1													
	HAAs	0-1													
Biological	HPC	0-1-3					0, 1, 3	0, 1, 3	0, 1, 3	0, 1, 3	0, 1, 3	0, 1, 3	0, 1, 3	0, 1, 3	0, 1, 3
	AOC	0-1-3					0, 1	0, 1	0, 1			0, 1			0, 1
	AOB	0-1-3					1	1	0, 1			0, 1			1
	TC/EC	0-1-3					1	1	1			1			1
	Aerobic Spores	0-1-3													0,1
	Biofilm-Pepa	3							3			3			3
	Biofilm-AOB	3							3			3			3
	Biofilm-HPC	3							3			3			3

Note: 0 = influent
1 = effluent
2 = copper loops
3 = coupon cradle

Table 3.22
Internal ports sampling plan for the period from 04/05/2003 to 07/04/2003

Parameter	Units	PDS 11	12	13	14	15	16	17	18
Ammonia	mg/L NH_3-N	x	x	x	x	x	x	x	x
NOx-N	mg/L NO_3-N	x	x	x	x	x	x	x	x
Nitrite	mg/L NO_2-N	x	x	x	x	x	x	x	x
Residual	mg/L Cl_2	x	x	x	x	x	x	x	x
D.O.	mg/L O_2	x	x	x	x	x	x	x	x
Temperature	^oC	x	x	x	x	x	x	x	x
pH	pH units	x	x	x	x	x	x	x	x
Alkalinity	mg/L as $CaCO_3$	x	x	x	x	x	x	x	x

Note: Samples dates were
05/09/2003, 05/30/2003, 06/20/2003 and 07/03/2003 for PDS 11 to 14
05/09/2003, 05/30/2003, and 06/20/2003 for PDS 15 to 18

Dynamic Behavior of Inorganic Scale Formation

A kinetic pipe coupon study for 6 months was performed in order to investigate dynamic behavior of pipe surface scales under various water qualities. CI-Cu-Pb coupons were placed in the four coupon holders (i.e. cradle) receiving 100% RO, 100 % S1 and 100 % G1 and their blend throughout Phase V and VI. A total of three sets of CI-Cu-Pb coupons were installed in the pipe cradles of PDS 15 to 18 at the beginning of Phase V.

Each set of CI-Cu-Pb coupons was collected at 1, 2, and 6 months for surface characterization by XRD and XPS. No coupons were collected after three months since pipe coupons had already been analyzed for 3 month during Phase I to V. The surface characterization analysis matrix for PDSs 15 to 18 is presented in Table 3.23. Sampling frequencies for cradle water quality are shown in Table 3.24.

Table 3.23
Analysis matrix for kinetic surface characterization study of coupons

Pipe Materials	PDS	Phase	Incubation	XPS	XRD	EDS	SEM
Cast Iron	15,16,17,18	V Final	1 month	√√	√√		
Cast Iron	15,16,17,18	V Final	6 months	√√	√√		
Copper	15,16,17,18	V Final	1 month	√√	√√		
Copper	15,16,17,18	V Final	2 months	√√	√√		
Copper	15,16,17,18	V Final	6 months	√√	√√		
Lead	15,16,17,18	V Final	1 month	√√	√√		
Lead	15,16,17,18	V Final	2 months	√√	√√		
Lead	15,16,17,18	V Final	6 months	√√	√√		
Cast Iron	1, 2, 3, 4	VI Final	3 months	√√	√√		
Copper	1, 2, 3, 4	VI Final	3 months	√√	√√		
Lead	1, 2, 3, 4	VI Final	3 months	√√	√√		

Table 3.24
Sampling plan for surface kinetic study

Parameter/Event	P5 & P6-Surface	Parameter/Event	P5 & P6-Surface
Cradles	15 to 18	Cradles	15 to 18
Field Laboratory		UCF Laboratory	
Inorganics		Ca	X
pH	X	Mg	X
Total Cl_2	X	Na	X
UV_{254}	X	Cu	X
Alkalinity	X	Pb	X
Turbidity	X	Fe (total)	X
Other		Fe (dissolved)	X
D.O.	X	Al	X
pH	X	SiO_2	X
Temperature	X	Br	0
		Cl	X
		SO_4	X
Surface Characterization-UCF		Aggregates	
SEM	SEM	True Color	X
XPS	XPS	Alkalinity	X
XRD	XRD	Conductivity	X
EDS	EDS	NPDOC	X

Chlorine – Chloramine Comparison Study

The effect of free chlorine versus chloramine residual on water quality was studied using PDS lines 1 through 8. The three principal water sources used during the previous year of field work: G1, S1, and RO were produced in both a free chlorine and chloramine version. Conventional finished waters (G1, S1, and RO) and a G1/S1/RO blend at a free chlorine residual of 4 mg/L fed PDS 1 to 4 and conventional finished waters (G1, S1, and RO) and a G1/S1/RO blend at a chloramine residual of 4 mg/L fed PDS 5 to 8. An HRT of two days was used during the study. The type of water and respective percentage fed to each PDS during the disinfectant comparison study is presented in Table 3.25.

Water quality parameters were analyzed weekly for the influent and effluent of the PDS as well as the coupon cradles effluent for the duration of the study. In addition, free chlorine and chloramine measurements were taken at the internal sample ports of the PDSs. Table 3.26 shows the different parameters analyzed during the study and the sampling frequency.

Bulk liquid HPC was measured weekly and all other microbiological parameters were measured during the last week of the phase. Pipe surface characterization using XPS and XRD techniques were performed on cast iron, copper and lead coupons. In this set of experiments, the pipe coupons were exposed to RO, SW and GW with free chlorine, analyzed and compared with those exposed to monochloramine characterized during Phases I to IV (see Table 3.12). The selected pipe coupons were installed in the corresponding cradles and incubated for three months prior to surface characterization as presented in Table 3.25.

Table 3.25
Water sources by PDS for free chlorine versus chloramine study

PDS #	HRT	G1 Free	S1 Free	RO Free	G1 Comb	S1 Comb	RO Comb
1	2 days	100%					
2	2 days		100%				
3	2 days			100%			
4	2 days	62%	27%	11%			
5	2 days				100%		
6	2 days					100%	
7	2 days						100%
8	2 days				62%	27%	11%

Table 3.26
Analytical plan for disinfectant comparison study

Influent, Effluent &Cradle
PDS 1- 8

	Start 04/05/03		Cycle 1-4 wk				Cycle 2-4 wk				Cycle 3-4wk			
	Parameters	Ports SP#	4/11/2003 week 1	4/18/2003 week 2	4/25/2003 week 3	5/2/2003 week 4	5/9/2003 week 1	5/16/2003 week 2	5/23/2003 week 3	5/30/2003 week 4	6/6/2003 week 1	6/13/2003 week 2	6/20/2003 week 3	6/27/2003 week 4
Field	Temperature	0-1-3	0-1-3	0-1-3	0-1-3	0-1-3	0-1-3	0-1-3	0-1-3	0-1-3	0-1-3	0-1-3	0-1-3	0-1-3
	pH	0-1-3	0-1	0-1	0-1	0-1-3	0-1	0-1	0-1	0-1-3	0-1	0-1	0-1	0-1-3
	Total Cl_2	0-1-3	0-1-3	0-1-3	0-1-3	0-1-3	0-1-3	0-1-3	0-1-3	0-1-3	0-1-3	0-1-3	0-1-3	0-1-3
	Free Cl_2	0-1-3	0-1-3	0-1-3	0-1-3	0-1-3	0-1-3	0-1-3	0-1-3	0-1-3	0-1-3	0-1-3	0-1-3	0-1-3
	DO	0-1-3	0-1	0-1	0-1	0-1-3	0-1	0-1	0-1	0-1-3	0-1	0-1	0-1	0-1-3
	NH_3-N	0-1-3				0-1-3				0-1-3				0-1-3
	NO_2-N	0-1-3				0-1-3				0-1-3				0-1-3
	NO_3-N	0-1-3				0-1-3				0-1-3				0-1-3
	Alkalinity	0-1-3	0-1	0-1	0-1	0-1-3	0-1	0-1	0-1	0-1-3	0-1	0-1	0-1	0-1-3
	UV-254	0-1-3	0-1	0-1	0-1	0-1	0-1	0-1	0-1	0-1	0-1	0-1	0-1	0-1
	Apparent Color	0-1-3	0-1	0-1	0-1	0-1	0-1	0-1	0-1	0-1	0-1	0-1	0-1	0-1
	Conductivity	0-1-3	0-1	0-1	0-1	0-1	0-1	0-1	0-1	0-1	0-1	0-1	0-1	0-1
	Turbidity	0-1-3	0-1	0-1	0-1	0-1	0-1	0-1	0-1	0-1	0-1	0-1	0-1	0-1
Laboratory	TKN	0-1-3												0-1-3
	NPDOC	0-1-3				0-1				0-1				0-1-3
	Ca	0-1-3		0,1		0-1		0,1		0-1		0,1		0-1
	Mg	0-1-3		0,1		0-1		0,1		0-1		0,1		0-1
	Cu	2	2	2	2	2	2	2	2	2	2	2	2	2
	Pb	2	2	2	2	2	2	2	2	2	2	2	2	2
	Fe(Total)	0-1-3		0,1				0-1-3				0,1		0-1-3
	Fe(dissolved)	0-1-3		0,1				0-1-3				0,1		0-1-3
	Si	0-1-3	1	1	1	1	1	1	1	1	1	1	1	1
	Na	0-1-3	1	1	1	1	1	1	1	1	1	1	1	1
	Cl	0-1-3	1	1	1	1	1	1	1	1	1	1	1	1
	SO_4	0-1-3	1	1	1	1	1	1	1	1	1	1	1	1
	Br	0-1				0-1				0-1				0-1
	THMs	0-1				0-1				0-1				0-1
	HAAs	0-1				0-1				0-1				0-1
Biological	HPC	0-1-3	0-1-3	0-1-3	0-1-3	0-1-3	0-1-3	0-1-3	0-1-3	0-1-3	0-1-3	0-1-3	0-1-3	0-1-3
	AOC	0-1-3												0-1-3
	AOB	0-1-3												0-1-3
	TC/EC	0-1-3												0-1-3
	Aerobic Spores	0-1-3												0-1
	Biofilm-Pepa	3												3
	Biofilm-AOB	3												3
	Biofilm-HPC	3												3

Note: 0 = influent
1 = effluent
2 = copper loops
3 = coupon cradle

QUALITY CONTROL

This portion of the final report summary is intended to provide a comprehensive overview of quality control procedures, analytical results and supplementary discussion of PDS sampling events and process sampling events associated the entirety of this project.

The total number of samples, spiked samples, duplicates and blind analyses varies by parameter. The following section addresses analytical QA/QC issues and other laboratory function and information issues pertinent to those samples that were analyzed in the field as well as those that were brought back to the main ESEI lab facility for analysis.

QC Methods

The following describes quality assurance and quality control (QA/QC) measures taken in order to insure data of known quality throughout the entire study. QA/QC plan for all data has followed guidelines as presented in Standard Methods (APHA, 1995). Furthermore, biological QA/QC measures were taken as recommended in reports by Verner (1990) and Najm et al. (2000).

Chemical Data- UCF Main Laboratory

The procedures used and their analytical specifications for the duration of this project are summarized in Table 3.27 and Table 3.28. Some parameters contain multiple method listings as relevant to field applications and/or multiple concentration ranges of quantification.

Table 3.27
Methods and reporting limits for aggregate properties and NPDOC measured in the main laboratory during project

Parameter	Method	Reporting Limit
Color	SM 2120A pages 2-2 to 2-4 UV Absorption at 254 nm	1 cpu
Conductivity	SM 2510B pages 2-45 to 2-46 Laboratory Method	0.1 μmho/cm
NPDOC	SM 5310C, pages 5-19 to 5-21	0.1 mg C/L
Solids (TDS)	SM 2540 pages 2-53 to 2-58 Gravimetric Method	1 mg/L
	Estimation of TDS by major ion sum	1 mg/L
Turbidity	SM 2130B pages 2-9 to 2-11 Nephelometric Method	0.01 NTU
UV-254	SM 5910 pages 5-60 to 5-62 UV Absorption at 254 nm	0.0001 cm^{-1}

*All techniques designated with "SM" referenced in Standard Methods are from the 19th Edition (1995).

Table 3.28

Methods and reporting limits for inorganic analysis in the main laboratory during project

Parameter	Method	Reporting Limit
Aluminum	SM 3113 pages 3-22 to 3-27 Electrothermal Atomic Absorption	0.02 mg/L
Bicarbonate	SM 2320B, pages 2-26 to 2-28 Titration Method	5 mg/L
Bromides	SM 4110 pages 4-2 to 4-6, Ion Chromatography with Chemical Suppression of Eluent Conductivity	0.1 mg/L
Calcium	SM 3111B pages 3-13 to 3-15 Direct Air/Acetylene Flame AAS	0.1 mg/L
Chloride	SM 4500-Cl B pages 4-49 to 4-50 Argentometric Method	1.0 mg/L
	SM 4110 pages 4-2 to 4-6, Ion Chromatography with Chemical Suppression of Eluent Conductivity	0.1 mg/L
Copper	SM 3113 pages 3-22 to 3-27 Electrothermal Atomic Absorption	0.01 mg/L
	SM 3111B pages 3-13 to 3-15 Direct Air/Acetylene Flame AAS	0.3 mg/L
Iron	SM 3111B pages 3-13 to 3-15, Direct Air/Acetylene Flame AAS	0.3 mg/L
	SM 3113 pages 3-22 to 3-27 Electrothermal Atomic Absorption	0.01 mg/L
Lead	SM 3113 pages 3-22 to 3-27 Electrothermal Atomic Absorption	0.01 mg/L
Magnesium	SM 3111B pages 3-13 to 3-15 Direct Air/Acetylene Flame AAS	0.1 mg/L
Manganese	SM 3113 pages 3-22 to 3-27 Electrothermal Atomic Absorption	1 ppb
Nitrate	SM 4110 pages 4-2 to 4-6, Ion Chromatography with Chemical Suppression of Eluent Conductivity	0.1 mg/L
Nitrite	SM 4110 pages 4-2 to 4-6, Ion Chromatography with Chemical Suppression of Eluent Conductivity	0.1 mg/L
Nitrogen (NH3,TKN)	SM 4500-Norg pages 4-92 to 4-94 Macro-Kjeldahl Method	0.1 mg/L
pH	SM 4500-H+ B pages 4-65 to 4-69 Electrometric Method	± 0.1 pH units
Phosphorus	SM 4500-P C, pages 4-111 to 4-112 VM Colorimetric Method	0.3 mg/L
	SM 4500-P E. pages 4-113 to 4-114 Ascorbic Acid Method	0.01 mg/L
Silica	SM 4500-Si D, pages 4-118 to 4-120 Molybdosilicate Method	1 mg/L
Sodium	SM 3500-Na B, page 3-96 to 3-98 Flame Emission Method.	0.1 mg/L
Sulfate	SM 4500-SO42- pages 4-136 to 4-137, Turbidimetric Method	1.0 mg/L
	SM 4110 pages 4-2 to 4-6, Ion Chromatography with Chemical Suppression of Eluent Conductivity	0.1 mg/L

Chemical Data - Field Laboratory

Maintenance of good analytical practice in the field was carried out according to the same relevant statistical principles as were used in the main laboratory. While facilities available in a field laboratory are somewhat less sophisticated than those in a full-scale laboratory, the same guidelines are applied to obtain the maximum quality possible from the data gathered in the field. The methodologies used in the field are summarized in Table 3.29. Many of these techniques are the same as those used in the laboratory (such as the alkalinity determination). Others are modifications of known standard laboratory methods that are commercially available for application to field usage (such as the Hach application for chlorine and for nitrogen species). In all cases involving water quality analysis during the course of this project, the commercial modifications cited and used here were widely accepted and available applications of existing standard protocols described in Standard Methods for Water and Wastewater Analysis. They are not novel or untested methods, and they are broadly used by utilities all over the US.

Statistics (accuracy, precision, etc.) were monitored in the same fashion as in the main laboratory, where applicable. Some parameters (UV254, pH, etc.) can not be monitored for accuracy under field conditions (i.e. can not be accurately spiked with a known increment of standard). In cases as these, either standard is not applicable (turbidity, color), or is not sufficiently stable (DO, sulfide). Precision was monitored for these parameters. Surrogate or secondary standards were used for alkalinity, turbidity, and pH to further assist with assessment.

Table 3.29
Methods and reporting limits for inorganic analysis in the field laboratory during project

Parameter	Method	Method Reference	Approx. Range
Alkalinity	Titration	SM 2320 B	5 - 500 ppm
Ammonia-N	Membrane Probe Method	SM 4500-NH3 C	0.1 - 3 ppm
Chloride	Argentometric Titration	SM 4500-Cl^- B	1 – 100 mg/L
Chlorine, free	DPD colorimetric	SM 4500-Cl G or Hach 8021	0.1 - 2 ppm
Chlorine, total	DPD colorimetric	SM 4500-Cl-G or Hach 8167	0.1 - 2 ppm
Color, apparent	Visual Comparison (by spectrometer)	SM 2120 B	1 - 50 cpu
Conductivity	Conductivity Bridge	SM 2510 B	variable
Hardness (total, calcium)	EDTA Titration	SM 2340 C	5 - 500 mg/L
Nitrate	Cadmium reduction	Hach 8192	0.1 - 0.5 mg/L
Nitrite	Diazotization	Hach 8507	0.1 - 0.3 mg/L
Oxygen, Dissolved (DO)	Membrane probe	SM 4500-O G	0.1 - 20 mg/L
pH	Electrometric	SM 4500-H^+ B	2 - 13
Sulfide	Iodometric	SM 4500-S^{2-} F	0.1 - 5 mg/L
Temperature	Direct reading		0 - 100 deg C
Turbidity	Nephelometric	SM 2130 B	0.02 - 200 ntu
UV-254	UV spectrometry	SM 5910 A	0-0.5200 cm^{-1}

Biological Data

In order to insure quality of the biological data of this study, several measures have been taken to minimize variability and analytical errors in the bioassays. These measures can be divided in three categories: (i) sampling protocol, (ii) preparation of reagents and microorganisms, and (iii) analysis of the samples. In each category, the fate of both bulk liquid and biofilm samples is further discussed.

For all the bulk liquid assays (e.g. AOC, HPC, TC/*E.Coli.*, FISH AOB and AOB MPN), samples were taken using sterile gloves and sterile containers (either sterile disposable 100-mL Coliform test sampling containers or 1-L Pyrex bottles that had been muffled at 525°C for 5 hrs). After collection, sample bottles were capped with persulfate washed caps. Bulk liquid samples were then placed in ice for transport and brought back to the laboratory upon sampling, where they were stored in a 4°C walk-in cooler until preservation and/or analysis. Storage of bulk liquid samples did not exceed 24 hrs prior to analysis unless proper preservation measures were taken. For biofilm assays (e.g. PEPA, BFHPC, FISH AOB and AOB MPN), coupons affixed on PVC holders were handled with sterile gloves and transported in air-tight PVC containers, in which moisture level was maintained with sponges soaked with corresponding incubation water (to prevent desiccation of the biofilms). Upon arrival at the UCF main laboratory, these containers were stored in a 4°C walk-in cooler until analysis was carried out (no later than 24hrs after sampling).

For the preparation of reagents in contact with bulk liquid samples and/or microorganisms, strict aseptic techniques were used. Sterilization of agar and reagents (such as distilled water, dilution buffer and control reagents) was performed by autoclaving at 121°C and 15 psi for 20 minutes. Preservation of AOC samples was carried out by pasteurization in a water bath at 70°C for 1 hr or with alternate pasteurization technique described by Escobar and Randall (2000). for extended storage. Aliquots of AOC samples were added to 40-mL EPA vials, previously muffled at 525°C for 5 hrs, and capped with persulfate washed caps. PVC containers used for incubation of coupons in PEPA assay were sterilized by autoclave as previously described. To prevent contamination, handling and manipulating of samples and reagents requiring aseptic conditions were carried out under a laminar flow hood, previously wiped with bactericidal detergent and irradiated with a UV lamp for 5 minutes. For biofilm samples, sterilized (with 70% ethanol solution) cell scraper and tissue homogenizer were used for cell detachment and homogenization of suspended materials.

Analysis of samples was mainly carried out under laminar flow hood (kept sterile as described above), except for PEPA, which was carried out in a fume hood. In order to check aseptic conditions and proper sterility of the reagents, control measures included the following three steps: (i) laminar flow hood atmosphere was checked using open agar plate during spreading, (ii) dilution buffer sterility was checked by spreading one plate with dilution buffer alone as a blank during the same time as the spreading of the actual samples, and (iii) sterility of the agar was verified by inclusion of one covered, non-inoculated plate in each run. These three plates were incubated along with the samples. A glass spreader (i.e. bent glass rod) was sterilized with 70% ethanol solution and flamed with a Bunsen burner. Sterile disposable pipette tips, sterile gloves and sterilized dilution vials were used to prevent contamination during experiments. For AOC, sterile needles and syringes were used to inoculate samples with Pseudomonas fluorescens P17 and Spirillum NOX. Purity and accuracy of inocula were also monitored. Incubation of samples when required was performed in a temperature-controlled atmosphere. For biofilm samples analyzed by PEPA, North Viton® gloves were used for handling carcinogenic reagents under fume hood conditions. For bioassays involving plate spreading techniques (HPC, BFHPC and AOC), serial dilutions were always performed in duplicate and plate counts in the range of 30-300 cfu were selected whenever possible. However in the case of low (less than 30 cfu at 10^{-1} dilution) or high (above 300 cfu at highest dilution) density samples, plate counts out of this range have been used for interpretation. Since the sample analysis and assessment required 7 to 10 days to complete, this renders additional duplication (i.e. re-collection of an identical duplicate of a sample in question) practically impossible.

The quality control measures on biological assays described above proved proper aseptic conditions and analysis of the samples with respect to Standard Methods guidelines for handling and analyzing biological samples throughout the entire study.

QC Assessment

Chemical Data – UCF Main Laboratory

The tables on the following pages provide a summary of the accuracy and precision of laboratory analysis during the entire project.

Accuracy. The mean recoveries for the inorganic parameters ranged from about 94% to 103%, as shown in Table 3.30. Metal recoveries ranged from 90 to 103% recovery. Similarly, anion recovery ranged from 94 to 103%. NPDOC and the other inorganic recoveries fell within 98 to 102% recovery. Individual spike measurements for all parameters were within the NIST required 80 to 120% recovery range, or the value was rejected and another spike was performed. Most warning limits were similar to or within these NIST ranges.

Spike recoveries were tracked using Shewart control charts in accordance with Standard Methods. Table 3.31 summarizes the warning and control limit information gathered in the accuracy control charts for all applicable parameters analyzed during the entire project.

Precision. Average RPD values for all inorganic parameters analyzed in the laboratory during the entire project are shown in Table 3.32. The precision control chart information for inorganics and NPDOC is provided in Table 3.33. A similar summary is provided in Table 3.34 for aggregate parameters measured in the laboratory. Aluminum and phosphorus were nearly always at or below detection levels. This trend was one of a few recurring circumstances that gave rise to high RSD values and broad ranges. Values that are near/at the detection limit for a given technique and can not be accurately resolved further (for example, TKN. measurement to + 0.01 mg/L) will inevitably give rise to a high RPD. This is responsible for the bulk of the large values shown in the "High" column of Table 3.30.

Table 3.30
Accuracy summary for NPDOC and inorganic parameters in the UCF main laboratory during the entire project

Parameter	Technique	Units	# Spikes	Mean %Rec	RPD	High	Low
High Metals							
Calcium	Flame AAS	ppm	224	98.2	8.3	120	80
Magnesium	Flame AAS	ppm	180	97.3	6.2	113	82
Sodium	Flame AE	ppm	195	99.1	3.8	104	89
Trace Metals							
Aluminum	GFAAS	ppm	98	99.0	8.3	120	80
Copper	GFAAS	ppb	117	100.4	6.3	116	81
Iron, diss	GFAAS	ppb	170	103.1	8.5	118	81
Iron, total	GFAAS	ppb	216	98.5	2.2	103	93
Lead	GFAAS	ppb	122	99.9	2.1	105	93
Manganese	GFAAS	ppb	14	90.3	4.8	120	80
Anions							
Bromide	IC	ppm	109	100.6	9.7	120	80
Chloride	Titration	ppm	142	97.6	12.1	120	80
Fluoride	IC	ppm	96	95.0	10.1	120	80
Ortho-P	IC	ppm	90	101.8	8.7	120	80
Sulfate	IC	ppm	142	94.1	10.0	120	80
Other Inorganics							
TKN	Digest/Distill	ppm	64	102.0	9.6	120	80
Alkalinity	Titration	ppm	140	97.3	7.1	117	80
Total-P	Vis spec	ppm	40	103.8	8.8	118	81
Silica	Vis spec	ppm	139	100.2	3.2	117	83
Organics							
NPDOC	UV/persulfate	ppm	140	98.9	8.2	118	80

Table 3.31
Accuracy - spike violations in the main UCF laboratory for the entire project

Parameter	Total Spikes	No. > WL	No > CL	Total ≤ WL	Total ≤ CL	Pct. ≤ WL	Pct. ≤ CL
High Level Metals							
Calcium	224	8	0	216	224	96.4	100.0
Magnesium	180	7	0	173	180	96.1	100.0
Sodium	195	10	0	185	195	94.9	100.0
Trace Metals							
Aluminum	98	3	0	95	98	96.9	100.0
Copper	117	7	0	110	117	94.0	100.0
Iron, dissolved	170	3	0	167	170	98.2	100.0
Iron, total	216	3	0	213	216	98.6	100.0
Lead	122	3	0	119	122	97.5	100.0
Manganese	14	NA	NA	NA	NA	NA	NA
Anions							
Bromide	109	5	0	104	109	95.4	100.0
Chloride	142	3	0	139	142	97.8	100.0
Fluoride	98	3	0	95	98	96.9	100.0
o-Phosphate	90	3	0	87	90	96.7	100.0
Sulfate	142	7	0	135	142	95.1	100.0
Other Inorganics							
TKN	64	3	0	61	64	95.3	100.0
Alkalinity	147	6	3	141	144	95.9	98.0
Phosphorus, total	40	NA	NA	NA	NA	NA	NA
Silica	139	7	1	132	138	95.0	99.3
Organics							
NPDOC	142	7	0	135	142	95.1	100.0

Table 3.32
Precision summary for inorganic parameters and NPDOC analysis in the UCF main laboratory during the entire project

Parameter	Technique	Units	# Dupes	Mean RPD	%RSD	High	Low
High Metals							
Calcium	Flame AAS	ppm	228	1.0	1.3	9.3	< 1
Magnesium	Flame AAS	ppm	180	1.0	1.2	9.2	< 1
Sodium	Flame AE	ppm	191	1.2	1.3	6.6	< 1
Trace Metals							
Aluminum	GFAAS	ppm	107	4.4	5.7	25.4	< 1
Copper	GFAAS	ppb	117	2.8	3.3	25.5	< 1
Iron, diss	GFAAS	ppb	168	4.1	4.9	34.1	< 1
Iron, total	GFAAS	ppb	221	3.2	2.0	9.1	< 1
Lead	GFAAS	ppb	120	4.0	4.1	25.4	< 1
Manganese	GFAAS	ppb	14	2.7	3.5	9.6	< 1
Other Inorganics							
Bicarbonate	Titration	ppm	147	2.5	4.7	39.5	< 1
Bromide	IC	ppm	86	6.7	11.4	59.3	< 1
Calcium	Flame AAS	ppm	228	1.0	1.3	9.3	< 1
Chloride	Titration	ppm	150	1.3	1.3	7.5	< 1
Fluoride	IC	ppm	95	2.5	9.2	59.3	< 1
Ortho-P	IC	ppm	90	NA	NA	NA	NA
Total-P	Vis spec	ppm	40	NA	NA	NA	NA
Silica	Vis spec	ppm	138	1.4	2.6	12.4	< 1
Sodium	Flame AE	ppm	191	1.2	1.3	6.6	< 1
Sulfate	IC	ppm	146	3.6	6.1	8.8	< 1
TKN	Digest/Distill	ppm	63	8.2	10.4	47.4	< 1
Organics							
NPDOC	UV/persulfate	ppm	151	3.7	4.8	37.3	< 1

[a] All phosphate and alumimum samples analyzed for QA/QC were below detection (spikes and dupes supportive)

Table 3.33
Inorganics and NPDOC precision and duplicate violations in the UCF main laboratory during the entire project

Parameter	Total Duplicates	Total > WL	Total > CL	Pct. < WL	Pct. < CL
High Level Metals					
Calcium	228	19	13	91.7	94.3
Magnesium	180	6	5	96.7	97.2
Sodium	191	13	7	93.2	96.3
Trace Metals					
Aluminum	107	8	4	92.5	96.3
Copper	117	6	2	94.9	98.3
Iron, dissolved	168	13	7	92.3	95.8
Iron, total	222	19	13	91.4	94.1
Lead	123	11	11	91.1	91.1
Manganese*	14	NA	NA	NA	NA
Anions					
Bromide	111	10	9	91.0	91.9
Chloride	148	12	8	92.0	94.7
Fluoride	96	12	12	87.5	87.5
o-Phosphate	90	3	3	96.7	96.7
Sulfate	148	14	6	90.4	95.9
Other Inorganics					
TKN	64	3	0	88.9	92.1
Alkalinity	147	6	3	94.5	96.6
Phosphorus, total	40	NA	NA	NA	NA
Silica	139	7	1	84.8	87.7
Organic					
NPDOC	142	7	0	93.4	95.4

*Insufficient data to construct control chart for this parameter.

Table 3.34
Precision summary for aggregate parameters for UCF main laboratory during the entire project

Parameter	Technique	Units	# Dupes	Mean RPD	Std dev	Max.	Min.
Color, App	Spectrometric	CPU	31	NA	NA	NA	NA
Color, True	Spectrometric	CPU	138	2.7	11.3	59.3	< 1
TDS	Ion Sum	ppm	NA	NA	NA	NA	< 1
Conductivity	Probe	μmho/cm	148	0.8	1.1	8.9	< 1
UV-254	Abs @ 254 nm	/cm	70	0.8	0.7	3.1	< 1

As an added measure of precision, blind duplicates were collected on site with every sampling event. With each work order submitted, at least two (usually more) blind duplicates were also collected. A summary of the blind precision assessment is found in Table 3.35. These samples were submitted along with their respective sample batch, submitted in the same work order as this batch, and analyzed accordingly. After the results for each parameter were submitted to the lab manager for the entire work order, the blind duplicate results were assessed.

Table 3.35
Blind precision summary for all parameters for UCF main laboratory during the entire project

Parameter	Technique	Units	# Dupes	Mean RPD	%RSD	High	Low
High Metals							
Calcium	Flame AAS	ppm	185	2.5	4.4	28.9	< 1
Magnesium	Flame AAS	ppm	183	2.7	4.1	27.3	< 1
Sodium	Flame AE	ppm	158	2.4	3.6	26.4	< 1
Trace Metals							
Aluminum	GFAAS	ppm	94	7.6	12.8	67.3	< 1
Copper	GFAAS	ppb	115	4.2	6.0	48.3	< 1
Iron, diss	GFAAS	ppb	125	16.1	23.0	100	< 1
Iron, total	GFAAS	ppb	148	7.9	9.4	50.0	< 1
Lead	GFAAS	ppb	92	8.7	15.9	100	< 1
Other Inorganics							
Bicarbonate	Titration	ppm	107	4.8	5.8	23.0	< 1
Bromide	IC	ppm	78	7.1	15.7	66.7	< 1
Chloride	IC	ppm	149	2.8	4.0	23.3	< 1
Ortho-P	IC	ppm	104	0.8	5.6	48.9	NA
Total-P	Vis spec	ppm	NA	NA	NA	NA	NA
Silica	Vis spec	ppm	108	4.7	6.0	26.7	< 1
Sulfate	IC	ppm	149	3.6	6.1	40.0	< 1
TKN	Digest/Distill	ppm	27	15.0	13.7	66.7	< 1
Aggregates							
Color, True	Vis spec	cpu	127	19.5	31.2	200	< 1
Conductivity	Probe	µmho/cm	79	2.2	9.0	18.9	< 1
UV-254	UV spec	cm^{-1}	110	4.9	10.1	100	< 1
Organics							
NPDOC	UV/persulfate	ppm	119	9.8	10.9	61.8	< 1

*Comparatively few or no blind PDS samples analyzed for ortho-phosphate, lead, or aluminum during this time ever showed values above detection limit. Only one blind sample showed detectable aluminum values; both it and its blind dupe were measure at 30 ppb or below.

Chemical Data - Field Laboratory

Table 3.36 and Table 3.37 provide a summary of the accuracy (when applicable) and precision of laboratory analysis during the entire project.

Table 3.36
Accuracy summary for field parameters during the entire project

Parameter	Technique	Units	# Spikes	Mean % Rec	RPD	High	Low
Metals							
Ca Hardness	Titration	ppm	73	102.1	5.8	119.5	86.4
Total Hardness	Titration	ppm	70	100.6	4.3	119.9	90.0
Iron, total	Vis spec	ppb	45	100.9	10.4	104.6	94.8
Anions							
Chloride	Titration	ppm	54	103.5	5.4	115.7	90.0
Nitrite-N	Vis spec	ppm	156	100.3	6.7	120.0	80.0
Nitrate+Nitrite-N	Vis spec	ppm	150	95.7	24.5	120.0	80.0
Other Inorganics							
Alkalinity	Titration	ppm	176	99.4	4.2	114.7	93.3
Ammonia-N	Probe	ppm	224	97.3	11.8	120.0	80.0
Chlorine, free	DPD color	ppm	533	97.2	1.7	101.3	94.5

Table 3.37
Precision summary for field parameters during the entire project

Parameter	Technique	Units	# Dupes	Mean Range	%RSD	High	Low
Metals							
Ca Hardness	Titration	ppm	76	1	1	4	< 1
Total Hardness	Titration	ppm	72	1	1	4	< 1
Iron, total	Vis spec	ppm	44	0.003	0.004	0.015	< 1
Anions							
Chloride	Titration	ppm	51	1	1	11	< 1
Nitrite-N	Vis spec	ppm	177	0.01	0.03	0.23	< 0.01
Nitrate+NitriteN	Vis spec	ppm	129	0.04	0.06	0.35	< 0.01
Other Inorganics							
Alkalinity	Titration	ppm	240	1	1	3	< 1
Ammonia-N	Probe	ppm	235	0.01	0.01	0.11	< 0.01
Chlorine, free	DPD color	ppm	695	0.02	0.02	0.03	< 0.01
Chlorine, total	DPD color	ppm	776	0.06	0.15	0.54	< 0.01
Oxygen, diss	Probe	ppm	515	0.03	0.02	0.44	< 0.1
Aggregates							
Color, apparent	Vis spec	cpu	433	1	1	5	< 1
Conductivity	Probe	µmho/cm	211	2	2	34	< 1
ORP	Probe	mV	75	4.2	3.0	68.7	0.1
pH	Probe		612	0.02	0.01	0.44	< 1
Turbidity	Nephelometer	ntu	557	0.05	0.04	0.37	< 1
UV-254, filtered	UV spec	cm^{-1}	159	0.0008	0.0011	0.0082	<0.0001
UV-254, unfilt	UV spec	cm^{-1}	129	0.0007	0.0006	0.0027	<0.0001

Biological Data

For plate counting techniques (HPC, BFHPC), all samples were run in duplicate. For each individual within each pair, the counts were assessed. The logarithm of each count was taken, and the range (difference) of the log values for each pair was calculated. The range was expected to be less than one, i.e. the counts were expected to be within one order of magnitude of each other (USEPA 1989). Out of 320 HPC analyses, the range varied from 0 to 0.64. Out of 324 BFHPC analyses throughout the project, the range varied from 0 to 0.75. In either case, the range never exceeded 1. This means that all duplicate analyses of HPC as well as BFHPC were acceptable.

Yield controls were used to assess AOC performance. Two separate solutions of carbon (100 µg/L of acetate carbon in mineral salts buffer) are inoculated with P17 and with NOX respectively Growth yields were assessed as cfu/µg C and compared to published values (Van der Kooij 1990). The difference between the logarithm of the value obtained experimentally versus the published (theoretical) value were expected to be within 1. There were a total of 11 P17 yield controls and 11 NOX controls. All of the NOX controls complied, whereas 9 of the 11 P17 yield controls complied; 2 of these were high (about 1.3-1.5). It was found upon inspection of the logs that the P17 inocula were slightly higher for 6/20/02 and 8/01/02, resulting in a higher yield control.

For PEPA analysis as carried out in this project (i.e. on coupons), biofilm structure and pattern of colonization can not be reproduced identically on two apparently identical coupons of the same material (Van der Kooij and Veenendaal 1992). Therefore, as long as the range of logs of the biofilm inventory of duplicate pairs were within 2, (i.e. the inventories were within two orders of magnitude), duplication was deemed acceptable. The range varied from 0 to 1.4 over the entire project (64 pairs); no range value ever exceeded 2. The majority of ranges (62 of 64) were below 1.

Bulk liquid FISH-AOB and MPN results correlated well with the sum of nitrite and nitrate concentration for a given sample location, referred to as NOx ($R^2 = 0.36$ and $R^2=0.25$, respectively). For FISH-AOB, two different probes were used to detect nitrifiers (NSO 190 and NSO 1225). Results from these two probes correlated well with each other ($R = 0.7$). It should be noted that this FISH-AOB technique was tailor-designed for this project, therefore no standard selection criteria or literature reference exist for comparison. In addition, the AOB-MPN counts correlated well with FISH-AOB results.

EQUIVALENT PIPE LENGTH

The results shown in this report were obtained using a pilot distribution system as previously described. Mass transfer models for apparent color release, total iron release, turbidity release and chloramine or free chlorine dissipation were developed so that the findings of this research project could be applied to a full size distribution system. These models are presented in the chapter on results and discussion, and it is important to note that these results can not be accurately applied directly to actual distribution systems from the information shown in results and discussion. The following discussion on equivalent pipe lengths illustrates that point.

The pilot distribution systems are 100 ft long and were made as four single material lines and 14 hybrid lines as previously described. The PDSs were operated at a 5 day and 2 day hydraulic residence time (HRT). The flow velocities in the PDS were 0.000232 and 0.000579 ft/sec, respectively. HRT selection was a group effort that included extensive discussion and final approval by the UCF project team, the AwwaRF project advisory committee and the TBW technical advisory committee. A conscious decision was made to study worst case conditions in actual distribution systems, which were thought to be represented initially by a 5 day HRT. As will be shown in results and discussion, the HRT was reduced from 5 to 2 days because a chloramine residual could not be maintained in the PDS over 5 days.

It is necessary to assume an actual distribution system HRT and velocity in order to determine equivalent pipe lengths in an actual distribution system based on the 2 and 5 day HRT in the PDSs. An actual distribution system velocity was estimated by assuming a 1 ft/sec velocity from 6 AM to 9 AM, noon to 3 PM and 6 PM to 8 PM, and zero at all other times. Hence the equivalent daily average velocity would be 8/24 x 1 ft/sec or 0.33 ft/sec. Equation (3.1) was developed and solved for equivalent distribution system pipe lengths.

$$\frac{L_{eq}}{V_{eq}} = \frac{L_{PDS}}{V_{PDS}} \tag{3.1}$$

Where:
L_{eq} = Equivalent pipe length, ft actual distribution system
V_{eq} = Equivalent velocity, ft/sec actual distribution system
L_{PDS} = PDS pipe length, ft pilot distribution system
V_{PDS} = PDS velocity, ft/sec pilot distribution system

The equation as shown requires the V_{eq}, L_{PDS} and V_{PDS} to be known in order to determine the L_{eq}. The L_{PDS} was 100 ft. V_{PDS} was calculated by dividing 100 ft. by 2 or 5 days. Consequently, the L_{eq}s can be directly solved for those two conditions, and were found to be 10.8 and 27 miles for a 2 and 5 day HRT. Additional L_{eq}s were determined as shown in Table 3.38, by assuming different PDS HRTs and calculating different V_{PDS}s to use in equation (3.1).

As shown in Table 3.38, the equivalent length varies from 189 to 0.27 miles in an actual DS for HRTs of 14 to 0.05 days. Whereas actual reaches of pipe in distribution systems might extend to 40 miles for some member governments, actual reaches of UCI or G pipe would seldom extend more than 1000 feet. Consequently, direct application of water quality for single material lines in the distribution system can be made to 6" PVC, LCI or UCI lines or a 2" G line receiving the same water quality at 10.5 or 27 miles from the water treatment plants (assuming 2 or 5 day HRT, respectively). Those instances will probably never exist, so to make application of the results in this study to actual distribution systems for apparent color release, total iron release, turbidity release, residual concentrations and nitrification requires assumption or knowledge of pipe geometry, pipe material, pipe hydraulics and water quality. If that is done, then reasonable estimation of apparent color release, total iron release, turbidity release, residual concentrations and nitrification can be made. Whereas exact mass transfer models were developed for apparent color release, total iron release, turbidity release and residual concentrations, they were not developed for nitrification. However, nitrification was linked and controlled by chloramine residual in this work. Hence residual modeling can be use to control nitrification if distribution system dead zones are controlled or eliminated.

Table 3.38
Equivalent pipe length for comparison between PDS and actual distribution systems

Assumed PDS HRT (days)	Equivalent Actual DS Pipe Length using 2 Day PDS HRT (ft)	(miles)	Equivalent Actual DS Pipe Length using 5 Day PDS HRT (ft)	(miles)
14	399168	75.6	997920	189.00
7	199584	37.8	498960	94.50
5	142560	27	356400	67.50
2	57024	10.8	142560	27.00
1	28512	5.4	71280	13.50
0.5	14256	2.7	35640	6.75
0.25	7128	1.35	17820	3.38
0.1	2851	0.54	7128	1.35
0.05	1426	0.27	3564	0.68

CHAPTER 4
RESULTS AND DISCUSSION

IRON RELEASE

Identification of Iron Corrosion Products on Aged Pipe Materials

XRD and XPS were performed to characterize the corrosion products formed on the internal surface of the four different pipe materials which were exposed to conventionally treated GW (GW) for several decades in the distribution system. Figure 4.1 to Figure 4.2 show the XRD patterns of the corrosion products on UCI and G. The observed crystalline compounds on the UCI coupons included α-FeOOH, β-FeOOH, γ-Fe_2O_3, Fe_3O_4, $FeCO_3$ and SiO_2. The primary crystalline compounds on the G coupon, shown in Figure 4.2, were α-FeOOH, β-FeOOH, γ-Fe_2O_3, Fe_3O_4, and ZnO. Here, β-FeOOH and ZnO were the major peaks observed. As for the LCI and PVC, no clear crystalline phases were found, indicating there was no iron corrosion or minimal iron corrosion deposits.

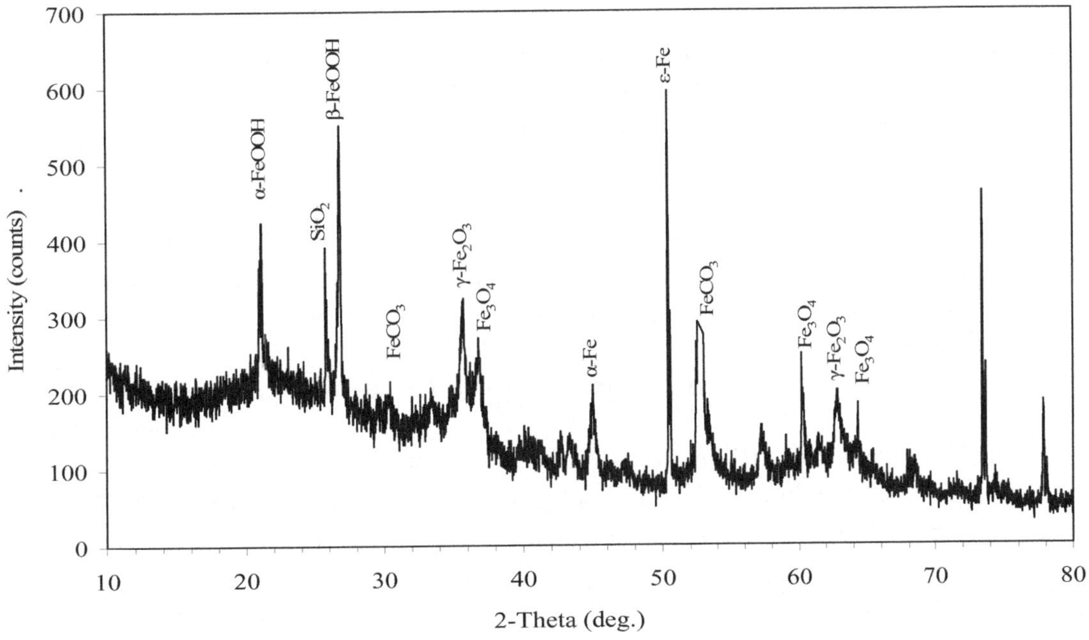

Figure 4.1 Representative XRD patterns of corrosion products on aged UCI pipe materials exposed to GW water environment for several decades

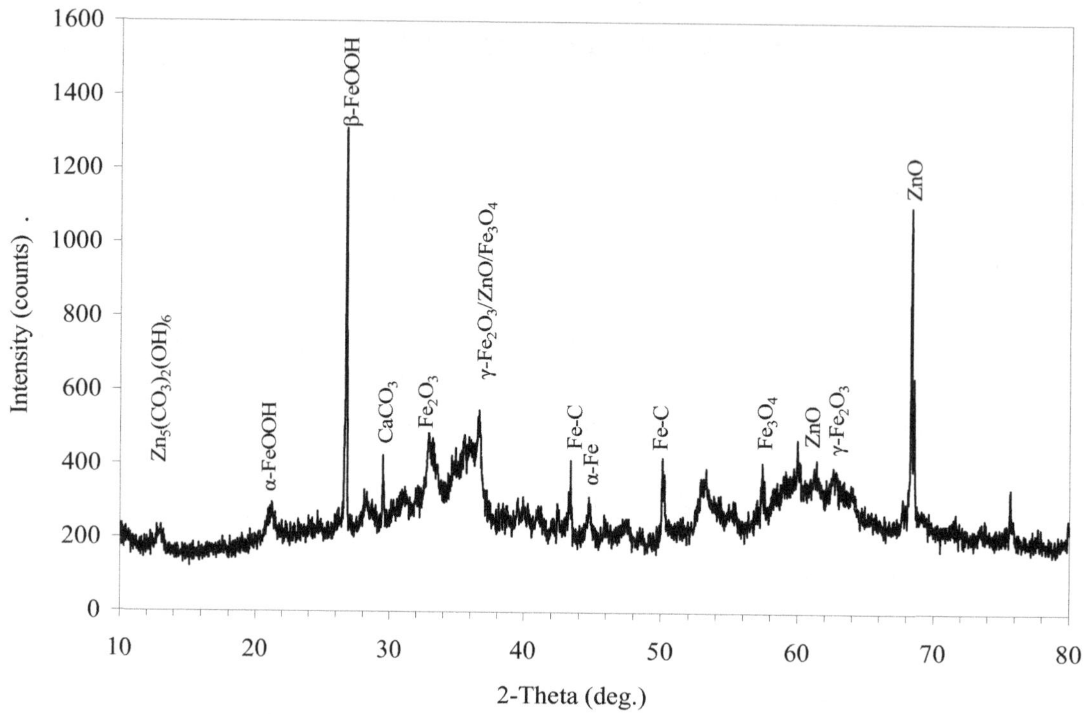

Figure 4.2 Representative XRD patterns of corrosion products on aged G pipe materials exposed to GW water environment for several decades

The XPS survey spectra from UCI showed that the primary elements observed were C, O, Fe, Si, and N. For G, a significant amount of Zn was also found in the XPS survey scan. A high resolution scan of the Zn2p3/2 peak indicated the presence of ZnO and $Zn(OH)_2$ on the surface of the corrosion products, verifying the XRD results shown in Figure 4.2. As expected, for LCI and PVC coupons, no or very small iron signals were observed in the XPS scan spectra.

Identification of Corrosion Products on Coupons after Three-Month Incubation

Figure 4.3 shows the XRD patterns of the corrosion products on an UCI coupon after 3-month exposure to GW/SW/RO (23/45/32%) water environment. As shown, the crystalline compounds observed were α-FeOOH, β-FeOOH, γ-Fe_2O_3, Fe_3O_4, $FeCO_3$, and SiO_2. Compared to the coupons before incubation, more γ-Fe_2O_3, Fe_3O_4 and $FeCO_3$, were found in the corrosion scale, showing continued corrosion. Lastly, background noise was more noticeable, implying more amorphous phases existed in the corrosion scales due to the 3-month exposure to new water. In other words, the corrosion process was ongoing.

The XRD pattern of the corrosion products on G coupons indicates that α-FeOOH, β-FeOOH, Fe_2O_3, Fe_3O_4, ZnO and $Zn_5(CO_3)_2(OH)_6$ were primary corrosion products. After 3-month exposure to GW/SW/RO environment, both α-FeOOH and β-FeOOH intensity decreased, and Fe_2O_3 and Fe_3O_4 became the major corrosion products. Unlike the cast iron pipes, $FeCO_3$ was not found in the XRD analysis, and ZnO and $Zn_5(CO_3)_2(OH)_6$ were observed. The XRD analyses of LCI and PVC coupons show no obvious crystalline compounds, with exception of small amounts of Ca and Fe precipitates such as $CaCO_3$ and γ-Fe_2O_3. Lastly, the XPS analysis performed on four different pipe coupons of single material PDSs showed the results similar to those of XRD analysis as described previously.

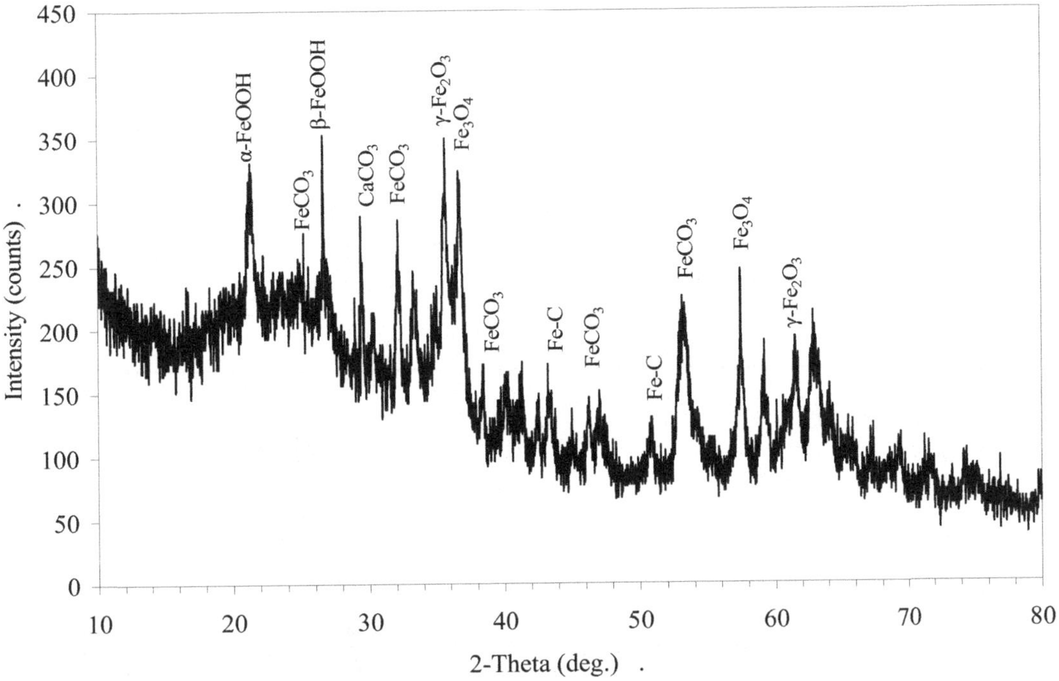

Figure 4.3 Representative XRD patterns of corrosion products formed on an UCI coupon when exposed to GW/SW/RO (23/45/32%) water for three months

Correlation between Surface Characteristics and Iron Release

In order to further investigate the effect of water quality on pipe surface characteristics and iron release, treated ground, surface, and saline waters were blended at various ratios and fed to hybrid PDSs consisting of all of these four pipe materials. The study focused on six blending scenarios including GW(100%), SW(100%) RO(100%), and GW/SW (55/45%), GW/RO (68/32%) and GW/SW/RO (62/27/11%).

The XRD patterns of cast iron exposed to six different blended water environments showed little difference, except that GW exhibited the lowest $FeCO_3$ content based on the relative $FeCO_3$ peak intensity. The XRD spectra of G also showed no significant variations among the pipe coupons incubated under different water quality conditions. Since the XRD was done on powder samples, three-month incubation might not significantly change the characteristics of the corrosion scales on aged pipe materials.

The XPS technique is very surface sensitive and could provide information on chemical changes that have occurred at the interface between water and inorganic film on the pipe surface. Figure 4.4 presents the detailed scan and the deconvolution of Fe2p3/2 peak of an UCI aged coupon exposed to GW(100%) water environment. The deconvolution of Fe(2p3/2) peak revealed four separate peaks, which are corresponding to Fe_3O_4, Fe_2O_3, FeOOH, and the satellite peak. The deconvolution of Fe2p3/2 peak of G coupons presented also showed the presence of

Fe_2O_3, Fe_3O_4, and FeOOH. According to the XPS results, the peak area percentage, which stands for the relative amount of corrosion products in the scale, varied with the different water quality conditions. The correlations (R^2) between Fe_2O_3 peak area percentages of cast iron pipes and total and dissolved iron levels observed in the pilot study, presented in Figure 4.5, showed that dissolved and total iron release increased with increasing Fe_2O_3. It seems that Fe_2O_3 plays an important role in iron release in water. However, Fe_2O_3 solubility is very low and actual dissolved iron concentration is much higher than Fe_2O_3 solubility. Since $FeCO_3$ could decompose easily and transform to Fe_2O_3 if exposed to air (Seal et al. 2000, Heuer and Stubbins 1999), a large portion of Fe_2O_3 peak area percentage observed in the XPS study may come from the transformation of $FeCO_3$. It suggests that there is a positive linear correlation between $FeCO_3$ content in the surface layer and the iron concentration. Therefore, $FeCO_3$ might be the controlling solid phase of iron release in an aqueous environment.

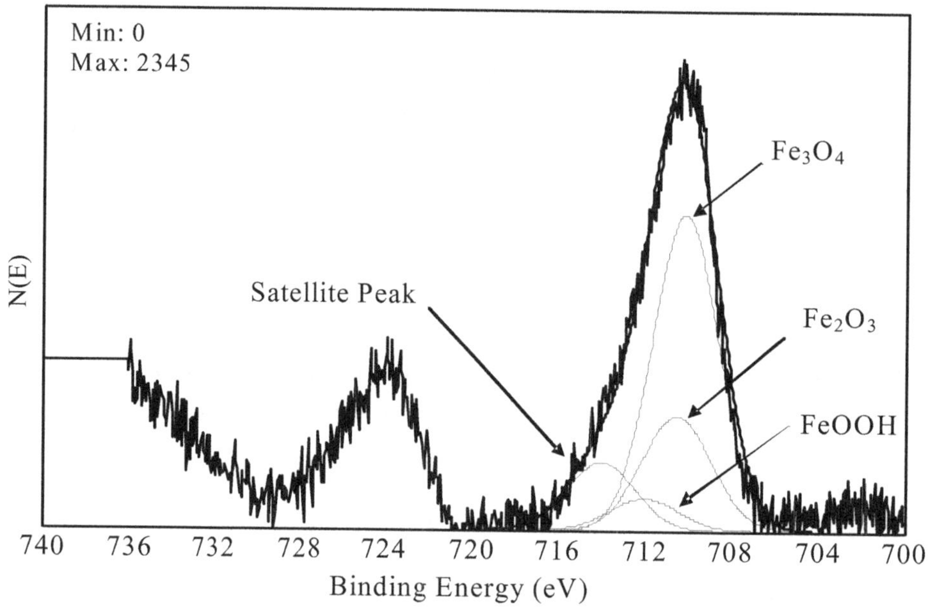

Figure 4.4 Representative high-resolution XPS scan and the deconvolution of $Fe_2p_{3/2}$ peak of corrosion products formed on UCI coupon during three-month incubation under GW source water

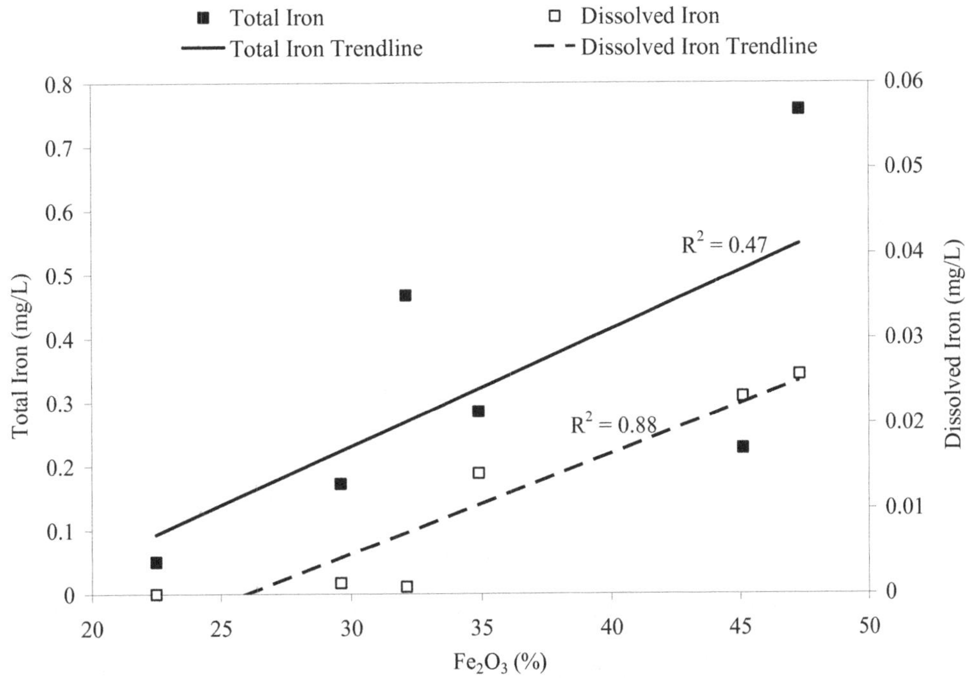

Figure 4.5 Correlation between Fe_2O_3 peak area percentage of XPS scan and actual total and dissolved iron release observed from hybrid PDSs receiving various blended source waters

Lastly, the correlations between surface characteristics of aged G pipes and iron release were also examined. No clear relationship was found between different iron species and iron release levels. This is not surprising since elevated iron levels observed were primarily due to the disruption and release of chemical films formed on UCI pipes.

Thermodynamic Equilibrium Iron Release Model

The iron release from aged pipe surfaces may be predicted simply by the solubility of solid phases formed within corrosion scales. This thermodynamic approach is limited to definition of the soluble fraction of iron, and would be expected to underestimate total iron concentrations due to the presence of particulate iron forms. In this study, siderite ($FeCO_3$) is assumed to be the solid phase controlling iron ion concentration in water, based on surface characteristics of cast iron pipes, determined by XRD and XPS analyses. In this thermodynamic model, the free iron ion concentration is first calculated assuming iron ions and the controlling solid phase are in equilibrium. Subsequently, the concentrations of other dissolved complexes (e.g. carbonate, chloride, and sulfate) are estimated according to the related reactions. Finally, the summation of dissolved iron ion and dissolved complexes are the total iron concentration in bulk water as shown in equation (4.1) where all units are in moles/L.

$$Fe_T = [Fe^{+2}]+[FeOH^+]+[Fe(OH)_2]+[Fe(OH)_3]+[FeHCO_3^{\ +}]+[FeCO_3]+[FeSO_4]$$
$$= \frac{10^{10.89}\times[H^+]}{[HCO_3]\times10^{10.3}}\left(1+\frac{10^{9.5}}{[H^+]}+\frac{10^{20.6}}{[H^+]^2}+\frac{10^{32}}{[H^+]^3}+10^2[HCO_3]+10^{4.38}[HCO_3]\times\frac{10^{10.3}}{[H^+]}+10^{2.25}\right) \quad (4.1)$$

It should noted that the model was developed based on the assumption that ORP is 0.38V (average of TBW water quality) and temperature is 25°C. According to the ORP values measured in the field, Fe^{+2} is much more than Fe^{+3}. Thus ferric species are ignored in the model development.

The siderite model developed was used to predict the iron release according to water qualities measured in the pilot study. Figure 4.6 shows the predicted iron levels as a function of pH and alkalinity.

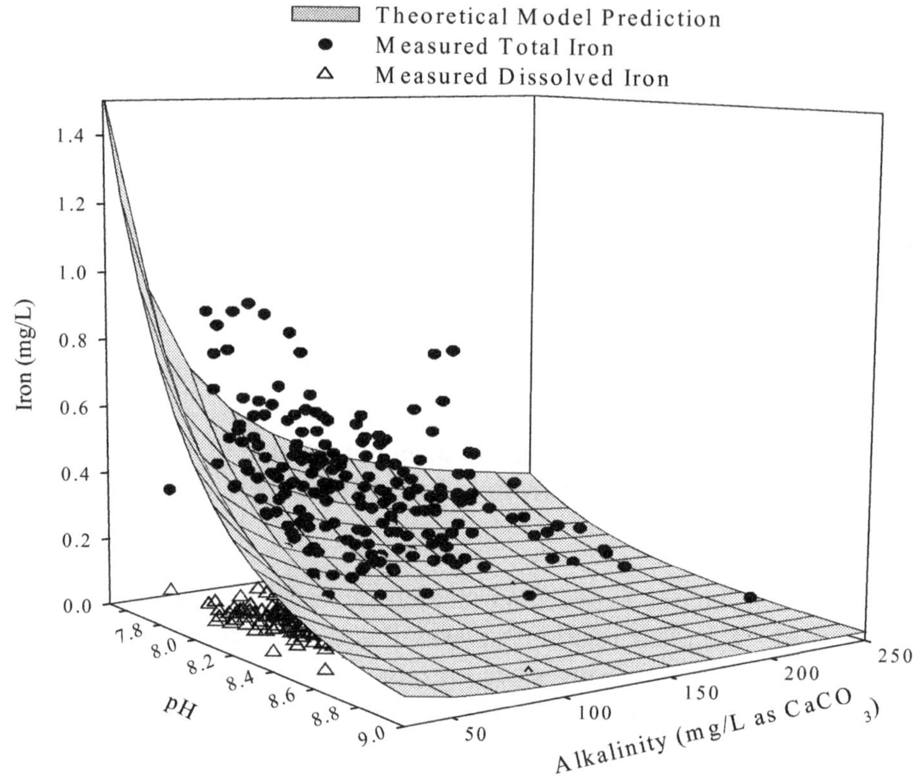

Figure 4.6 Siderite model predictions versus actual total and dissolved iron observed during one-year study of hybrid PDSs receiving various blended source waters

The actual dissolved and total iron concentrations observed in the pilot study are presented in Figure 4.6 for comparison. As shown, the siderite model predicts iron release to decrease as pH and alkalinity increase. However, it must be noted that the iron values predicted by the siderite model are larger than actual dissolved iron concentrations and less than actual total iron concentrations.

Lastly, to compare with the $FeCO_3$ model, other equilibrium models were developed assuming $Fe(OH)_2$ or $Fe(OH)_3$ as the controlling solid phase, respectively according to Pourbaix diagram. The predicted values of the $Fe(OH)_2$ model are much higher (1000 times) than actual TBW pilot total iron data. The $Fe(OH)_3$ model grossly under-predicts iron release and the predicted values are 200,000 times lower than actual TBW pilot dissolved iron data. Thus, compared with $Fe(OH)_2$ and $Fe(OH)_3$ models, the $FeCO_3$ model gives the closest concentrations when compared to actual data.

Development of a Statistical Model for Total Iron Release

During the equilibration period, it was determined that effluent values for apparent color were well correlated with turbidity and total iron concentrations (Cullen, 2002). Phase average effluent data for the hybrid PDSs during the first 12 months of operation are presented in Figure 4.7 and Figure 4.8 that confirm this observation for turbidity and total iron, respectively. Accordingly, apparent color data has been selected for this discussion to characterize water quality effects associated with disruption of surface films. More precisely, the change in apparent color (effluent minus influent apparent color) was utilized as the response variable. Summary data for the rise in apparent color is provided in Table 4.1 for Phases I to V (12/8/01 to 4/4/03).

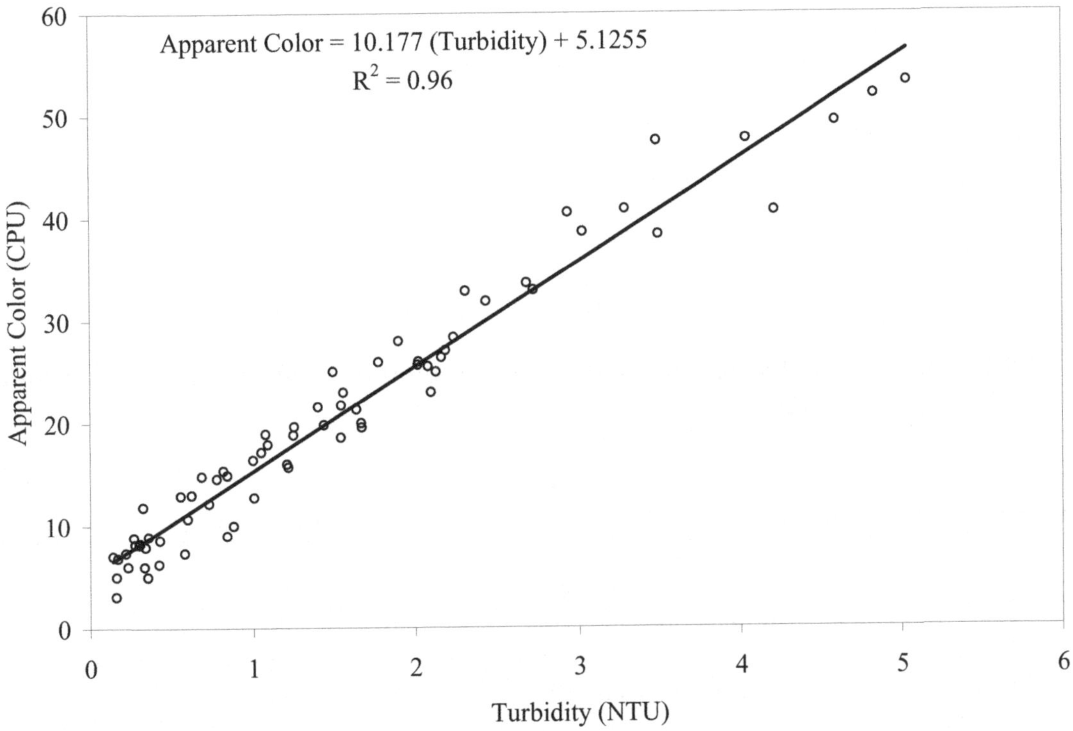

Figure 4.7 Relationship between turbidity and apparent color for average PDS data from phase I to IV (12/8/01 to 12/13/02)

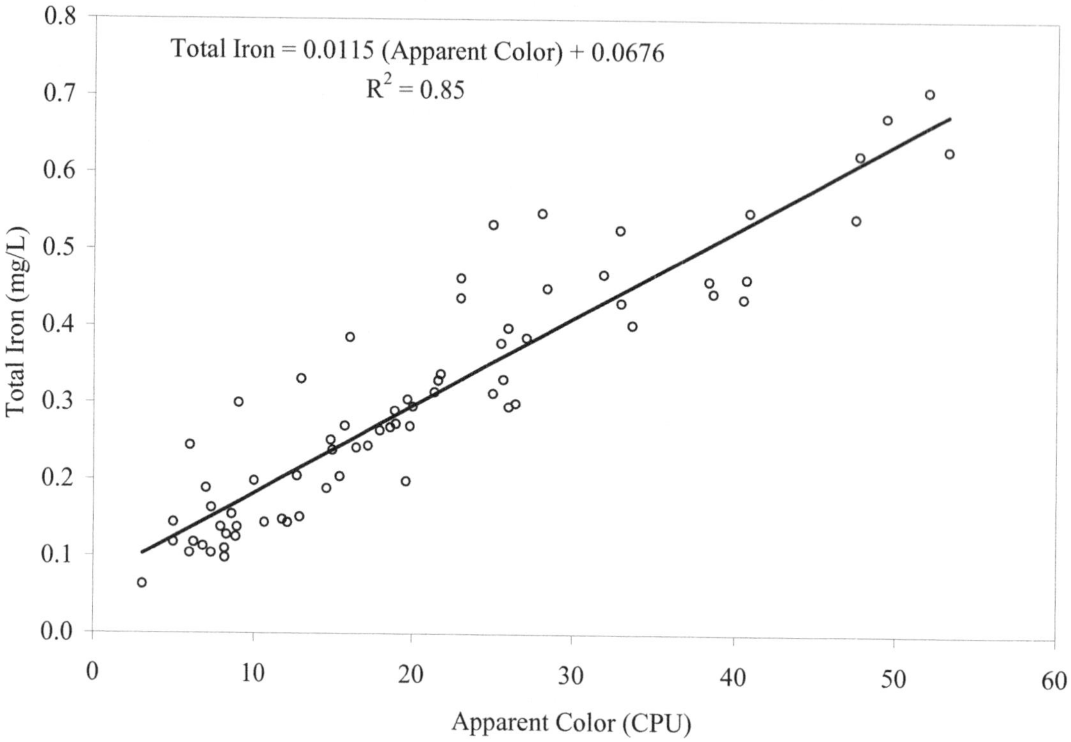

Figure 4.8 Relationship between apparent color and total iron for average data from phase I to IV (12/8/01 to 12/13/02)

Table 4.1
Average delta apparent color (CPU) response of PDS for different periods of study

Date	PDS 01	02	03	04	05	06	07	08	09	10	11	12	13	14	15	16	17	18
12/8/01 – 3/15/02	3	6	32	12	20	14	13	22	2	16	16	40	18	13	36	1	0	3
3/16/02 – 6/14/02	9	5	37	33	45	12	36	15	11	28	29	25	47	13	29	1	0	10
6/15/02 – 9/13/02	4	5	45	19	38	11	21	21	9	23	21	49	37	10	36	0	0	7
9/14/02 – 12/13/02	1	1	15	5	21	4	4	4	3	11	9	11	19	7	12	1	0	7
12/24/02 – 1/24/03	3	1	6	2	2	2	20	5	2	6	Nitrification Experiments				Flux Experiments			
1/25/03 – 2/7/03	5	4	5	6	4	0	20	9	5	8								
2/8/03 – 2/21/03	4	3	13	8	10	7	50	12	7	11								
2/22/03 – 3/7/03	6	3	20	12	22	12	34	26	5	17								
3/8/03 – 3/21/03	12	6	16	20	24	16	14	19	5	7								
3/22/03 – 4/4/03	11	6	13	11	15	11	7	16	2	5								

Note: Empirical model was developed on data available from 12/6/01 to 12/14/02. Model verification was done on data from PDS 01 to 10 from 1/1/03 to 4/4/03.

Pipe Equilibration

The pipe materials were obtained from the Member Governments during the spring and summer of 2001. The pipes were maintained in a moist condition after excavation during transport to the field site. Upon arrival at the study site, treated GW was provided to maintain the interior films in their original condition. The water supplied during equilibration (GW) was believed to be a close match for the historic water that had been exposed to the pipes. Construction of the pilot PDSs was completed in July 2001.

The harvesting, transport, and re-assembly of the aged pipes would be expected to impose some disruption of the interior films. It was recognized that a period of equilibration would be required to restore the pipe surface characteristics after construction of the pilot distribution system. Elevated (relative to the treated GW feed) levels of apparent color, turbidity, and iron (total) were experienced during the equilibration period (7/16/01 to 12/07/01). The dynamics were described as a first-order impulse response (Cullen 2002). An equilibration period of three months was required to achieve return of the effluent conditions to values equal to the input (Cullen 2002). A typical response for a hybrid PDS during the equilibration period is presented in Figure 4.9. Introduction of different source water blends was initiated in December 2001.

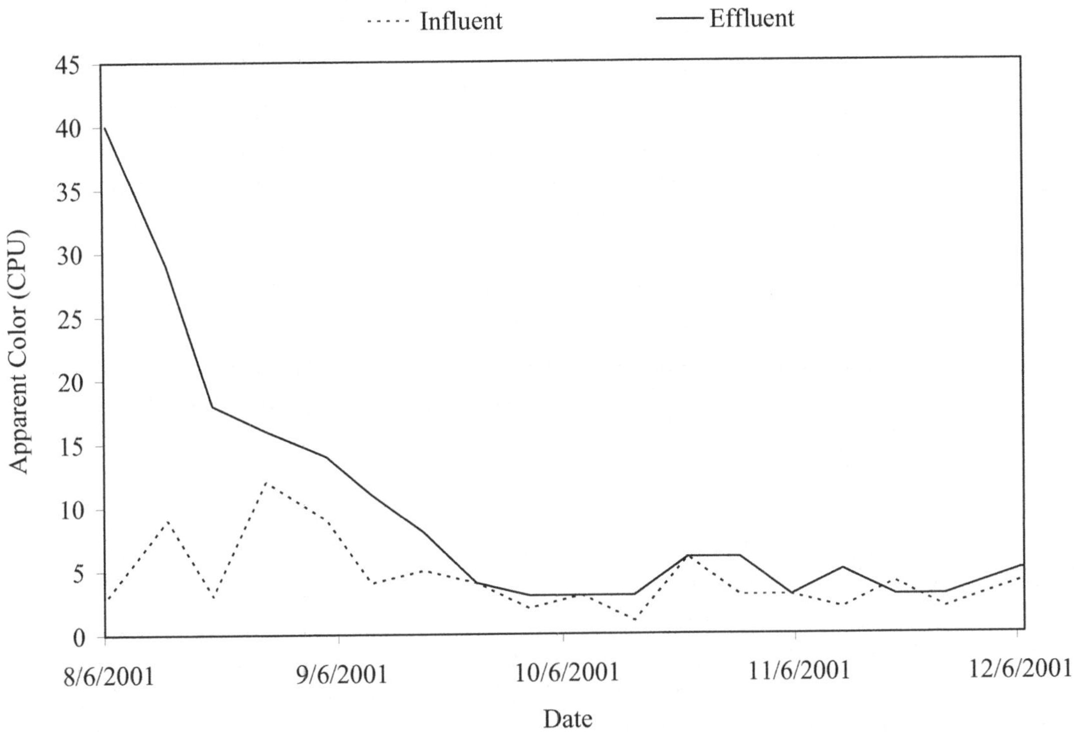

Figure 4.9 Typical apparent color release behavior during equilibration of hybrid PDS (PDS 03)

Empirical Model Development

Changed conditions during phase I to phase IV (12/8/01 to 12/13/02). Several significant changes were experienced in conditions in the pilot distribution systems during the first twelve months of operation (December 2001 to December 2002). These factors are all related in some way to seasonal trends associated with temperature cycles and/or wet weather impacts on SW quality:

- An increase in temperature during summer conditions was experienced, with PDS values approaching 30°C (86°F). Although extreme, this range of temperature is typical of worst case conditions in the Member Governments distribution systems.

- A decline in chlorine residual (chloramine and free chlorine) was documented to coincide with the elevation in temperature. The effect was more pronounced with the UCI and G pipe materials. Only the lined iron and PVC PDSs were able to carry a modest residual during the high temperature conditions.
- A decline in DO concentration was also noted during summer conditions. This effect is partially attributable to a reduction in saturation values at the elevated temperatures as evidenced by a reduction in feed water DO levels. A greater consumption of oxygen was also experienced due to microbiological and/or chemical reactions in the pilot distribution system.
- Oxidation of reduced nitrogen forms (measured by TKN), coupled with the emergence of oxidized nitrogen forms (nitrite and nitrate) was documented by the end of Phase II (3/16/02 to 6/14/02) and continued into Phase III (6/15/02 to 9/13/02), suggesting that a nitrifying population had become established in several of the pilot distribution PDSs. The nitrification coincided with periods of high temperature and low chloramine residual. Reduction of the hydraulic residence time from five days to two days (8/31/02 to 12/13/02) provided some mitigation of these effects (low chloramine residual and presence of nitrification). Implementation of the change in HRT occurred during the final two weeks of Phase III (after 8/30/02).

Source water summary. The data base, as defined by Phase I to IV (12/8/01 to 12/13/02) results, includes confounding effects between factors that may be relevant to release of corrosion products. Specifically, the water source(s) with the highest alkalinity (GW and blends enriched in GW) also contain the lowest chlorides. The water source(s) with the greatest sulfate (SW and blends enriched in SW), exhibited low alkalinity. Similarly, the water source(s) with the highest chlorides (RO and blends enriched in RO), also report low alkalinity. Collection of data through the traditional rainy season in Florida (June to September representing Phase III and IV) was useful to permit development of a data source for SW with reduced mineral content. Development of the full matrix of data for a twelve-month period is necessary to properly evaluate these interactions. Additional studies were completed in Phase V to isolate the effect of alkalinity, chloride, and sulfate by addition of various salts to modify blend composition.

Water quality alteration in the pilot distribution system – Summary. The majority of parameters exhibit minimal difference between the feed and effluent values. As noted during equilibration studies, degradation of water quality attributed to release of surface constituents during PDS construction was noted for apparent color, turbidity, and iron. Similar effects were observed for certain blends and specific pipe materials. Elevation of apparent color values were reported previously in Table 4.1. The results from the first 12 months of operation are summarized below:

- The majority of color and iron were associated with particulates, suggesting a release of corrosion products from the pipe surface. The elevation of turbidity confirmed the presence of particulate materials.
- The pipe materials were not uniformly affected. The UCI (PDS15), and to a lesser degree G pipe (PDS18), were associated with release of corrosion products (color, iron, and turbidity). Lined iron and PVC did not experience a comparable release of corrosion products.

- The source water characteristics were also an important determinant of water quality. Impairment of water quality was widespread, but not universal, among the hybrid PDSs. Application of water (G1or G2) with similar characteristics to the historic water source for the pipe materials did not result in pronounced water quality changes. These waters are characterized by elevated alkalinity, depressed chloride, and depressed sulfate. Introduction of source waters with characteristics that varied significantly from the historic water source resulted in significant increases in apparent color, iron, and turbidity. Departures from the historic water quality included reduced alkalinity (SW, RO, and blends of SW and RO), increased chloride (RO, SW, and blends of RO and SW), and increased sulfate (SW and blends of SW).

Association of Color Release with Water Quality Parameters – Model Development

The assembly of data for the twelve-months of operation (Phase I to IV, 12/8/01 to 12/13/02) permitted development of a non-mechanistic empirical model. The data was restricted to the fourteen hybrid PDSs to maintain uniformity in pipe material for the data set. Average values for each phase were utilized in the analysis to avoid complications with missing values (the sampling frequency for all water quality parameters is not identical). There are fourteen pilot distribution systems that were operated in five distinct phases: Phase I (5 day HRT, 12/8/01 to 3/15/02), Phase II (5 day HRT, 3/16/02 to 6/14/02), Phase III (5 day HRT, 6/15/02 to 8/30/02), Phase III (2 day HRT, 8/31/02 to 9/13/02), and Phase IV (2 day HRT, 9/14/02 to 12/13/02). Thus, the resultant database represents 70 sets of average observations.

The increase in apparent color (effluent minus influent apparent color) was selected as the response variable. A large number of water quality parameters were considered for inclusion in the model: alkalinity, calcium, conductivity, pH, sulfate, chloride, sodium, silica, UV-254, DO, and temperature. The PDS hydraulic residence time was also considered. A log-variant non-linear model form was examined as shown in equation (4.2).

$$\begin{aligned} \Delta Apparent\ Color = {} & 10^{B_1}(Alkalinity)^{B_2}(Sulfate)^{B_3}(Chloride)^{B_4}(Sodium)^{B_5}(Ca)^{B_6} \\ & (Conductivity)^{B_7}(pH)^{B_8}(Silica)^{B_9}(UV_{254})^{B_6}(DO)^{B_{11}}(Temperature)^{B_{12}}(HRT)^{B_{13}} \end{aligned} \qquad (4.2)$$

The empirical database represents a range of values for the independent variables that spans the anticipated water quality range in the Member Governments' distribution system and the units used during the modeling effort are noted in Table 4.2.

Analysis of variance procedures (ANOVA) were applied to determine the individual significance (alpha = 0.05) of the independent variables. More than 4000 combinations of variables were examined. The resulting model is provided as equation (4.3).

Table 4.2
Study variable range for empirical model development (hybrid PDS)

Parameter	Minimum	Maximum
Alkalinity (mg/L as $CaCO_3$)	51	210
Calcium (mg/L)	23	88
Chloride (mg/L)	19	100
Sulfate (mg/L)	4	232
Sodium (mg/L)	12	65
Silica (mg/L)	1	16
DO (mg/L)	6.9	12.2
UV-254 (cm^{-1})	0.0233	0.0688
Conductivity (μS/cm)	331	707
pH	7.8	8.4
Temperature (°C)	16.8	27.0
HRT (days)	2	5

$$\Delta Apparent\ \ Color(CPU) = 0.04775(Alkalinity)^{-0.912}(Sulfate)^{0.118}(Chloride)^{0.485} \times (Sodium)^{0.561}(DO)^{0.967}(Temperature)^{0.813}(HRT)^{0.836} \qquad (4.3)$$

The modeling results indicate that alkalinity, sulfate, chloride, sodium, DO, temperature, and hydraulic residence time significantly affect the release of corrosion products from unlined cast iron and G pipe materials. Resolution of the singular effect of each variable is complicated by confounding effects in the database. A correlation matrix of the seven remaining variables is presented in Table 4.3.

Specific confounding effects of potential concern are noted between alkalinity and chloride, alkalinity and sodium, alkalinity and sulfate, sodium and chloride, and temperature and DO. These difficulties result from differential characteristics of the sources (for example, GW sources are high in alkalinity but low in chloride), requirements for electroneutrality (for example, sodium is coupled with chloride in the sea salt addition to RO permeate), and/or seasonal changes (for example, high temperatures are associated with low DO due to solubility characteristics). The experimental efforts in Phase V (12/14/02 to 4/04/03) addressed the confounding effects between alkalinity, chloride, and sulfate. The current model achieved an impressive agreement with the data from the first twelve months of operation (R^2 = 0.83) as illustrated in Figure 4.10.

Table 4.3
Correlation matrix for empirical model variables, hybrid PDS – phase I to IV (12/8/01 to 12/13/02)

	Alkalinity	Chloride	DO	Sodium	Sulfate	Temperature
Alkalinity	1.00	-0.28	0.21	-0.47	-0.29	-0.01
Chloride	-0.28	1.00	-0.21	0.56	-0.27	-0.04
DO	0.21	-0.21	1.00	-0.27	-0.16	-0.33
Sodium	-0.47	0.56	-0.27	1.00	0.14	0.14
Sulfate	-0.29	-0.27	-0.16	0.14	1.00	-0.09
Temperature	-0.01	-0.04	-0.33	0.14	-0.09	1.00

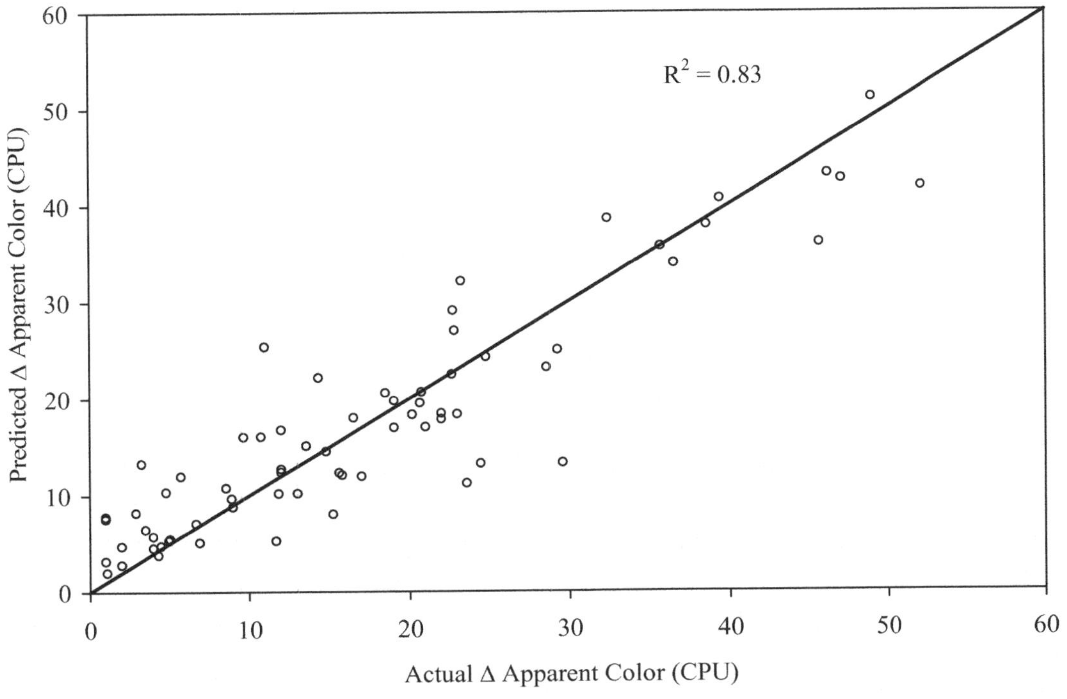

Figure 4.10 Performance of empirical model (hybrid PDS) for averaged water quality data from phase I through IV (12/8/01 to 12/13/02)

It is cautioned that any consideration of mitigation of release of corrosion products must recognize the complexities and electroneutrality constraint associated with alteration of the finished water chemistry. For example, the empirical model suggests that an increase in alkalinity would be beneficial. However, supplementation of alkalinity by addition of sodium bicarbonate would also elevate the sodium ion concentration. Based on the empirical model, an increase in sodium may offset the intended benefit achieved by increase of alkalinity. Other approaches to boost alkalinity that do not add sodium may be more desirable (for example, lime plus carbon dioxide). It is noted that TBW practices alkalinity supplementation by addition of lime plus carbon dioxide.

A brief discussion of limitations of this type of empirical modeling effort is provided to guide application of the resulting empirical models. The procedure for selection and/or removal of variables from the final model was an analysis of variance (ANOVA). Two forms of a model were compared: a version that included a potential variable in addition to the shared variables, and a version that included all variables (the shared variables) except the potential variable. If the expanded model, including the potential variable, achieved a statistically significant improvement in ability to describe the data relative to the model with shared variables only, then the variable was judged to be significant and would remain in the model. Thus, inclusion (or exclusion) of variables from the final model would be based on criteria other than a single goodness-of-fit statistic such as the correlation coefficient. This approach is fundamentally different from a mechanistic model that would presume the significance of variables in advance of parameter estimation.

The empirical approach may be useful to suggest mechanistic relationships, but the presence of confounding effects between study variables may obscure proper identification of such mechanistic relationships. The present study mandated examination of three specific water sources, with the result that the dataset contained an inherent correlation between certain water quality parameters (Table 4.3). Limitations associated with field operations precluded independent absolute control of all study variables (for example, temperature and associated solubility of dissolved oxygen). Although these factors may hinder a mechanistic interpretation of the data, it should be noted that the exact same type of confounding issues will be present in the prototype distribution system when varying blend ratios of the three principal sources are present. This approach would be expected to provide a satisfactory description of events that would be experienced in the Member Governments' distribution systems when subjected to variation in water quality associated with blends of groundwater, surface water, and desalination sources.

Application of any empirical model should recognize the range of values of the study variables in the dataset that was used to develop the empirical model. For the present study, a wide range of water qualities were provided (Table 4.2) as feed to the hybrid PDSs. This robust data represented the full spectrum of potential blends that could be experienced in the Member Governments' distribution systems, ranging from 100% of any single source to various blends of multiple sources. Neverthesss, use of the empirical model to describe water quality not represented by the data would be inappropriate. Examples of such an inappropriate use of the model would include description of anaerobic systems, unstabilized waters (with respect to calcium carbonate solubility), and waters with a low pH relative to the data range defined in Table 4.2.

Color Release Model for G pipes

The preceding model development (equation (4.3)) utilized water quality variations that encompassed the greatest range of values, as hybrid PDS inputs included all seven finished waters plus various blends. The corresponding range of water quality parameters for PDS 18 is reported Table 4.4.

The hybrid system response (color, iron, or turbidity) was dominated by the impact of the UCI pipe, relative to other pipe materials. A complementary analysis of data using the single material G PDS (PDS 18) was undertaken to develop an empirical model that would be specific to the G material. It is acknowledged that the data base for the single material PDSs is less robust than that available for the hybrid PDSs, owing to the reduced variability in input water quality to the single G PDS in comparison to the 14 hybrid PDS. Specifically, for the first year of data collection, the input to the single material PDSs consisted only of blends of the three principal sources (GW, SW, and RO). The resulting empirical model is reported as equation (4.4). The agreement between the empirical model and the PDS18 data is reviewed in Figure 4.11.

$$\Delta App.\ Color(CPU) = 0.71(Alkalinity)^{-1.63}(Silica)^{-0.55}(Sulfate)^{1.12}(Temperature)^{1.06} \qquad (4.4)$$

Table 4.4
Study variable range for empirical model development (G PDS)

	Minimum	Maximum	Average
Alkalinity (mg/L as $CaCO_3$)	92	200	126
Chlorides (mg/L)	35	66	48
Silica (mg/L)	3	14	8
Sodium (mg/L)	28	85	46
Sulfate (mg/L)	59	102	86
DO (mg/L)	6.5	11.4	8.4
pH	7.8	8.2	8.0
Temperature (°C)	14.5	28.6	22.6

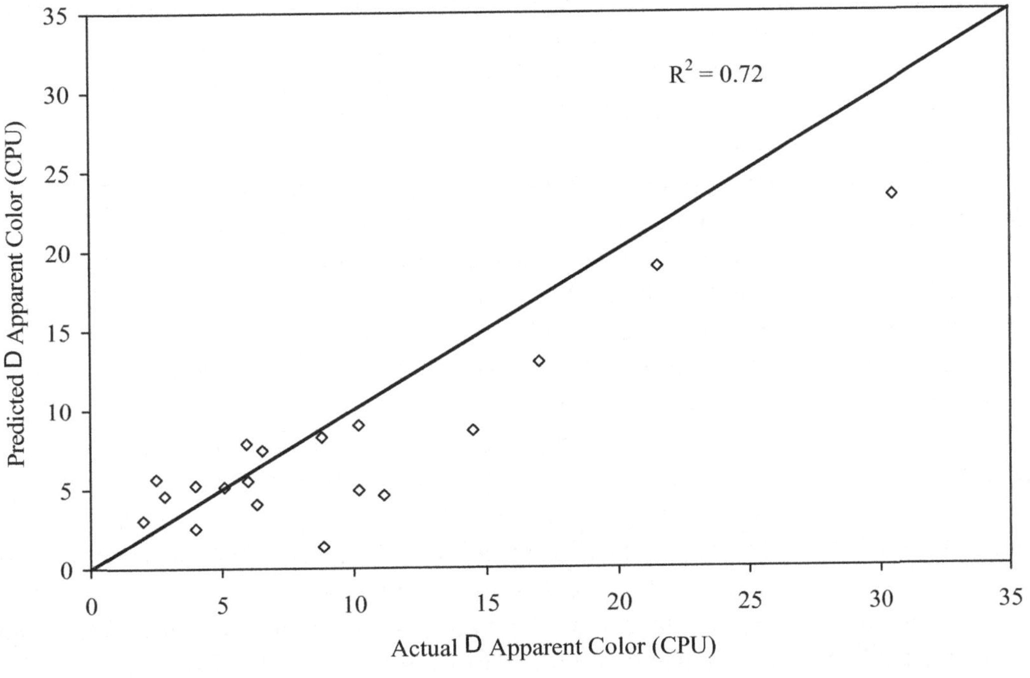

Figure 4.11 Performance of empirical model (G PDS) for monthly averaged water quality data from phase I through V (12/8/01 to 4/4/03)

The resulting empirical model for the G PDS includes many, but not all, of the same variables that were shown to be statistically significant for the hybrid PDS. For both models, the release of corrosion products is positively correlated with sulfate and temperature. For both models, alkalinity exhibits a strong negative correlation with release of color. The presence of silica was determined to provide a mitigating effect for the G material.

Empirical Model Verification.

Data collection. Verification of the empirical model for the hybrid PDS was undertaken by conducting a fifth independent phase of study to identify the singular effects of the major water quality parameters: alkalinity, chlorides and sulfates. The verification studies provided a wider range of alkalinity, chlorides and sulfates than was available from Phase I to IV (12/8/01 to 12/13/02) to support model development. The source waters were prepared by the addition of alkalinity in the form of sodium bicarbonate, chlorides from sea salt, and sulfates from calcium sulfate. The purpose of these additions was to simulate a particular water quality rather than a specific source water or treatment process. PDS 1 through 6 were used for determining the effects of various anions. For instance, in the first month of Phase V (12/24/02 to 2/7/03), alkalinity was varied by PDS and sulfates and chlorides were kept to a constant level (Table 4.5). The water quality and the color response during this period are shown in Figure 4.12. The second month (2/8/03 to 3/7/03) focused on the evaluation of sulfates while fixing alkalinity and chlorides. In the third month (2/22/03 to 3/21/03) chlorides were varied while alkalinity and sulfates were kept constant. The PDS input water quality and associated color response are reported for the second and third month of Phase V in Table 4.6 and Table 4.7 respectively. These results are also presented in Figure 4.12, Figure 4.13 and Figure 4.14 for the alkalinity, sulfate, and chloride experiments, respectively. For all of these experiments, manipulation of the anion (bicarbonate, sulfate, or chloride) concentration was achieved by addition of the sodium salt. Thus, in order to maintain electroneutrality, sodium concentrations also varied in proportion to the increase in the specific anion concentration. It is not possible to truly examine the singular effect of an anion due to electroneutrality considerations. In this discussion, reference to the singular effect of a parameter (alkalinity, sulfate, or chloride) implies that other anion concentrations were controlled at near constant concentrations.

The results indicate that supplementation of alkalinity has a beneficial effect in reducing effluent color. The supplementation of alkalinity in Phase V was achieved with addition of sodium bicarbonate. As indicated by the empirical model (equation (4.3)), elevation of sodium concentrations is not desired. Measures to boost alkalinity that do not simultaneously elevate sodium (for example, addition of lime plus carbon dioxide) may produce a more favorable result. Increasing sulfates show a trend of increasing color as predicted by the nonlinear empirical model (equation (4.3)). However, the effect of increasing concentrations of chloride on color release is not apparent. The empirical model predicts that increasing the concentrations of chloride would lead to more corrosive waters. However, this trend of increasing color release with increasing chloride concentrations is not reflected in Figure 4.14. The Phase V results for chloride supplementation do not provide definitive support of the empirical model. This finding is not believed to warrant modification of the empirical model, as many other historical studies and corrosion indices (for example the Larson Ratio) have documented a positive association between iron corrosion rates and chloride (and sulfate) concentration (Imran 2003).

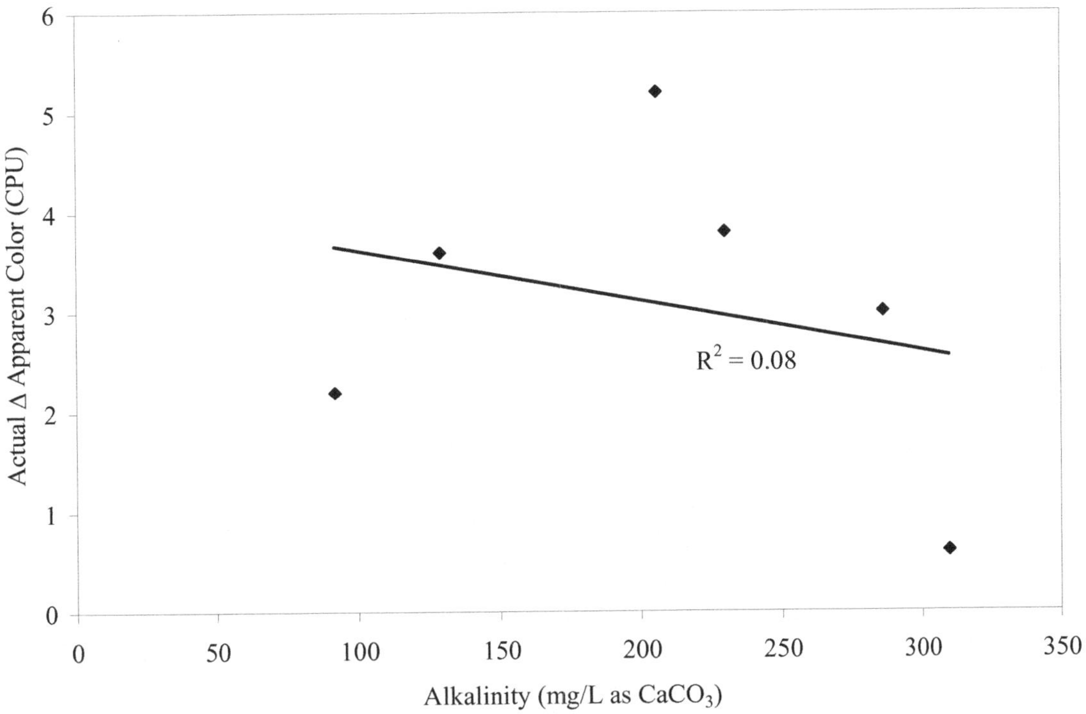

Note: Data represents average response for PDS 1 – 6. (Values indicate average alkalinity in mg/L as $CaCO_3$)

Figure 4.12 Phase V study to evaluate singular effects of alkalinity (month 1: 12/24/02 to 2/7/03)

Table 4.5
Phase V study to evaluate singular effects of alkalinity (month 1: 12/24/02 to 2/7/03)

	PDS					
	01	02	03	04	05	06
Alkalinity (mg/L as $CaCO_3$)	92	130	207	230	286	310
Chlorides (mg/L)	32	34	32	30	33	35
Sulfates (mg/L)	21	15	13	15	13	12
Sodium (mg/L)	30	49	78	87	114	124
DO (mg/L)	9	9	9	10	9	9
Temperature (°C)	18	18	18	18	17	17
HRT (Days)	2	2	2	2	2	2
Apparent Color (CPU)	4	2	5	4	3	1

Note: Data represents average water quality and response for PDS 1 – 6.

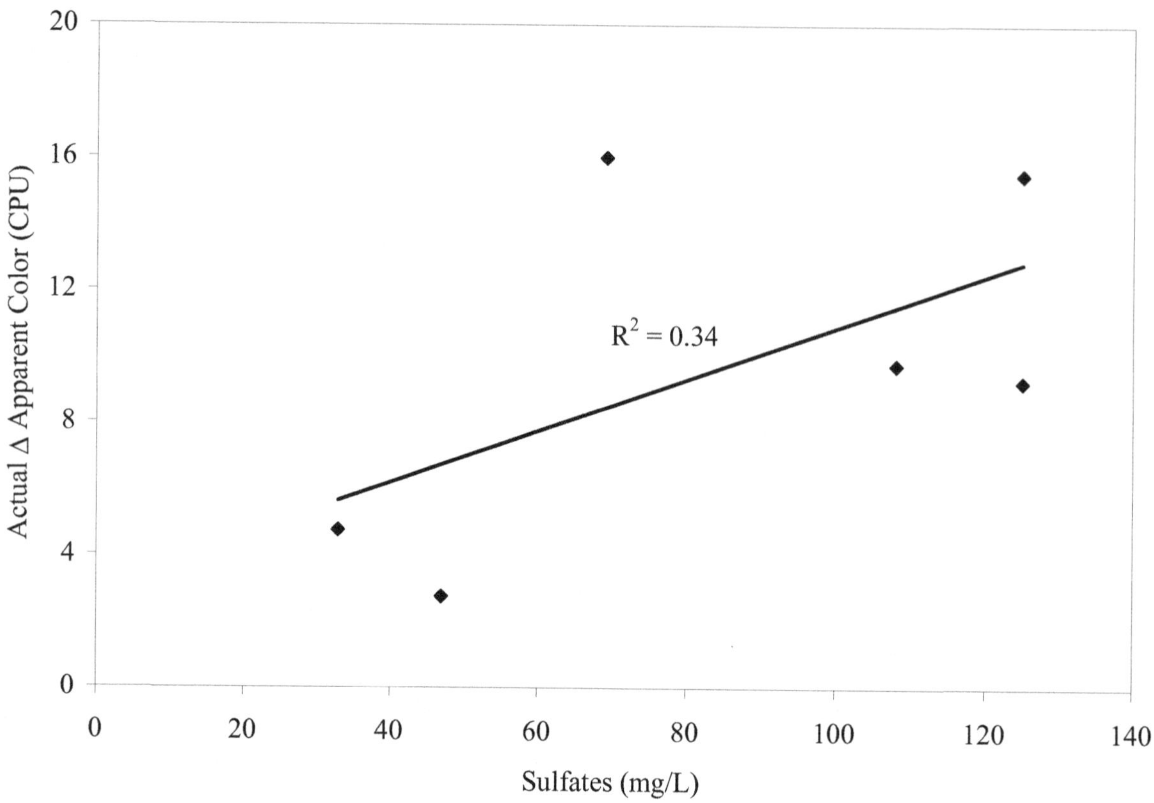

Note: Data represents average response for PDS 1 – 6. (Values indicate average sulfate in mg/L)

Figure 4.13 Phase V study to evaluate singular effects of sulfates (month 2: 2/8/03 to 3/7/03)

Table 4.6
Phase V study to evaluate singular effects of sulfates (month 2: 2/8/03 to 3/7/03)

	PDS 01	02	03	04	05	06
Alkalinity (mg/L as $CaCO_3$)	82	81	78	80	83	98
Chlorides (mg/L)	36	35	34	33	32	31
Sulfates (mg/L)	33	48	69	108	125	125
Sodium (mg/L)	36	45	55	66	79	80
DO (mg/L)	9	9	9	9	9	9
Temperature (°C)	21	21	22	22	22	22
HRT (Days)	2	2	2	2	2	2
Apparent Color (CPU)	5	3	16	10	16	9

Note: Data represents average water quality and response for PDS 1 – 6.

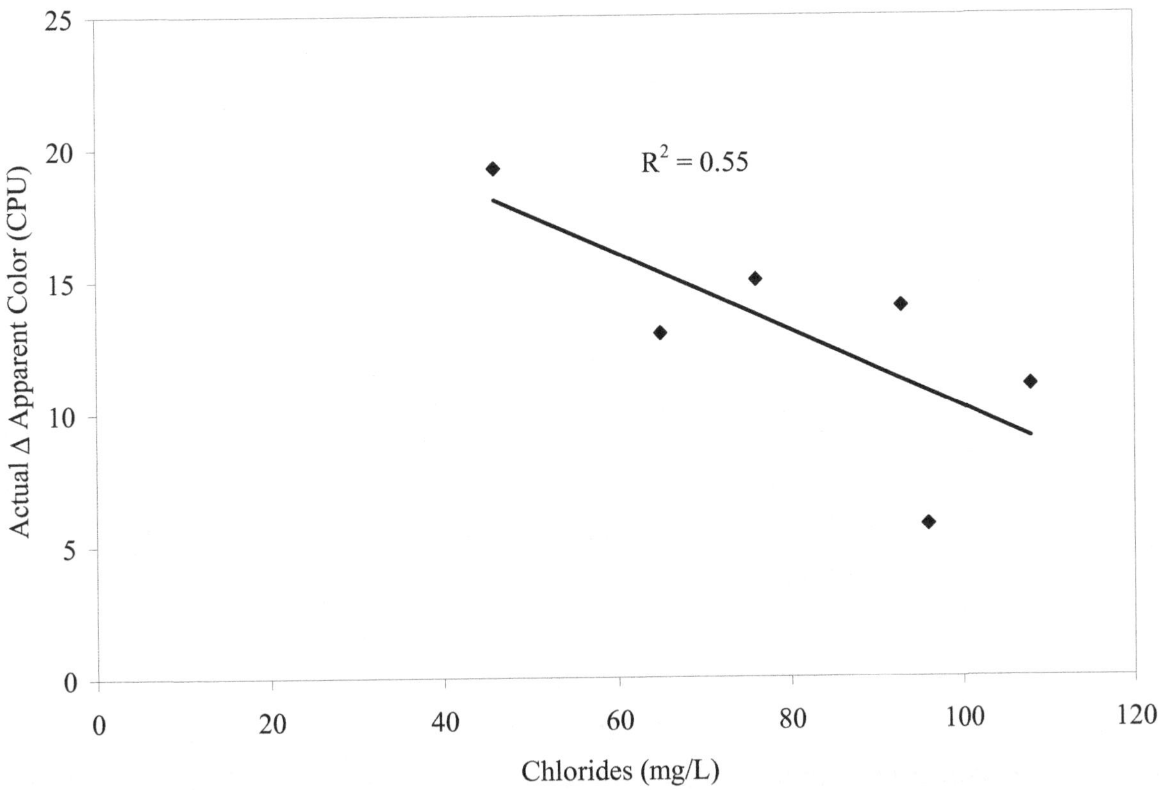

Note: Data represents average response for PDS 1 – 6. (Values indicate average chlorides in mg/L)

Figure 4.14 Phase V study to evaluate singular effects of chlorides (month 3: 3/8/03 to 4/4/03)

Table 4.7
Phase V study to evaluate singular effects of chlorides (month 3: 3/8/03 to 4/4/03)

	PDS 01	02	03	04	05	06
Alkalinity (mg/L as $CaCO_3$)	72	72	68	73	82	73
Chlorides (mg/L)	108	96	93	77	65	46
Sulfates (mg/L)	20	19	15	24	17	46
Sodium (mg/L)	69	65	64	60	53	49
DO (mg/L)	8	8	8	8	8	8
Temperature (°C)	25	25	25	25	25	25
HRT (Days)	2	2	2	2	2	2
Apparent Color (CPU)	11	6	14	15	19	13

Note: Data represents average water quality and response for PDS 1 – 6.

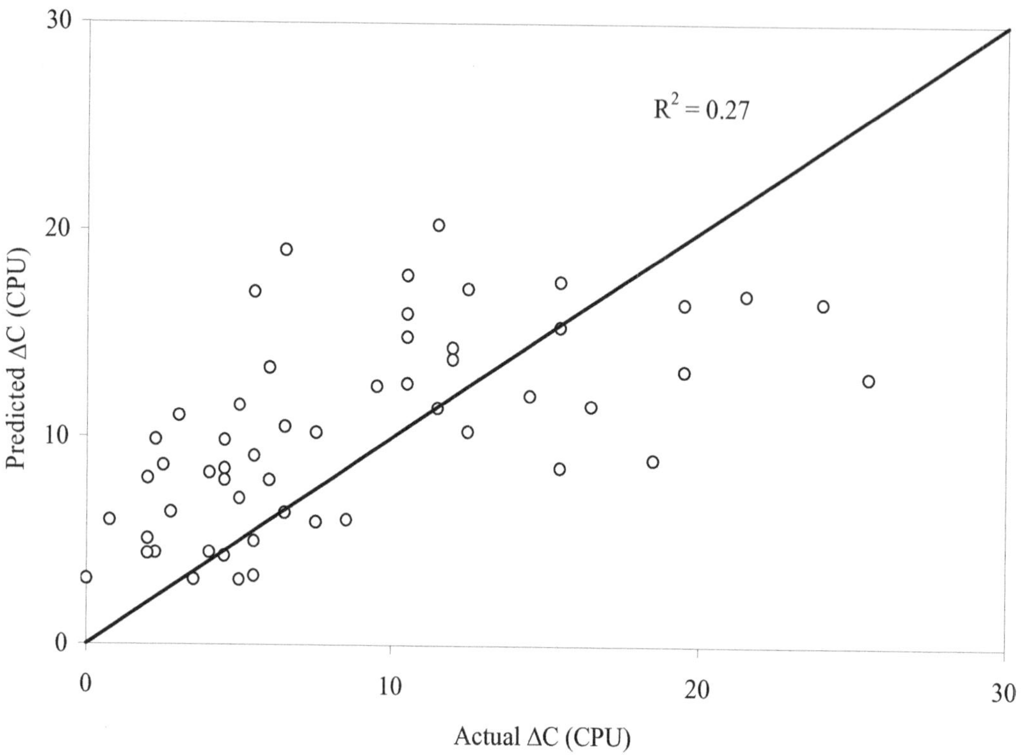

Figure 4.15 Verification of model by independent data from phase V (12/14/03 to 4/4/03)

An additional set of verification experiments was pursued using PDS 8 through 10, with blend changes every two weeks to adjust alkalinity within a range of 60 to 250 mg/L as $CaCO_3$. As can be observed from Figure 4.15, the performance of the model was consistent when verified on the independent data set from Phase V.

Mitigation measures suggested by the empirical model. From the empirical model for the hybrid PDS, it is apparent that the only water quality parameter that can be controlled by chemical addition with some success is alkalinity. It is important to note that elevation of alkalinity is also reported to mitigate release of corrosion products using the empirical model for the G PDS. The other parameters are inherent in the processes that produce the source waters. For instance, RO water will have an elevated chloride level that can only be reduced with modifications to the RO process. SW will have elevated sulfates due to ferric sulfate coagulation. Temperature is based on seasonal (and sometimes even diurnal) variations. HRT is governed by the dynamics of water demand. The response of the hybrid PDS to changes in alkalinity during Phase I to IV (12/8/01 to 12/13/02) of the study are presented in Figure 4.16. The field data are reported for both operating values of HRT (two and five days). The release of color is inversely related to alkalinity. A clear line of demarcation can be seen at an alkalinity of 80 mg/L as $CaCO_3$. There is a drastic reduction in the release of color above an alkalinity of 80 mg/L as $CaCO_3$. With the exception of a few source waters (incidentally with elevated in sulfate and chloride and circled in Figure 4.16), waters with alkalinity greater than 80 mg/L as $CaCO_3$ tended to be less corrosive. The empirical model developed from Phase I to IV (12/8/01 to 12/13/02) data as well as the verification data from PDS 1 to 6 and PDS 8 to 10 from Phase V suggest that an increase in alkalinity is desirable to reduce the release of color. However, chemical measures of alkalinity supplementation should be practiced with care. For instance, an increase in alkalinity by addition of sodium bicarbonate would also increase the sodium content of the water. Based on the empirical model, this corresponding increase in sodium may offset the intended benefit of increased alkalinity. Whenever practicable, alkalinity supplementation is recommended based on the empirical model.

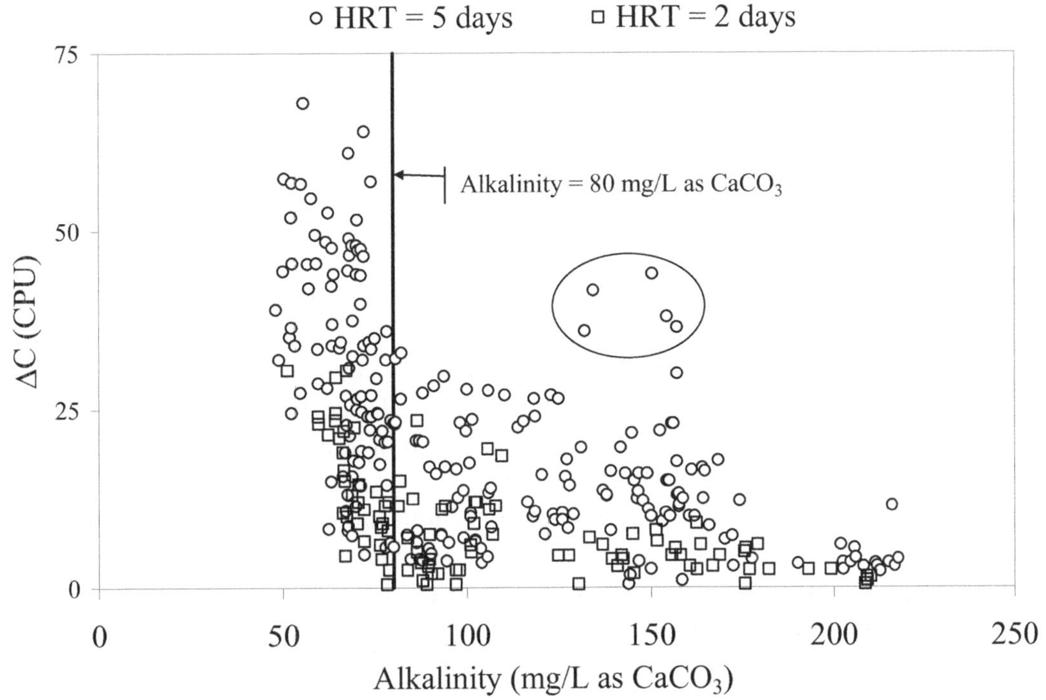

Note: Points enclosed within the circle represent high sulfates (> 100 mg/L) and chlorides (> 40 mg/L)

Figure 4.16 Dependence of release of color on alkalinity of source water

Iron Release Summary

An empirical model was developed for the hybrid PDS, using a robust data set. The resulting model is dominated by the effect of the UCI pipe material. For this hybrid system model:

- The variation in levels for the model independent variables in the Phase I to Phase IV (12/8/01 to 12/13/02) dataset encompasses a wider range that would be expected in the feed water and/or feed water blends in the Member Governments' distribution system.
- Alkalinity, chloride, sulfate, sodium, DO, temperature, and hydraulic residence time were identified by statistical procedures as having a significant effect on release of corrosion products. Confounding in the experimental design complicates absolute resolution of the singular effect of each variable.
- Alkalinity exhibited a significant negative correlation with color release.
- Chloride, sulfate, sodium, DO, temperature, and hydraulic residence time exhibited a significant positive correlation with color release.

- Calcium and pH were not demonstrated to have a significant effect on color release for this dataset. This observation may be attributable to stabilization of the waters (with respect to CaCO3 solubility).
- The empirical model numerical prediction of color release is specific to the geometry and hydraulic conditions of the hybrid pilot distribution systems. Although the PDS operation was limited to extremely low velocities (laminar flow), additional short-term experiments were completed to develop data for turbulent conditions as described in a subsequent section of the report. A more robust model is ultimately developed that incorporates pipe geometry and hydraulics to complement the evaluation of water quality provided by the empirical model.
- The relative value of the flux of corrosion products can be assessed for different sources (or blends) based on their individual water chemistry. The relative color release can be determined for the three principal source waters (GW, SW, and RO) using the empirical model, coupled with a set of assumed water quality parameters for each source. The assumed characteristics of the finished water from the three principal sources are presented in Table 4.8. The corresponding release of corrosion products (expressed as an increase in apparent color) is also noted based on the empirical model prediction. All color release quantities are expressed in relation to the historic GW source. It is evident that both the surface and RO sources present a potential concern for release of corrosion products. This adverse effect is attributed to altered water chemistry in comparison to the GW source. Specific differences include a reduction in alkalinity (both surface and RO sources), an increase in chloride (RO), an increase in sodium (both sources, especially RO), and an increase in sulfate (both sources, especially SW).
- The adverse effect of chloride and sulfate cited above is consistent with the literature, specifically with definition of corrosion indices such as the Larson Ratio. Evaluation of various corrosion indices listed below was pursued as an alternative to the reported empirical model.

  Langelier Saturation Index
  Calcium Carbonate Precipitation Potential
  Riddick Index
  Ryznar Index
  Driving Force Index
  Momentary Excess
  Aggressiveness Index
  Larson Ratio
  Feigenbaum, Galor, and Yahalom Index
  pH Buffer Intensity
  Casil Index

  It was determined that the empirical model provided a superior description of the data (Imran 2003).

Table 4.8
Relative release of corrosion products by water source

	Alkalinity (mg/L as $CaCO_3$)	Chloride (mg/L)	Sodium (mg/L)	Sulfate (mg/L)	Silica (mg/L)	Relative Δ Apparent Color UCI (CPU)	Relative Δ Apparent Color G (CPU)
GW	225	15	10	10	14	1	1
SW	50	10	15	180	10	6	356
RO	50	50	30	30	2	15	116
Blend	155	17	13.5	63	9	2	18

A similar empirical model was developed for the single material G PDS to isolate the effect of the G pipe material. The data base for this effort was less robust, as much smaller variations in input water quality were provided for the single material PDSs.

- The G model (equation (4.4)) indicated a similar effect of sulfate, temperature, and alkalinity. In addition, the presence of silica was indicated to be beneficial in mitigation of release of corrosion products for the G source. The color release for the three principal sources and the blend, as predicted by both empirical models, is reported in Table 4.8.

Development of a Flux Model – Mass Balance Approach

Review of the data obtained at variant hydraulic residence times during Phase III (8/15/02 to 9/13/02) suggested that the release of corrosion products (for example iron) increases in proportion to the hydraulic residence time. This observation is consistent with the occurrence of red water in low flow regions of a distribution system (dead ends). A simple, steady-state model can be developed from a material balance in which a constant release of material from the pipe surface is postulated. This release, also referred to as a flux, would vary with water chemistry and hydraulic conditions in the bulk liquid. For a given set of water chemistry and hydraulic conditions, however, the flux of corrosion products is assumed to remain constant. This mechanism would be inconsistent with diffusion driven processes that would experience a reduction in flux in response to an increase in the bulk phase concentration. This mechanism would also be inconsistent with equilibrium processes (for example, chemical precipitation) that would approach a single equilibrium condition irrespective of the hydraulic residence time. In view of earlier results that have demonstrated the majority of the corrosion products (apparent color or iron) are present in a particulate form, it may not be unreasonable to examine alternatives to diffusion- and equilibrium-controlled processes for description of release of corrosion processes. Development of a mass balance for iron is presented in equations (4.5) and (4.6).

$$Accumulation = Input - Output + Release\ from\ pipe\ surface \quad (4.5)$$

$$0 = QC_{FEED} - QC_{EFFLUENT} + K_m SA \quad (4.6)$$

where Q = flow rate (L/day)
C_{FEED} = iron concentration in feed (mg/L)
$C_{EFFLUENT}$ = iron concentration in effluent (mg/L)
K_m = iron flux (mg/m^2-day)
SA = pipe interior surface area (m^2)

After consideration of the pipe geometry, it can be shown that the rise in concentration is proportional to the hydraulic residence time and inversely proportional to the pipe diameter. The factor of 1000 is required to convert m^3 to liters.

$$SA = \pi DL \quad (4.7)$$

$$HRT = \frac{Volume}{Flow} = \frac{1000 \ \pi D^2/4 \ L}{Q} \quad (4.8)$$

$$C_{EFFLUENT} = C_{FEED} + \frac{4\,K_m\,HRT}{1000\,D} \quad (4.9)$$

where D = pipe diameter (m)
L = pipe length (m)
HRT = hydraulic residence time (days)

The proportional relation to HRT is intuitive, assuming a constant flux. Consider a batch reactor that receives a constant input of iron from the reactor wall. The increase in bulk phase concentration would increase linearly with time in response to the constant rate of mass input from the reactor wall. The data for feed and effluent total iron concentrations were used to calculate iron flux values using equations (4.10) and (4.11). Data are actually reported as the product of the flux (K_m) and the pipe interior surface area (SA), also referred to as the release rate term in the mass balance:

$$K_m = \frac{Q\ \Delta C}{Surface\ Area} = \frac{1000\ D\,\Delta C}{4\ HRT} \quad (4.10)$$

$$Release\ rate = K_m\ SA \quad (4.11)$$

It is noted that this area is the same for all hybrid PDSs, thus the calculated release of material is proportional to the flux. Results are summarized in Table 4.9. For purposes of this analysis, only data for the hybrid PDSs were used. The analysis suggests that the calculated values for iron flux (Km) are dependent on water quality but independent of hydraulic residence time. For example, the calculated release values for PDS01 (GW feed) and PDS02 (G2 feed) are much smaller than the calculated release values for PDS03 (SW feed), PDS04 (G4 feed), PDS05 (RO feed), PDS12 (G3 feed), or PDS13 (S2 feed). The release values are comparable, however, when 2-day and 5-day HRT data are examined for a constant water quality associated with a particular blend. A comparison of release values for individual pilot distribution systems is presented in Figure 4.17. The average ratio of 5-day to 2-day release values for the fourteen hybrid PDSs is 0.93. Because this ratio is nearly equal to one, the data support a hypothesis of a constant iron flux. It should be recognized that the entire PDS operation for this data represents very low velocity conditions.

Table 4.9
Calculated total iron flux values, phase III (6/15/02 to 9/13/02) data

PDS	Blend	5 day HRT			2 day HRT		
		C_{Feed} (mg/L)	$C_{Effluent}$ (mg/L)	Release (mg/day)	C_{Feed} (mg/L)	$C_{Effluent}$ (mg/L)	Release (mg/day)
1	G1 (100)	0.06	0.11	3	0.07	0.09	3
2	G2 (100)	0.01	0.13	7	0.02	0.08	9
3	SW (100)	0.03	0.78	47	0.07	0.30	36
4	G4 (100)	0.01	0.21	12	0.01	0.14	20
5	RO (100)	0.01	0.47	29	0.01	0.25	37
6	GW (55) SW (45)	0.04	0.26	14	0.06	0.14	12
7	GW (68) RO (32)	0.04	0.32	17	0.08	0.17	14
8	GW (23) SW (45) RO (32)	0.03	0.41	24	0.03	0.24	33
9	GW (60) S2 (30) RO (10)	0.03	0.16	8	0.04	0.08	6
10	G2 (50) SW (50)	0.02	0.31	18	0.03	0.20	26
11	G2 (62) SW (24) RO (14)	0.01	0.35	21	0.02	0.21	29
12	G3 (100)	0.01	0.69	42	0.01	0.30	45
13	S2 (100)	0.01	0.49	30	0.01	0.28	42
14	GW (62) SW (27) RO (11)	0.03	0.20	11	0.05	0.11	9

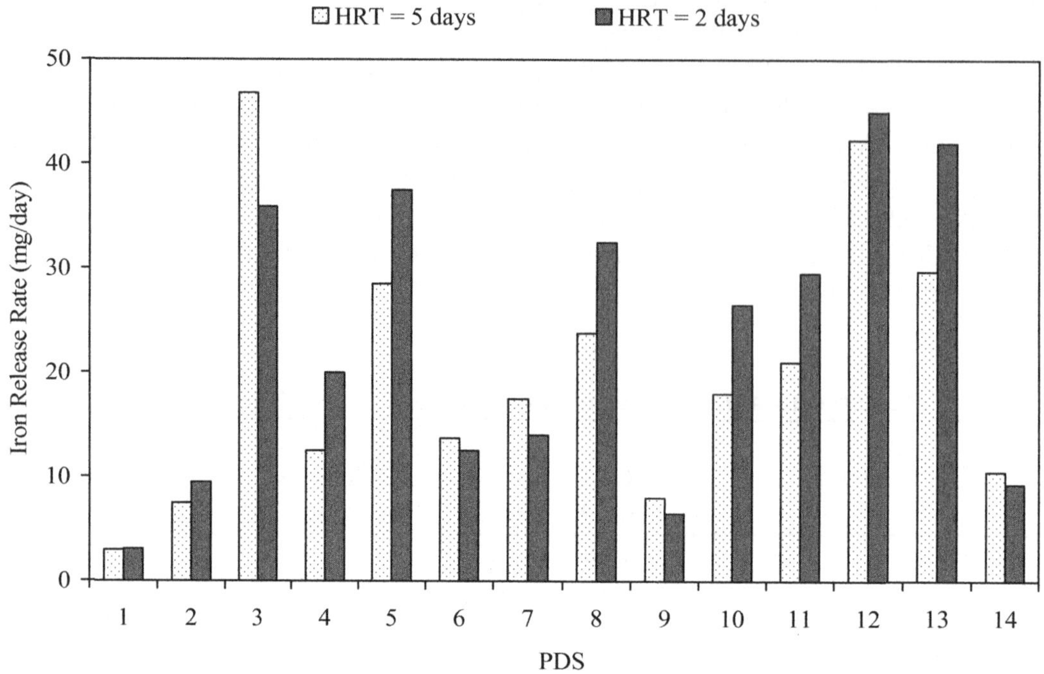

Figure 4.17 Iron release dependence on HRT. Phase III (6/15/02 to 9/13/02) data

Effect of hydraulic conditions on iron flux (K_m)

Experimental procedure. Normal operation of the pilot distribution systems was limited to extremely small velocities in order to achieve the desired hydraulic residence times (two or five days). Consequently, the database developed from normal PDS operation would not be suitable for evaluation of the effect of variant hydraulic conditions (for example, Reynolds Number) on the iron flux values. A series of experiments were completed to investigate the effect of Reynolds Number on the flux values. These experiments were conducted using PDS 15 (UCI) and PDS 18 (G) to define values for each pipe material.

The flux experiments were carried out by feeding a blend of treated SW – SW (30%), conventionally-treated groundwater – GW (60%) and desalinated water - RO (10%) sources at selected Reynolds Numbers in the UCI and G pilot scale distribution system pipes. Water quality data for the blend used in the flux experiment is provided in Table 4.10. The feed and outlet water quality were measured to determine the color and turbidity at various time intervals.

Iron release model results. Total iron and turbidity are highly correlated, supporting the use of turbidity as surrogate for iron measurements for during flux experiments. The best-fit model describing the change in particulate iron concentration is reported in equation (4.12).

$$\Delta [Fe] \left(\frac{\mu g}{L}\right) = \frac{\Delta Turbidity\ (NTU)}{0.0073} \tag{4.12}$$

Table 4.10
Average feed water quality

	Turbidity (NTU)	Total Fe (mg/L)	Ca^{2+} (mg/L)	Mg^{2+} (mg/L)	Silica (mg/L)	Na^{+} (mg/L)	pH	Alkalinity (mg/L as $CaCO_3$)	Conductivity (μS/cm)	TDS (mg/L)	Cl^{-} (mg/L)	SO_4^{2-} (mg/L)
Feed	0.306	0.045	68	6.4	9	31	7.96	148	557	377	38	66

The UCI pipe flux experiments were conducted using the same total water volume of 4.54 m^3 (1200 gallons), but run time was varied for each Reynolds Number. This volume is sufficient to flush the pipe contents with 9.5 pipe volumes of fresh feed. A characteristic response curve was generated for each experiment, showing an initial increase in turbidity to a maximum value, followed by a decline to a steady-state value that is characteristic for that particular flow rate. A typical time-series response curve generated for one of the experiments is given in Figure 4.18. The flow velocity for the above response curve was 0.058 m/s (0.190 ft/s) giving a hydraulic retention time of 6.6 minutes in the 26.3 m (86.3 ft) length of pipe. On the basis of this response curve, it was shown that at least three pipe volumes are required before steady-state turbidity values can be expected.

The flux experiment data for the UCI and G pipes are reported in Table 4.11 and Table 4.12, respectively.

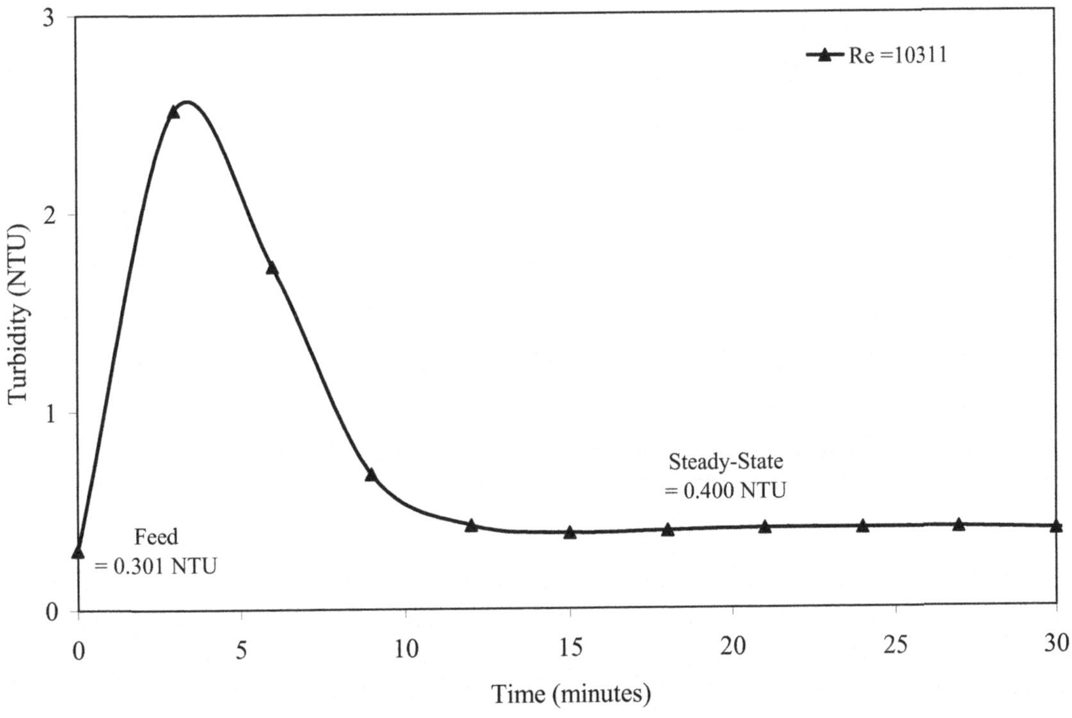

Figure 4.18 Turbidity response time series graph obtained during flux experiment

Table 4.11
UCI pipe flux experiment results

Flow Rate m^3/min (gpm)	Flow Velocity m/s (ft/s)	HRT (Days)	Reynolds Number	Log Reynolds Number	Δ Turbidity (NTU)	Δ Iron μg Fe/L	Flux, K_m mg/m^2/day (mg/ft^2.day)
0.00008 (0.02)	0.00007 (0.0002)	4.40	10.8	1.03	2.782	381.1	3.30 (0.31)
0.0002 (0.05)	0.00017 (0.0006)	1.76	27.0	1.43	1.128	154.5	3.34 (0.31)
0.0007 (0.18)	0.00066 (0.0022)	0.460	103	2.01	0.260	35.62	2.93 (0.27)
0.0011 (0.29)	0.0010 (0.0033)	0.303	157	2.19	0.058	8.00	1.0 (0.90)
0.0018 (0.48)	0.00163 (0.005)	0.187	254	2.40	0.093	12.71	2.59 (0.24)
0.0028 (0.74)	0.0026 (0.009)	0.117	405	2.61	0.079	10.82	3.51(0.33)
0.0035 (0.92)	0.003 (0.011)	0.095	502	2.70	0.225	30.75	12.38 (1.15)
0.0056 (1.48)	0.0051 (0.017)	0.060	794	2.90	0.143	19.55	12.44 (1.16)
0.0057 (1.51)	0.0052 (0.017)	0.059	810	2.91	0.033	4.52	2.94 (0.27)
0.0081 (2.14)	0.0074 (0.024)	0.041	1161	3.06	0.053	7.29	6.79 (0.63)
0.0088 (2.32)	0.008 (0.026)	0.038	1253	3.10	0.020	2.74	2.75 (0.26)
0.011 (2.91)	0.010 (0.033)	0.030	1566	3.19	0.042	5.72	7.18 (0.67)
0.0117 (3.09)	0.011 (0.035)	0.028	1674	3.22	0.054	7.40	9.93 (0.92)
0.0142 (3.75)	0.013 (0.042)	0.024	2019	3.31	0.016	2.23	3.60 (0.33)
0.0155 (4.1)	0.014 (0.044)	0.021	2213	3.35	0.070	9.59	17.0 (1.6)
0.0235 (6.21)	0.021 (0.070)	0.014	3353	3.53	0.018	2.50	6.72 (0.6)
0.0341 (9.0)	0.031 (0.102)	0.010	4859	3.69	0.079	10.82	42.2 (3.9)
0.0435 (11.5)	0.040 (0.130)	0.0058	8260	3.92	0.057	7.77	51.5 (4.8)
0.0579 (15.3)	0.053 (0.174)	0.0046	10311	4.01	0.082	11.16	92.3 (8.58)
0.0723 (19.1)	0.066 (0.217)	0.0029	16466	4.22	0.082	11.16	147.4 (13.7)
0.1154 (30.5)	0.105 (0.346)	0.0022	21918	4.34	0.082	11.16	196.2 (18.2)
0.1537 (40.6)	0.140 (0.465)	0.0016	29476	4.47	0.089	12.19	288.2 (26.7)

The difference between influent and effluent turbidity (ΔNTU), was calculated from the average of initial turbidity values as well as the averages of the steady-state turbidity values measured after passage of three-pipe volumes. With knowledge of the change in iron concentration, flow rate and pipe geometry; the flux can be calculated using equation (4.13).

$$K_m = \frac{\Delta[Fe] \times Flow}{Surface\ \ Area} \qquad (4.13)$$

Table 4.12
G pipe flux experiment results

Flow Rate m^3/min (gpm)	Flow Velocity m/s (ft/s)	HRT (days)	Reynolds Number	Log Re Number	Δ Turbidity (NTU)	Δ Iron (μg/L)	Flux, K_m mg/m^2-d (mg/ft^2-d)
0.000012 (0.0032)	0.0000016 ($5.4x10^{-6}$)	4.49	5.1	0.71	2.802	383.84	1.28 (0.12)
0.00003 (0.008)	0.0000041 ($1.4x10^{-5}$)	1.78	13.0	1.11	1.568	214.79	1.81 (0.17)
0.00038 (0.1)	0.000052 ($1.7x10^{-4}$)	0.142	162.1	2.21	0.115	15.73	1.65 (0.15)
0.00087 (0.23)	0.00012 ($3.9x10^{-4}$)	0.062	372.9	2.57	0.017	2.34	0.56 (0.05)
0.0016 (0.42)	0.00022 ($7.1x10^{-4}$)	0.034	680.9	2.83	0.021	2.93	1.29 (0.12)
0.003 (0.8)	0.00042 ($1.4x10^{-3}$)	0.018	1297.0	3.11	0.026	3.49	2.94 (0.27)
0.0034 (0.9)	0.00047 ($1.5x10^{-3}$)	0.0158	1459.1	3.16	0.014	1.92	1.81 (0.17)
0.0038 (1.0)	0.00052 ($1.7x10^{-3}$)	0.0142	1621.2	3.21	0.022	3.01	3.17 (0.29)
0.0045 (1.19)	0.00062 ($2.0x10^{-3}$)	0.0120	1929.2	3.29	0.055	7.53	9.42 (0.88)
0.0051 (1.34)	0.0007 ($2.3x10^{-3}$)	0.0106	2172.4	3.34	0.020	2.71	3.81 (0.35)
0.0065 (1.72)	0.00089 ($2.9x10^{-3}$)	0.0083	2793.3	3.45	0.038	5.26	9.52 (0.89)
0.0189 (5.0)	0.0026 ($8.5x10^{-3}$)	0.0028	8106.0	3.91	0.043	5.95	31.2 (2.9)
0.034 (8.98)	0.0047 ($1.5x10^{-2}$)	0.0016	14558.3	4.16	0.045	6.14	57.9 (5.4)
0.058 (15.3)	0.0079 ($2.6x10^{-2}$)	0.0009	24820.5	4.39	0.056	7.70	124 (11.5)

The data include the Reynolds Number at which experiments were conducted, the flow velocity, hydraulic retention time, average change in turbidity, calculated corresponding change in iron concentration (using equation (4.12)), and calculated flux values (using equation (4.13)).

The calculated flux values for the UCI and G are presented as a function of Reynolds Number in Figure 4.19. The coefficients of determination (R^2) for both the G and the UCI were equal to 0.98 when all experiment results were considered. A semi-log plot, Figure 4.20, is more revealing for definition of the relation between flux and Reynolds Number.

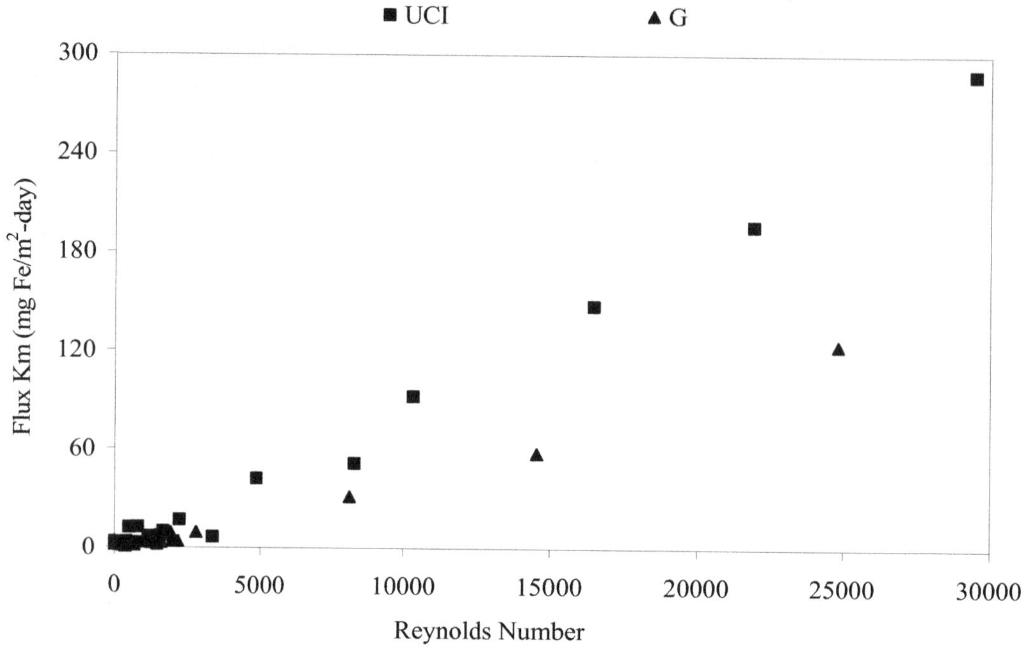

Note: Conversion factor: (mg/m^2-day) ÷10.7639 = mg/ft^2-day

Figure 4.19 Flux (K_m) as a function of Reynolds Number, G and UCI pipe

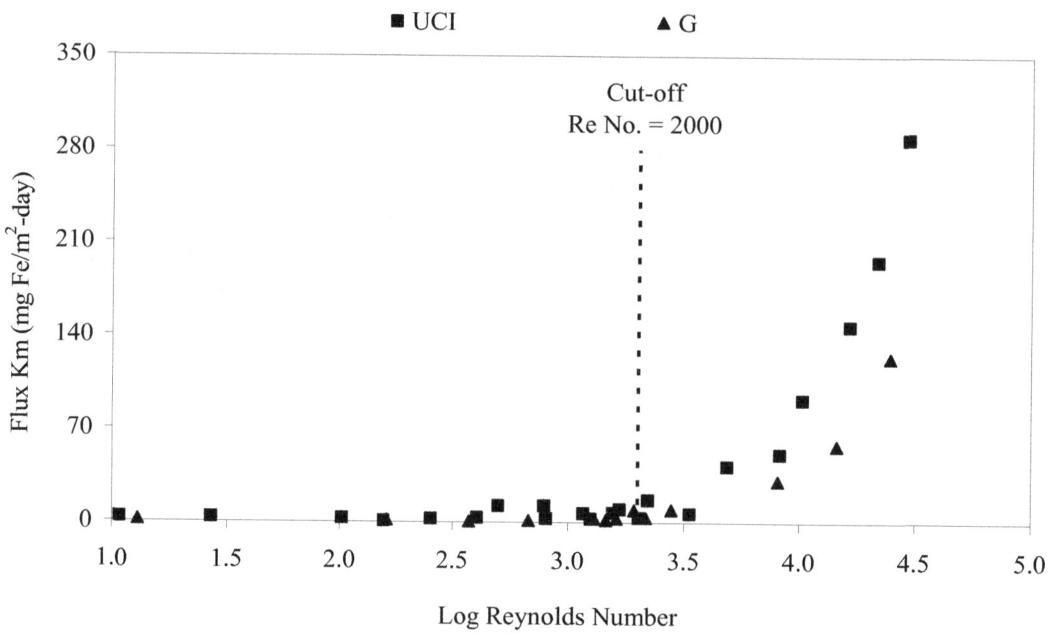

Note: Conversion factor: (mg/m^2-day) ÷10.7639 = mg/ft^2-day

Figure 4.20 Flux – log Reynolds Number plot, for the G and UCI pipes

The semi-log plot indicates a distinct change in the flux relationship at a Reynolds Number equal to 2000. The point coincides with the traditional definition cut point of a laminar flow regime. Thus a different relation exists between Reynolds Number and Flux under laminar flow conditions than under turbulent flow conditions. To correctly describe this relationship, a piecewise model is required. One relation is used to represent stagnant to laminar flow, and a second relation is appropriate for transitional to fully turbulent flow conditions. The flux-Reynolds Number relation under both laminar and turbulent flow conditions is represented in Figure 4.21.

For laminar flow conditions, the mean flux values for the G and UCI pipes were 1.99 and 4.16 mg Fe/m^2-d (0.185 and 0.387 mg Fe/ft^2-d) respectively. These uniform rates are appropriate for Reynolds Numbers less than 2000. For Reynolds Numbers greater than 2000, a linear regression was conducted that was constrained to force the flux value to equal the average constant flux values obtained for the laminar flow when the Reynolds Numbers equal or less than 2000. The resulting flux models for the turbulent region are given in Table 4.13.

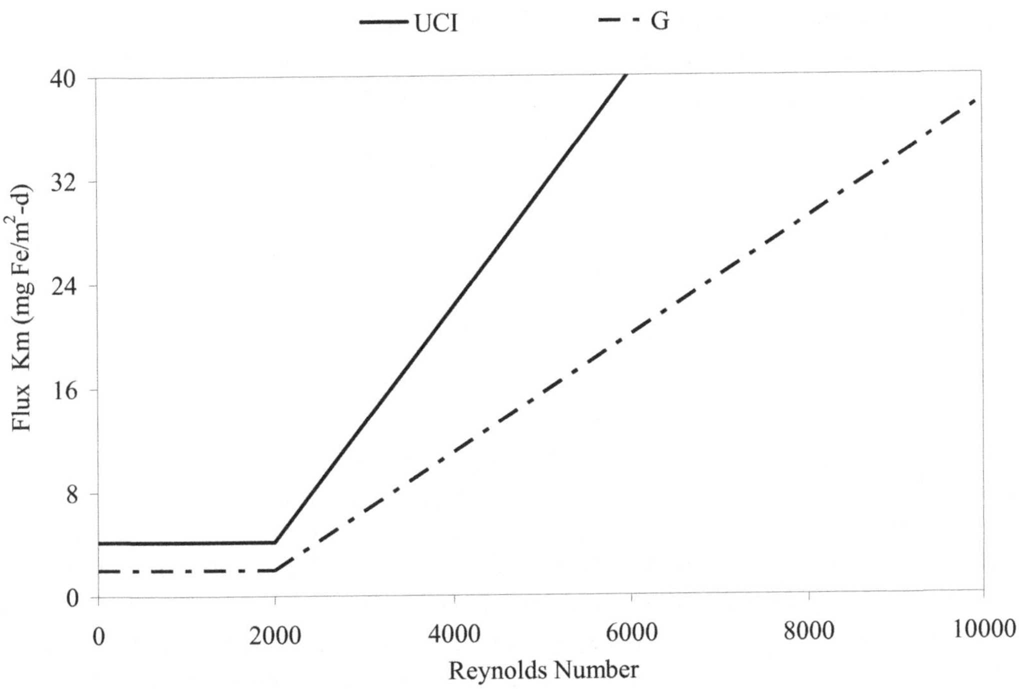

Note: Conversion factor: (mg/m^2-day) ÷10.7639 = mg/ft^2-day

Figure 4.21 Flux (K_m) as a function of Reynolds Number, for the G and UCI pipes

Table 4.13
Flux models under turbulent flow (fixed water chemistry)

Material	Flux, K_m (mg Fe/m^2.d)
G	$K_m = 4.5x10^{-3}$ (Re-2000) + 1.99
UCI	$K_m = 9.0x10^{-3}$ (Re-2000) + 4.16

The resulting models are of the form shown in equation (4.14), where α and β (and thus, K_m) would be a function of water quality and pipe material/condition.

$$K_m = \alpha + \beta\ Re \tag{4.14}$$

Modeling iron release for varying hydraulic conditions. Incorporation of the prediction of iron flux into the mass balance equation permits prediction of the effect of hydraulic conditions (for example, hydraulic residence time) on the impairment of water quality associated with the release of corrosion products. It has been previously reported that total iron, apparent color, and turbidity are highly correlated in PDS effluent samples. Accordingly, it would be acceptable to consider any of these three variables as qualitatively equivalent for the response to a change in hydraulic conditions.

For systems that operate at low flow conditions, the iron flux value (K_m) may be considered to be independent of Reynolds Number. (It is noted that K_m would be expected to depend on water chemistry and pipe material.) If a constant value for K_m is assumed, the mass balance approach described previously is applicable, and the resulting increase in total iron concentration (or apparent color, or turbidity) would be proportional to the hydraulic residence time and inversely proportional to the pipe diameter:

$$Accumulation = Input - Output + Release\ from\ pipe\ surface \tag{4.15}$$

$$0 = QC_{FEED} - QC_{EFFLUENT} + K_m SA \tag{4.16}$$

$$C_{EFFLUENT} - C_{FEED} = \Delta C = \frac{4\ K_m\ HRT}{D} \tag{4.17}$$

The preceding result is consistent with observations that adverse water quality is often associated with dead ends or other locations in the distribution system that experience extended hydraulic residence times. The mathematics suggest that for extreme conditions (very long HRT), adverse water quality is a possibility regardless of the water chemistry. This statement depends only on an assumption of a constant, positive value for K_m and a large enough hydraulic residence time to express an undesired increase in total iron (or apparent color or turbidity). To the extent that water chemistry and/or pipe material and condition causes an increase in K_m, the adverse water quality would be experienced at less severe hydraulic residence times.

For turbulent operation (Reynolds Number > 2000), the relation for K_m must be modified:

$$K_m = \alpha + \beta\ Re \tag{4.14}$$

Equation (4.16) could them be written as equation (4.18).

$$0 = QC_{FEED} - QC_{EFFLUENT} + (\alpha + \beta\ \mathrm{Re})\ SA \tag{4.18}$$

Simplification of the mass balance is possible with the following relations:

$$Re = VD/v \tag{4.19}$$

where V = velocity
ν = kinematic viscosity.

The increase in concentration can be defined as equation (4.20):

$$C_{EFFLUENT} - C_{FEED} = \Delta C = \left(\alpha + \beta \frac{VD}{\nu}\right)\left(\frac{\pi DL}{Q}\right) = \left(\alpha + \beta \frac{VD}{\nu}\right)\left(\frac{4L}{VD}\right) \quad (4.20)$$

For large values of Reynolds Number (β Re >> α), the mass balance equation simplifies to yield prediction of a constant concentration increase for a particular reach (constant length) of pipe that is independent of hydraulic residence time and pipe diameter. The increase would be proportional to the length of pipe.

$$C_{EFFLUENT} - C_{FEED} = \Delta C = \left(\beta \frac{VD}{\nu}\right)\left(\frac{4L}{VD}\right) = \frac{4\beta\beta}{\nu} \quad (4.21)$$

Integration of water quality effects into the flux model. It was shown (Figure 4.7 and Figure 4.8) that both color and turbidity are reasonable surrogates for iron measurement in drinking water distribution systems. Thus for a measured iron flux value $K_{m(1)}$, a new flux value $K_{m(2)}$ can be obtained by adjustment of the original flux value based on changes in color release predicted by the hybrid PDS empirical model (equation (4.3)) and/or G model (equation (4.4)) to account for variation in water chemistry as shown in equation (4.22). Use of the hybrid PDS model is recommended for applications with UCI pipe.

$$K_{m(2)} = K_{m(1)} \times \frac{\Delta Color_{(2)}}{\Delta Color_{(1)}} \quad (4.22)$$

Characteristics of the water blend utilized for the flux experiments were presented previously in Table 3.17. The color release ($\Delta Color_{(1)}$) predicted by the empirical model for the hybrid PDS for this water blend is 8.76 cpu (assuming a 5-day hydraulic retention time, 7.5 mg/L average DO and 20°C (68°F) water temperature). This prediction is valid for the hybrid PDS line geometry. Similarly for the G PDS model, the predicted color release for this experimental blend (Table 3.17) is 2.7 CPU.

Iron Release Simulations

The set of equations developed previously can be combined to enable simulation of water quality in an arbitrary distribution system for any arbitrary water quality and hydraulic condition. Completion of the calculations requires knowledge of the pipe material, pipe geometry (length and diameter), velocity (or flow rate), and water chemistry (parameters included in the empirical water quality model, equation (4.3)). The elevation of iron concentration is cumulative, so that distribution networks can be analyzed by sequential addition of incremental increases in iron concentration for pipe reaches with variant conditions (material, geometry, and velocity). The sequence of calculations is described below.

For each pipe reach, the velocity (or flow rate) and pipe geometry are used to calculate the Reynolds Number and hydraulic residence time.

For the specific pipe material, a value for the iron flux (Km) is calculated using the relationship in Table 4.13 for turbulent flow. For laminar flow, the constant values for flux cited in Table 4.13 are appropriate. It is noted that these flux values are specific for the water quality used in the flux experiments defined in Table 3.17.

The K_m values obtained previously are adjusted for the actual water quality using equation (4.22). For this calculation, the unadjusted K_m value is defined as $K_{m(1)}$ and the water-chemistry adjusted value is defined as $K_{m(2)}$. A value for $\Delta Color_{(2)}$ was calculated with the empirical model as shown in the equation using the actual water quality for unlined iron and G pipes respectively.

The change in iron concentration in the single reach of pipe can then be calculated using a mass balance (equation (4.17)), with the HRT term being replaced by L / u (equation (4.26)). The calculation process is repeated for sequential reaches of pipe to yield equation (4.24).

$$K_m = \alpha + \beta \; Re \tag{4.23}$$

$$\Delta[Fe] = \sum_{1}^{m} \left. \frac{4K_{m(2)ij} L_{ij}}{u_{ij} D_{ij}} \right|_{UnlinedCas\,tIron} + \sum_{1}^{n} \left. \frac{4K_{m(2)kl} L_{kl}}{u_{kl} D_{kl}} \right|_{Galvanized} \tag{4.24}$$

where L = length in the respective pipe reaches ij and kl
D = diameter in the respective pipe reaches ij and kl
u = flow velocity in the respective pipe reaches ij and kl

The preceding assembly of equations was used to simulate water quality for varying Reynolds Number, pipe diameter, and pipe length. The impact of finished water quality was also evaluated by examination of water qualities representative of the three primary sources (GW, SW, and RO) and a blend of these sources. The chemistry of these sources is summarized in Table 4.14. The assumed pipe geometry is summarized in Table 4.15. It is emphasized that the simulation results are specific for the assumed pipe geometry and water chemistry. The methodology defined previously would permit analysis of other defined geometries and water chemistries.

Table 4.14
Source water quality for simulations

	GW	SW	RO	Blend
Alkalinity (mg/L as $CaCO_3$)	225	50	50	155
Chloride (mg/L)	15	10	50	17
Sodium (mg/L)	10	15	30	13.5
Sulfate (mg/L)	10	180	30	63
Silica (mg/L)	14	10	2	9
Temperature (°C)	20	20	20	20
DO (mg/L)	7.5	7.5	7.5	7.5
HRT (Days)	5	5	5	5
Relative Δ Apparent Color – UCI (CPU)	1	6	15	2
Relative Δ Apparent Color – G (CPU)	1	356	116	18

Impact of Reynolds number on iron concentration. The impact of Reynolds number on iron concentration for the UCI and G pipes is illustrated in Figure 4.22, Figure 4.23 and Figure 4.24. The pipe geometry assumed for these simulations is summarized in Table 4.15.

The simulation results for both pipe materials are displayed for laminar flow conditions in Figure 4.22. The presentation in Figure 4.22 was developed using the source blend described in Table 4.14. For a system of fixed geometry, a reduction in Reynolds number would be accompanied by a corresponding increase in the hydraulic residence time. For laminar flow conditions, the numerical value for the flux (K_m) remains constant. As developed previously in equation (4.17), the anticipated increase in iron concentration would be proportional to the hydraulic residence time.

$$C_{EFFLUENT} - C_{FEED} = \Delta C = \frac{4\, K_m\, HRT}{D} \tag{4.25}$$

Table 4.15
Pipe geometry used in iron release simulation

Pipe Material	UCI	G
Length, m (ft)	1609.3 (5280)	1609.3 (5280)
Diameter, m (in)	0.1524 (6)	0.0508 (2)

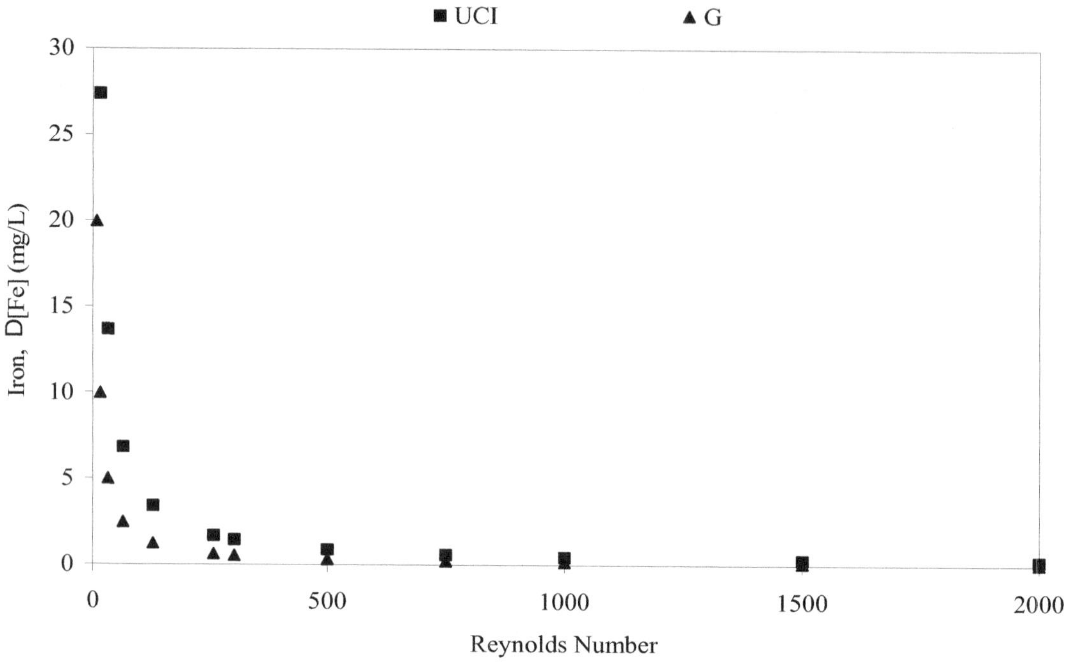

Figure 4.22 Predicted iron concentration under laminar flow conditions for the G and UCI pipes

Alternately, the increase in iron concentration would be inversely proportional to the Reynolds Number for a fixed geometry system under laminar flow conditions. This result is evident in Figure 4.22. It is noted that these conditions may be approximated by distribution system dead ends (excessive hydraulic residence times).

Under turbulent flow regimes, the iron concentration approaches an asymptote at a large value of the Reynolds Number. This relation was reported previously in equation (4.21).

Simulation results for the cast iron pipe geometry specified in Table 4.15 and the four sources defined in Table 4.14 are reported in Figure 4.23. Similar results would be obtained for simulation of the G pipe behavior. The relative iron release for the different source waters follows the predictions of the empirical model (equation (4.4)) that were reviewed previously in Table 4.8. The GW source (high alkalinity, low sulfate, and low chloride) exhibits minimal release of iron corrosion products. The GW source matches the chemistry of the historical water exposure that produced the corrosion products on the pipe surfaces. In contrast, a water quality similar to the treated S1 (low alkalinity, high sulfate) is predicted to result in elevated iron concentrations. Finished waters of similar composition to the RO source (low alkalinity, high sodium, and high chloride) pose the greatest potential for water quality impairment.

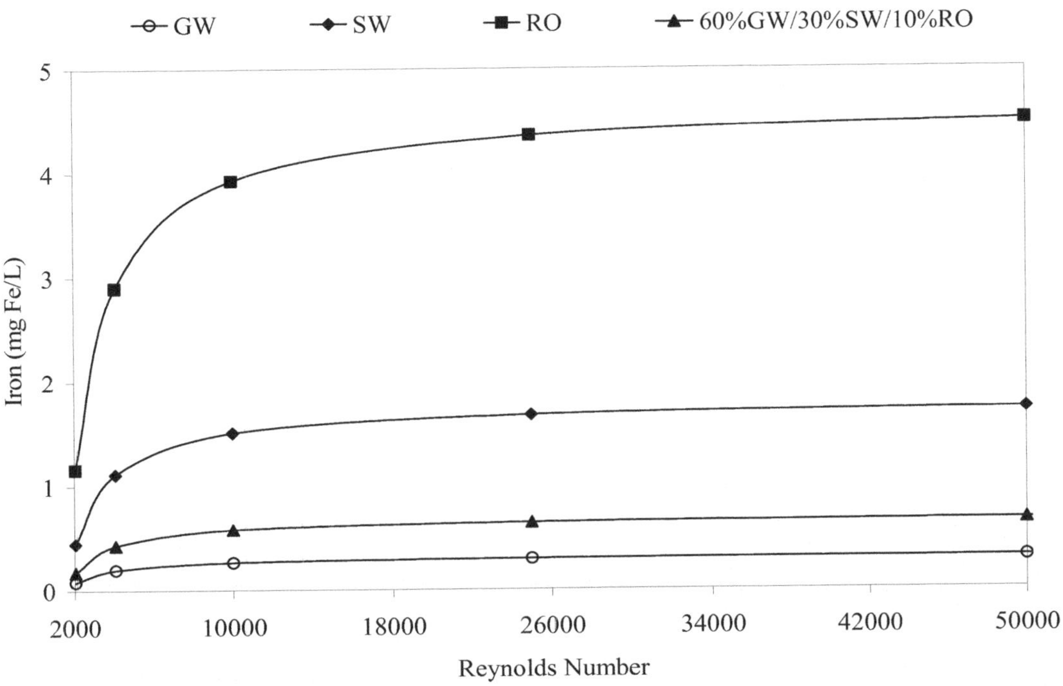

Figure 4.23 Predicted iron concentration, UCI pipe, turbulent flow conditions

The results for laminar and turbulent conditions are combined in Figure 4.24 for the UCI pipe. The levels of iron concentration are an order of magnitude higher in the laminar flow region compared to that in the turbulent flow region, supporting the observation that high iron concentration is problematic in drinking water distribution system dead ends.

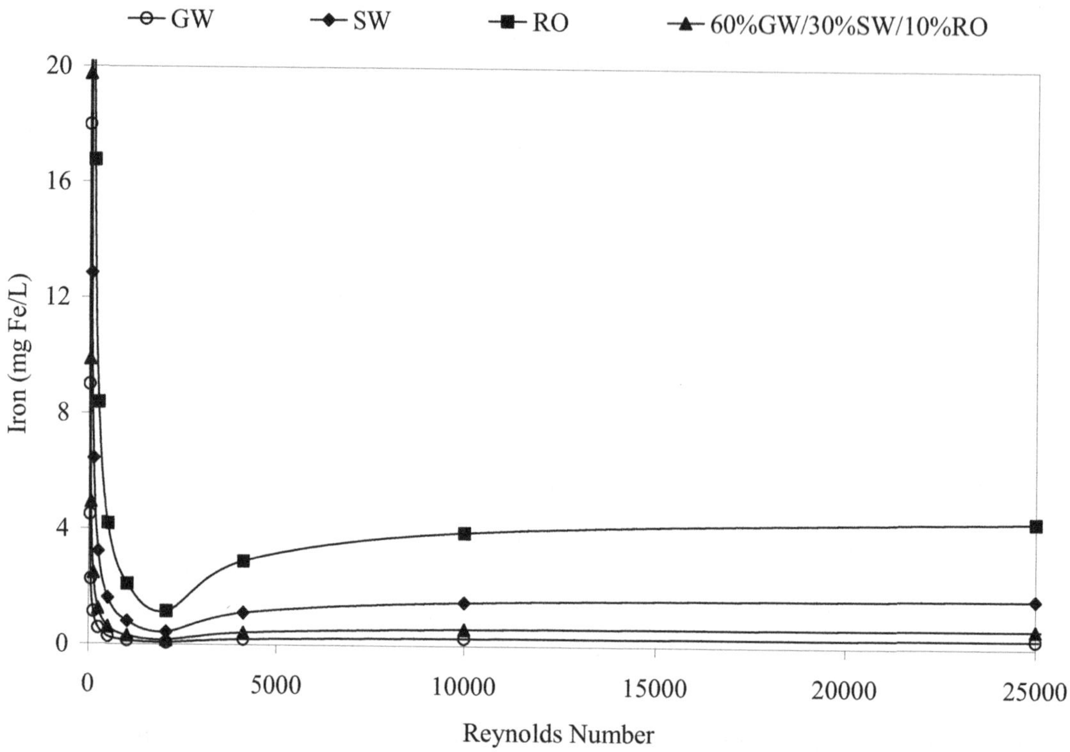

Figure 4.24 Iron concentration predictions, UCI pipe, laminar and turbulent flow

Impact of pipe diameter. The limiting behavior at low and high Reynolds number (laminar and turbulent conditions, respectively) was defined in equations (4.17) and (4.21), respectively.

These equations suggest that greater iron concentrations would be expected for smaller diameter pipes under laminar flow conditions. As the flow conditions approach fully turbulent, the equations predict a constant water quality impact that is independent of diameter. These results are confirmed by simulations for both pipe materials shown in Figure 4.25 and Figure 4.26. These simulations were developed for an assumed pipe length of 1 mile, water chemistry defined by the source blend in Table 4.14. and fixed flow velocities of 6.096×10^{-5} m/s (0.0002 ft/s) and 0.1524 m/s (0.5 ft/s) for laminar and turbulent flow conditions, respectively.

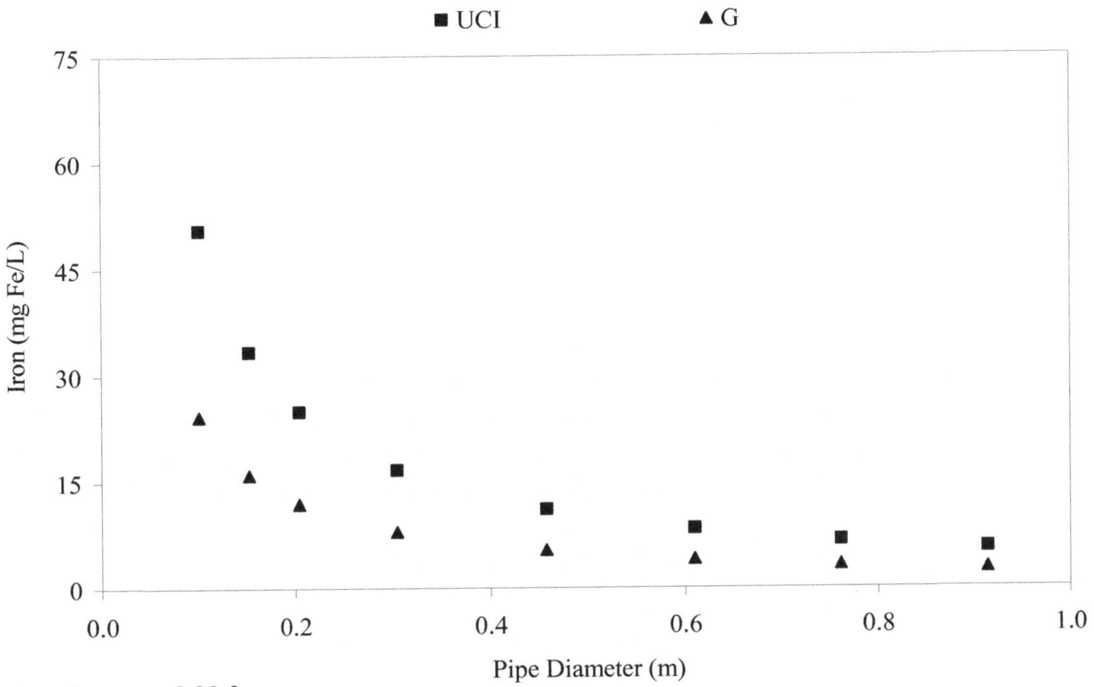

Note: 1 meter = 3.28 ft

Figure 4.25 Impact of pipe diameter under laminar flow conditions for the G and UCI pipes

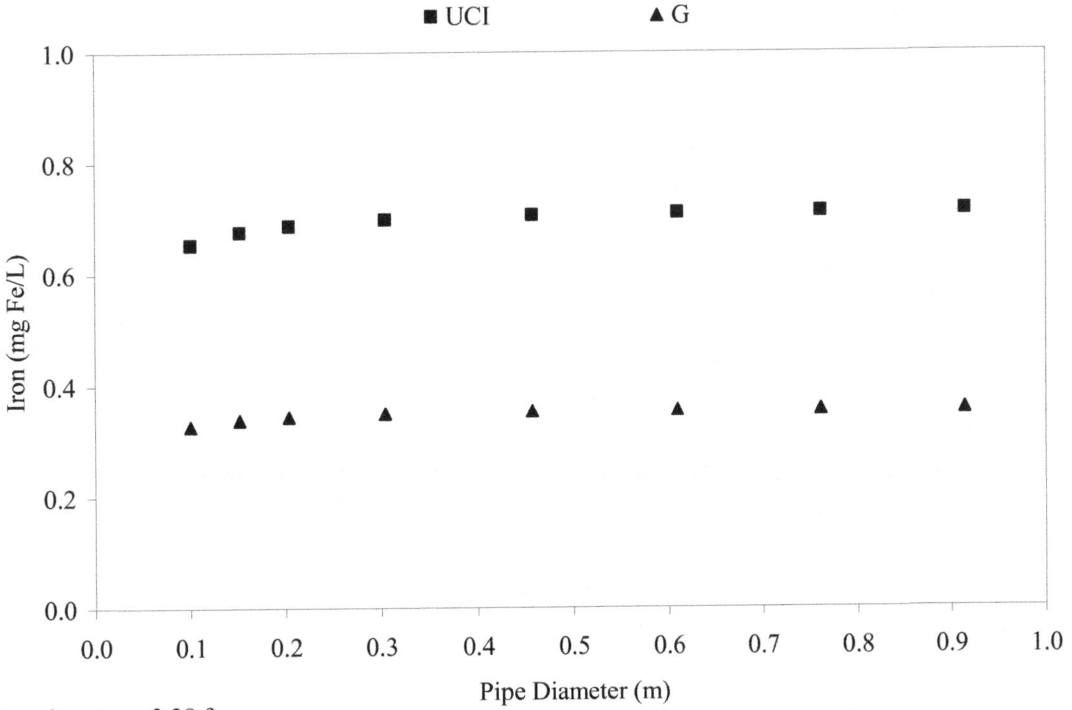

Note: 1 meter = 3.28 ft

Figure 4.26 Impact of diameter under turbulent flow conditions for the G and UCI pipes

Impact of pipe length. The limiting behavior at low and high Reynolds Number (laminar and turbulent conditions, respectively) was defined in equations (4.17) and (4.21), respectively. The hydraulic residence time (HRT) is equal to the pipe length (L) divided by the velocity (u), thus equation (4.17) could also be developed in terms of the pipe length (equation (4.26)).

$$C_{EFFLUENT} - C_{FEED} = \Delta C = \frac{4\,K_m\,L}{u\,D} \tag{4.26}$$

It is evident from these equations that the impairment in water quality defined by the increase in iron concentration is proportional to the length of pipe. A similar conclusion is supported by equation (4.24). This result is confirmed by simulations for both pipe materials shown in Figure 4.27 and Figure 4.28. These simulations were developed for assumed pipe diameters of 6 inches (UCI) and 2 inches (G), Reynolds Number of 10 (laminar flow), Reynolds Number of 5000 (turbulent flow), and water quality defined by the source blend in Table 4.14.

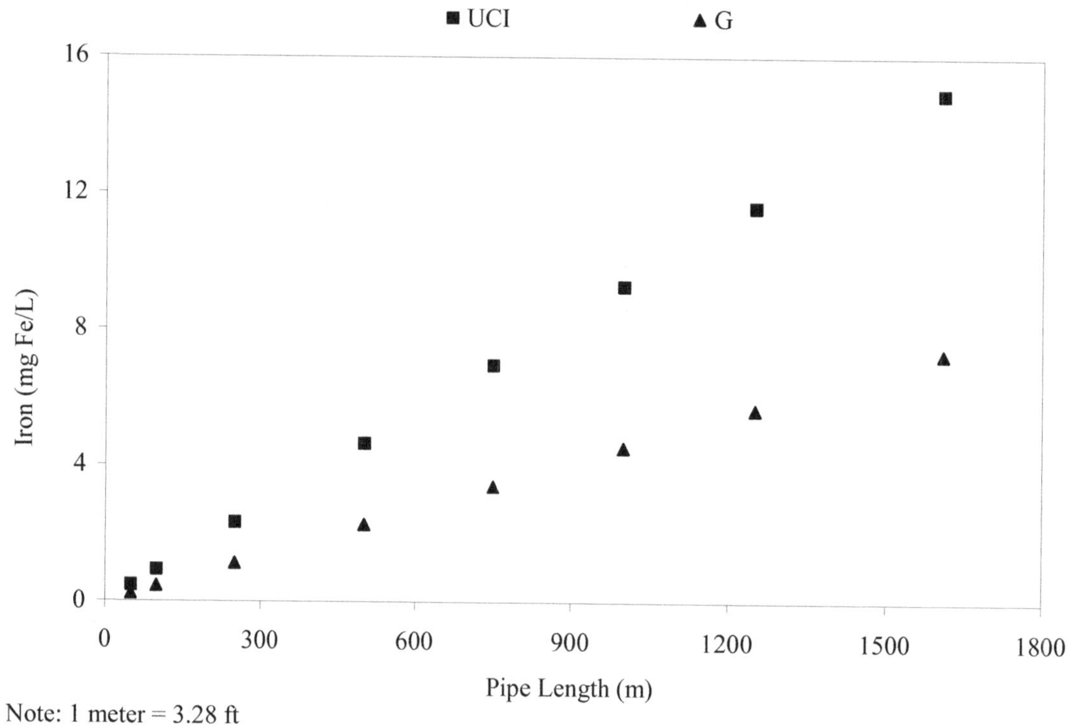

Figure 4.27 Impact of pipe length under laminar flow conditions for the G and UCI pipes

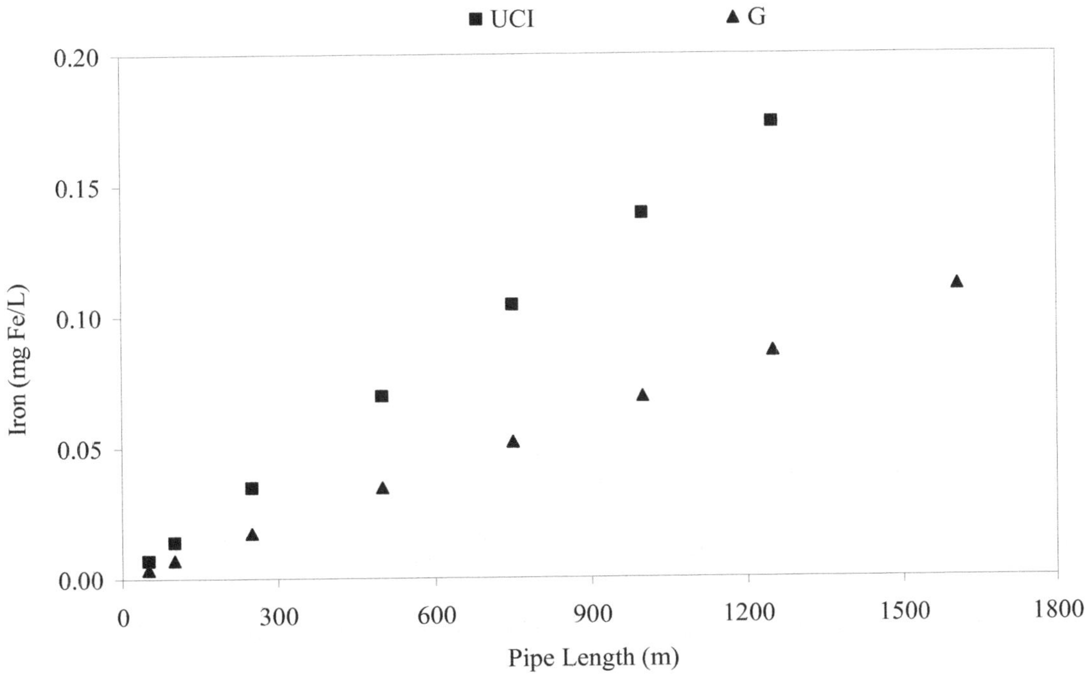

Note: 1 meter = 3.28 ft

Figure 4.28 Impact of pipe length under turbulent flow conditions for the G and UCI pipes

Practical Application of Iron Release Models – User Manual

Prediction of water quality impairment in galvanized and unlined cast iron pipe materials (release of corrosion products and associated apparent color) may be pursued as detailed in the previous section of the report. Application of these robust models requires knowledge of pipe material, pipe geometry (length and diameter), hydraulics (flow rate or velocity), and water quality. Application of the comprehensive models would be hindered by the lack of specific information to define pipe geometry, hydraulics, and/or water quality. It is recognized that the models could also be used to develop less precise predictive relationships that do not require the full knowledge of system parameters (pipe geometry, hydraulics, and water quality). It is believed that these simplified models, in spite of a reduction in accuracy, may still have significant value for the Member Governments due to their ease of application.

This discussion outlines the procedures for application of the comprehensive models with incorporation of complete knowledge of system parameters. This discussion also details a simplified approach that defines a proportional relationship between water quality impairment (measured as increase in apparent color) and pipe length. It is cautioned that this simplified approach would not provide a suitable description of stagnant regions in the distribution system. Such stagnant regions may include dead ends and dead spots that are characterized by laminar flow conditions, low velocities, and excessive hydraulic residence times.

Examples are provided to determine water quality impacts (apparent color) associated with different pipe materials, pipe geometries (length and diameter), hydraulics, and water quality. Specific calculations are presented using customary units of measurement.

Summary of Predictive Models

The basic equation (4.27) for prediction of change in iron concentration is obtained from a material balance on a pipe of length = L:

$$\Delta C = \frac{4\,Km\,L}{V\,D} = \frac{4\,Km\,HRT}{D} \qquad (4.27)$$

where DC = increase in iron concentration ($mass/L^3$)
Km = iron flux ($mass/L^2$-time)
L = pipe length (L)
V = fluid velocity (L/time)
D = pipe diameter (L)
HRT = hydraulic residence time (time)

The flux term (Km) depends on pipe material, water quality, and hydraulic conditions (Reynolds Number, Re). Different predictive equations are required for laminar (equation (4.29)) and turbulent (equation (4.30)) hydraulic conditions, corresponding to Reynolds Numbers (equation (4.28)) less than 2000 and greater than 2000, respectively.

$$\mathrm{Re} = \frac{V\,D}{\nu} \qquad (4.28)$$

$$\mathrm{Km} = \alpha \qquad \text{(Laminar Flow, Re < 2000)} \qquad (4.29)$$

$$\mathrm{Km} = \alpha + \beta\,(\mathrm{Re} - 2000) \qquad \text{(Turbulent Flow, Re > 2000)} \qquad (4.30)$$

Where ν = kinematic viscosity (L^2/time)
α = flux constant term ($mass/L^2$-time)
β = flux slope term ($mass/L^2$-time)

The kinematic viscosity is dependent on temperature, but may be assumed equal to 0.00001 ft^2/sec.

The flux constant and slope terms (α, β) depend on water quality and pipe material. Experiments were completed with four pipe materials (PVC, lined cast iron, unlined cast iron, and galvanized) to permit definition of the flux constant and slope terms. It was determined that the PVC and lined cast iron pipes do not experience release of iron corrosion products. Quantification of the release from galvanized and unlined cast iron requires knowledge of the water quality.

Galvanized pipe. An empirical model (equation (4.4)) was developed to relate change in color to water quality in galvanized pipe materials. Knowledge of the alkalinity, silica, sulfate, and temperature is required. Use of the empirical model requires use of the specified units of measurement for all water quality parameters.

$$\Delta C_{galv} = 10.71\,(Alk)^{-1.63}\,(Si)^{-0.55}\,(SO_4)^{1.12}\,(Temp)^{1.06} \qquad (4.4)$$

where ΔC_{galv} = increase in apparent color, galvanized (cpu)
Alk = alkalinity (mg/L as $CaCO_3$)
Si = silica (mg/L as SiO_2)
SO_4 = sulfate (mg/L)
Temp = temperature (°C)

Values for the flux terms can be calculated (equation (4.31) and equation (4.32)) that are specific for the galvanized pipe material and experienced water quality:

$$\alpha_{galv} = (0.822)\,\Delta C_{galv} \tag{4.31}$$

$$\beta_{galv} = (0.00186)\,\Delta C_{galv} \tag{4.32}$$

where α_{galv} = flux constant term, galvanized pipe (mg/m^2-day)
β_{galv} = flux slope term, galvanized pipe (mg/m^2-day)

Unlined cast iron pipe. A separate empirical model was developed to relate change in color to water quality in unlined cast iron pipe materials as shown in equation (4.3). Knowledge of the alkalinity, sulfate, chloride, sodium, dissolved oxygen, and temperature is required. For application of the model to define Km, the HRT should be set equal to 5 days. Use of the empirical model requires use of the specified units of measurement for all water quality parameters.

$$\Delta C_{UCI} = 0.04775(Alk)^{-0.912}(SO_4)^{0.118}(Cl)^{0.485}(Na)^{0.561}(DO)^{0.967}(Temp)^{0.813}(HRT)^{0.836} \tag{4.3}$$

where ΔC_{UCI} = increase in apparent color, unlined cast iron (cpu)
Alk = alkalinity (mg/L as $CaCO_3$)
SO_4 = sulfate (mg/L)
Cl = chloride (mg/L)
Na = sodium (mg/L)
DO = dissolved oxygen (mg/L)
Temp = temperature (°C)
HRT = hydraulic residence time (days)

Values for the flux terms can be calculated (equations (4.33) and equation (4.34)) that are specific for the unlined cast iron pipe material and experienced water quality:

$$\alpha_{galv} = (0.411)\,\Delta C_{UCI} \tag{4.33}$$

$$\beta_{galv} = (0.000888)\,\Delta C_{UCI} \tag{4.34}$$

where α_{UCI} = flux constant term, unlined cast iron pipe (mg/m^2-day)
β_{UCI} = flux slope term, unlined cast iron pipe (mg/m^2-day)

Relationship between iron and color. An empirical relation was defined to relate increases in apparent color to increases in iron concentration as shown in equation (4.35):

$$\Delta Color = 87\, \Delta Iron \tag{4.35}$$

where ΔColor = increase in apparent color (cpu)
ΔIron = increase in iron concentration (mg/L)

Simplified predictive equation – Laminar flow. For laminar flow conditions, the flux term is constant and equal to α. For laminar conditions, the predicted increase in iron concentration is determined with equation (4.36).

$$\Delta C = \frac{4\alpha L}{VD} = \frac{4\alpha HRT}{D} \tag{4.36}$$

It is observed that for a fixed length of pipe, the impairment in water quality becomes more severe as the velocity decreases and/or the pipe diameter decreases. This relation is consistent with the observation that red water problems are often associated with stagnant regions in the distribution system. The equation can be rewritten (equation (4.37) and equation (4.38)) to predict elevation of apparent color per mile of pipe length for both pipe materials.

For galvanized pipe:

$$\gamma = \frac{\Delta Color}{L} = \frac{0.689 \Delta C_{galv}}{VD} \tag{4.37}$$

For unlined cast iron pipe:

$$\gamma = \frac{\Delta Color}{L} = \frac{0.344 \Delta C_{UCI}}{VD} \tag{4.38}$$

where γ = increase in apparent color per unit length of pipe (cpu/mile)
ΔColor = increase in apparent color (cpu)
L = pipe length (miles)
ΔC_{UCI} = increase in apparent color for specified water quality (cpu)
ΔC_{galv} = increase in apparent color for specified water quality (cpu)
V = velocity (ft/sec)
D = pipe diameter (inches)

Simplified predictive equation – Fully turbulent flow. For fully turbulent flow conditions with very large values for the Reynolds Number, the flux (Km) can be approximated by the relation defined by equation (4.39).

$$\begin{aligned} Km &= \alpha + \beta(Re - 2000) = (\alpha - 2000\,\beta) + \beta Re \\ Km &\cong \beta Re \end{aligned} \tag{4.39}$$

The change in iron concentration is determined with equation (4.40), reflecting the simplification to the flux defined in equation (4.39).

$$\Delta C \cong \frac{4\,\beta\,\mathrm{Re}\,L}{V\,D} = \frac{4\,\beta\left(\frac{V\,D}{\nu}\right)L}{V\,D} = \frac{4\,\beta\,L}{\nu} \qquad (4.40)$$

It is observed that for a fixed length of pipe, the impairment in water quality would be independent of hydraulic conditions (such as velocity) or pipe diameter in the limit as the Reynolds Number becomes very large. The equation can be rewritten (equation (4.41) and equation (4.42)) to predict elevation of apparent color per mile of pipe length for both pipe materials.

For galvanized pipe:

$$\gamma = \frac{\Delta \mathrm{Color}}{L} = 12.98\,\Delta C_{galv} \qquad (4.41)$$

For unlined cast iron pipe:

$$\gamma = \frac{\Delta \mathrm{Color}}{L} = 6.19\,\Delta C_{UCI} \qquad (4.42)$$

where γ = increase in apparent color per unit length of pipe (cpu/mile)
ΔColor = increase in apparent color (cpu)
L = pipe length (miles)
ΔC_{galv} = increase in apparent color for specified water quality (cpu)
ΔC_{UCI} = increase in apparent color for specified water quality (cpu)

Rigorous approach for turbulent conditions with dependence of Km on hydraulics. For situations where turbulent conditions are present and the velocity is known, it is possible to incorporate the specific velocity into the analysis. This approach may provide some improvement in accuracy relative to the previous approach that assumed fully turbulent conditions.

The analysis relies on the same equations noted previously.

$$\Delta C = \frac{4\,Km\,L}{V\,D} \qquad (4.27)$$

$$\Delta Color = 87\,\Delta Iron \qquad (4.35)$$

The mass balance equation and the empirical relation between apparent color and total iron are combined to yield equation (4.43).

$$\gamma = \frac{\Delta Color}{L} = \frac{4\,(87)\,Km}{V\,D} \qquad (4.43)$$

A value for Km must be obtained that is specific to the Reynolds Number and water quality for the specific location.

$$Km = \alpha + \beta\,(Re - 2000) \qquad (4.30)$$

$$\alpha_{galv} = (0.822)\ \Delta C_{galv} \tag{4.31}$$

$$\beta_{galv} = (0.00186)\ \Delta C_{galv} \tag{4.32}$$

$$\alpha_{UCI} = (0.411)\ \Delta C_{UCI} \tag{4.33}$$

$$\beta_{UCI} = (0.000888)\ \Delta C_{UCI} \tag{4.34}$$

The relationships are then combined to yield equation (4.44). The approach is general for prediction of γ and could be used for both pipe materials, any pipe geometry, any hydraulic condition, or any water quality. The effect of hydraulics and water quality are reflected in the value for Km.

$$\gamma = \frac{\Delta Color}{L} = \frac{0.837\ Km}{V\,D} \tag{4.44}$$

where γ = increase in apparent color per unit length of pipe (cpu/mile)
ΔColor= increase in apparent color (cpu)
L = pipe length (miles)
Km = flux for specified water quality and Reynolds Number (mg/m^2-day)
V = velocity (ft/sec)
D = pipe diameter (inches)

Water quality specifications. Use of the prediction models requires estimation of the water quality in the respective pipes. Production of the source waters in the field study included stabilization with respect to calcium carbonate solubility. Thus application of the empirical water quality model should be restricted to situations where stable waters are anticipated. Site-specific water quality should be used when available. In the absence of such data, anticipated characteristics of the three principal water sources are defined in Table 4.16. This water quality matches the assumed water quality used in earlier simulations and reported in Table 4.11 with the exception of alkalinity. The alkalinity values in Table 4.16 for the SW and RO sources were increased to 100 mg/L as $CaCO_3$ to match current TBW operating practices. A temperature of 28 ^{o}C was selected to represent summertime conditions. Values are also provided for DCgalv and DCUCI for each source as shown in Table 4.16. For anticipated blends of the three principal sources, the characteristics of the blend may be calculated as a weighted average for each source.

Table 4.16
Water quality parameters by sources and predicted apparent color release for G and UCI pipe

Parameter	Groundwater (GW)	Surface Water (SW)	Desal (RO)
Alkalinity (mg/L as $CaCO_3$)	225	100	100
Chloride (mg/L)	15	10	50
Sodium (mg/L)	10	15	30
Sulfate (mg/L)		10	180
Silica (mg/L as SiO_2)	14	10	2
Dissolved Oxygen (mg/L)	7.5	7.5	7.5
Temperature (°C)	28	28	28
ΔC_{galv} (cpu)	0.166	19.0	6.21
ΔC_{UCI} (cpu)	2.46	7.46	19.4

Examples

The following examples are intended to illustrate application of the preceding steady-state equations for estimation of color release under various distribution system conditions. As reviewed in a subsequent section of this report ("Copper, Lead and Iron Transition"), the complete response to alteration in bulk water quality associated with source blend changes in not immediate. The numerical values reported in these examples represent changes that would only be realized if a sufficient duration of exposure to the altered water quality was experienced to reach a new steady state. For iron (or apparent color) release, the duration required to reach a new steady state is approximately one month.

1. Determine the release of color from a 1000 foot length of 2-inch galvanized pipe for a blend of the three principal water sources: GW (60%), SW (30%), and RO (10%). Assume laminar flow conditions with a velocity = 0.01 ft/sec. Repeat for turbulent flow conditions with a velocity = 1 ft/sec. Repeat for fully turbulent flow conditions.

Determine water quality of the blend by weighted average of the three sources:

$$Alk_{blend} = Alk_{GW}\,Fraction_{GW} + Alk_{SW}\,Fraction_{SW} + Alk_{RO}\,Fraction_{RO}$$
$$= (225\text{ mg/L})\,(0.60) + (100\text{ mg/L})\,(0.30) + (100\text{ mg/L})\,(0.10)$$
$$= 175\text{ mg/L as }CaCO_3$$

The resulting characteristics of the blend are summarized in Table 4.17

Table 4.17
Blend water quality

Parameter	Blend WQ
Alkalinity (mg/L as $CaCO_3$)	175
Chloride (mg/L)	17
Sodium (mg/L)	13.5
Sulfate (mg/L)	63
Silica (mg/L as SiO_2)	11.6
Dissolved Oxygen (mg/L)	7.5
Temperature (°C)	28

For this blend:

$$\begin{aligned}\Delta C_{galv} &= 10.71\,(Alk)^{-1.63}\,(Si)^{-0.55}\,(SO_4)^{1.12}\,(Temp)^{1.06}\\ &= 10.71\,(175)^{-1.63}\,(11.6)^{-0.55}\,(63)^{1.12}\,(28)^{1.06}\\ &= 2.17 \text{ cpu}\end{aligned}$$

It should be noted that due to non-linearities in the empirical equation for determination of ΔC_{galv} and ΔC_{UCI}, the value of ΔC_{galv} and ΔC_{UCI} can not be determined as a weighted average of the values for the individual sources. In this case:

$$\begin{aligned}\Delta C_{blend} &= 2.17 \text{ cpu}\\ &\neq DCGW \text{ x Fraction GW} + DCSW \text{ x Fraction SW} + DCRO \text{ x Fraction RO}\\ &= (0.166 \text{ cpu})(0.60) + (19.04 \text{ cpu})(0.30) + (6.21 \text{ cpu})(0.10) = 6.43 \text{ cpu}\end{aligned}$$

<u>For laminar flow with a velocity = 0.01 ft/sec</u>

$$\Delta Color = \frac{0.689\,\Delta C_{galv}}{V\,D}\,L = \frac{0.689\,(2.17)}{(0.01)\,(2)}\,(0.189) = 14 \text{ cpu}$$

<u>For turbulent flow with a velocity = 1 ft/sec</u>

In this example, the velocity is known. A value for the flux terms (α and β) will be calculated that are specific to both the water quality and the Reynolds Number.

Determine the Reynolds Number:

$$Re = \frac{V\,D}{\nu} = \frac{(1\,\frac{ft}{sec})\,(2\,in)}{(0.00001\,\frac{ft^2}{sec})\,(12\,\frac{in}{ft})} = 16{,}667$$

Determine a value for Km for the specific conditions:

$$\begin{aligned}\alpha_{galv} &= (0.822)\,\Delta C_{galv} = (0.822)\,(2.17) = 1.78 \text{ mg/m}^2\text{-day}\\ \beta_{galv} &= (0.00186)\,\Delta C_{galv} = (0.00186)\,(2.17) = 0.00404 \text{ mg/m}^2\text{-day}\\ Km &= \alpha + \beta\,(Re - 2000) = 1.78 + (0.00404)\,(16667 - 2000) = 60.98 \text{ mg/m}^2\text{-day}\end{aligned}$$

Calculate the increase in apparent color:

$$\Delta \text{Color} = \gamma\, L = \frac{0.837\, \text{Km}}{\text{V D}} L = \frac{0.837\,(60.98)}{(1)\,(2)} (0.189) = 4.8\, \text{cpu}$$

For fully turbulent flow

$$\Delta \text{Color} = \gamma\, L = (12.98)\,(\Delta C_{galv})\, L = (12.98)\,(2.17)\,(0.189) = 5.3\ \text{cpu}$$

It is observed that assumption of complete turbulence results in an error (overestimate) of approximately 10% for these specific conditions (5.3 cpu vs. 4.8 cpu).

2. Determine the release of color for all pipe links in the hypothetical subdivision layout shown in Figure 4.29. The hydraulic parameters (velocities, flows, and head loss) were determined using a standard software package. The water quality was assumed to match the blend of groundwater (60%), surface water (30%), and desalination (10%) described in previous problems. The large diameter pipes (> 4 inches, P-35, P-30, P-22, and P-34) are PVC. The smaller diameter pipes (2 and 4 inch) are galvanized

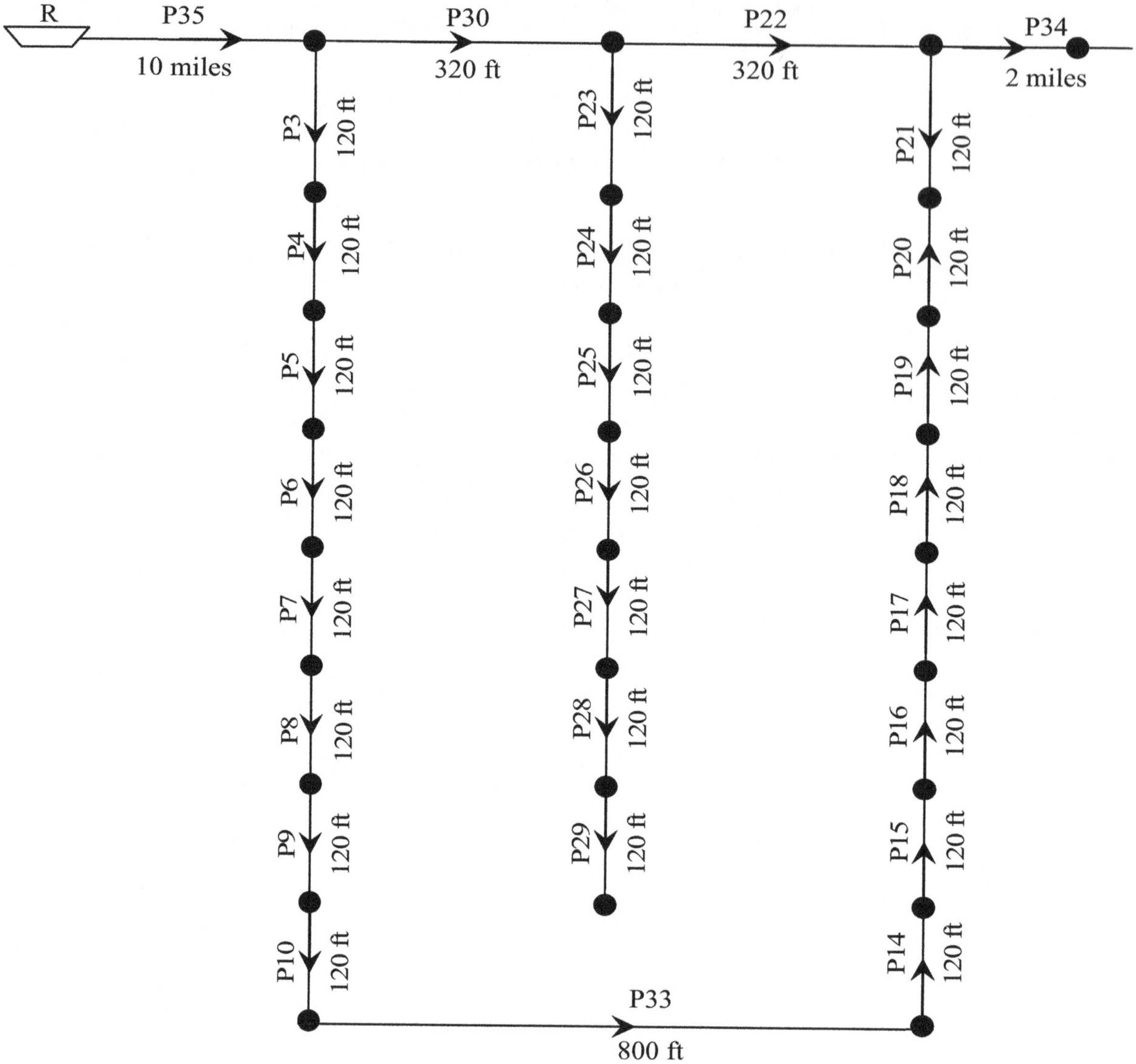

Figure 4.29 Example drinking water distribution system

The results are summarized in Table 4.18, Table 4.19, and Table 4.20.

Table 4.18
Linear pipe segment

Pipe	Material	Length ft	Diameter Inches	Velocity Ft/sec	ΔColor CPU	Σ ΔColor CPU	HRT hrs	Σ HRT hrs
P23	Galv	120	2	0.3475	0.47	0.47	0.10	0.10
P24	Galv	120	2	0.2979	0.44	0.91	0.11	0.21
P25	Galv	120	2	0.2482	0.40	1.31	0.13	0.34
P26	Galv	120	2	0.1986	0.34	1.64	0.17	0.51
P27	Galv	120	2	0.1489	0.24	1.88	0.22	0.73
P28	Galv	120	2	0.0993	0.17	2.05	0.34	1.07
P29	Galv	120	2	0.0496	0.34	2.39	0.67	1.74

Table 4.19
Pipe loop segment – counterclockwise path

Pipe	Material	Length Ft	Diameter Inches	Velocity Ft/sec	ΔColor CPU	Σ ΔColor CPU	HRT hrs	Σ HRT hrs
P3	Galv	120	4	0.1913	0.48	0.48	0.17	0.17
P4	Galv	120	4	0.1789	0.47	0.96	0.19	0.36
P5	Galv	120	4	0.1665	0.46	1.42	0.20	0.56
P6	Galv	120	4	0.1540	0.45	1.86	0.22	0.78
P7	Galv	120	4	0.1416	0.43	2.29	0.24	1.01
P8	Galv	120	4	0.1292	0.41	2.70	0.26	1.27
P9	Galv	120	4	0.1168	0.38	3.08	0.29	1.56
P10	Galv	120	4	0.1044	0.35	3.44	0.32	1.88
P33	Galv	800	4	0.0920	2.10	5.54	2.42	4.29
P14	Galv	120	4	0.0796	0.26	5.80	0.42	4.71
P15	Galv	120	4	0.0672	0.19	5.99	0.50	5.21
P16	Galv	120	4	0.0548	0.15	6.15	0.61	5.81
P17	Galv	120	4	0.0423	0.20	6.35	0.79	6.60
P18	Galv	120	4	0.0299	0.28	6.63	1.11	7.72
P19	Galv	120	4	0.0175	0.48	7.12	1.90	9.62
P20	Galv	120	4	0.0051	1.66	8.78	6.54	16.16

Table 4.20
Pipe loop segment – clockwise path

Pipe	Material	Length Ft	Diameter Inches	Velocity Ft/sec	ΔColor CPU	Σ ΔColor CPU	HRT hrs	Σ HRT hrs
P21	Galv	120	4	0.0073	1.16	1.16	4.57	4.57

It is of interest to note that the presence of a "dead zone" or region of low velocity occurs in the pipe loop in the vicinity of pipe links 20-21. The corresponding cumulative HRT in the galvanized material (16 hours) and associated increase in apparent color (8.78 cpu) is greater at this point in the pipe loop than is encountered in the linear pipe segment (pipe links 23 to 29, HRT of 2 hours, cumulative color increase of 2.39 cpu).

Summary Graphs

Results are summarized in a series of graphs to display the relationship between the increase in apparent color and flow velocity. Three graphs are presented in Figure 4.30 to Figure 4.32 for the following pipes:

1. 2-inch diameter galvanized pipe (Figure 4.30)
2. 6-inch diameter unlined cast iron pipe (Figure 4.31)
3. 24-inch diameter unlined cast iron pipe (Figure 4.32)

A blend of 60% groundwater, 30% surface water, and 10% desalination was assumed for the simulations. Similar relationships could be defined for other water quality conditions using the methods described in this section.

The color release (γ) is quantified as a rate per mile of pipe length, i.e. as cpu per mile. Because the increase in apparent color is proportional to the length of pipe, predictions of the increase in apparent color may be obtained by multiplication of the rate per mile of pipe length by the pipe length:

$$\Delta\text{Color (cpu)} = \gamma\text{ (cpu/mile) L (miles)}$$

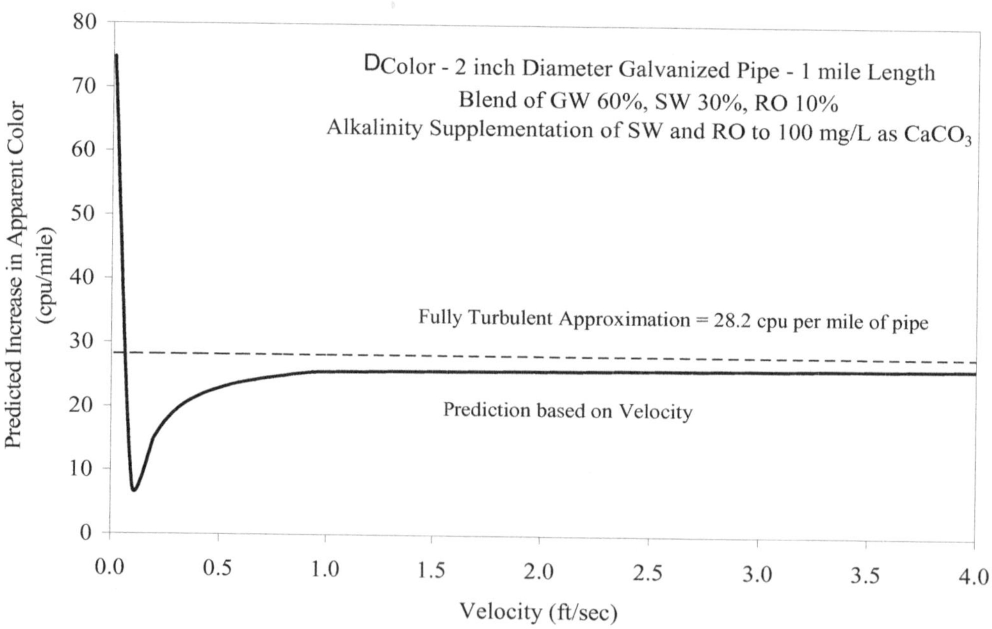

Figure 4.30 Predicted increase in apparent color for 2" G pipe for blended water quality with alkalinity supplement

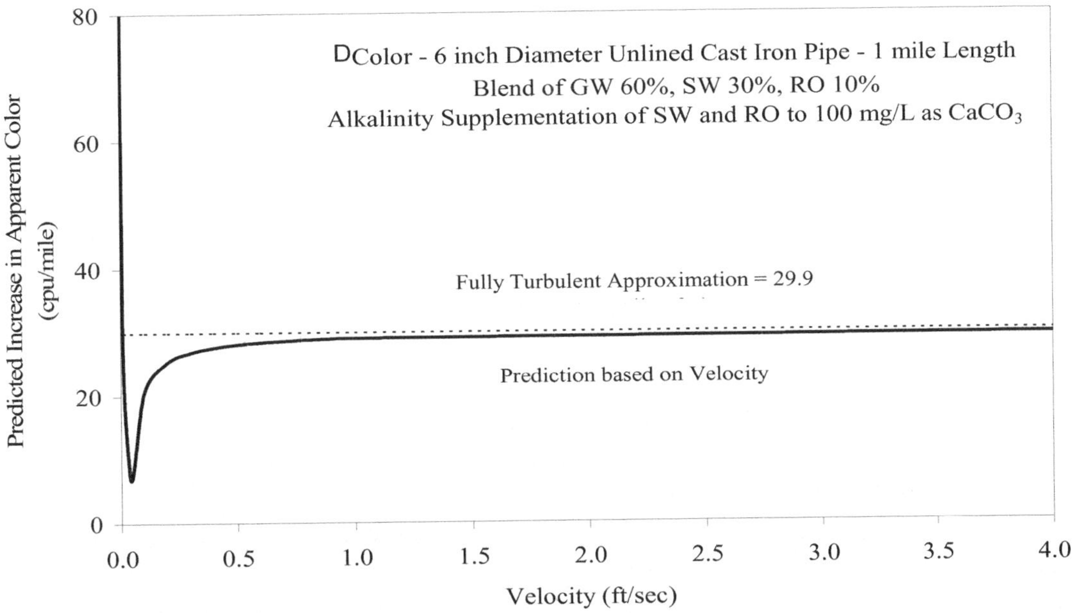

Figure 4.31 Predicted increase in apparent color for 6" UCI pipe for blended water quality with alkalinity supplement

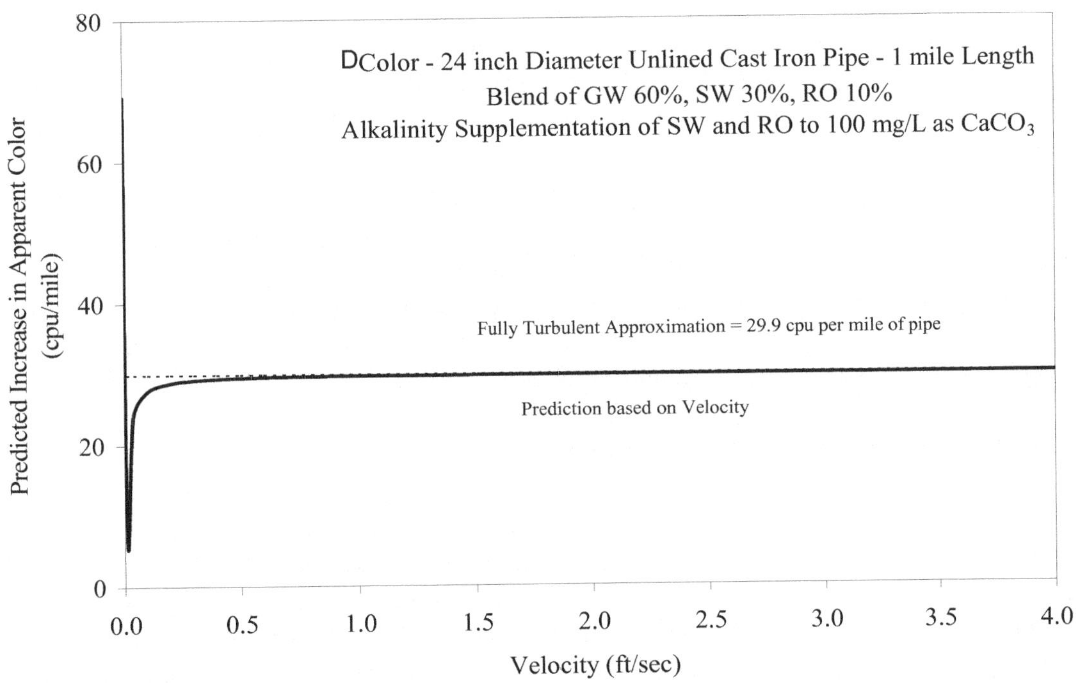

Figure 4.32 Predicted increase in apparent color for 24" UCI pipe for blended water quality with alkalinity supplement

It can be demonstrated from the curves that a simplifying assumption of complete turbulence introduces relatively minor errors as long as turbulent conditions are present. The simplified equations for definition of γ are stated below.

For galvanized pipe:

$$\gamma = \frac{\Delta Color}{L} = 12.98\,\Delta C_{galv}$$

For unlined cast iron pipe:

$$\gamma = \frac{\Delta Color}{L} = 6.19\,\Delta C_{UCI}$$

where γ = increase in apparent color per unit length of pipe (cpu/mile)
ΔColor = increase in apparent color (cpu)
L = pipe length (miles)
ΔC_{galv} = increase in apparent color for specified water quality using the empirical model for galvanized pipe (cpu)
ΔC_{UCI} = increase in apparent color for specified water quality using the empirical model for unlined cast iron pipe (cpu)

If flow conditions are in the laminar region, however, accurate predictions are not possible without accurate definition of the hydraulics (either velocity or flow rate). In view of the relatively short reaches of unlined cast iron and galvanized pipe that remain in the distribution system, the most severe impairment of water quality will likely occur in stagnant portions of the distribution system.

COPPER RELEASE

Copper release was investigated using the corrosion loops described in the Methods and Procedure Chapter exactly as lead release was investigated. Actual copper release came from common copper pipe that was the same as copper pipe used for home plumbing systems. Copper scale samples came from copper coupons that were carefully controlled for surface area exposure and were identical in each of the eighteen corrosion loops. Specifically, surface characterization of copper scale was done to investigate the controlling solid copper phase. Thermodynamic and empirical models for copper release as a function of bulk water quality at steady state were developed to gain an understanding of copper release in home plumbing systems. Finally, copper transition resulting from water quality change was modeled as a function of time and steady state total copper release for changing water quality.

Identification of Copper Corrosion Products

Identification of corrosion products on coupons was done to determine if the water quality variations from ground, surface and saline sources would change the controlling solid phases. Hence, the discussion of coupon scales is common to all source waters. SEM/EDS, XRD, and XPS were used to identify the composition of the corrosion scale layer on the copper coupons. Copper coupons were incubated in PVC pipe cradles for three months in each PDS for each phase during year one. Copper coupons were harvested after three months and analyzed by surface techniques. Copper coupons taken from a GW environment had a dark-brown color, which was indicative of cuprite (Cu_2O) film. Copper coupons taken from either surface or RO finished water had a blue-green color, which was indicative of a cupric salt film. Corrosion scale formed in GW was more compact that the films formed in surface and RO water. Scale on coupons from a surface and RO water environment could be easily removed by wiping. In some cases, the copper coupons exposed in GW were also covered by blue-green deposits. However all copper coupons had the dark brown cuprite film next to the copper metal, which was always revealed by initial visual inspection or after wiping away the top film. EDS analysis revealed the main surface area elementary composition was copper, carbon, oxygen, with minor amounts of sulfur, chlorine and calcium. EDS showed major amounts of copper and oxygen in the surface film, which illustrated the oxide or hydroxide copper compounds dominated the scale.

The corrosion scale layer was thin and did not provide enough mass to make a powder for XRD analysis. Only metallic copper and cuprite (Cu_2O) were found by XRD detection, there were no other crystalline structures such as nantokite (CuCl) or tenorite (CuO). Figure 4.33 shows the representative XRD spectra obtained by grazing X'Pert (Philips XRD). Cuprite was always a major scale constituent. The overall smooth spectra and sharp peaks in Figure 4.33 indicate that there were no other amorphous cuprous or cupric compounds formed directly on the metallic copper. Like other cuprous compounds, the solubility of cuprite is extremely low, it is readily precipitated following oxidation and is quickly oxidized to the cupric state.

Copper concentration in bulk water was too high to be explained by cuprite solubility. Hence, more soluble copper phases existed on the cuprite layer, which could not be detected by XRD as XRD is a technique for analyzing composition in scales that are approximately 50 Å to 10,000 Å thick.

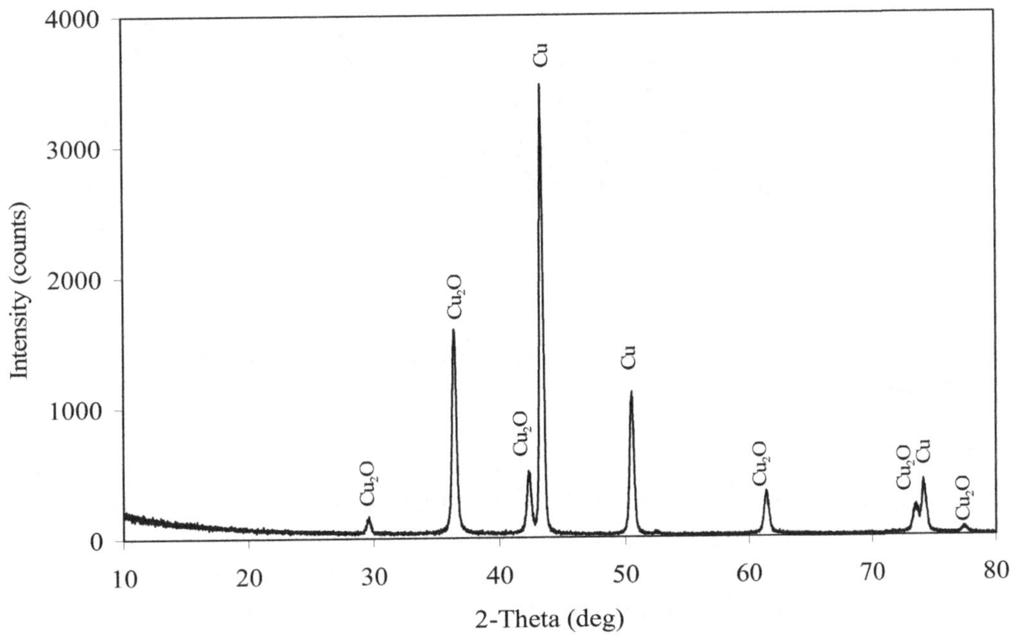

Figure 4.33 Representative XRD spectra scanned by Philips X'Pert.

XPS was used to identify chemical compounds formed in the top layer of copper scales. Considering an inelastic mean free path for LMM Auger electrons of 15Å, XPS is more surface sensitive and can detect composition in the first fifty angstroms of scale layers. Spectra analysis showed that only oxide or hydroxide copper compounds were formed. The broad Cu(2p3) peak indicated that there were several copper compounds on the immediate top scale layer. Curve fitting analysis was performed to determine the different chemical states of copper. Figure 4.34

shows a representative XPS multiplex spectra analysis from copper coupon exposed to GW for three months.

Curve fitting (Table 4.21) revealed that metallic copper existed in the top layer, which indicated a low copper corrosion rate in the corrosion loops. Cu(II) was not present in the top level of the surface scale. Cu_2O was found and confirmed cuprite was a major corrosion component, which was also supported by XRD analysis. Tenorite (CuO) was also confirmed. XRD did not clearly differentiate between cupric hydroxide ($Cu(OH)_2$) or cupric carbonate ($CuCO_3$) by peak values. However, equilibrium water chemistry significantly reduces the possibility of cupric carbonate ($CuCO_3$), which indicated cupric hydroxide was present.

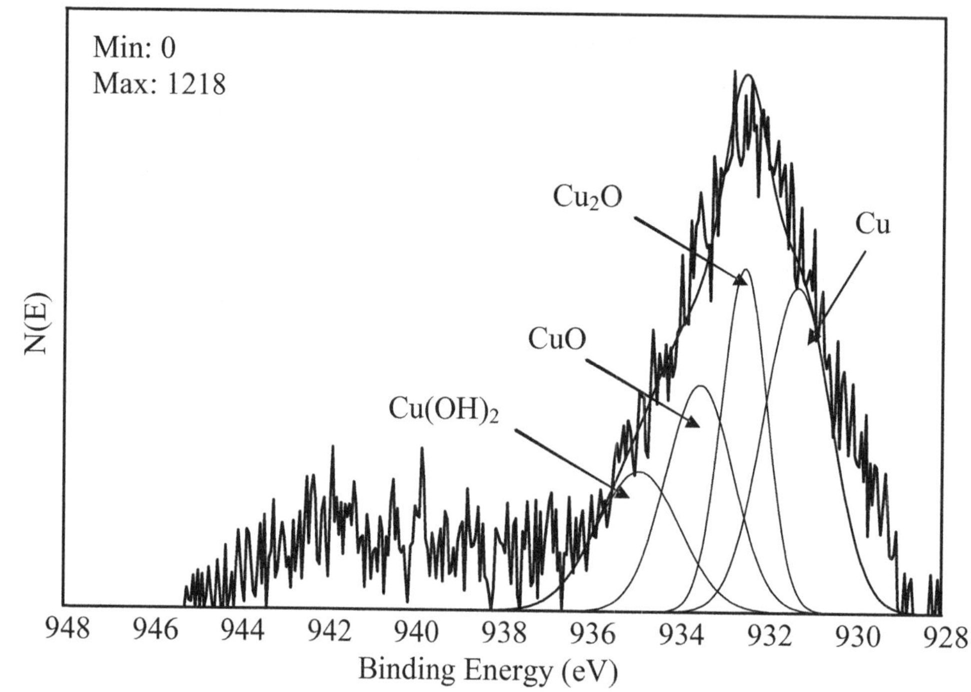

Figure 4.34 Representative XPS multiplex spectra from copper coupon incubated in GW for three months.

XRD and XPS analyses indicated that cuprite (Cu_2O) is the major corrosion scale formed on metallic copper. Cuprite was covered by Tenorite (CuO) and cupric hydroxide ($Cu(OH)_2$ (Cu(II) species and controlled the release of copper to the bulk water. Tenorite and cupric hydroxide always coexisted, which was caused by dehydration of cupric hydroxide. Although Malachite ($Cu_2(CO_3)(OH)_2$) had been found in other studies, none was detected here in any water quality environment.

Table 4.21
Copper XPS curve fit summary

Position	Intensity	Area	FWHM	Area	Percent	Component
1	931.31	730	1.80	1399	34.1%	Cu
2	932.52	773	1.20	987	24.1%	Cu_2O
3	933.52	510	1.80	977	23.8%	CuO
4	934.90	315	2.20	738	18.0%	$Cu(OH)_2$

Thermodynamic Model Development for Copper Release

It is acknowledged that the copper regulation is based on total copper, which is the sum of particulate and dissolved copper and thermodynamic models only predict dissolved copper species. However, fundamental models can sometimes discern among release mechanism and can be modified into semi-empirical models that are useful for predicting metal release.

Considering the major speciation affecting the copper release and hypothetical equilibrium between free Cu(II) ion and the controlling solid phase CuO or $Cu(OH)_2$, the following items listed in the equation (4.45) account for 99.1% of all the dissolved total copper and were used as the basis for the development of the thermodynamic model for copper release.

$$\begin{aligned}[Cu]_T &= [Cu^{2+}] + [CuOH^{+}] + [Cu(OH)_2{}^{0}] + [Cu(OH)_3{}^{-}] + [CuHCO_3{}^{-}] \\ &+ [CuCO_3{}^{0}] + [Cu(CO_3)_2{}^{2-}] + [Cu(OH)_2CO_3{}^{2-}]\end{aligned} \tag{4.45}$$

Within a typical pH range of 7.5-8.5 and alkalinity of 50-200mg/L $CaCO_3$, it is clear that the aqueous $Cu(CO_3)^0$, $CuHCO^{3+}$, and $CuOH^+$ dominate the composition of total copper. Thus, pH and alkalinity are the most important and sensitive factors affecting the release of dissolved copper. The other anions that are not included in equation (4.45) because they do not contribute to dissolved copper in drinking water environments. The only difference between the CuO and $Cu(OH)_2$ models was the solubility constants for these species, which are pK_{sp} 7.98 for CuO and pK_{sp} 8.89 for $Cu(OH)_2$. Thus, the final total dissolved copper species are approximately 8 times for a $Cu(OH)_2$ controlling solid phase than CuO controlling solid phase. The predicted and actual copper versus alkalinity and pH are illustrated in Figure 4.35. The $Cu(OH)_2$ model slightly underestimated the copper release. Given that the field copper release consisted of dissolved and particulate species, the $Cu(OH)_2$ model is reasonable. However, $Cu(OH)_2$ may not be the controlling solid phase. As shown, copper release is more sensitive to alkalinity when pH decreases and to pH when alkalinity increases. Thus, one approach for preventing excessive total copper release is to maintain a relatively high pH with low alkalinity.

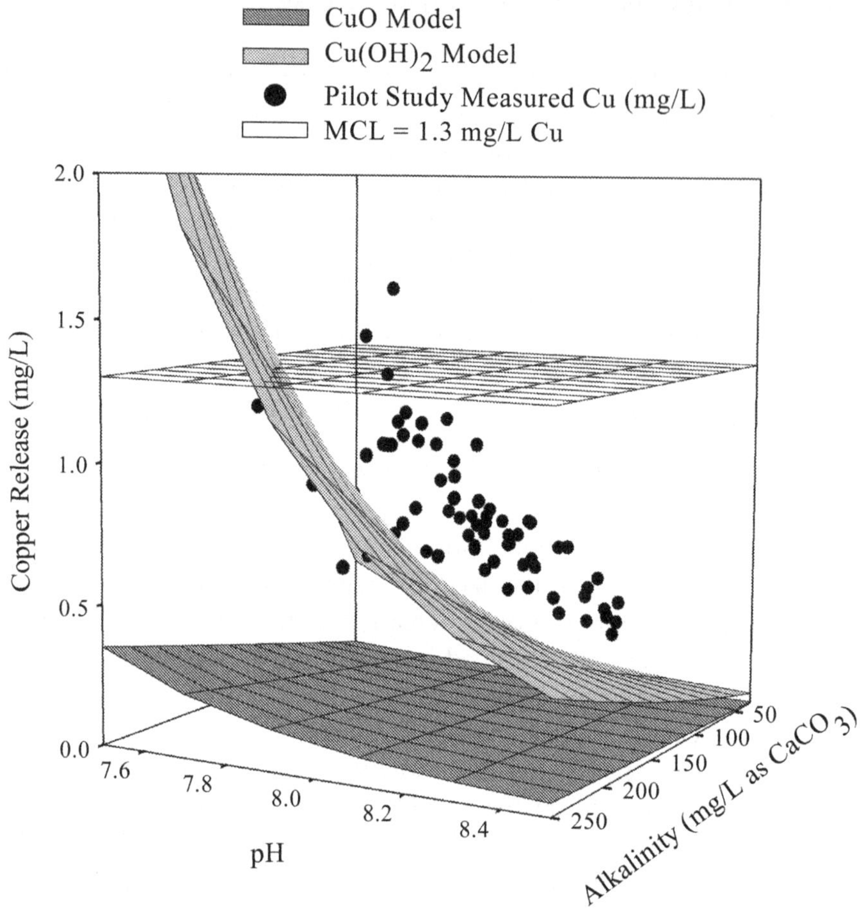

Figure 4.35 Comparison of theoretical model prediction with averaged actual copper release

Effects of Water Quality on Copper Release

In this study, the pH ranged between 7.71 and 8.28. The average copper release throughout an experimental period of 12 months versus pH is shown in Figure 4.36. Clearly, two qualitative trends are present. Copper release decreases as pH increases. In order to distinguish the role of pH from confounding effects caused by alkalinity, released copper data was classified into three groups, corresponding to low, moderate, and high alkalinity. The groups were categorized by the smallest Euclidean distance standard within each group and this produced approximately 36.1% and 66.7% percentile of alkalinity delimitation corresponding to alkalinity ranging from 51 to 87, 88 to 127, 128 to 212 mg/L as CaCO3, respectively. The effect of alkalinity on pH grouping with respect to total copper release can be visualized by passing imaginary trend lines through the grouped data in Figure 4.36. The release of total copper versus pH increases as with the grouped alkalinity.

The second trend is the release of total copper for increasing alkalinity. The slope of the linear trend line reflects the sensitivity of the pH effect on total copper release at different alkalinity levels. The slope within the grouped alkalinity data actually increased as the alkalinity increased, which indicates that copper corrosion release is more sensitive to pH variation at higher alkalinity. Even though the pH effect is rather obvious at low pH, the pH range of this study was limited from 7.5 to 8.5. Due to this narrow range, the pH effect is not always clear, especially if there is interference from other species.

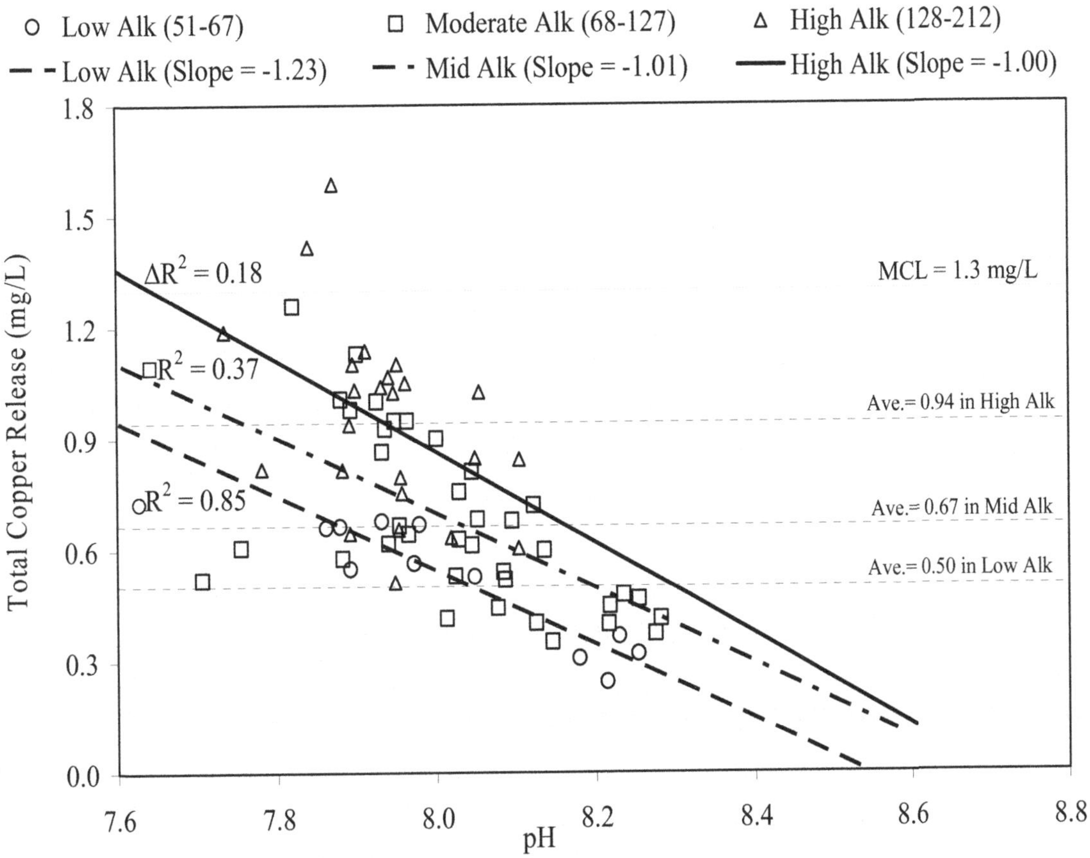

Figure 4.36 pH effect on total copper release

The alkalinity range was 51mg/L to 212 mg/L as $CaCO_3$. For the pilot scale operations, the effect of alkalinity with the interference of other varying water quality parameters was investigated, as shown in Figure 4.37. Within the range of alkalinities evaluated in this study, the general trend is that higher alkalinities enhanced copper release. Even though the impact of other confounding factors cannot be eliminated within the pilot-scale context, alkalinity produced increased copper release.

Similarly as in the pH effect analysis, pH was classified into three groups designated as low pH, ranging from 7.71 to 7.93, moderate pH ranging, from 7.94 to 8.05, and high pH, ranging from 8.06 to 8.28. The slope of trend lines show total copper release increases as pH decreases as shown in Figure 4.37. The slope of copper release versus alkalinity. It was noted that in the experiment, a lower pH corresponded to a relatively high alkalinity, while a higher pH corresponded to relative low alkalinity. Note, in this work, alkalinity was relatively independent of pH (R^2=0.16), which confirmed that high alkalinity produced high copper release.

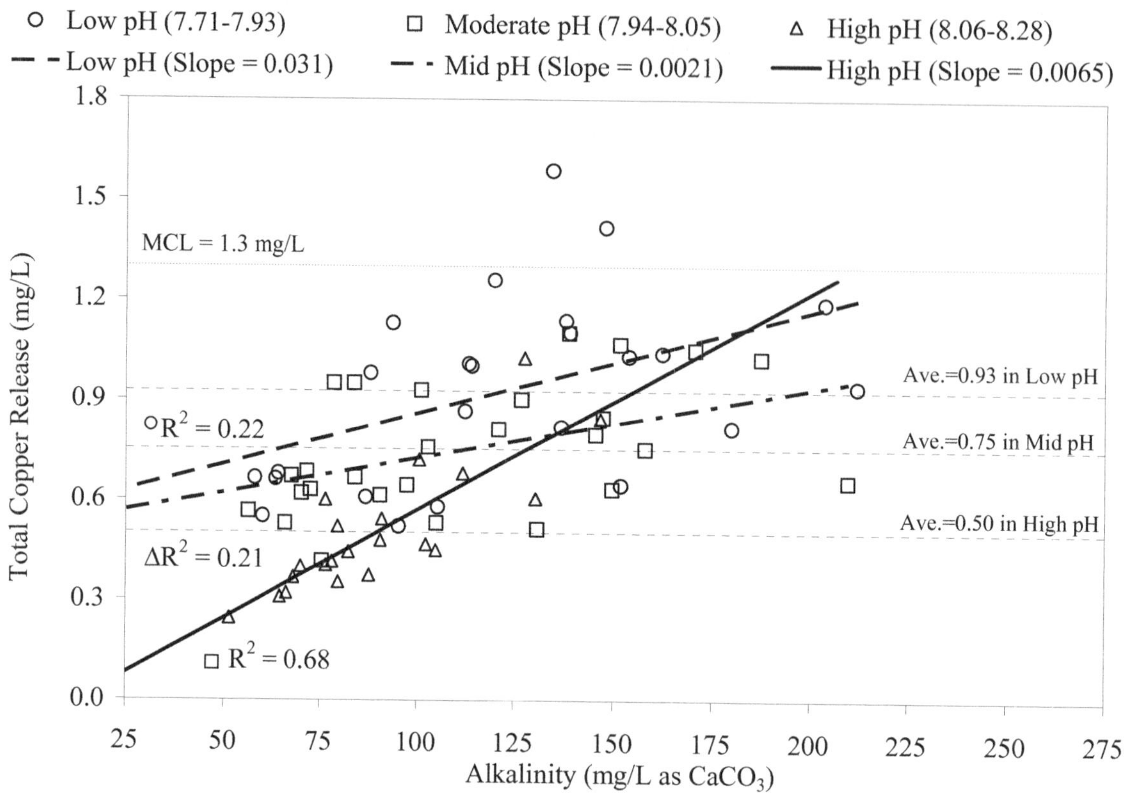

Figure 4.37 The effect of alkalinity variation on total copper release with sulfate as constant

As noted in the literature review, conflicting conclusions have been published regarding the effect of chloride. Some researchers have suggested that chloride would increase the dissolved copper concentration due to soluble chloro-complexes formation, but others had the opposite results. In this work, chlorides have trivial effects on total copper release, which is probably due to the strong confounding effects of alkalinity and pH. However chlorides were normalized with respect to alkalinity and pH and still did not exhibit any effects on copper release for the conditions of this work. Calcium has been viewed as a protective factor that reduces metal release following formation of a solid calcium carbonate layer. Calcium was always positively correlated with alkalinity (R^2=0.60) in this work and was normalized with respect to alkalinity. Still no obvious relationship between calcium and copper release was found. There is no doubt that scaling calcium carbonate can reduce metal release. The stabilization of all waters in the treatment process limited combinations of alkalinity and calcium to adequately demonstrate the inhibitive metal release effects of calcium carbonate scaling.

Initially silica did not show any effect on copper release, but following normalization with respect to alkalinity, silica did show inhibition of copper release. DO and chloramines are oxidants and can accelerate metal release. However DO and chloramines did not affect total copper release in this work as was found by statistical analysis of all data gathered in this project.

Controlled Study of Water Quality Effects on Total Copper Release

A one-month experiment was conducted to determine the confounding effects of chlorides and sulfates on the effect of alkalinity on copper release. In order to isolate water quality influences, certain PDSs were fed varying alkalinity concentrations while holding sulfate at 13.6mg/L (+2.3/-2.0mg/L) and chloride concentration at 32.7mg/L (+6.9/-5.1mg/L). As had been previously observed, copper release was enhanced as alkalinity increased. However, pH increased from 8.09 to 8.44 as alkalinity increased, which indicates that the pH effect is not the reason for the increased copper level at increasing alkalinity as ascending pH had been clearly shown to decrease copper release. This observed alkalinity effect was consistent with aquatic chemistry as carbonate complexes dominated the composition of dissolved copper speciation.

In order to reduce the interference of alkalinity on the investigation of the effects of sulfates and chlorides on the pH effect on copper release, controlled experiments for alkalinity were conducted. Sulfate was controlled at 17.9mg/L (+4.7/-3.7mg/L) and alkalinity was controlled at 71 mg/L as $CaCO_3$ (+18/-9 mg/L as $CaCO_3$) when the effect of chloride was investigated. Chloride was controlled at 32.9mg/L (+3.0/-4.6mg/L) and alkalinity was controlled at 81 mg/L as $CaCO_3$ (+7/-5 mg/L as $CaCO_3$) when the effect of sulfate was examined. pH varied little (pH 7.89-8.13 for the sulfate work, 7.87 –8.06 for chloride work) and was essentially constant.

Visual inspection of the general trend of copper release versus chloride concentration was slightly negative in that increasing chlorides appeared to reduce copper release. Visual inspection of the general trend of copper release versus sulfate concentration was slightly positive in that increasing sulfate appeared to increase copper release and was opposite to the general chloride trend. These trends were only visual and not strongly apparent, which limited any conclusion regarding the effects of sulfates and chlorides on copper release.

Statistical Copper Release Model

Empirical models for total copper release as a function of steady-state water quality were developed. The non-linear predictive models for the 50th and 90th percentile copper concentration are shown in equations (4.46) and (4.47), respectively. The range of the independent variables used to develop these models is shown in Table 4.2. Caution should be taken when applying these models outside the range of the independent variables used to develop these models.

$$[Cu]_{T-50} = Temperature^{0.72} \times Alkalinity^{0.73} \times pH^{-2.86} \times Sulfate^{0.10} \times Silica^{-0.22} \quad (4.46)$$

$$[Cu]_{T-90} = 0.28 + Temperature^{0.72} \times Alk^{0.73} \times pH^{-2.86} \times Sulfate^{0.1} \times Silica^{-0.22} \quad (4.47)$$

where: $[Cu]_{T-50}$ = total copper release corresponding to 50th percentile, mg/L
$[Cu]_{T-90}$ = total copper release corresponding to 90th percentile, mg/L
Alkalinity = mg/L as $CaCO_3$
Temperature = °C
pH = -Log $\{H^+\}$
Sulfate = mg/L
Silica = mg/L as SiO_2

A brief discussion of limitations of this type of empirical modeling effort was provided previously in the iron section to guide application of the resulting empirical models. The procedure for selection and/or removal of variables from the final model was an analysis of variance (ANOVA). Two forms of a model were compared: a version that included a potential variable in addition to the shared variables, and a version that included all variables (the shared variables) except the potential variable. If the expanded model, including the potential variable, achieved a statistically significant improvement in ability to describe the data relative to the model with shared variables only, then the variable was judged to be significant and would remain in the model. Thus, inclusion (or exclusion) of variables from the final model would be based on criteria other than a single goodness-of-fit statistic such as the correlation coefficient. This approach is fundamentally different from a mechanistic model that would presume the significance of variables in advance of parameter estimation.

The empirical approach may be useful to suggest mechanistic relationships, but the presence of confounding effects between study variables may obscure proper identification of such mechanistic relationships. The present study mandated examination of three specific water sources, with the result that the dataset contained an inherent correlation between certain water quality parameters (Table 4.3). Limitations associated with field operations precluded independent absolute control of all study variables (for example, temperature and associated solubility of dissolved oxygen). Although these factors may hinder a mechanistic interpretation of the data, it should be noted that the exact same type of confounding issues will be present in the prototype distribution system when varying blend ratios of the three principal sources are present. This approach would be expected to provide a satisfactory description of events that would be experienced in the Member Governments' distribution systems when subjected to variation in water quality associated with blends of groundwater, surface water, and desalination sources.

Application of any empirical model should recognize the range of values of the study variables in the dataset that was used to develop the empirical model. For the present study, a wide range of water qualities were provided (Table 4.2) as feed to the hybrid PDSs. This robust data represented the full spectrum of potential blends that could be experienced in the Member Governments' distribution systems, ranging from 100% of any single source to various blends of multiple sources. Neverthesss, use of the empirical model to describe water quality not represented by the data would be inappropriate. Examples of such an inappropriate use of the model would include description of anaerobic systems, unstabilized waters (with respect to calcium carbonate solubility), and waters with a low pH relative to the data range defined in Table 4.2.

The model predicted copper concentration increase with increasing alkalinity, temperature, and sulfate. Increasing pH and silica were observed to decrease copper concentration for the conditions of this investigation. As noted previously, these predictive equations should be used with caution outside the experimental range of the independent variables in equations (4.46) and (4.47).

These variables undoubtedly affect dissolved copper concentration but their effect on total copper is difficult to theoretically predict because of the effects of particulate copper. The empirical model did identify the same water quality parameters that affected the thermodynamic models previously described. Consequently, these non-linear models are representative of fundamental copper release phenomena. The predicted average copper release using the empirical model (equation (4.46)) versus the average actual copper released is shown in Figure

4.38. The coefficient of correlation (R^2) for predicted versus actual total released copper was 0.59. While R^2 is not high, the independent variables shown to increase copper in these models have been shown to increase copper in the TBW MG copper samples and in other national works, and can be used by the MGs to proactively control copper release.

The action level for copper and lead is evaluated based on 90th percentile data. Conversion of the average copper release (equation (4.46)) to the 90th percentile value (equation (4.47)) is achieved by assuming a normal data distribution. The data variability is characterized by a sample standard deviation (s). The 90th percentile value is calculated using the sample mean, the sample standard deviation, and the z-statistic per equation (4.48):

$$[Cu]_{T-90} = [Cu]_{T-50} + Z \times s \qquad (4.48)$$

where Z = Z statistic for conversion to 90th percentile, 1.28

S = Standard deviation Cu (mg/L)

The standard deviation (s) of the observations of total copper release for the conditions of this study was 0.22 mg/L. Hence, equation (4.48) is converted to equation (4.49)

$$[Cu]_{T-90} = [Cu]_{T-50} + 0.28 \qquad (4.49)$$

The 90th percentile copper release based on these observed standard deviations can be calculated from equation (4.47).

Model verification was done using data that was not used for model development. This data was collected over a six month period following the one-year period of data collection. The predicted versus the actual copper concentrations during the verification phase for the non-linear model is presented in Figure 4.39.

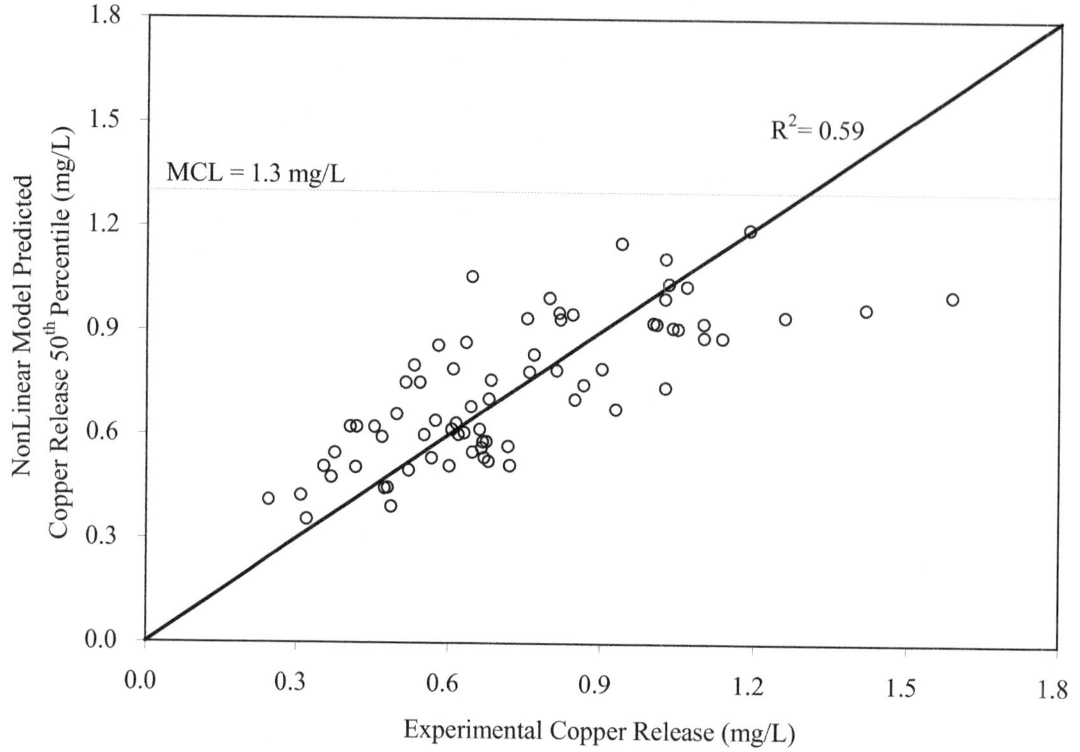

Figure 4.38 Actual versus predicted 50th percentile copper concentrations using non-linear model

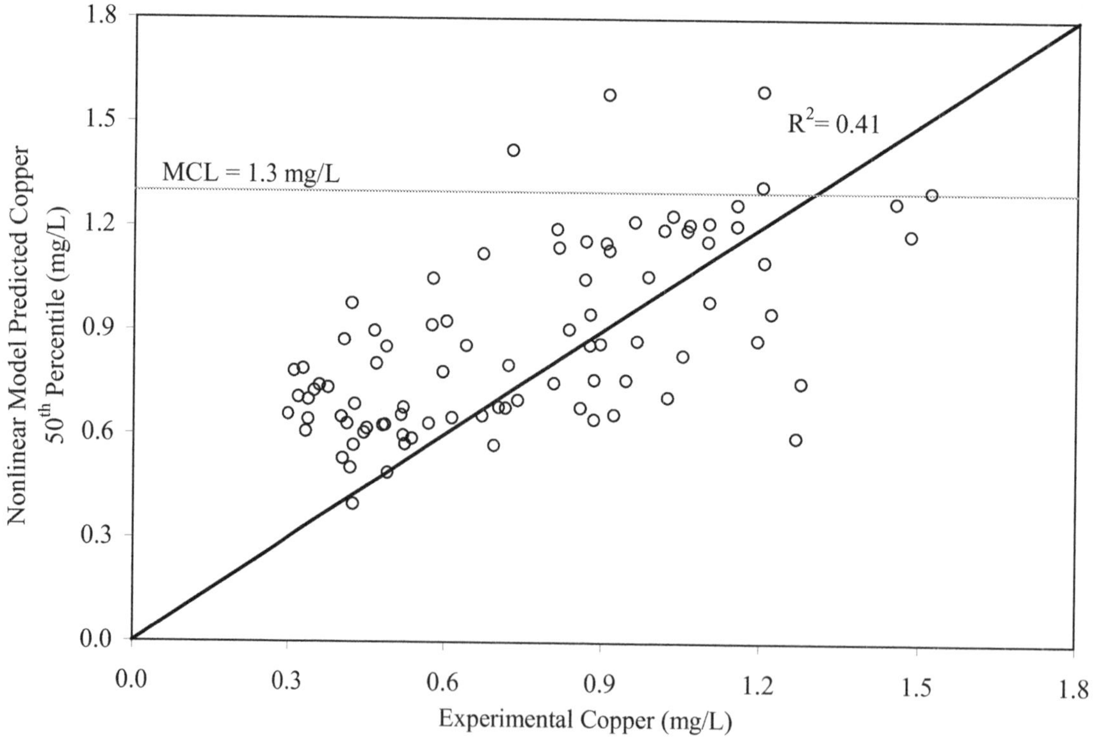

Figure 4.39 Verification plot of nonlinear model

Simulation of Copper Rule Compliance

The effect of blended water quality on copper concentration was assessed graphically for blends of GW, SW and RO. The resulting ternary plot is shown in Figure 4.40 using the 90th percentile copper model shown in equation (4.47). This plot provides a desktop estimation of 90th percentile copper concentrations for any combination of ground, surface or RO finished water based in the water quality in Table 4.22. Reading the Ternary plots can be difficult. Hence the procedure for reading the Ternary plot for total copper is presented.

- Within the ternary plot region, the summation of three source water ratio is 1.0, and any point represents a unique blend.
- Assume a blend of 62 % GW, 27 % SW and 11 % RO.
- Locate 0.62 on the GW Ratio axis.
- Draw a line parallel to the SW axis that passes through the 0.62 GW Ratio point. (Note the parallel axis will always be the first axis that is clockwise to the axis containing the blend ratio point. Hence, SW is clockwise from GW and is the parallel axis of reference for this line.).
- Locate 0.27 on the SW Ratio axis.
- Draw a parallel line to the RO Ratio axis that passes through the 0.27 SW Ratio point. The intersection of these two lines defines the predicted 90th percentile copper concentration for this blend using the linear model to be 1.29 mg/L, which violates the copper action level.
- Locate 0.11 on the RO Ratio axis.
- Draw a parallel to the GW axis through the 0.11 RO Ratio axis point. (The intersection of this line passes through the 1.29 mg/L point found in step 6 and confirms the graph was used correctly).

This procedure can be used to develop and predict singular water quality parameters for any combination of blended water quality from three sources given the models for prediction of water quality are available. The desktop technique can easily be used by engineering scientific, operational, regulatory and managerial personnel to proactively manage water quality.

Table 4.22
Water quality range for development of ternary diagram

Variable	Unit	Range	Concentration (mg/L)
Alkalinity	mg/L as $CaCO_3$	Maximum	225
		Minimum	50
pH		Maximum	8.30
		Minimum	7.60
Temperature	°C	Maximum	35
		Minimum	15
Sulfate	mg/L	Maximum	120
		Minimum	10
Silica	mg/L	Maximum	15.05
		Minimum	0.59

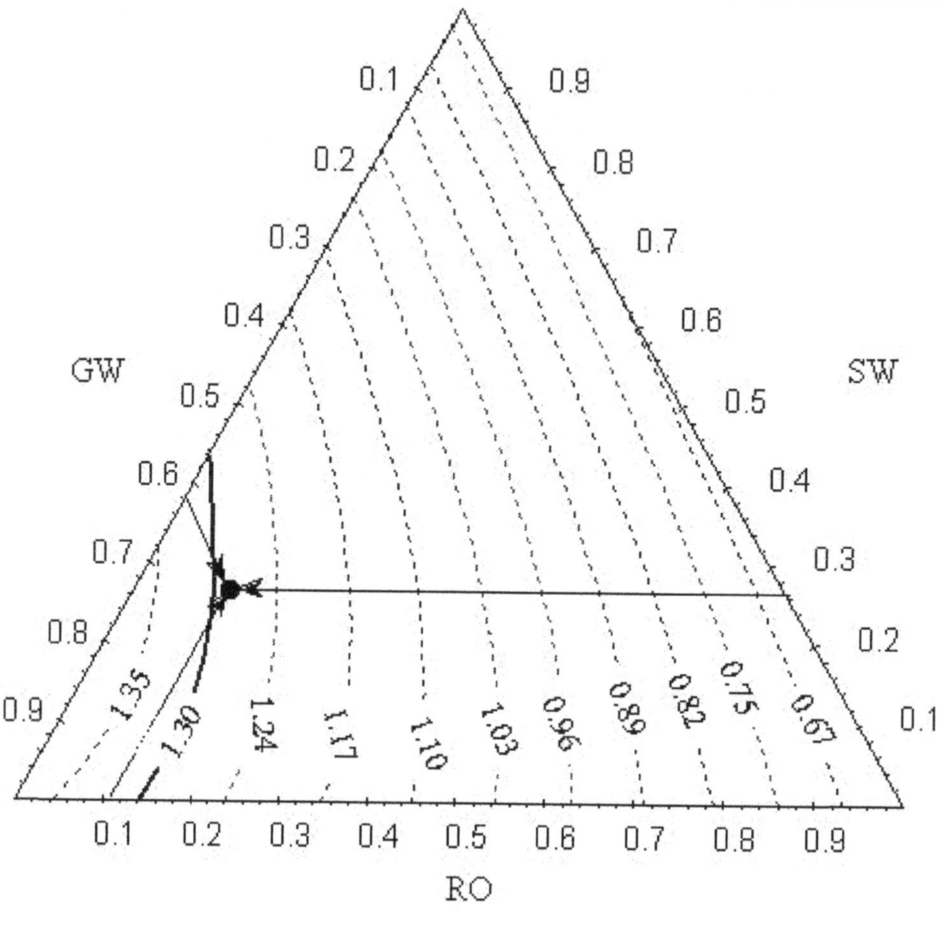

Figure 4.40 Ternary diagram showing predicted 90th percentile copper concentration for blends of finished ground, surface and RO water using non-linear model

Sensitivity Analysis for Copper Model

The 90th percentile copper release model, as a function of feed water chemistry, has been described previously in equation (4.47). The sensitivity of each parameter shown in equation (4.47) depends on the initial feed water quality as described by the nonlinear relationship with individual water quality parameters. For instance, during the study, GW had maximum alkalinity and low to moderate sulfate content. Conversely SW had maximum sulfate and low alkalinity. Therefore, field evaluation of a high sulfate, high alkalinity water was not possible based on blends of these two sources. Though the empirical model was not developed from such extreme water chemistries, quantitative predictions of total copper release can be made for those extremes using the empirical model.

Three single water sources GW, SW, and RO were used as baselines to evaluate the sensitivity of copper release model to water quality changes permitted by Exhibit D, the water quality agreement among TBW and the MGs. Water quality used in steady-state model is listed in Table 4.14. The sensitivity was evaluated by changing a particular parameter for a given source, while keeping the other parameters constant.

Figure 4.41 shows the change in copper release for the entire range of alkalinity for the three source waters listed in Table 4.14. RO never violates the action level for copper at any alkalinity to 250 mg /L as $CaCO_3$. Total copper release will violate the action level at alkalinities exceeding 217 mg/L as $CaCO_3$ in GW, and 166 mg/L as $CaCO_3$ in SW. The action level for copper is violated at a lower alkalinity in SW than in GW because of the impact of sulfate concentrations in SW (180mg/L) relative to GW (10mg/L).

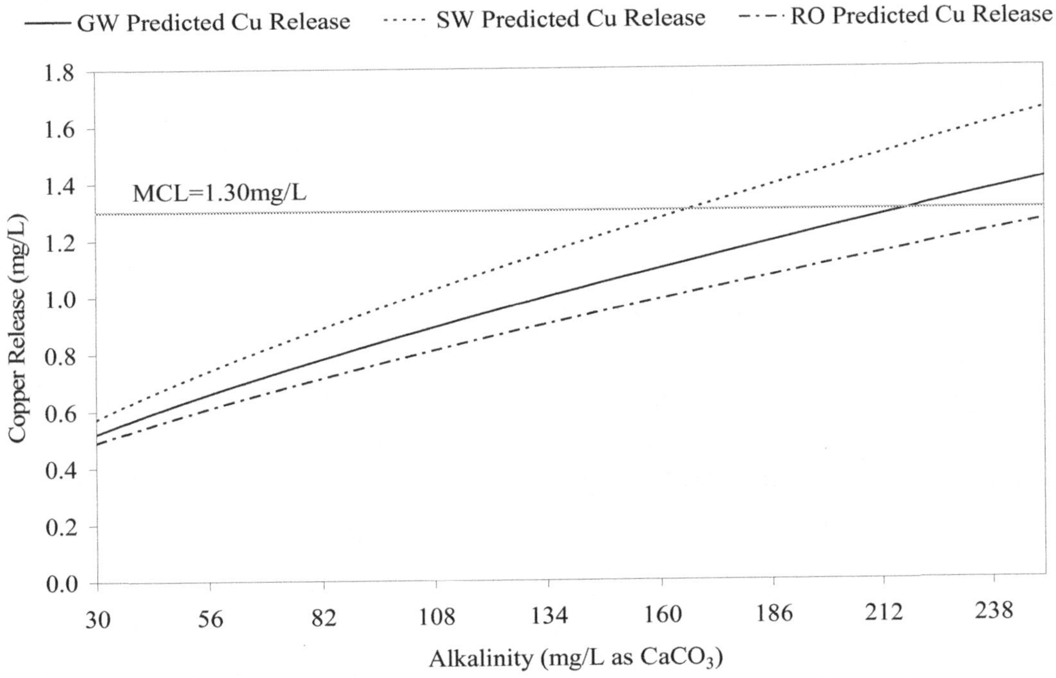

Figure 4.41 Sensitivity of copper release to alkalinity

Total copper release for varying pH is shown in Figure 4.42 for the expected pH range of the three source waters. As shown in Figure 4.42, pH will not cause a violation of the copper action level for the expected ranges for RO and SW. However, pHs of 7.67 or less in GW will cause violations of the copper action level. The change of total copper released with respect to pH is not linear, and will increase more rapidly at lower pH. Sulfate concentrations are high in SW (180 mg/L) but since alkalinity is relatively low, sulfates do not cause violation of the copper action level in SW or RO finished water

As shown in Figure 4.43, more than 8 mg/L of sulfates will cause the total copper release to exceed the 1.30 mg/L action level in GW. Sulfate in GW is approximately 10 mg/L. Based on the baseline GW quality, GW would violate the copper action level at temperatures greater than 25°C (71°F). Silica retards total copper release. The steady state copper model predicts that maintaining a silica concentration in GW higher than 22.5 mg/L as SiO_2 will prevent violation of the copper action level. Because of the low alkalinity in surface and RO water, silica variations do not significantly affect violation of the copper action level in these sources.

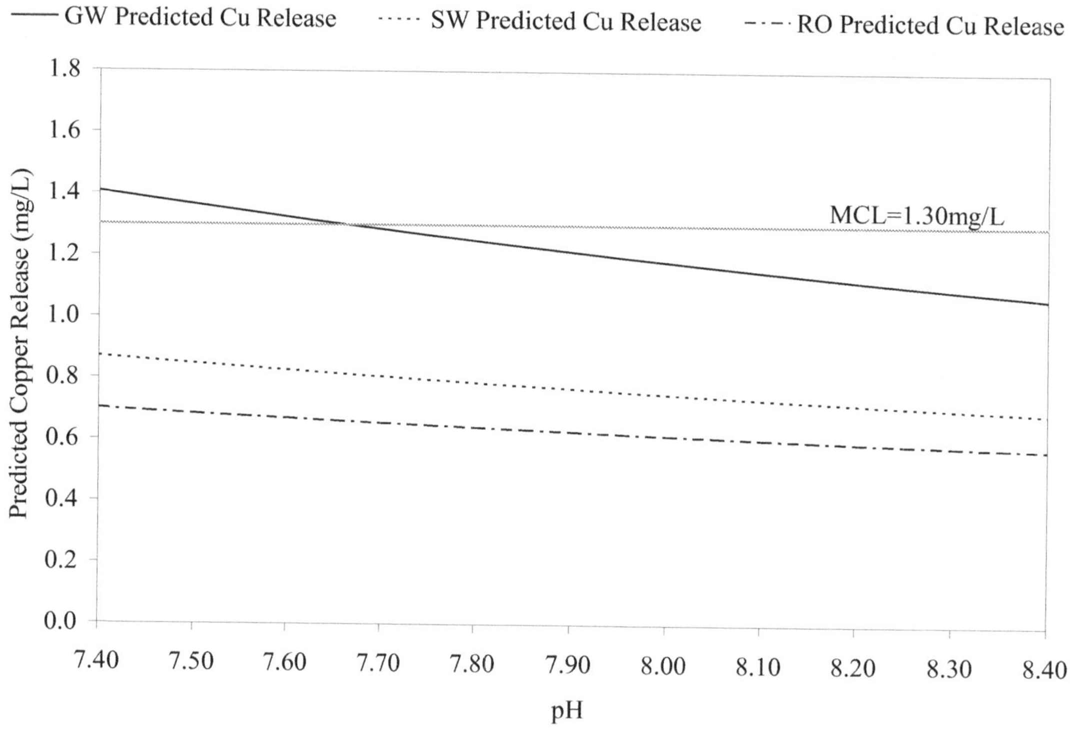

Figure 4.42 Sensitivity of copper release to pH

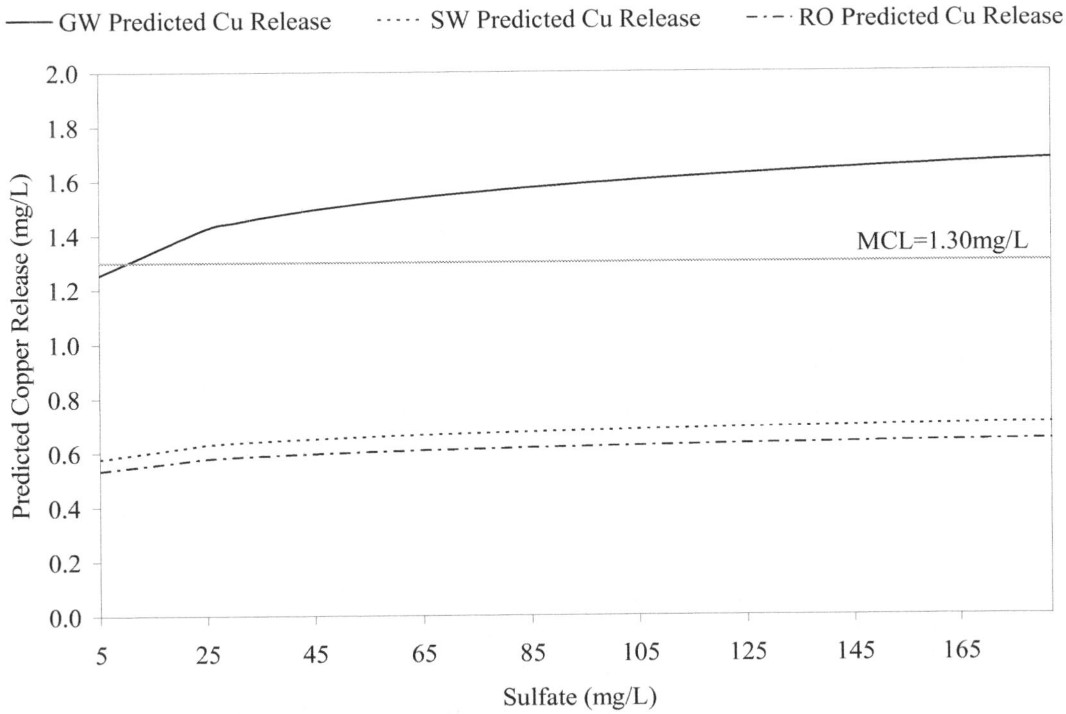

Figure 4.43 Sensitivity of copper release to sulfate

Summary for Copper Release

In this work, eighteen copper loops were installed connecting with eighteen PDS lines individually to simulate the kitchen plumbing system for LCR compliance study. A thermodynamic model was developed to assist evaluating the significance of water quality parameters, and experimental copper release was statistically modeled as a function of water quality to investigate the blending water quality effects in the PDS, and a Ternary Diagram was developed to help assess the potential copper release from blending treated ground, surface, and RO waters.

Experimental data and empirical models developed from non-linear regression showed the release of total copper depended on alkalinity and pH, and also to a lesser extent on sulfates, silica and temperature for the conditions of this investigation. Increasing alkalinity will produce more total copper release than any other water quality parameters; high alkalinity is typically observed in the regional groundwaters. Additionally, pH_s decreases as alkalinity increases, which also increases total copper release if waters are stabilized. Hence, the copper release model can be used to anticipate the total copper release from various waters, including the blended waters.

High concentrations of sulfate increased total copper release, but not as much as high alkalinities and low pH for the water quality ranges of this study. The high sulfates were present in SW due to ferric sulfate coagulation, and sulfates increased in blended waters as the percentage of SW increased. Chloride did not significantly affect total copper release in this study. Total copper release decreased as silica increased, presumably because silica functioned as an inhibitor to prevent copper release, which might not be obvious because increasing alkalinity and silica both came from GW. However, normalization of silica (silica versus alkalinity) demonstrated the capability of silica to reduce copper release in this investigation.

A nonlinear statistical model was developed from one year of total copper release data that was collected for varying source water quality and season. The model found total copper released increased as alkalinity and temperature increased, and decreased as pH, sulfates and silica increased. This model was verified using data that was not used for model development. The model was used to develop a 90th percentile ternary diagram for total released copper that showed total copper released for finished waters composed of RO, SW and GW. The results indicate that GW will violate the copper action limit, and that SW and RO finished water will not violate the copper action limit at baseline water qualities. Blends that contain more than 86 % GW are predicted to violate the copper action. Reducing alkalinity in high alkalinity finished waters can mitigate release of total copper; however, this will increase iron release to some extent.

LEAD RELEASE

Lead release was investigated using the corrosion loops described in the Methods and Procedure Chapter. Actual lead release came from lead coupons that were carefully controlled for surface area exposure and were identical in each of the eighteen corrosion loops. Specifically, surface characterization of lead scale was done to investigate the controlling solid lead phase. Thermodynamic and empirical models for lead release as a function of bulk water quality at steady state were developed to gain an understanding of lead release in home plumbing systems. Finally, lead transition resulting from water quality change was modeled as a function of time and steady state total lead release for changing water quality.

Identification of Lead Corrosion Products

After a three month incubation period, the corrosion products formed on the lead coupon surface were identified by XRD and XPS. Figure 4.44 presents a typical XRD result of lead coupons incubated in G1 water. The XRD patterns indicate that $Pb_3(OH)_2(CO_3)_2$, $PbCO_3$, PbO, and PbO_2 were the main crystalline compounds of corrosion products. There were also Pb peaks from the substrate of coupon material. Among these corrosion products, $Pb_3(OH)_2(CO_3)_2$ and $PbCO_3$ peaks were more pronounced. It should be noted that a large amount of amorphous phases were present in the corrosion products and increased the background noise, which made it difficult to analyze quantitatively.

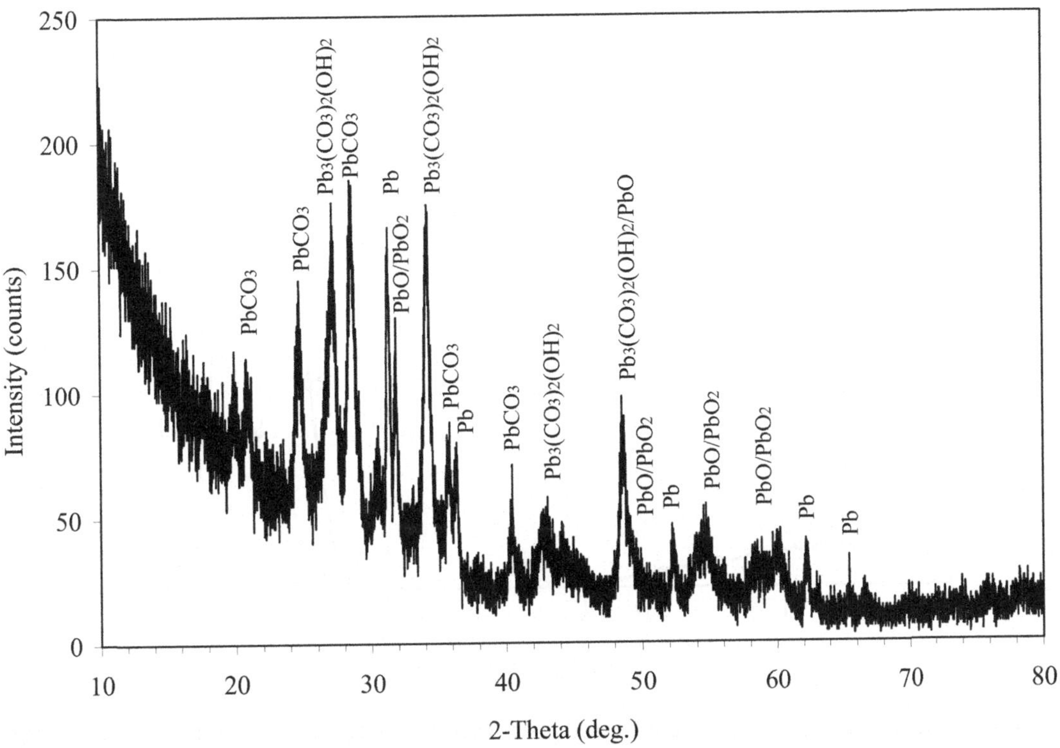

Figure 4.44 Representative XRD patterns of corrosion products on lead coupons exposed to GW (G1) water environment for three months

The XRD results were verified by XPS scans. XPS survey scans of lead coupons indicated that the main elements in corrosion layers included Pb, C, O, Na, N, Cl, and Si. Detailed high resolution scans and deconvolution of main element peaks were performed to identify chemical structures of lead corrosion scales. As shown in Figure 4.45, XPS identified the main corrosion products in the outer scale layers as PbO, PbO_2, $Pb_3(OH)_2(CO_3)_2$, and $PbCO_3$. The results agreed very well with XRD observations.

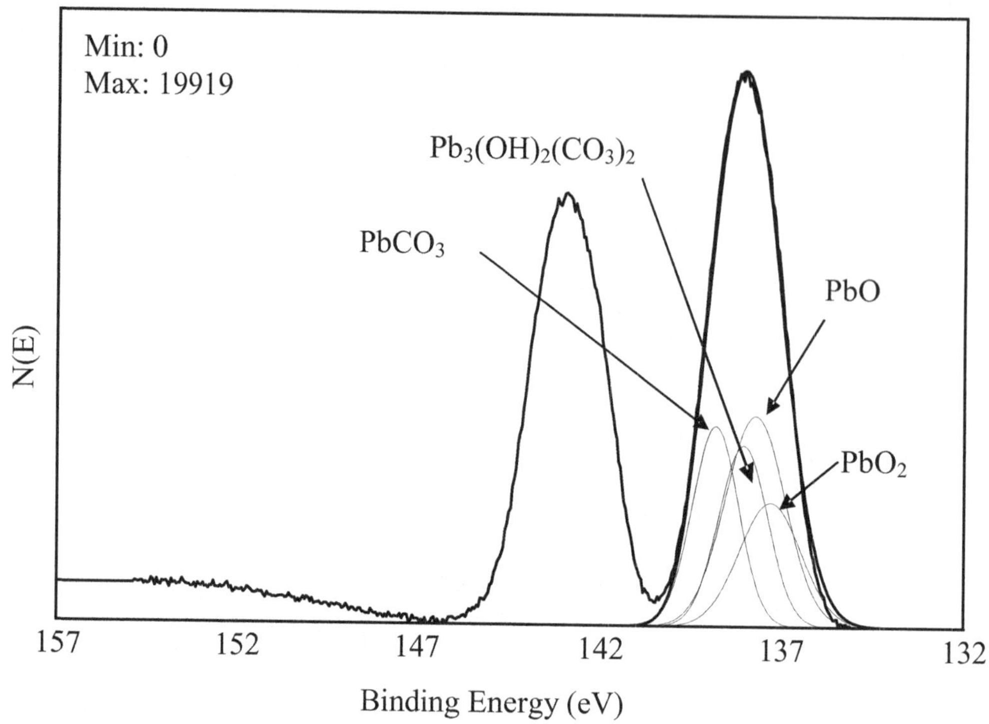

Figure 4.45 Representative high-resolution and the deconvolution result of Pb(4f7) peaks of XPS scan of corrosion products formed during 3-month incubation under GW (G1) source water

Correlation between Surface Characteristics and Lead Release

The relationship between surface scales and lead release was established through XPS analysis. Specifically, the lead peak area % obtained from deconvolution curve fitting of Pb(4f7) peaks was correlated to lead release observed for six blended waters. In addition, the peak area % of lead species was plotted against lead concentrations of six blended waters. As shown in Figure 4.46, a positive linear relationship ($R^2 = 0.73$) was found between lead release and peak area % of $Pb_3(OH)_2(CO_3)_2$. Higher area % of $Pb_3(OH)_2(CO_3)_2$ corresponded to higher lead release meaning larger surface coverage by $Pb_3(OH)_2(CO_3)_2$ resulted in higher lead release. This observation suggested that $Pb_3(OH)_2(CO_3)_2$ was the solid phase controlling lead release through dissolution/precipitation.

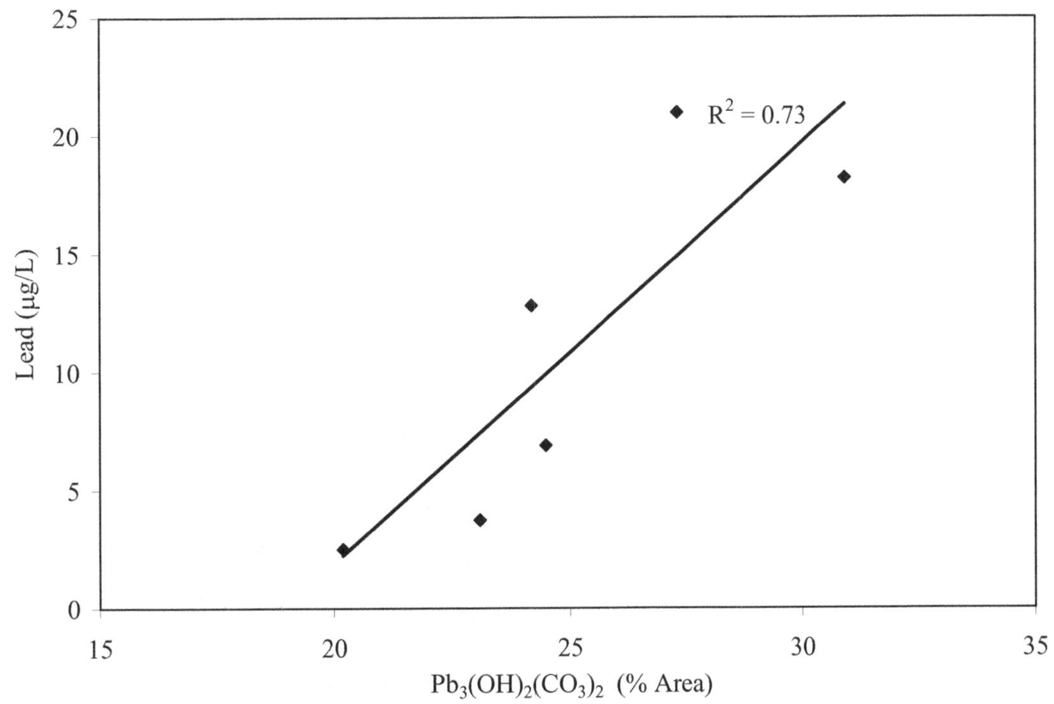

Figure 4.46 Total lead release versus XPS identified $Pb_3(OH)_2(CO_3)_2$ peak area percent

Thermodynamic Model Development for Lead Release

In the previous sections, it was demonstrated that hydrocerussite ($Pb_3(OH)_2(CO_3)_2$) might be the main corrosion product determining lead concentrations in drinking water. Therefore, a thermodynamic equilibrium model was developed assuming $Pb_3(OH)_2(CO_3)_2$ was the solid phase controlling lead release. Dissolved lead concentration was first calculated assuming chemical equilibrium was established between lead ions and the controlling solid phase (i.e. $Pb_3(OH)_2(CO_3)_2$). Then, the concentrations of other dissolved complexes (e.g. hydroxyl, carbonate, chloride, and sulfate) were estimated based on the related reactions. The total lead concentration in water (DIPb) was determined by summing dissolved lead ions and complexes as follows:

$$\begin{aligned} DIPb = {} & [Pb^{2+}] + 4[Pb_4(OH)_4^{4+}] + 2[Pb_2OH^{3+}] + [PbOH^{+}] + 3\{Pb_3(OH)_4^{2+}] \\ & + 6\{Pb_6(OH)_8^{4+}] + [Pb(OH)_2^{0}] + [Pb(OH)_3^{-}] + [Pb(OH)_4^{2-}] + [PbCO_3^{0}] \\ & + 3[Pb_3CO_3^{4+}] + 2[Pb_2CO_3^{2+}] + [Pb(CO_3)_2^{2-}] + [PbHCO_3^{+}] + [PbSO_4^{0}] \\ & + [Pb(SO_4)_2^{2-}] + [PbCl^{+}] + [Pb(Cl)_2^{0}] + [Pb(Cl)_3^{-}] + [Pb(Cl)_4^{2-}] \end{aligned} \tag{4.50}$$

The lead concentrations predicted by the developed models are presented as a function of pH and alkalinity in Figure 4.47. The $Pb_3(OH)_2(CO_3)_2$ model shows that increasing pH decreases lead release. The model also shown when alkalinity is less than 75 mg/L as $CaCO_3$, increasing alkalinity decreases lead release. However, for higher alkalinity (>100 mg/L as $CaCO_3$), lead release increases with increasing alkalinity.

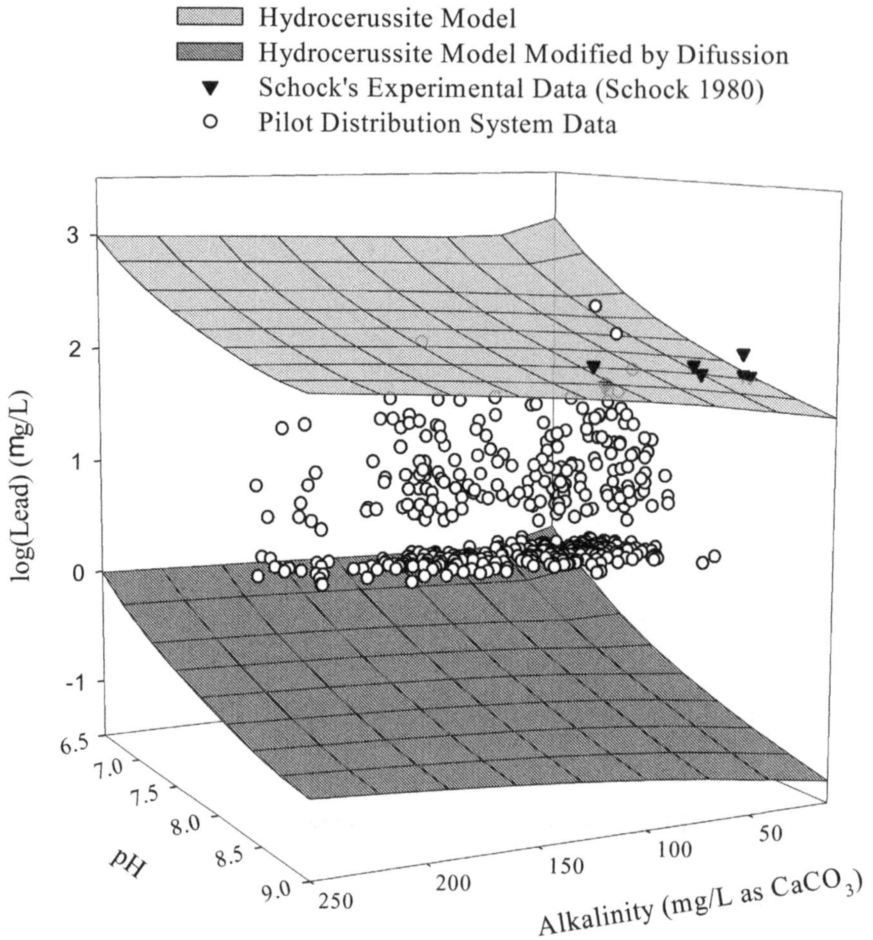

Figure 4.47 Lead–Alkalinity–pH diagram for hydrocerussite ($Pb_3(OH)_2(CO_3)_2$ model

In Figure 4.47, the thermodynamic model predictions are compared with the lead concentrations observed during the first year of the pilot study. In most cases, the values predicted by theoretical models were much higher than actual lead measurements. However, experimental results reported by Schock (1980, 1983), presented as triangles in Figure 4.47, showed a good agreement with the lead levels predicted by the hydrocerussite model in this study. This finding may suggest that the developed theoretical model can provide maximum lead concentrations in the drinking water as a function of water quality when equilibrium is achieved. In this study, a small lead coupon was used in the long copper tubing, and equilibrium could not be reached for the entire water volume. In this case, dissolved lead concentration was controlled by diffusion. Based on diffusion theory, C_x (lead concentration at the location which has distance x from the lead source) can be calculated by equation (4.51).

$$C_x = C_0 (1 - erf \frac{x}{\sqrt{4 D_{eff} t}}) \tag{4.51}$$

The copper tubing used in this study was 9 m (30 ft) long, which contained 1.8 L of water. A lead coupon was located near the outlet of the tubing. A 1L sample was taken, which corresponded to the volume of 5-meter tubing, thus the maximum distance, x, was 5 meters. The stagnation time, t, was six hours. Note that D_{eff} is the diffusion coefficient that is 10^{-9} m^2s^{-1} for lead. Based on experimental configurations and diffusion theory, the lead concentration of a 1 L sample was estimated to be 0.001 C_0 (lead concentration at the lead source, that is the equilibrium concentration predicted by the thermodynamic model described previously) and was presented in Figure 4.47 as the bottom plate. The resulting lead concentrations were 0.2 μg/L for typical pilot conditions, which were consistent with the majority of pilot observations that were below the detection limit of AAS (5μg/L).

The thermodynamic model predicts a small lead increase when sulfate increases, but our pilot experiments showed the opposite trend. Oliphant (1983) reported that lead corrosion was decreased by sulfate. Currently there is no clear explanation for this phenomenon. The speculation is that sulfate may alter the physical structure of lead corrosion scales, which help prevent lead from corroding. SEM scan pictures supported this hypothesis. The flat and compact corrosion scales were observed for lead surface exposed to S1 water, while rough and needlelike structures were found on the lead surface incubated under RO water. The flat and compact scales may prevent lead from corroding, while lead particulates can get easily released from rough surfaces. The release of lead particulate also explains why lead concentration had high variability.

Effects of Water Quality on Lead Release

Blending of different source waters caused significant changes in distribution system water quality. The alteration of water quality in distribution systems imposes potential for lead release in drinking water by creating chemical conditions favorable for lead corrosion, and/or by disrupting chemical film formed under historical ground water quality.

Controlled Study of Water Quality Effects on Lead Release

The effect of blended water quality on lead release was investigated for confounding effects of sulfates, bicarbonate alkalinity and chloride. A series of experiments were performed to address confounding effects as described in the Empricial Model Verification section for the steady state iron model. The water quality for those studies was reported in Table 4.6, Table 4.7 and Table 4.8. In this study, only one water quality parameter was varied systematically while others were kept constant. The changes of chlorides, alkalinity and sulfates were accomplished by varying blends of GW, SW, RO and RO permeate without the addition of sea salt, and by adding sulfuric acid and sodium bicarbonate. Chloramines were used for residual maintenance during the studies of water quality effects on lead release. The variation of chloride levels was achieved by adjusting sea salt added to RO water.

As shown in Figure 4.48, increasing sulfates decreased lead, while increasing chlorides increased lead. These observations suggest that lead concentration will increase as RO blending increases, since finished water produced by RO has the highest chloride content. Conversely, lead concentrations in finished water produced from surface water (S1) are predicted to have lower lead content because of the additional sulfates from coagulation.

In this study, higher alkalinity was related to lower pH because all water were stabilized with respect to $CaCO_3$ before discharge to the PDSs. Alkalinity was highest in finished groundwater, G1, which had the lowest pH. Chloride concentration was highest in RO finished water, and sulfate concentration was greatest in S1, which was produced by ferric sulfate coagulation of SW. The range of chloride and alkalinity was three to four fold, while the range of sulfate concentration was ten fold. Finished water quality was dependent on source and treatment process, and was investigated for confounding effects on lead release.

As shown in Figure 4.49, lead concentration was reduced as pH and alkalinity increased. pH and alkalinity varied simultaneously because sulfuric acid and sodium bicarbonate were added to control alkalinity, which affected pH. Consequently, the alkalinity effect was difficult to demonstrate. Schock (1999) presented lead equilibrium diagrams as a function of pH and dissolved inorganic carbon assuming $PbCO_3$ and $Pb(CO_3)_2(OH)_2$ as controlling solid phases. These diagrams showed soluble lead species decreased with pH and increased with alkalinity over smiliar pH and alkalinity ranges as used in this study. While the lead trend with alkalinity is consistent with the expected trend for soluble lead at equilibrium, the lead trend with pH is not. Lead release here includes particulate lead, which may have affected the observations of the confounding effects study.

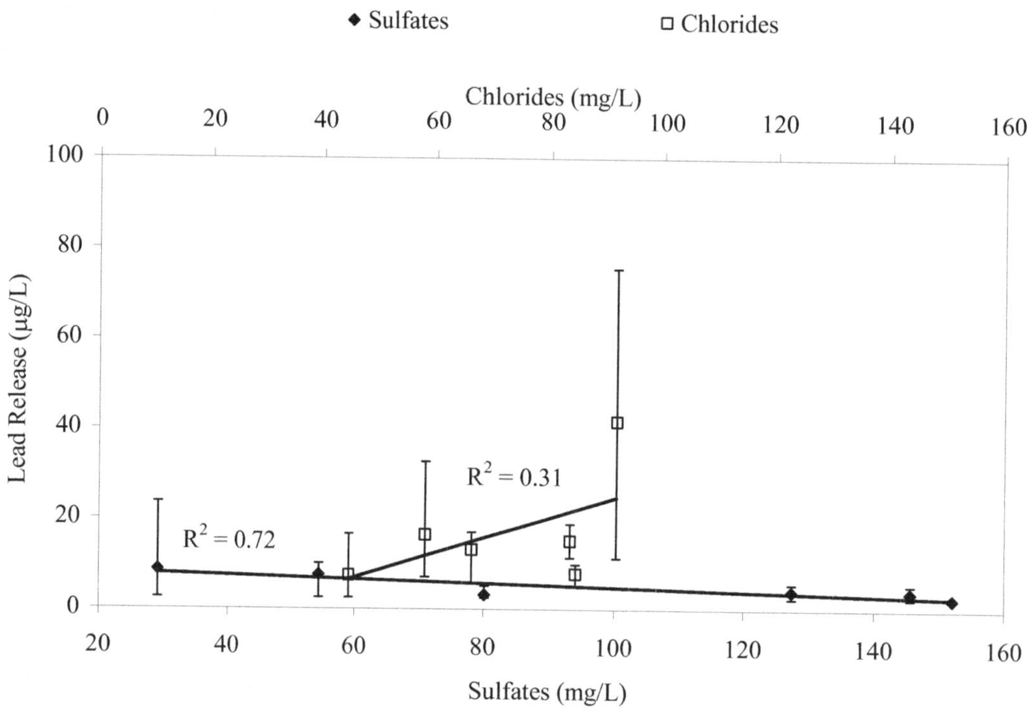

Figure 4.48 Effect of sulfate and chloride on maximum, minimum and average lead release

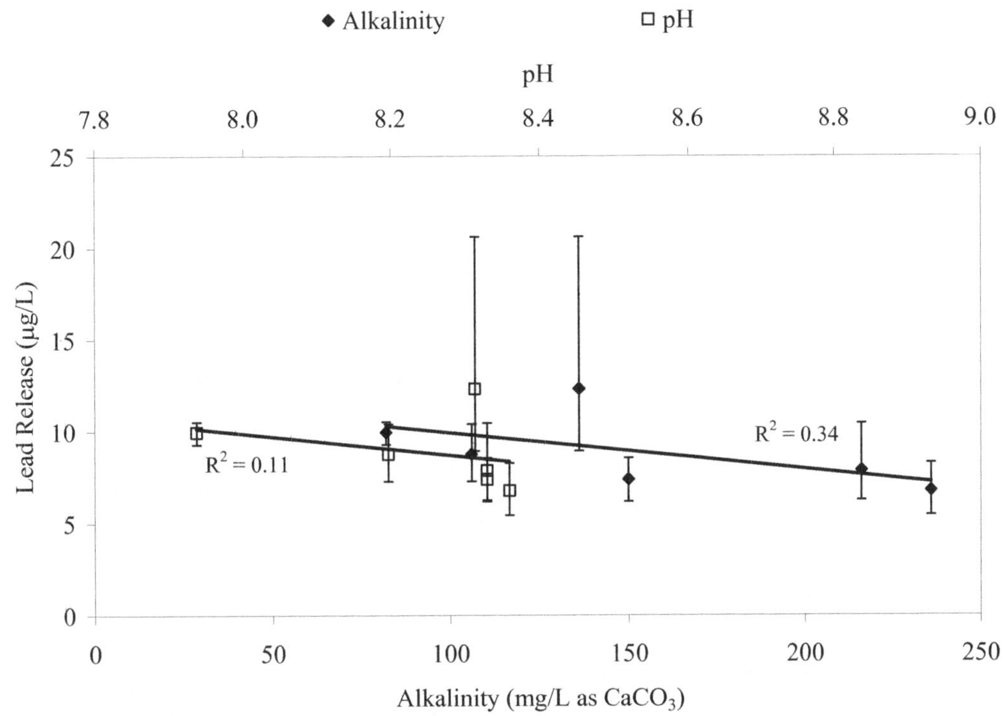

Figure 4.49 Effect of alkalinity and pH on maximum, minimum and average lead release in controlled experiments

Statistical Lead Release Model

Release of total lead corresponding to PDS water quality was regressed to develop empirical models for predicting the release of total lead for varying water quality. Regressions were conducted based on 90 percent data points chosen randomly. Seasonal average data was used for the one year pilot study and monthly average data was used for controlled study in the model development, because certain parameter measurements were missing for some data sets. According to pilot observations, ten major parameters were chosen for multiple regressions, e.g. alkalinity, pH, chloride, sulfate, DO, temperature, calcium, SiO_2, conductivity, UV-254, and chloramine. A variety of regression functions were examined and evaluated. The fiftieth and ninetieth percentile models shown in equation (4.52) and (4.53) were statistically significant to the 0.05 confidence interval and utilized the same water quality parameters as the thermodynamic model. The predicted variation of total lead release with water quality is the same as was observed in the confounding effects study except for alkalinity, which increased total lead release and was consistent with the predicted effect of alkalinity on soluble lead (Schock 1999).

$$[Pb]_{T-50} = 1.027^{(T-25)} \times Alkalinity^{0.677} \times pH^{-2.726} \times Chloride^{1.462} \times Sulfate^{-0.228} \quad (4.52)$$

$$[Pb]_{T-90} = 1.027^{(T-25)} \times Alkalinity^{0.677} \times pH^{-2.73} \times Chloride^{1.46} \times Sulfate^{-0.23} + 8.96 \quad (4.53)$$

where $[Pb]_{T\text{-}50}$	=	total lead release corresponding to 50th percentile (μg/L)
$[Pb]_{T\text{-}90}$	=	total lead release corresponding to 90th percentile (μg/L)
Alkalinity	=	mg/L as $CaCO_3$
Chloride	=	mg/L
Sulfate	=	mg/L
T	=	Temperature, °C

A brief discussion of limitations of this type of empirical modeling effort was provided previously in the iron section to guide application of the resulting empirical models. The procedure for selection and/or removal of variables from the final model was an analysis of variance (ANOVA). Two forms of a model were compared: a version that included a potential variable in addition to the shared variables, and a version that included all variables (the shared variables) except the potential variable. If the expanded model, including the potential variable, achieved a statistically significant improvement in ability to describe the data relative to the model with shared variables only, then the variable was judged to be significant and would remain in the model. Thus, inclusion (or exclusion) of variables from the final model would be based on criteria other than a single goodness-of-fit statistic such as the correlation coefficient. This approach is fundamentally different from a mechanistic model that would presume the significance of variables in advance of parameter estimation.

The empirical approach may be useful to suggest mechanistic relationships, but the presence of confounding effects between study variables may obscure proper identification of such mechanistic relationships. The present study mandated examination of three specific water sources, with the result that the dataset contained an inherent correlation between certain water quality parameters (Table 4.3). Limitations associated with field operations precluded independent absolute control of all study variables (for example, temperature and associated solubility of dissolved oxygen). Although these factors may hinder a mechanistic interpretation of the data, it should be noted that the exact same type of confounding issues will be present in the prototype distribution system when varying blend ratios of the three principal sources are present. This approach would be expected to provide a satisfactory description of events that would be experienced in the Member Governments' distribution systems when subjected to variation in water quality associated with blends of groundwater, surface water, and desalination sources.

Application of any empirical model should recognize the range of values of the study variables in the dataset that was used to develop the empirical model. For the present study, a wide range of water qualities were provided (Table 4.2) as feed to the hybrid PDSs. This robust data represented the full spectrum of potential blends that could be experienced in the Member Governments' distribution systems, ranging from 100% of any single source to various blends of multiple sources. Neverthesss, use of the empirical model to describe water quality not represented by the data would be inappropriate. Examples of such an inappropriate use of the model would include description of anaerobic systems, unstabilized waters (with respect to calcium carbonate solubility), and waters with a low pH relative to the data range defined in Table 4.2.

The statistical models indicate that increasing pH and sulfate will decrease lead release, while high alkalinity and chloride content and conductivity will increase lead release. The models indicate a similar trend for main parameters as the controlled study, except for alkalinity. It should be noted there were correlations between pH and alkalinity in the controlled study since both of them varied with addition of acid or base: increasing pH came with increasing alkalinity, while low pH corresponded with low alkalinity. So the effect of alkalinity on lead release, shown in the controlled study, could have been caused by pH and the statistical model probably reveals the actual trends.

Using a total of 69 water quality data points, that were not used in model development, the applicability of the statistical model was verified and showed that the model agreed well with the actual data.

The model for total lead release (equation (4.52)) was developed using average total lead concentrations and average water quality concentrations by phase and PDS. The model for total lead release (equation (4.52)) was adjusted to define the 90th percentile concentrations (equation (4.53)) in the same manner as previously described for the steady-state total copper release model.

Calibration of the Total Lead Release Model

The corrosion loops were constructed with copper piping material and an inserted lead coupon. The lead coupon was sized to simulate input of lead from household plumbing fixtures and soldered joints. The actual lead coupon surface area was 21.8 cm^2 (3.38 in^2). This surface area may not accurately reflect lead impacts from all individual member governments' customers due to differing construction methods and materials. It would be expected that the realized lead concentrations would be proportional to the lead surface area. Adjustment of the predicted lead concentrations is proposed to match actual member government historical monitoring data as described below. Specific data requirements for this model calibration would include historical lead concentration data (mean, standard deviation, and 90th percentile value) and water quality data for all parameters in the non-linear empirical lead prediction model (equation (4.53)). The relevant parameters are alkalinity, pH, sulfate, chloride, and temperature.

The empirical model would be used to predict a 90th percentile lead concentration for the specific member government water quality. This predicted total lead released would be representative of the lead coupon surface area used in the corrosion loops for this experimental study. The predicted lead value would be compared with the historical 90th percentile number defined by the specific member government monitoring program. Any discrepancy between these values may be partially attributable to differences in the lead surface area in the pilot study and member government households. Calibration of the predictive model (equation (4.53)) to the specific member government data is achieved by multiplication of the predicted value by the ratio of the measured 90th percentile value to the predicted 90th value. Assuming a historical monitoring 90th percentile value of 5 μg/L, the correction factor is calculated and applied as shown below.

$Pb_{90,Historical} = 5$ μg/L
$Pb_{90,Historical,predicted} = 18.7$ μg/L
Therefore correction factor (CF) = 5/18.7 = 0.27
Therefore, $Pb_{90,actual} = 0.27 * Pb_{90,predicted}$

This issue does not exist with copper because copper pipe was used in the corrosion loops and is used in MGs consumer homes. Hence, the area for copper pipe is the same in both instances. Unfortunately, that is probably not the case for lead since lead pipe is not used in the MG distribution systems or in MGs consumer homes.

Simulation of Lead Rule Compliance

The 90th percentile statistical model was applied to simulate and predict lead release based on various water qualities caused by different blending ratios of three source waters. As shown in Figure 4.50, the model predicted low lead release at a high percentage of SW, while high percentages of RO and GW (especially RO water) were corresponding to high lead release. The predicted lead release ranged from 9.4 to 18.0 μg/L at 25°C,(77°F) indicating no lead rule violation. At high temperature, however, lead concentration was elevated. The highest lead release occurred with GW/RO ≈ 40/60% (18.0 μg/L at 25°C (77°F) and 20.8 μg/L at 35°C (95°F)), which has low pH and sulfate content but high chloride concentration.

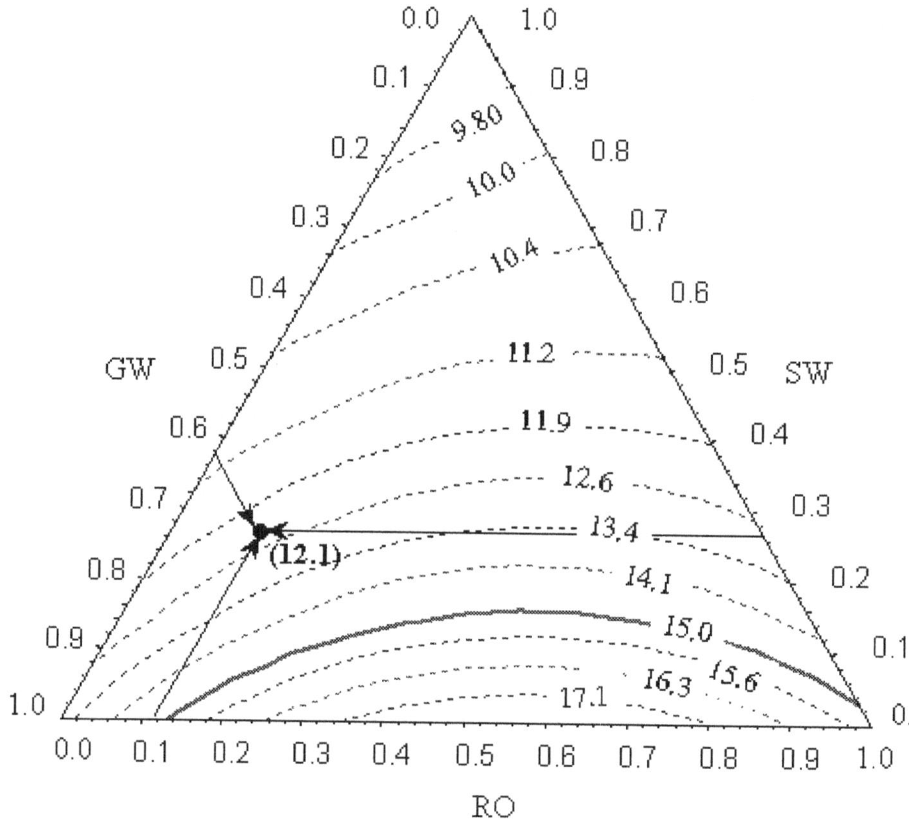

Note: Pick any G1, RO or S1 blend fraction on appropriate axis. Draw parallel line to the axis that is immediately clockwise to the selected reference axis and that passes through the selected reference blend. Pick any of remaining blend fractions and locate point on appropriate axis. Draw a line through this point that is parallel to the first axis that is clockwise from the axis that represents the second blend fraction. A third line can be which will intersect the intersection point of the previously drawn lines. The intersection of these lines represents the estimated lead concentration corresponding to that blend.

Figure 4.50 Simulation of lead release by the statistical model (25°C). Black point represents the lead concentration predicted by 62% GW, 27% SW, and 11% RO water blending

Sensitivity Analysis for Lead Model

The 90th percentile steady-state lead release model as a function of water quality has been described previously in equation (4.53). The sensitivity of total lead release to water quality can be determined by plotting lead release for a given parameter in equation (4.53) while holding other parameters constant. Lead release is sensitive to both alkalinity and chlorides of the source waters. GW has a high alkalinity and low chloride content, while RO water has low alkalinity and high chloride content. Blending constraints under actual field conditions did not permit the evaluation feed waters with high alkalinity and high chloride as these conditions could not be achieved with blends of the finished RO, SW and GW used in the project. However, extrapolation of the empirical model for lead release provides useful quantitative assessment of the effect of water chemistry changes in source waters on total lead release.

Figure 4.51 shows total lead release for varying alkalinity, given an assumed water quality for the three principal sources (GW, SW, and RO) reported previously in Table 4.14. As shown, increasing alkalinity to 250 mg/L as $CaCO_3$ will not cause a violation of the lead action level (15 µg/L) in finished ground or SW. However the lead action level is violated if 100 % finished RO water is used because of chlorides in RO finished water. This projection can be adjusted, if pure RO water was ever used, by the calibration method previously described.

As shown in Figure 4.52, variations of pH within the anticipated ranges of (7.2 – 9.0) of treated ground and SW will not cause a violation of the lead action level given the water qualities shown in Table 4.14. For RO finished water, violation of the lead action level will occur if pH is lower than 8.37. Although not shown, variations of temperature from 5 to 35°C will not cause a violation of the lead action level for SW; however, the total lead action level will be violated in 100 % GW and 100 % surface at the baseline water qualities shown in Table 4.14 at 33.5°C (92.3°F), and 24°C (75°F), respectively.

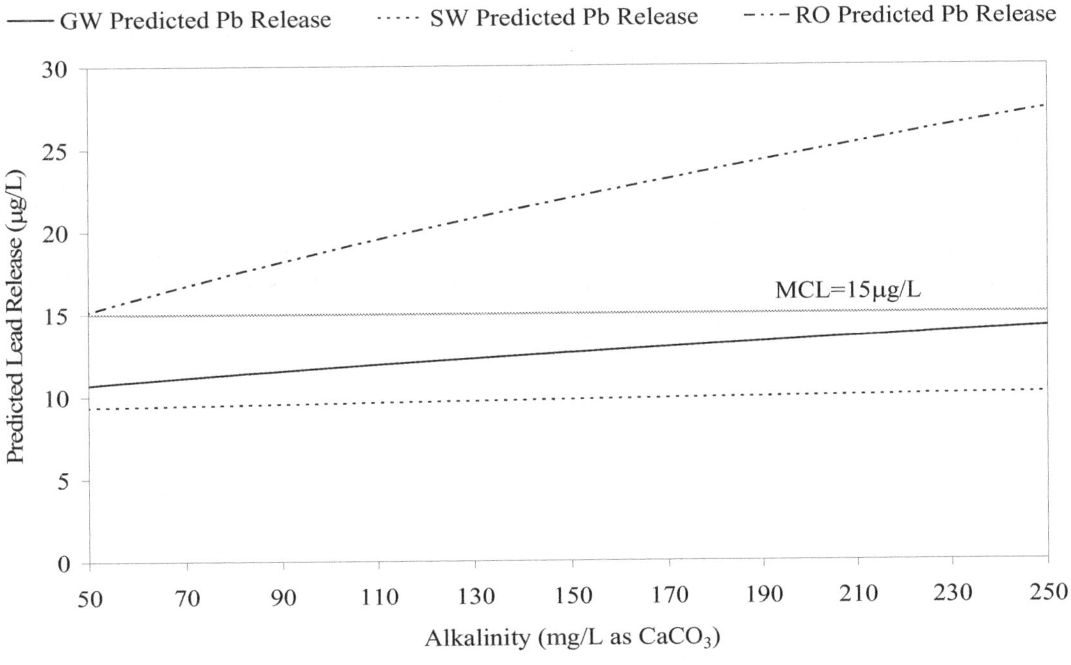

Figure 4.51 Sensitivity of lead release to alkalinity

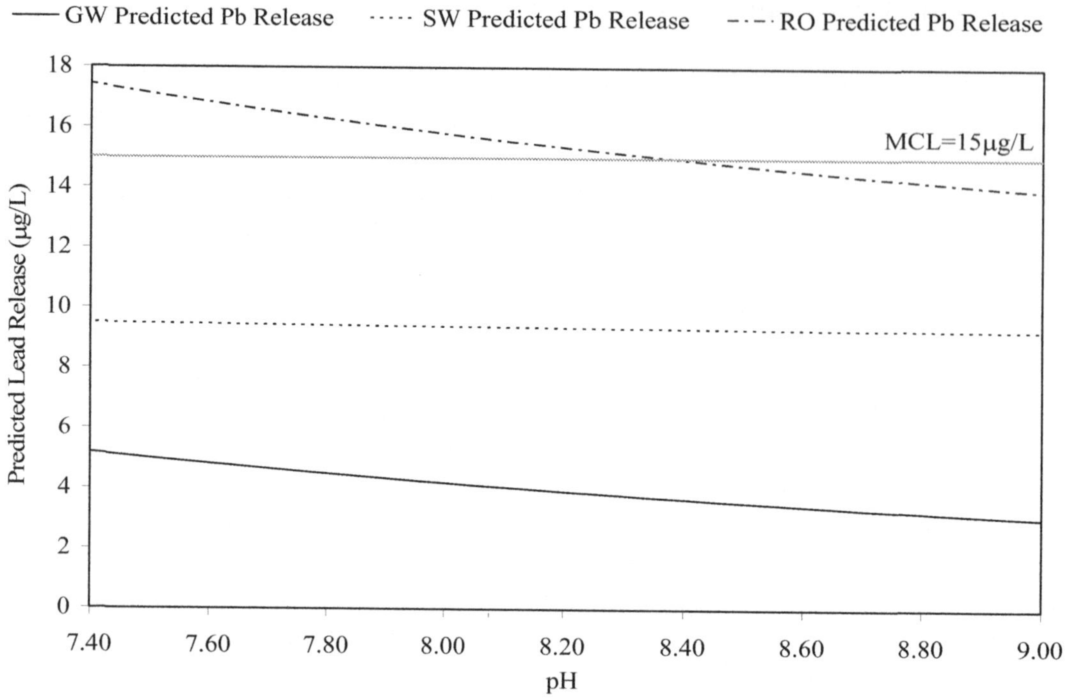

Figure 4.52 Sensitivity of lead with respect to pH

As shown in Figure 4.53, lead concentrations will violate the action level when chloride increases to 18 mg/L for GW, 49 mg/L for RO water, and 65 mg/L for SW. High chloride typically occurs in RO permeate because of the high concentration in the source. Blends of finished GW and RO water will potentially cause violation of the lead action level because of a high alkalinity and chlorides water quality. Hence, attention should be given to RO product water quality to maintain chloride concentrations at acceptable values to comply with the LCR.

Any consideration given for chloride concentrations in drinking water should also be accompanied by the consideration of the impact of chloride concentrations on lead release. Total lead release increases with increasing chlorides. However, as noted previously the lead release model developed in this work over predicted the MG government historical lead release. Consequently, the lead release models should be calibrated to the individual MGs 90th percentile lead concentrations as previously described before the impact of chlorides are considered. However, the potential of violation of the lead action level because of high chlorides in the RO finished water should not be dismissed.

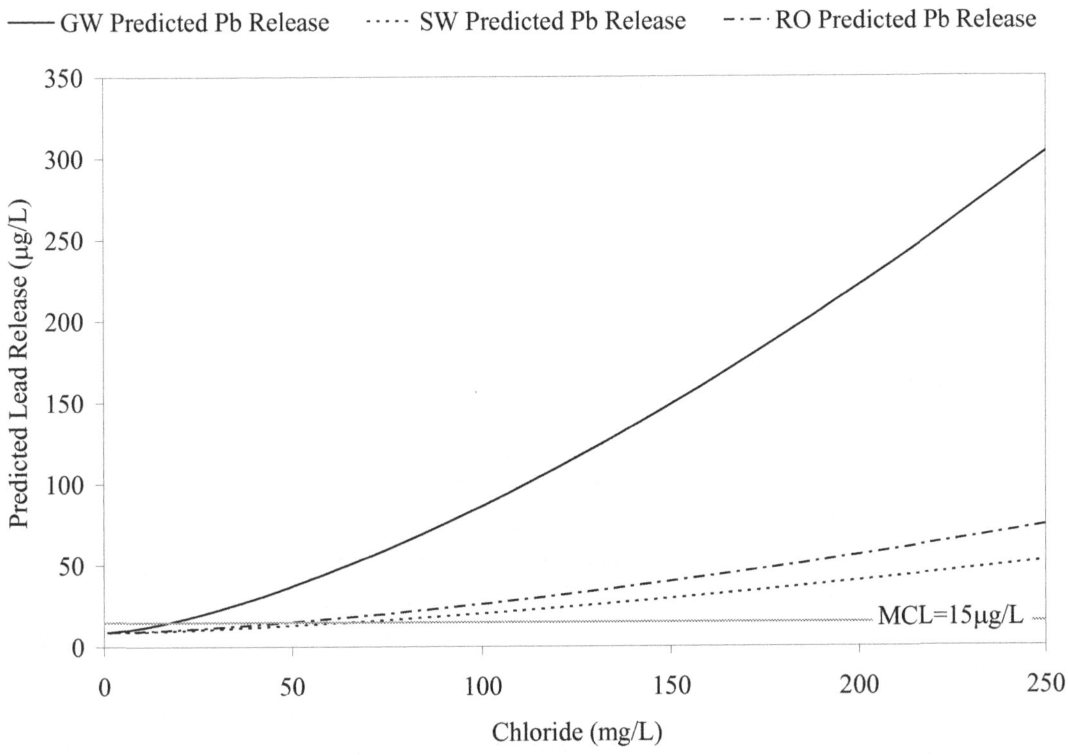

Figure 4.53 Sensitivity of lead with respect to chloride

Summary for Lead Release

In this work, eighteen copper loops were installed connecting with eighteen PDS lines individually to simulate the kitchen plumbing system for LCR rule compliance study. In each of these copper loops, one lead coupon was incubated according to the specific lead area (relative to unit water volume). A thermodynamic model was developed to assist evaluating the significance of water quality parameters, and experimental lead release was statistically modeled as a function of water quality to investigate the blending water quality effects in the PDS, and a tool was made to help access the potential lead release from blending treated ground, surface, and RO waters.

Experimental data indicated that lead corrosion by-product releases depended heavily on water quality, especially on chloride, alkalinity, pH, and also to a lesser extent on sulfate and temperature for the conditions of this investigation. Considering the confounding water quality effects of blending, increasing chloride will produce more lead release, and this high chloride is primarily caused by a high ratio of RO water in a water blend. Hence, consideration should be given to blended water with a high ratio of RO water. Lead release is more sensitive to the potential impacts of high chlorides and alkalinity, than the potential impact of high sulfates. The high sulfates associated with a high blending ratio of surface water would not offset the potential impact of high chlorides or alkalinity. However, nothing was generated in this study that would indicate a potential violation of the lead action level given the existing water quality of the singular and blended sources.

The lead area used in the corrosion loops was not likely the same average area of exposed lead in the MGs consumers' home plumbing systems. Calibration of the steady state lead model to the MGs 90th percentile can be easily done by linearly proportioning the 90th percentile lead concentrations of the MGs and the steady state lead model.

COPPER, LEAD AND IRON TRANSITION

The transition section describes the change of scale formation for copper, lead and iron using copper, lead and UCI coupons, and the release of total copper from common copper pipes, total lead from identical lead coupons in the corrosion loops and total iron from UCI or G pipe to the bulk water during changes in water quality from one phase to another.

Transition of Metal Corrosion Scale Composition

The transition of scale formation for copper, lead and iron using copper, lead and UCI coupons, and transition of total copper release from common copper pipes, total lead release from identical lead coupons in the corrosion loops and total iron release from UCI or G pipes to the bulk water during water quality phase changes was described and modeled as a function of water quality, time and temperature.

Metal coupons were exposed to a constant water quality over a six month, and coupons were taken from the cradles at regular intervals. Surface characterization techniques were used to quantify the changes in surface scales over a six month transitional period. The objective of the investigation was determination of the minimum required time of exposure to develop a constant and controlling scale.

Transitional Metal Release Response in the Aqueous Phase

The effect of water quality on the steady-state metal release was previously described for copper, lead and iron in G and UCI pipes in equations (4.46), (4.52), (4.3) and (4.4) respectively. These models have the ability to predict the steady-state metal release in distribution systems due to blend changes sustained over a period of several months. However, the transient metal release behavior immediately after a blend change (several weeks) is not adequately described. Member governments may expect temporary changes in blends due to operational and seasonal variations in demand and supply dynamics for the three principal source waters (GW, SW and RO). This section gives a quantitative method of determining the effect of short term water quality variations (due to blending changes) on the metal release behavior in the distribution systems.

The development of the general transition model for total metal release was initiated by defining the release of total metal as shown in equation (4.54), as the sum of present total metal release and change in total metal release.

$$C_{Total} = C_P + C_{Change} \tag{4.54}$$

where C_{Total} = total metal release

C_P = present total metal release

C_{Change} = change in total metal release

The change in total metal release was defined and the sum of the total unreleased metal and the total released metal as functions of time as defined in equation (4.55).

$$C_{Change} = C_{unrel} + C_{Rel} \tag{4.55}$$

where C_{unrel} = unreleased total metal at time, t
C_{Rel} = released total metal at time, t

The unreleased total metal at time zero was defined as equal to the released total metal at time infinity as shown in equation (4.56).

$$C_{unrel\ at\ t=0} = C_{Rel\ at\ t=\infty} \tag{4.56}$$

The unreleased total metal was defined as a first order differential equation as shown in equation (4.57), which was integrated with respect to time as shown in equation (4.58).

$$\frac{dC_{unrel}}{dt} = -kC_{unrel} \tag{4.57}$$

$$C_{unrel} = C_{unrel\ max} e^{-kt} \tag{4.58}$$

The released total metal as a function of time was determined by a mass balance and is shown in equation (4.59). The unreleased total metal maximum can be defined from the observed release of total metal over time, or predicted from the steady-state water quality equations for predicting total metal release.

$$C_{rel} = C_{unrel\ max}\left(1 - e^{-kt}\right) \tag{4.59}$$

The transition of total metal release from one steady state to another steady state as would be brought about by a change in blended distribution system water quality is shown in equation (4.60). In this equation, C_1 refers to the steady-state concentration corresponding with the original water quality. Similarly, C_2 refers to the steady-state concentration that would be experienced for the changed water quality if the altered water quality persisted for a sufficient duration to reach a new steady-state. The unreleased maximum (C_2 – C_1) was actually determined by the difference between the steady-state water concentration between phases determined visually from a plot of total metal concentration as a function of time from the end of one phase until a steady state developed for the following phase. As noted previously, this could also be approximated by using the steady-state water models for the two different water quality conditions to determine the difference or "driving force" between changes in steady-state total metal release. In summary, this driving force represents the maximum potential change in metal ion concentration that would be experienced for the specified change in water quality. A sufficient duration of exposure to the new water quality would be required to reach steady state.

$$C_t = C_1 + (C_2 - C_1)(1 - e^{-kt}) \tag{4.60}$$

Equation (4.61) was further modified by incorporating the effect of temperature on the rate constant (k) using the Arrhenius relationship as shown in equation (4.61).

$$C_t = C_1 + (C_2 - C_1)\left(1 - e^{-k\theta^{(T-20)}t}\right) \tag{4.61}$$

where C_1 = steady-state copper release before water quality change (mg/L)
C_2 = steady-state copper release after water quality change (mg/L)
C_t = transient copper release at time 't' days after change
T = temperature (°C)
θ = temperature correction factor
k = rate constant (day^{-1})
t = time (days) after water quality change

All C, θ and k terms can be determined by non-linear regression of equation (4.61) against data set of total metal release during transition between phases. Typically, all C terms were identified visually from a plot of total metal release during transition and, θ and k were determined by regression. A model for total metal release during transition was developed by incorporation of steady-state estimates of C terms using the models that related the steady-state total metal release to steady-state water quality. The model is shown in equation (4.62) and can be used proactively to mitigate undesired total metal release in the distribution system.

$$C_{tp} = C_{1p} + (C_{2p} - C_{1p})\left(1 - e^{\left(-\theta^{(T-20)} k\, t\right)}\right) \tag{4.62}$$

where C_{1p} = predicted steady-state metal release before water quality change (mg/L)
C_{2p} = predicted steady-state metal release after water quality change (mg/L)
C_{tp} = predicted transient metal release at time 't' days after change

Copper Transition

Copper Scale Transition

Figure 4.54 presents the superimposed XRD patterns of the copper coupon samples after incubation in aerated GW water environments for 1, 2, 3, and 6 months. The water incubating the coupon shown in Figure 4.54 was characterized as high alkalinity; the average level was maintained at 200 to 210mg/L as $CaCO_3$.

Focusing on the qualitative analysis relationship between the incubation period and corrosion products formed, the spectra were superimposed for coupons exposed to each similar water matrix. As shown in the Figure 4.54, the common feature indicated by all of the spectra patterns is that the composition of crystal formed on copper coupons was relative simple and stable despite the significantly different water quality.

Further examination shows that the thickness of cuprite (Cu_2O) increased with a longer incubation period. The fact that the ratio of peak intensity of cuprite to substrate copper increased to around 2-Theta 75° indicated that the averaged relative amount of cuprite formed increased if the exposure time was longer. The cuprite layer thickness increase was obviously not caused by the water quality variation because this increase was observed through all the water conditions in the experiment.

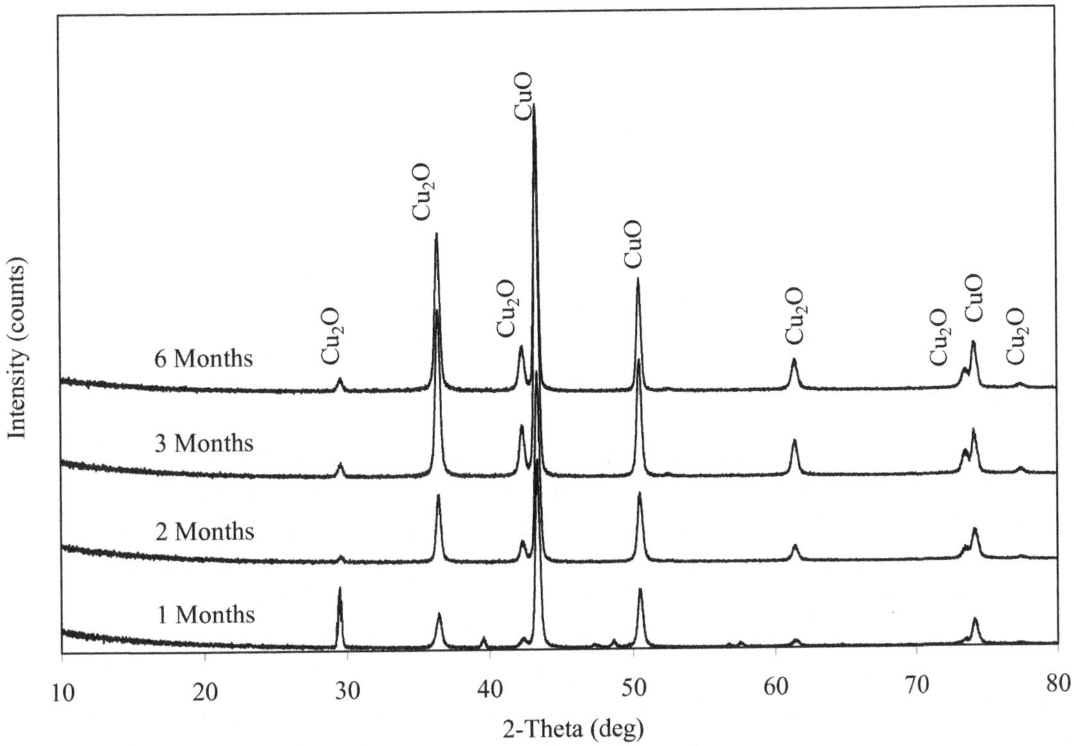

Figure 4.54 XRD pattern from Cu coupon incubated in 100% groundwater (G1)

In order to verify the uninfluenced effect of water quality on the kinetic variation of subsurface corrosion products, the XRD spectra of four individual coupons incubated in four different water qualities for 3 months were also checked. There was no observable difference between each spectrum in terms of the cuprite to copper ratio. This fact indicates the ratio variation through time span could not be caused by water quality changes, but the function of time during which copper corrosion continues to proceed. With the consideration of fixed penetration depth of XRD and the associated absence of cupric phases among all samples, the ratio indicating the relative amount variation of cuprite (Cu_2O) and copper was straightforward.

The normalized percentage variation of cuprite formed in different water context occupying the entire top surface by using XPS was examined. It is noted that the cuprite content decreased first until the minimum at around 2 or 3 months and then increased to the previous level. The common variation trends among the content of cuprite formed in four different waters implicated that the cuprite formation was not attributed to the water quality. Cu(II) species followed the opposite trend. The analysis performed on tenorite (CuO) and cupric hydroxide might not be as informative as cuprite. There is no general traceable trend that can be extracted. But, it was noted that the cupric hydroxide content tends to be stable with a longer incubation period, which might be the signal of approaching to equilibrium, but this needs further verification. The data points obtained at the first two months, however, accordingly increased indicating the thermodynamic equilibrium has not been reached in terms of dissolution/precipitation models.

Total Copper Transition in the Aqueous Phase

Influent water quality was changed in all PDS and consequently the corrosion loops at the start of each phase. Minor influent water quality variations occurred during a particular phase due to pump adjustments, fluctuations of some parameters in raw source waters or during treatment. Temperature variations were seasonal and sometimes varied significantly during a particular day. In summer conditions the water temperature sometimes reached 30°C (86°F).

Figure 4.55 is a typical time series plot of change in effluent copper concentration with time elapsed from the change in water quality (phase change). It shows that the change in copper release was time dependent and slowly increased or decreased with time and reached a nearly stable value with increasing time. Inspection of Figure 4.55 indicates that the major changes in copper release were realized during the first 15-20 days from the beginning of the phase, which was defined by the change in PDS feed stream water quality. The copper concentration did not change significantly between 20-42 days.

Whether there is an increase or decrease in copper concentration following the change in influent water quality depends on the initial copper concentration and the water quality parameters affecting copper corrosion as described in equation (4.46). The water quality driving force determines the uptake or release of total copper following the change in water quality. If the initial copper is greater than the copper predicted by the driving force, the transition direction will always be down. If the initial total copper is less than the copper predicted by equation (4.46), the transition direction of total copper will be upwards. Hence, a positive driving force will not always increase total copper unless the actual copper is less than that predicted by the steady-state equation. The trend of total copper release or any metal is similar to changes in water quality when controlled by equilibrium. Changing from low alkalinity (RO) to high alkalinity (GW) will increase the driving force significantly and will almost always cause significantly higher copper concentrations. However, if the RO water is in contact with newly exposed copper surfaces and the steady-state total copper concentration is not achieved, then the total copper trend may actually decrease because the actual total copper is still higher than the steady-state total copper predicted at steady state with GW. The same is true for changes of any water quality variables as described by equation (4.46), which represents steady-state operation.

The rate constant (k) and the temperature correction factor (θ) were determined by non-linear regression of equation (4.61) against total copper release data from Phases II-III and III-IV. The driving force was visually determined from the steady-state total copper release during Phase II and Phase III. The steady-state release of total copper was determined to occur at 42 days after phase change as shown in Figure 4.55. Initial copper concentration (Cu_I) was determined as average of the last two total copper release observations before phase change. Ten percent of the data was kept aside for validating the transitional release model. The remaining dataset was regressed to find a single value of the rate constant (k) and the temperature correction factor (θ).

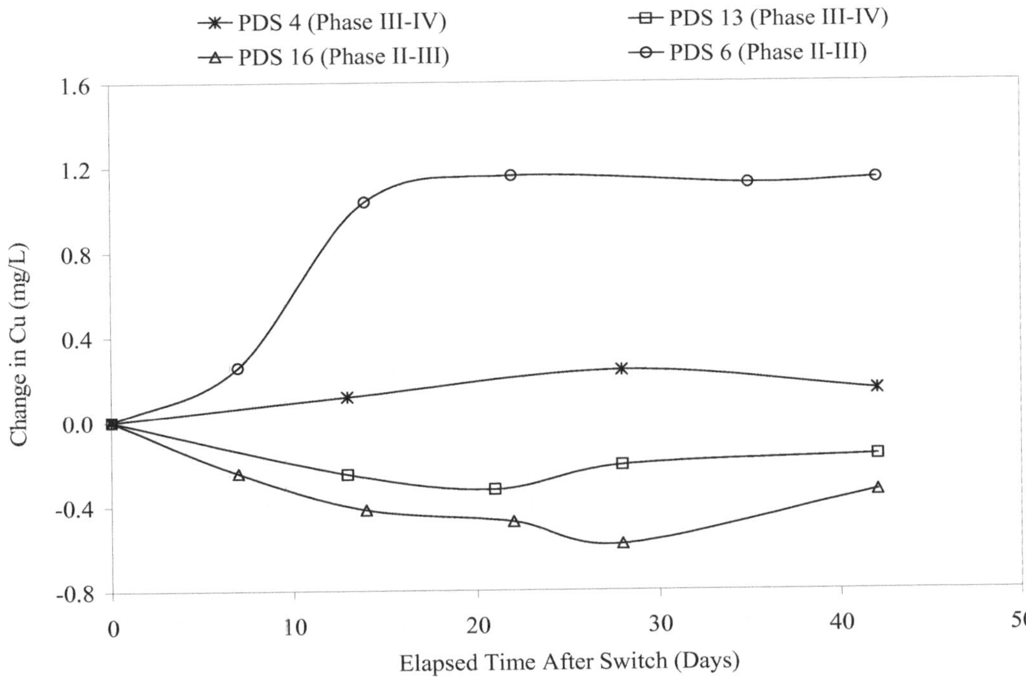

Figure 4.55 Typical response of change in copper concentration with time elapsed from the change in water quality

As determined by non-linear regression, the rate constant at 20°C ($k_{20^\circ C}$) was found to be 0.071 day^{-1} and the value of θ was found to be 1.02. The R^2 for the regression was 0.63. A paired t-test for difference in means using 10% of the data set, which was set aside for validation, found that there was no difference in actual and predicted total copper released during transition. The t-statistic was 1.22 and was less than the t-critical statistic (2.13) for a two-tailed t-test, 15 degrees of freedom and 95 % confidence interval. The results of the paired t-test found there was no significant difference between actual and predicted total released copper during transition. The first order model that was developed by regression of equation (4.61) against the transition data set for total copper release is shown in equation (4.63). A first order rate constant of 0.071 day^{-1} means that 62 % of the total copper release would be realized at approximately 14 days (20°C). Steady state corresponds to 95 % of the maximum total copper release and occurs at 42 days and is the same time for reaching steady state for total copper release as was visually determined from Figure 4.55.

$$Cu_t = Cu_1 + (Cu_2 - Cu_1)\left(1 - exp\left(\ 1.02^{(T-20)}\ (0.071)\ t\right)\right) \tag{4.63}$$

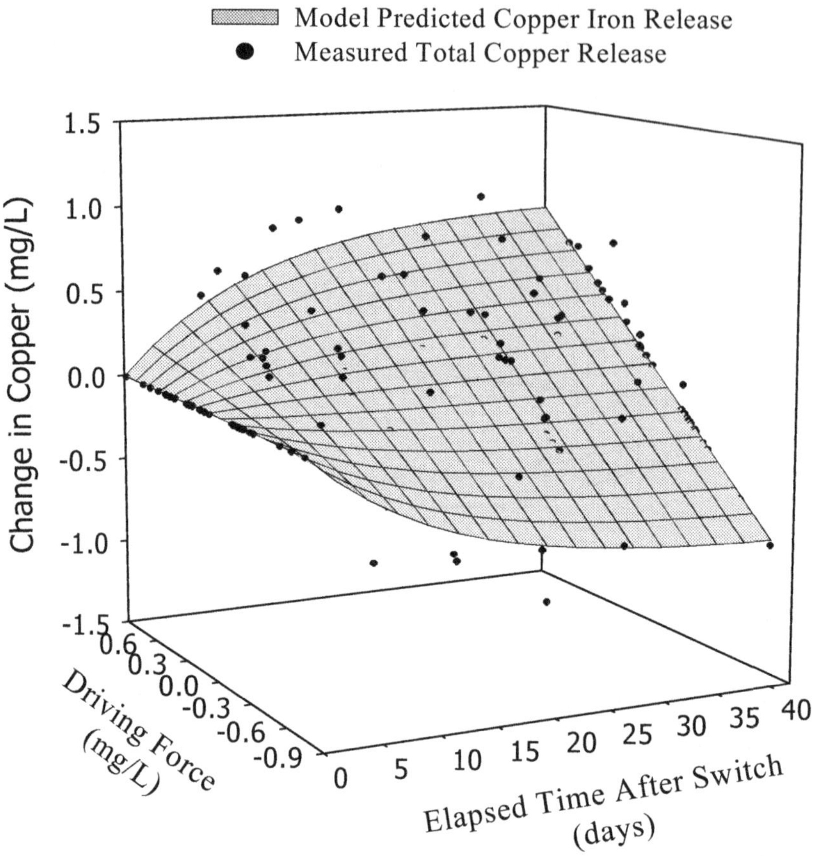

Figure 4.56 Total copper release versus driving force and time following a water quality change

The three dimensional plot shown in Figure 4.56 provides visualization of the first order response of total copper transition following a steady-state water quality change in the PDS and actual distribution systems. Actual total copper released is shown as dots in Figure 4.56 and the surface is the predicted plane of total copper release. Total copper release is shown on the Z axis for varying driving forces and transition times. The release or uptake of total copper increases with transition time as the driving force increases either positively or negatively. The three-dimensional plot demonstrates the relatively good fit of the actual and predicted copper release and shows the significance of the driving force and k. The exponential response is controlled by k as the reciprocal of k is 14 days, which represents the transition time when 62 % of the total response is realized. Hence, there is some time of release reaction and for utilities to mitigate adverse transient copper release caused by water quality change. A high driving force will generally increase the total copper reaching the consumer while a small driving force will not cause a major increase in total copper, and the driving forces can be predicted by the difference between the actual copper and predicted copper using equation (4.62). The same method of predicting total metal release can be used for total lead and total iron.

Incorporation of Steady-State Release Model in Transient Release Model

The driving force used to determine the rate constant (k) and the temperature correction factor (θ) were calculated based on the actual data available from this study. However, for proactive utility use, the driving force must be predicted using the steady-state model defined in equation (4.46). The proactive transitional model for total copper release is shown in equation (4.64), which is almost identical to equation (4.63), except that the driving force is predicted by the difference in the predicted copper before and after a water quality change using equation (4.52).

$$Cu_t = Cu_{ss,1} + (Cu_{ss,2} - Cu_{ss,1})\left(1 - e^{\left(-\theta^{(T-20)}\,0.071\,t\right)}\right) \tag{4.64}$$

where $Cu_{ss,1}$ = steady-state model predicted copper concentration before change

$Cu_{ss,2}$ = steady-state copper concentration predicted after change

Equation (4.64) was assessed by comparison of predicted and actual total release copper during transition for all eighteen copper corrosion loops during phase change. Typical predicted versus actual transition of total copper release is shown in Figure 4.57. A paired t-test was again used to evaluate differences between the predicted and actual transient copper release. The t statistic was 1.71 compared to the t-critical statistic for a 2 tailed t-test, 190 degrees of freedom at a 95 % confidence interval was 1.97, which was greater than the actual t statistic of 1.71. The paired t test found there was no significant difference between the predicted and actual total release copper during transition using equation (4.64). The t-test comparison indicates that the revised model accurately describes total released copper during transition and validates the use of the steady-state model (equation (4.46)) in the model for transition of total released copper as shown in equation (4.64).

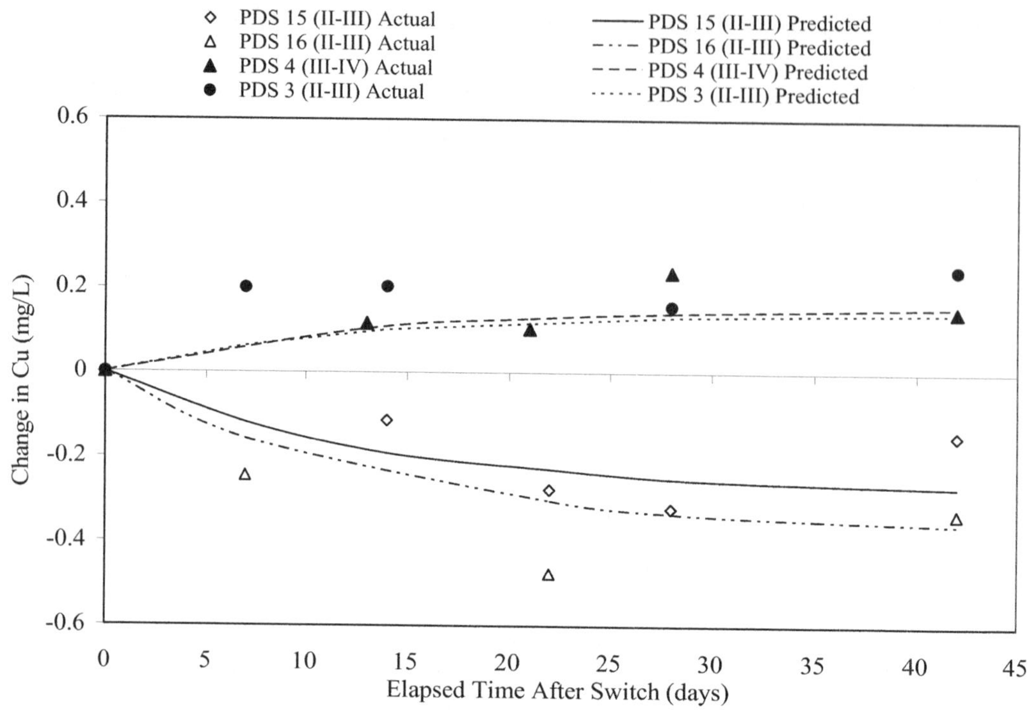

Figure 4.57 Actual versus predicted change in copper with time elapsed from the change in water quality

Application of the Model

The impact of the rate constant on the transition of total metal release can be demonstrated by plotting the fractional transition as a function of time using equation (4.65). As noted previously the reciprocal of a first order rate constant represents 62 % of the total response of the dependent variable. Consequently, 62 % of the transitional total copper release is released in 14 days. The fractional response for any change in total copper release driven by water quality changes is shown in Figure 4.58 at 20°C. The point of the Figure 4.58 is that a utility would potentially have some time to react before the maximum response is realized. The fractional release of total copper at 5 and 10 days is approximately 25 and 50 %, which means utilities may be able to change quality to reverse adverse impacts or that proactive actions can be taken to mitigate adverse impacts prior to water quality change.

$$\frac{Cu_t - Cu_1}{Cu_2 - Cu_1} = 1 - e^{\left(-1.02^{T-20} \, 0.071 t \right)} \tag{4.65}$$

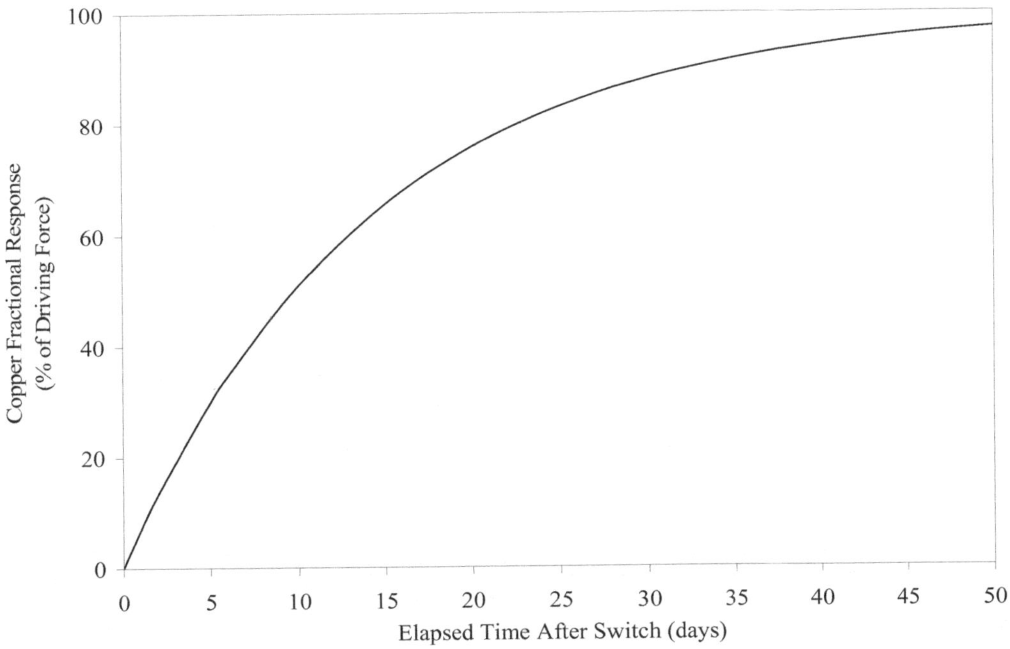

Figure 4.58 Fractional response in copper release with time elapsed from the change in water quality

As shown in Figure 4.58 one day after a water quality change in water quality only 7% of total copper release would be realized. At ten days, 50 % of the total copper release occurs. Thus these dynamics show that if the water quality change is mitigated after ten days the total copper release is reduced by half, which could possibly avert exceeding a regulatory action level. Although the magnitude of the total copper change is determined by the initial total copper and the difference in driving forces before and after steady-state water quality changes, the transition of the total copper release is dependent on the reciprocal of the reaction constant. Hence the fractional impact can always be described as a function of time and provides opportunities for reducing adverse impacts.

The transitional model developed previously (equation (4.64)) can be used to simulate total copper release following any changes in water quality. The driving forces, based on the assumed water quality for the principal finished water sources shown previously in Table 4.14, corresponding to the change from GW to RO, SW and Blend at 20°C are shown in Table 4.23. These driving forces are calculated as the difference between the predicted steady-state copper release due to SW, RO or blend and the predicted steady-state total copper release due to GW source.

Table 4.23
Driving forces corresponding to changing source waters for copper

Change in source	Driving force (mg/L)
GW to RO	-0.65
GW to SW	-0.54
GW to Blend (60%GW,30%SW, 10%RO)	-0.13

The 50th percentile model is suitable for comparing and validating the actual and predicted total copper release during the pilot study as the mean total copper released is what is expected to be seen in any one observation. However, a 90th percentile transition model is required for MGs to proactively use the total copper released model during transition as MG utilities are regulated at the 90th percentile. Therefore, MGs should use the 90th percentile copper release model for evaluating action level compliance.

As evident from Table 4.23 going from GW to either source or blend at steady states creates a negative driving force, which means total copper release will decrease. The magnitude of decrease depends on the number of days the new water quality is sustained in the distribution system. The simulations are shown in Figure 4.59 for changes from GW to the sources described in Table 4.23. The decreased total copper release following the change from GW to blend, SW and RO during transitions is shown in Table 4.24. As expected, the total copper released decreases following the change from GW to either of the source waters or blends.

Table 4.24
Decreased total copper release following change from GW to SW or RO

Elapsed time (days)	Decreased total copper release (%) GW to SW	GW to RO	GW to Blend
2	6	7	1
5	14	16	3
10	23	28	5
20	35	41	8

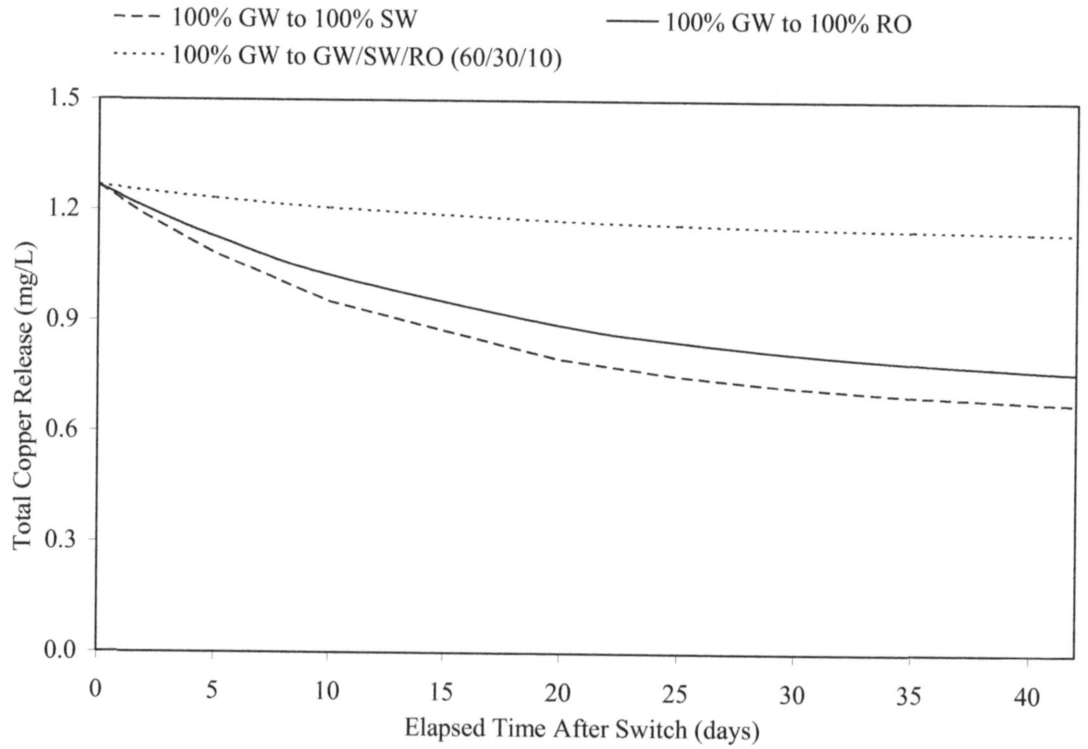

Figure 4.59 Transient copper simulation of change from GW to other sources/blends

Figure 4.59 and Table 4.24 reveal that a long period of exposure to a new source is required before an obvious change in copper concentration can be realized by the consumer. Blend impact on total copper release and LCR copper action levels can be anticipated and possibly mitigated before the transient response exceeds the regulatory limit. The 42 day period to attaining steady state does not imply the LCR would be violated after 42 days. For instance shifting from a 60/30/10 blend of GW/SW/RO to a GW high in alkalinity (300 mg/L as $CaCO_3$) would require less time (10 days) to result in a predicted violation of the copper action level than when shifting from low alkalinity water (RO) to GW (40 days).

Table 4.25
High frequency line blends (PDS 14)

Blend	Source
1	70% GW, 30% SW, 0% RO
2	85% GW, 0% SW, 15% RO
3	0% GW, 71% SW, 29% RO

The finished water feed stream to PDS 14 consisted of two of three blends, which were randomly varied every twelve hours to determine the effects of high frequency blend changes on distribution system water quality. The three blends are shown in Table 4.25. The high frequency data set for total metal release was developed over one year (12/7/01 – 12/10/02). Figure 4.60 shows the actual copper release in PDS 14 over the one year period. The steady-state predictions at 25°C and average source water qualities for the three blends identified in Table 4.25 are shown as horizontal lines. The time weighted composition of the three blends was 60% GW, 30% SW and 10% RO. The actual data shows that the copper release response in this high frequency line was approached thee horizontal total copper concentration represented by the average blend.

Simulations were performed using the actual blends composition, duration and temperatures. The data from 12/15/01 to 2/23/02 (Phase I) was used for simulation. The transitional release of total copper in the high frequency PDS was adequately described by the transition total copper release model.

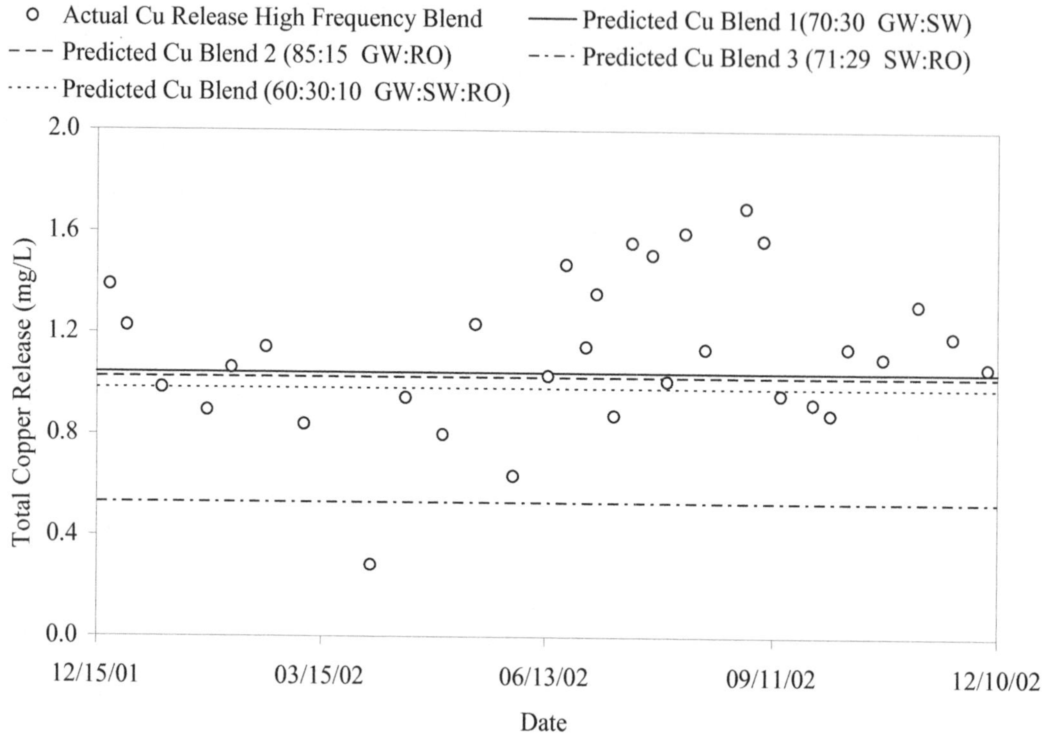

Figure 4.60 Copper release from the high frequency blend line (PDS 14) over the period of the study

The finished water feed stream water quality for PDS 14 was determined by mass balance. Typical blend water quality for PDS 14 is shown Table 4.26. Equation (4.64) was used to determine the driving forces for the transient release of total copper release. The rate constant was adjusted for actual water temperatures measured in the field.

The predicted transient total copper release shows a relatively constant release of total copper despite the rapid blend changes in source waters. The transition data shown in Figure 4.61 was taken from the beginning of the field investigation when the copper pipes were relatively new and not at steady state. Although the total copper release is relatively high, a negative trend is shown by the actual and predicted transition data because the initial copper was higher than predicted at steady state and the actual and predicted drivers were negative. The observations in the high frequency line are consistent with the rate constant (k) as 42 days is required to develop steady state.

Table 4.26
Transient water quality and simulation for copper release in high blend frequency line (PDS 14)

Blend	Time (Days)	pH	Alkalinity (mg/L $CaCO_3$)	Sulfate (mg/L)	Silica (mg/L)	Actual Temp (°C)	Steady-State Cu Prediction (mg/L)	Cu Actual (mg/L)	Transient Cu Predicted (mg/L)
2	0.5	7.9	186	22	22.0	12	0.88	1.77	1.77
1	0.5	8.0	163	91	22.0	13	0.90		1.74
2	0.5	7.9	186	22	22.0	12	0.88		1.72
3	0.5	8.2	62	175	22.0	8	0.48		1.68
2	0.5	7.9	186	22	22.0	12	0.88		1.65
3	0.5	8.2	62	175	22.0	8	0.48		1.61
3	1	8.2	62	175	22.0	8	0.48		1.54
1	0.5	8.0	163	91	22.0	13	0.90		1.52
1	1	8.0	163	91	22.0	13	0.90		1.48
1	1.5	8.0	163	91	22.0	13	0.90		1.43

The rate constant and the negative drivers essentially mitigate or decrease the initially high release of total copper as shown in Figure 4.61. However, it is not the frequency of change but the water quality drivers that would drive copper past the action level. The first order response means the rate constant provides some time for mitigation of adverse effects. A paired t-test of the actual and predicted total copper released during transition in PDS 14. The actual t-statistic was 2.47, which was greater than the 2.45 critical t-statistic for a 2 tailed t-test, 6 degrees of freedom and 95 % confidence interval. The paired t-test actually indicates there was a significant difference between the actual and predicted total copper release during transition in PDS 14. The result is not surprising given the distribution shown in Figure 4.60. However, new copper pipes will have elevated release and the close proximity of the t-critical and the actual t statistic indicate that the transition model, which incorporates the steady-state model as shown in equation (4.46) for predicting the driving forces ($Cu_{ss,2}$ and $Cu_{ss,1}$) can be used to describe total copper variations in the high frequency line.

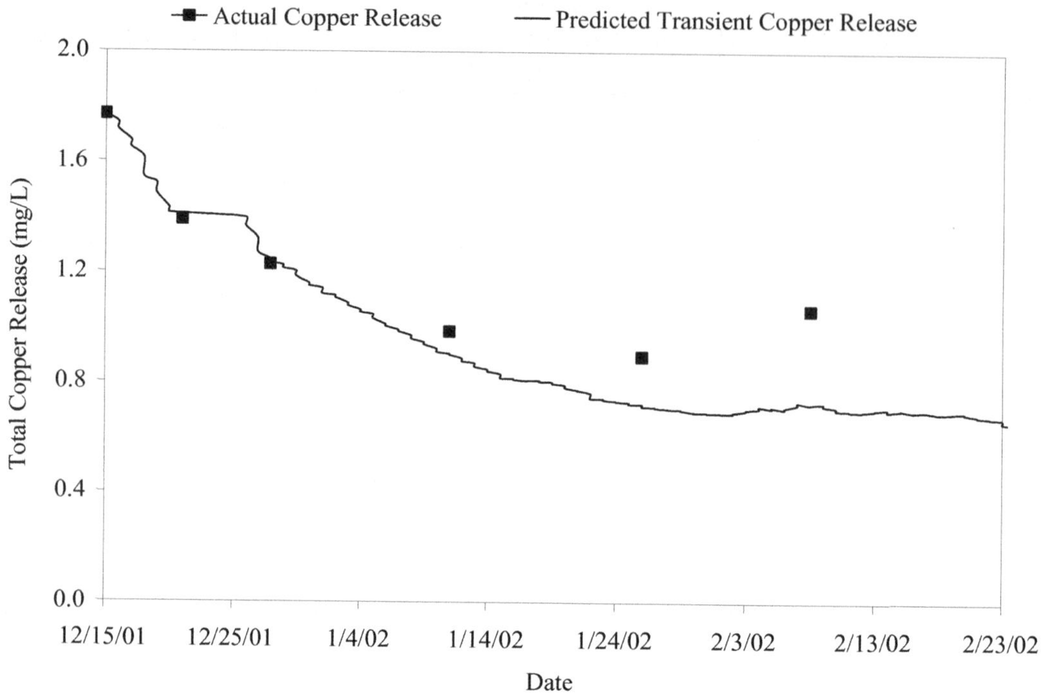

Figure 4.61 Transient copper release from corrosion loop fed by high frequency line (PDS 14)

Summary for Transient Copper Release

Visual inspection of copper release data indicated that the copper release versus time response were typical of an exponential relationship that was affected by water quality change.

A transient total copper release model was developed which was consistent with the observed exponential relationship of observed copper release following phase changes. The model developed was verified on 10% of data set aside for validation as well as for the steady-state model predicted driving forces.

The rate constant at 20°C ($k_{20^{\circ}C}$) for transient copper release model was found to be 0.071 day^{-1} and the value of θ was found to be 1.02. Forty-two days are required for total copper release to reach a steady state following a steady-state water quality change.

The total copper release from the high frequency blend change PDS was initially higher than predicted at steady state because steady state possibly had not been achieved in the newly exposed copper loop. The actual and predicted release of total copper in the high frequency loop showed a negative trend toward steady state. The frequency of water quality change did not adversely affect total copper release, the water quality drivers can adversely affect the steady state and transient total copper release.

If a utility can anticipate water quality changes, there is an opportunity to mitigate adverse release of total copper. Use of the transition models can provide a time frame, which would allow proactive measures to mitigate adverse total copper release.

Lead Transition

Lead transition studies were conducted on lead surfaces or scales change over time of contact with the bulk water. Lead transition studies were also conducted on the release of total lead in the bulk water in contact with the lead coupons in the corrosion loops.

Lead Scale Transition

To study the kinetic transition of lead surfaces, lead coupons were incubated in four feeding waters (G1 = 100%, S1 = 100%, RO = 100%, G1/S1/RO = 62/27/11%) for one, two, three, and six months. This study investigated the change in solid structures of lead scales with time.

The coupons were analyzed by XRD and the results indicated that the surface crystalline composition did not qualitatively change with time during the six month incubation period. $Pb_3(OH)_2(CO_3)_2$, $PbCO_3$, and PbO were the main crystalline compounds of corrosion products for all of these samples, regardless of the incubation period. Comparing the relative intensities in one sample, some trends are found: with the extension of incubation time, less PbO existed in the corrosion layer while more $Pb_3(OH)_2(CO_3)_2$ and $PbCO_3$ appeared. Since new pure lead coupons were used in this study, the results indicated that PbO appeared at the beginning of the corrosion process but could have transformed into $Pb_3(OH)_2(CO_3)_2$ and $PbCO_3$ gradually, which reveals the process of lead corrosion.

XPS scanning was done for the kinetic study samples and showed $Pb_3(OH)_2(CO_3)_2$, $PbCO_3$, PbO and PbO_2 were the primary corrosion products for all of these samples. From the previous study, $Pb_3(OH)_2(CO_3)_2$ was found to correlate well with lead concentration in water and may play an important role on controlling lead release. Therefore, the effects of incubation time on $Pb_3(OH)_2(CO_3)_2$ peak area% was investigated by XPS.

$Pb_3(OH)_2(CO_3)_2$) and $PbCO_3$ peak area percentage changed when the incubation time varied. For $PbCO_3$, there was no obvious trend. However there was trend for growth of $Pb_3(OH)_2(CO_3)_2$ as indicated by XPS analysis of the lead coupons. A high growth of $Pb_3(OH)_2(CO_3)_2$ as indicated by a high percentage of peak areas associated with $Pb_3(OH)_2(CO_3)_2$ occurred in one-month, three-month and six-month samples, but not in two month samples. The occurrence of high peak area percentages in XPS samples of $Pb_3(OH)_2(CO_3)_2$ suggests $Pb_3(OH)_2(CO_3)_2$ appears at the surface before other species. $Pb_3(OH)_2(CO_3)_2$ continues forming and needs additional time to reach equilibrium than other corrosion products. Therefore, after two months, the percentage of $Pb_3(OH)_2(CO_3)_2$ area shown by analysis of XPS samples increased. These results suggest that the equilibrium of lead surface solids phases was not reached in the six month period. This trend of $Pb_3(OH)_2(CO_3)_2$ was the same for all four of these waters. The results also indicate that the formation of $Pb_3(OH)_2(CO_3)_2$) and $PbCO_3$ on the surface is complicated, and additional work is needed to understand the growth of solid phases on lead surfaces.

It is obvious that the growth of $Pb_3(OH)_2(CO_3)_2$ on the surface of the lead coupon increased over the six-month period. The order of increasing percentage of $Pb_3(OH)_2(CO_3)_2$) on the coupon surface by source was RO>G1>G1/S1/RO>S1, which was consistent with previous results. XPS analysis of samples taken from the corrosion loops revealed that high $Pb_3(OH)_2(CO_3)_2$ peak areas were related to high lead release. The kinetic study and corresponding total lead release from the corrosion loops revealed that the order by source of the rate $Pb_3(OH)_2(CO_3)_2)$ growth as indicated by XPS analysis and total lead release were the same.

Total Lead Transition in the Aqueous Phase

Four corrosion loops were chosen to show the typical responses of the lead to changes in influent water quality (Figure 4.62). Although, lead concentration slowly increased or decreased with time following a water quality change before stabilizing; based on a visual inspection of the time series data in Figure 4.62, the major changes in lead concentration were realized within the first 15 or 25 days following the change in water quality. It is important to note that temperature variations were also experienced within the period shown in the plot and hence may have attributed to minor variations in lead concentrations. The source water in PDS 13 was changed from 100% RO to 100% S2 going from phase II-III, which essentially decreased chlorides by 30 mg/L and increased sulfates by 8 mg/L, which resulted in decreased total lead release as shown in Figure 4.62. The total lead release in PDS 13 increased when the source was changed from 100% S2 to 100% RO in the phase III-IV transition because of the resulting decreased sulfates and increased chlorides.

Equation (4.66) was developed for describing transitional release of total lead. Equation (4.61) was regressed against the transition of total lead between phases II-III and III-IV over time to develop k and θ for total lead. The driving force was visually determined from the steady-state total lead release before and after phase change. Ten percent of the data was kept aside for validating the transitional release model.

$$Pb_t = Pb_1 + \left(Pb_2 - Pb_1\right)\left(1 - e^{\left(-1.05^{(T-20)}\ 0.081\ t\right)}\right) \tag{4.66}$$

where Pb_2-Pb_1	=	driving force
Pb_1	=	steady-state lead release before water quality change (μg/L)
Pb_2	=	steady-state lead release after water quality change (μg/L)
Pb_t	=	transient lead release at time 't' days after change

The value of the rate constant at 20°C ($k_{20^\circ C}$) was found to be 0.081 day^{-1} and the value of θ was found to be 1.05. The R^2 for the regression was 0.72. A paired t-test was done for difference in predicted and actual total released lead using 10% of the data set, which was not used for model developed. The t-statistic was 0.69, which was less that the critical t statistic of 2.13 as determined from a two tailed t-test at 15 degrees of freedom and 95 % confidence interval. The results indicate there is no significant difference between actual and predicted total released lead during transition.

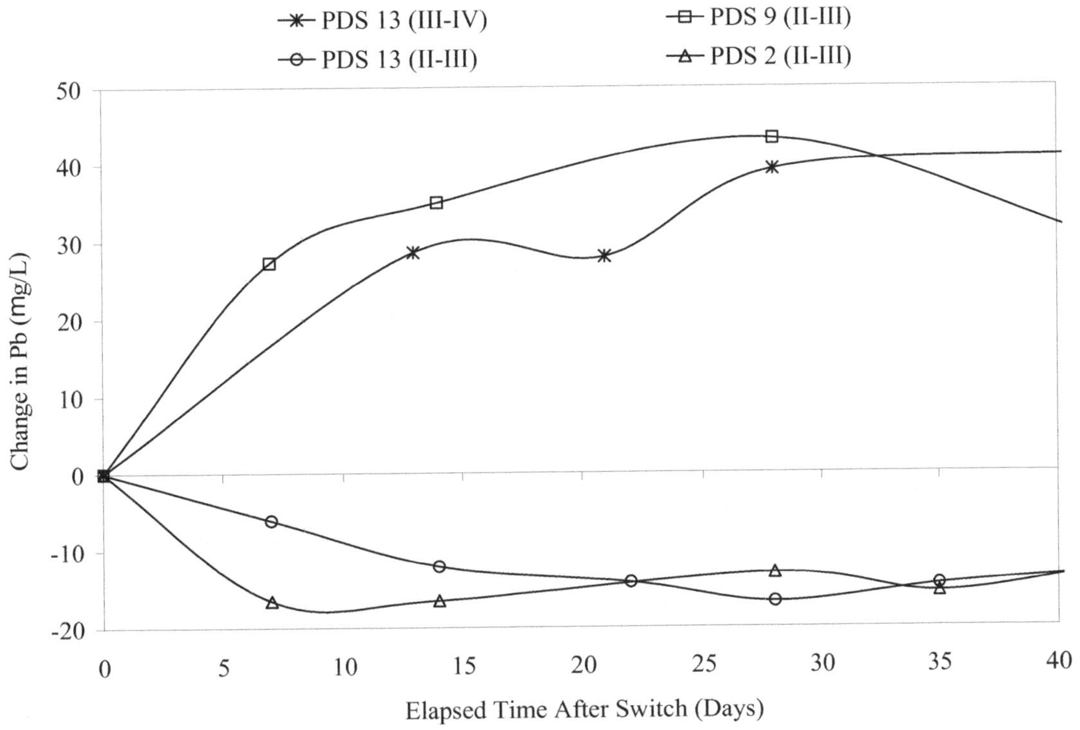

Figure 4.62 Typical response of change in lead concentration with time elapsed from the change in water quality

Given a rate constant of 0.081 day^{-1} and a specification that steady state would be reached when the product of the rate constant and time is equal to three (corresponding with 95% of maximum total lead release), thirty-seven days are required to reach steady state at 20°C following a water quality change. The actual data reported in Figure 4.62 support attainment of steady state within this same time frame. A three-dimensional plot of the actual and predicted total released lead versus time and driving force is presented in Figure 4.63. The distribution of the actual data and the plane of predicted total lead release show the model reasonably described total lead release during transition based on examination of Figure 4.63. The time to reach steady state for total lead is shown in Figure 4.63 as well as the effect of driving force. The change of total released lead increases as the driving force increases and significant time is required to reach steady state following a steady-state water quality change.

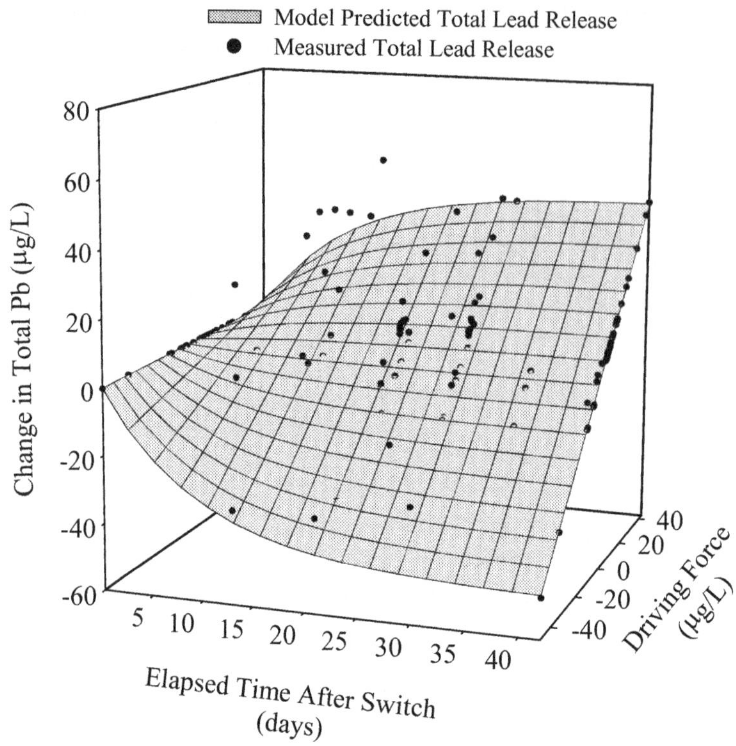

Figure 4.63 Three dimensional plot of change in lead, driving force and time elapsed following change in water quality.

Incorporation of Steady-State Release Model in Transient Release Model

The driving force used to determine the rate constant (k) and the temperature correction factor (θ) were calculated based on the actual data available from this study. In the absence of such data, initial and final steady-state concentrations ($Pb_{ss,1}$ and $Pb_{ss,2}$) can be predicted with steady-state model to define the driving force. The transitional model was modified by incorporating the steady-state model for predicting the driving force as shown in equation (4.67).

$$Pb_t = Pb_{ss,1} + (Pb_{ss,2} - Pb_{ss,1})\left(1 - e^{\left(-\theta^{(T-20)}\,k\,t\right)}\right) \tag{4.67}$$

where $Pb_{ss,1}$ = Initial steady-state model predicted concentration

$Pb_{ss,2}$ = Steady-state model predicted concentration due to changed water quality

To confirm the applicability of the steady-state copper release model $Pb_{ss,1}$ and $Pb_{ss,2}$ were predicted by equation (4.52) and used in equation (4.67) for all the 18 loops following both the phase changes. Figure 4.64 presents the predicted and actual total lead release during transition using the steady-state water quality model for total released lead (equation (4.52)) to predict driving force.

T-test statistics were used to evaluate the equality in means between the paired predicted and actual transient lead release. The t-stat was 1.54 compared to t-critical (2 tailed t-test, degree of freedom = 190, α = 0.05) of 1.97, which means the model is statistically validated and the steady-state model can be used for predicting the driving force.

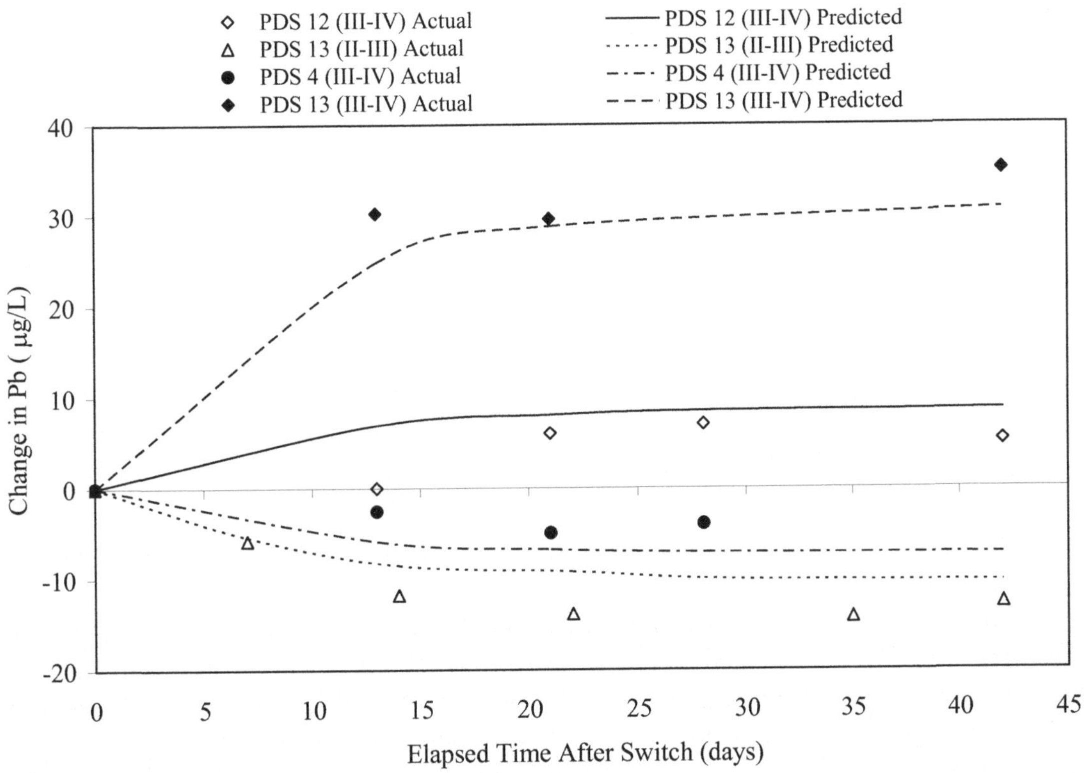

Figure 4.64 Actual versus predicted change in lead with time elapsed from the change in water quality

Application of the Model

Using the estimates of the rate constant (k) the proportion of the maximum change in lead concentration (driving force) that was realized can be calculated for various time intervals following the change in water quality. The fractional response to the changed conditions for the time duration following the change in water quality is defined by equation (4.68). The fractional response during transition at 20°C is shown in Figure 4.65.

$$\frac{Pb_t - Pb_1}{Pb_2 - Pb_1} = 1 - e^{\left(-1.05^{T-20}\, 0.081\, t\right)} \qquad (4.68)$$

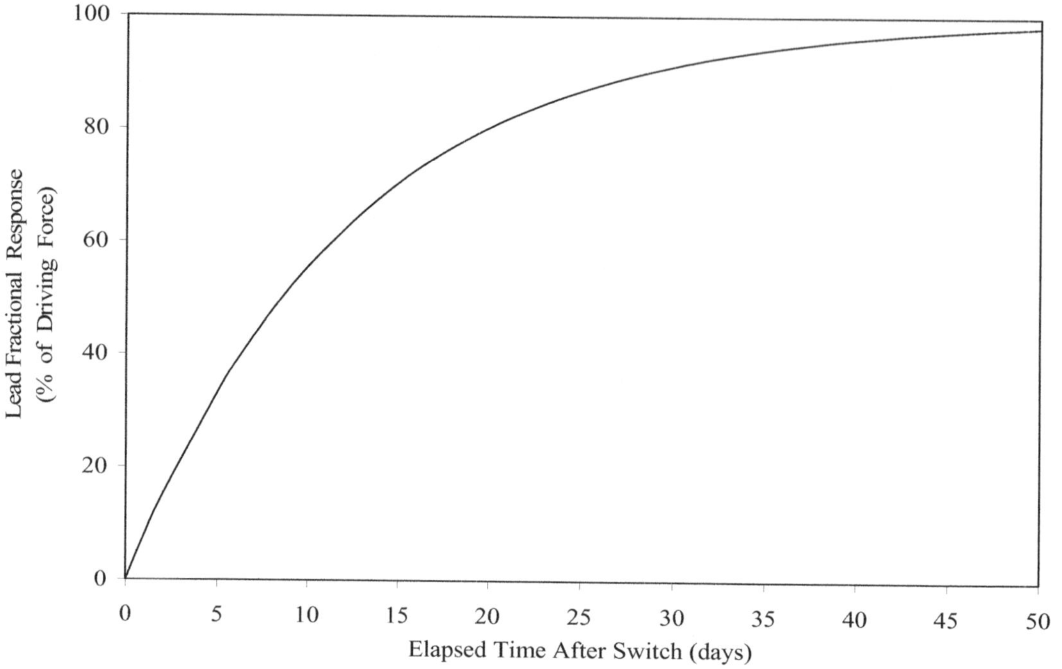

Figure 4.65 Percentage change in lead with time elapsed from the change in water quality

As shown in Figure 4.65 after a day elapsed from the change in water quality the lead concentration would have reached approximately 8% of the steady-state lead concentration corresponding to the new water quality. A 50% change in total lead release is realized in nine days. As with copper, these dynamics demonstrate that many days must past before the total released lead will reach steady state. The magnitude of the total released lead change is dependent only on water quality, but the percent change between the two steady states is only dependent on time as gauged by the reciprocal of k.

Table 4.27
Driving forces corresponding to changing source waters for lead

Change in source	Driving force (μg/L)
GW to RO	1.2
GW to SW	-3.9
GW to Blend (60%GW,30%SW, 10%RO)	-1.9

The transitional model developed previously (equation (4.67)) can be used to simulate change in lead concentration following any changes in water quality. The driving forces, based on representative water quality for the three principal finished waters reported previously in Table 4.14, corresponding to the change from GW to RO, SW and Blend are shown in Table 4.27.

As evident from Figure 4.66, changing from steady-state total lead release in GW to SW or blend water quality decreases total lead release to the consumer while changing to RO water quality slightly increases in total lead release.

It is evident from the plot that the lead action level is not violated during any of the changes described in Table 4.27. The change in total lead release following a change from GW to blend, SW and RO at varying times is shown in Table 4.28. The actual lead concentrations in the Member Governments' distribution systems would be somewhat lower than the values reported, due to the relative quantities of lead that are believed to be present in the Member Governments' distribution systems.

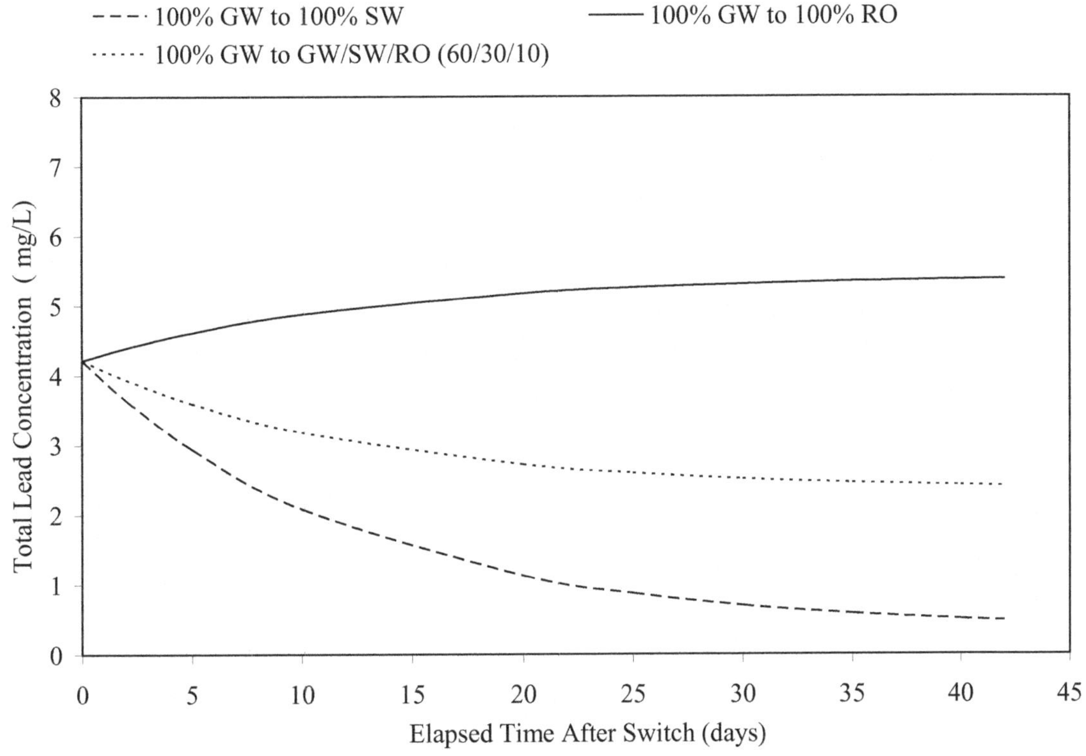

Figure 4.66 Transient lead simulation of change from GW to other sources/blends

Table 4.28
Percentage change in lead concentration following the change from GW to other source waters/blend

Elapsed time	% change in Pb concentration		
(days)	GW to SW	GW to RO	GW to Blend
2	-14	4	-7
5	-30	9	-15
10	-50	16	-24
20	-73	23	-35

However, the large value of the rate constant reveals that any change in water quality will cause a significant impact only if the new water is sustained for a long period of time. These results are supported by the lead data obtained from the high frequency line (PDS 14). Figure 4.67 shows the actual lead release in PDS 14. The predicted steady-state lead release corresponding to the three different blends is indicated by three horizontal lines. Most of the lead release data was below quantifiable detection limit (10 μg/L) for the analytical method used. However, consistent with the modeling approach, the values below detection limit are shown here as 5 μg/L to illustrate the muting effect of high frequency changes on lead release. Similar to copper, total lead release was simulated from 12/15/01 to 2/23/02 (Phase I). The water quality parameters, actual and simulated total released lead are shown in Table 4.29. The value of rate constant was adjusted for actual water temperature. A plot of predicted and actual transition of transient lead release is shown in Figure 4.68.

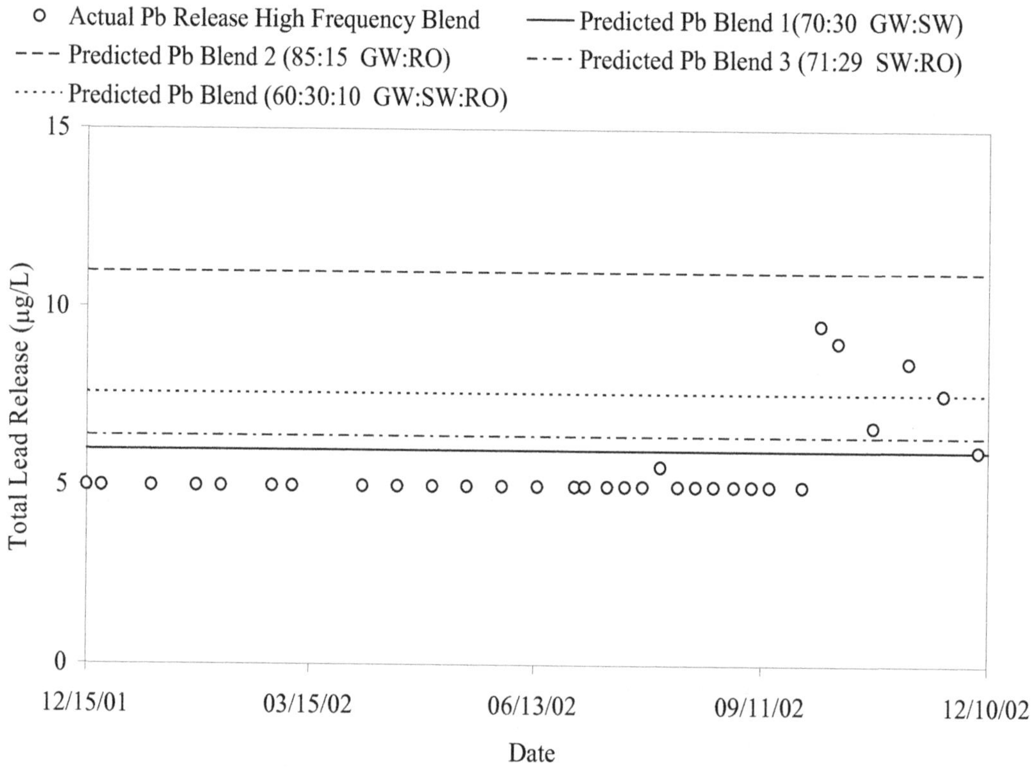

Figure 4.67 Lead release from the high frequency blend line (PDS 14) over the period of the study

Table 4.29
Transient water quality and simulation for lead release in high blend frequency line (PDS 14)

Blend	Time (Days)	pH	Alkalinity (mg/L $CaCO_3$)	Chloride (mg/L)	Sulfates (mg/L)	Temp (°C)	Steady-State Pb Prediction (μg/L)	Pb Actual (μg/L)	Transient Pb Predicted (μg/L)
2	0.5	7.9	186	34	22	22.0	10	10	10
1	0.5	8.0	163	30	91	22.0	5		10
2	0.5	7.9	186	34	22	22.0	10		10
3	0.5	8.2	62	60	175	22.0	6		10
2	0.5	7.9	186	34	22	22.0	10		10
3	0.5	8.2	62	60	175	22.0	6		9
3	1	8.2	62	60	175	22.0	6		9
1	0.5	8.0	163	30	91	22.0	5		9
1	1	8.0	163	30	91	22.0	5		9
1	1.5	8.0	163	30	91	22.0	5		8

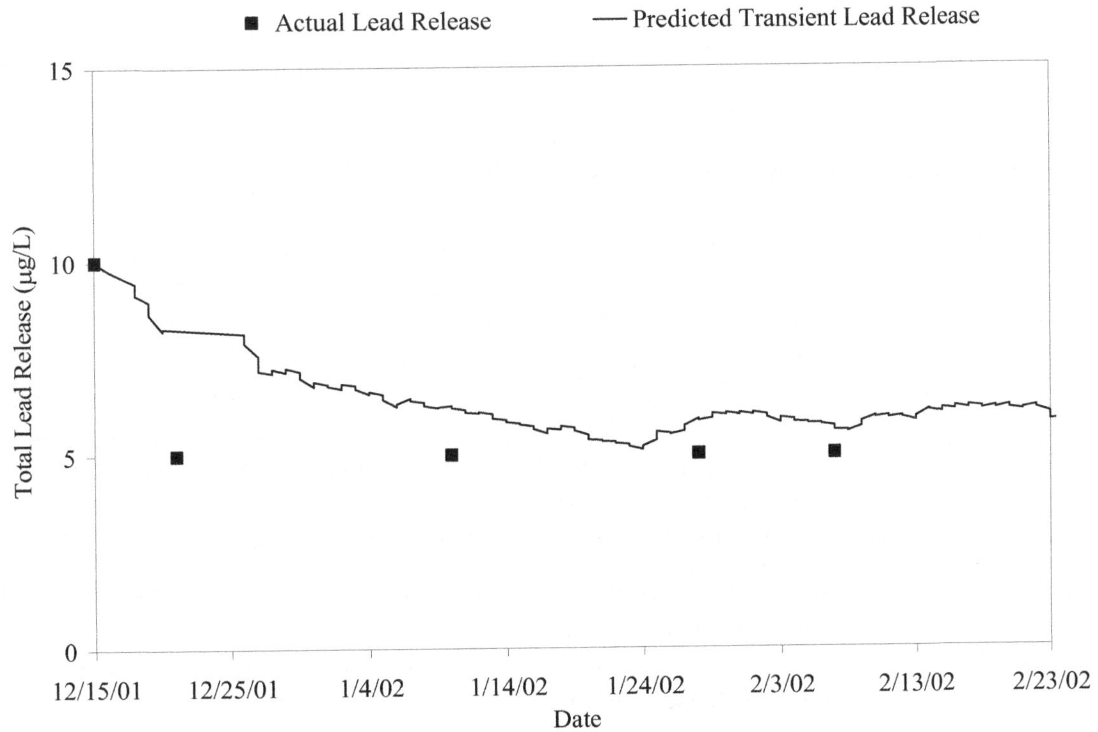

Figure 4.68 Transient lead release from corrosion loop fed by high frequency line (PDS 14) from 12/15/01 to 2/23/02

The simulation and data show a relatively constant total lead release despite of the rapid changes in PDS water quality. Given the 0.081/day rate constant, total lead released would reach just 4% of the steady-state total lead released in twelve hours following the change in water quality. As seen in Figure 4.66, the actual and simulated total released lead trend is negative, which corresponds to the difference between the predicted steady-state total lead release and the actual total lead release. The water in PDS 14 was over saturated with lead because the scales on the lead coupons were not at steady state and the total lead release was high. Hence it was decreasing. Also apparent in Figure 4.66 is the mitigating effect of time on a first order reaction when the reciprocal of the rate constant is significant. Changes every twelve hours make little difference when almost one thousand hours are required to reach steady state. Thus it can be concluded that due to the higher frequency of the change in blends any adverse effects resulting from the change in water quality were mitigated. Statistical hypothesis testing using a paired t test at the 95 % confidence interval was done to compare the actual and simulated transient total lead released. The results of the paired t test found there was no significant statistical difference between the actual and predicted values, hence the transient model using steady-state water quality to predict the total released lead driving forces was validated. There is a 3 ug/L error on 12/23/01; however, the transient model as shown in equation (4.68) did accurately predict time of transition and the final steady state lead concentration in a water quality environment that was changing frequently.

As with total released copper, the 50th percentile model is suitable for comparing the actual and predicted total lead release during the pilot study as average total released copper would be expected to be observed for any one observation. However, the 90th percentile model for total lead release during transition must be used to identify when the action level will be exceeded. The 90th percentile lead release model can be used as shown in this work or it could be modified to incorporate the 90th percentile and standard deviations of historical lead data collected by a MG.

Summary of Transient Lead Release in Aqueous Phase

Visual inspection of lead release data indicated that the release versus time response were typical of a first order exponential relationship that was driven by water quality change.

A first order transient total lead release model was developed by non-linear regression using an Arrhenius relationship to compensate for temperature and driving forces that were determined graphically. This model accurately described the actual total lead release following phase changes. The model was further developed using the steady-state water quality models to predict total release lead, which was used as the driving force. Both models were validated using 10% of the data set that was aside for validation.

The rate constant at 20°C ($k_{20^{\circ}C}$) for transient lead release model was 0.081 day^{-1} and θ was 1.05. Thirty-seven days were required for transient total lead release to achieve steady state following a steady-state water quality change.

The negative driving force and time required to achieve steady state for total lead release mitigated the transient total lead release in the high blend frequency PDS 14.

Member governments that anticipate a change in steady-state water quality that adversely impacts total lead release have the opportunity to mitigating actions such as chemical addition or mixing with other favorable sources to reduced total lead release. Additionally these actions can be taken following a steady state change in water quality as thirty-seven days are required before total release released achieves a new steady state.

Iron Transition

Transition of Iron Scale

To investigate the ongoing corrosion process, a kinetic study was conducted. The effects of incubation time on the surface characteristics were studied for four water blends (GW = 100%, SW = 100%, RO = 100%, GW/SW/RO = 62/27/11%). The coupons were incubated in these waters for one, three, and six months.

Three and six month samples were analyzed by XRD and Figure 4.69 presents the XRD pattern for the coupon incubated in GW (100%) water. The results indicated that the major corrosion products are almost the same for three and six month incubation, which means that incubation time doesn't affect surface crystalline composition qualitatively. Nevertheless, comparing the three and six month samples, they still showed some differences: significant SiO_2 appeared in the six month samples for GW and SW water, which indicated the continuing deposition of SiO_2 on the surface, that possibly alleviated iron release from the surface corrosion layer. It also showed that less crystalline FeOOH existed in six month samples than 3-month samples, which implied that crystalline FeOOH might have transformed into amorphous or other species.

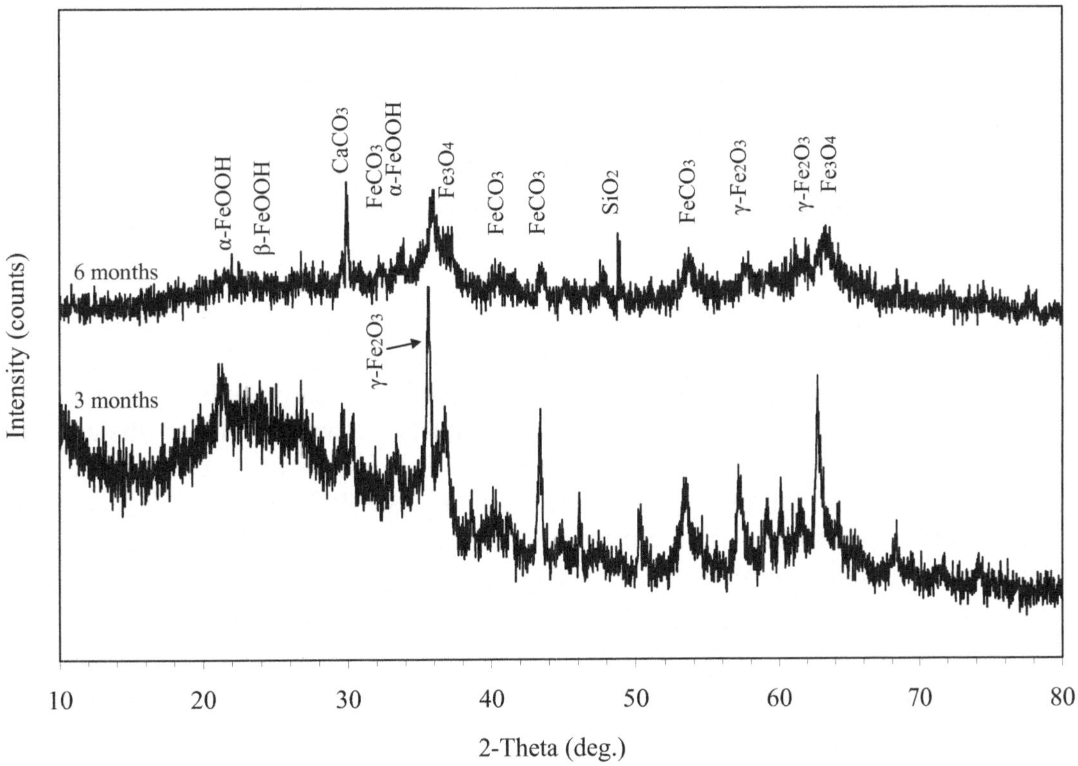

Figure 4.69 XRD pattern for UCI kinetic study incubated in 100% groundwater (G1)

XPS analysis was conducted for the one, three, and six month samples. The results showed that FeOOH, Fe_2O_3, and Fe_3O_4, are the major corrosion products in the top surface layer. From previous studies, Fe_2O_3 was found to correlate well with total and dissolved iron. Because the detected Fe_2O_3 might have been transformed from $FeCO_3$, and the observed dissolved iron concentrations were much higher than Fe_2O_3 solubility, the correlation of Fe_2O_3 with iron release might imply the correlation of $FeCO_3$ with iron release. So the effect of incubation time on Fe_2O_3 peak area % was investigated and shown in Figure 4.70.

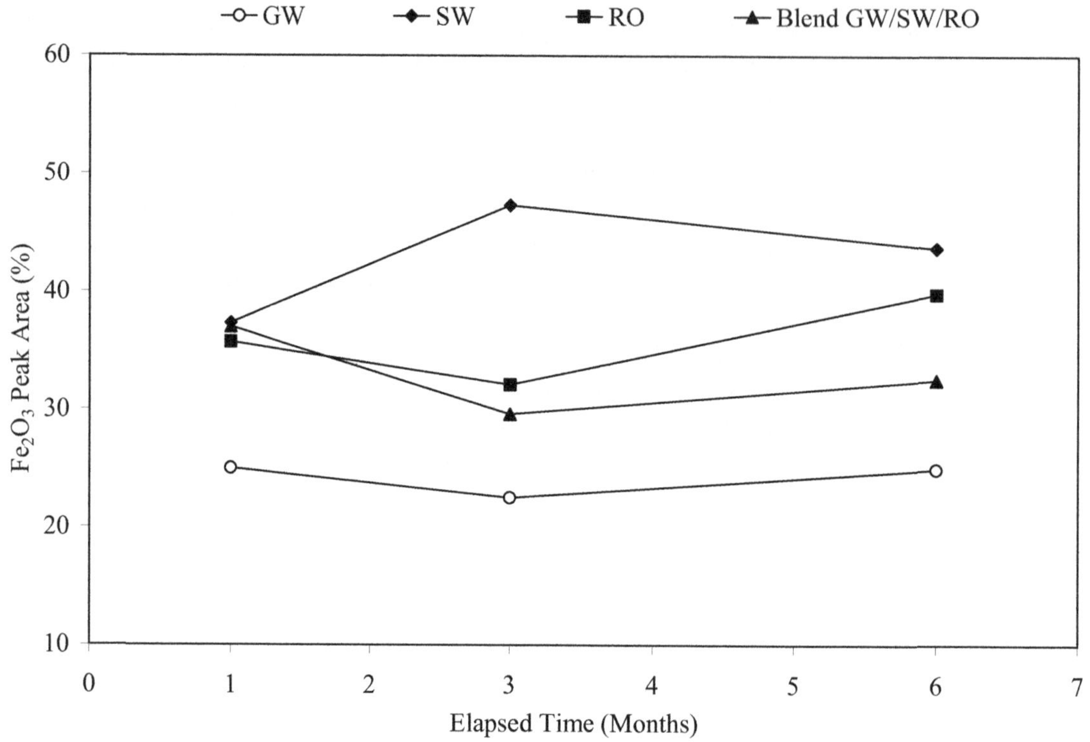

Figure 4.70 Effect of incubation time on surface characteristics (Fe2O3 Peak Area %)

Figure 4.70 shows that Fe_2O_3 peak area percentage does not change much when the incubation time was one, three, or six months, implying Fe_2O_3 ($FeCO_3$) content in the surface layer reached equilibrium quickly (in one month). Therefore, the iron release level will be stable in less than one month after the old cast iron pipe contacts the water. The results also show that even the Fe_2O_3 peak area % changed to some degree under various incubation times and different source waters, all of them follow the trend: SW>RO>GW/SW/RO>GW, which is consistent with what was observed previously. Since previous research found that higher Fe_2O_3 ($FeCO_3$) area percent was corresponding to higher iron release, the trend could explain why SW water showed the highest iron release, followed by RO. Note that all predicted iron, copper or lead release depends on water quality. The source water quality used to simulate iron release is shown in Table 4.14 resulted in higher predicted iron release from RO than SW using the empirical water quality model.

Transition of Total Iron Release in Aqueous Phase

Transitional model development for total iron release in G and UCI pipes is similar to the approach followed for the copper and lead transition. In addition to water quality, variability exists due to the effect of geometric and hydraulic characteristics of the distribution systems. An effort has been made to complement the transitional iron release model with the water quality and flux models developed previously. This would assist the member governments to identify potential problems based on the specific geometry and hydraulic conditions of their distribution systems.

Water quality and flux models were developed previously that have the ability to describe the steady-state total iron release due to change in water quality and hydraulic characteristics of the G (equations (4.4) and (4.21)) and UCI (equations (4.3) and (4.20)) distribution systems. Transient behavior of iron release between different steady states can be described by determining the θ and k terms in the transient response equation developed earlier (equation (4.62). The incorporation of flux model into the transient iron release model facilitates the visualization of the transient iron release in different pipe geometries and hydraulic regimes. However, since the primary development is for laminar flow conditions under high hydraulic retention times, application to turbulent conditions is qualitative.

The modification of input water quality occurred between each phase for all PDS. Examination of data during the phase changes for the single material UCI (PDS15) and G PDS (PDS18) permits an evaluation of the speed of the system response to alteration of water quality. The operation of PDS14, also known as the high frequency PDS, was slightly different. For PDS14, the blends were changed twice per day in a random pattern to simulate water quality that would be experienced with only two of the three principal sources (GW, SW, or RO) available. Thus, this section of the summary report is an investigation of the short term behavior (several weeks) with respect to apparent color, turbidity and iron release in distribution systems following a water quality change produced by blending ground, surface and RO finished waters.

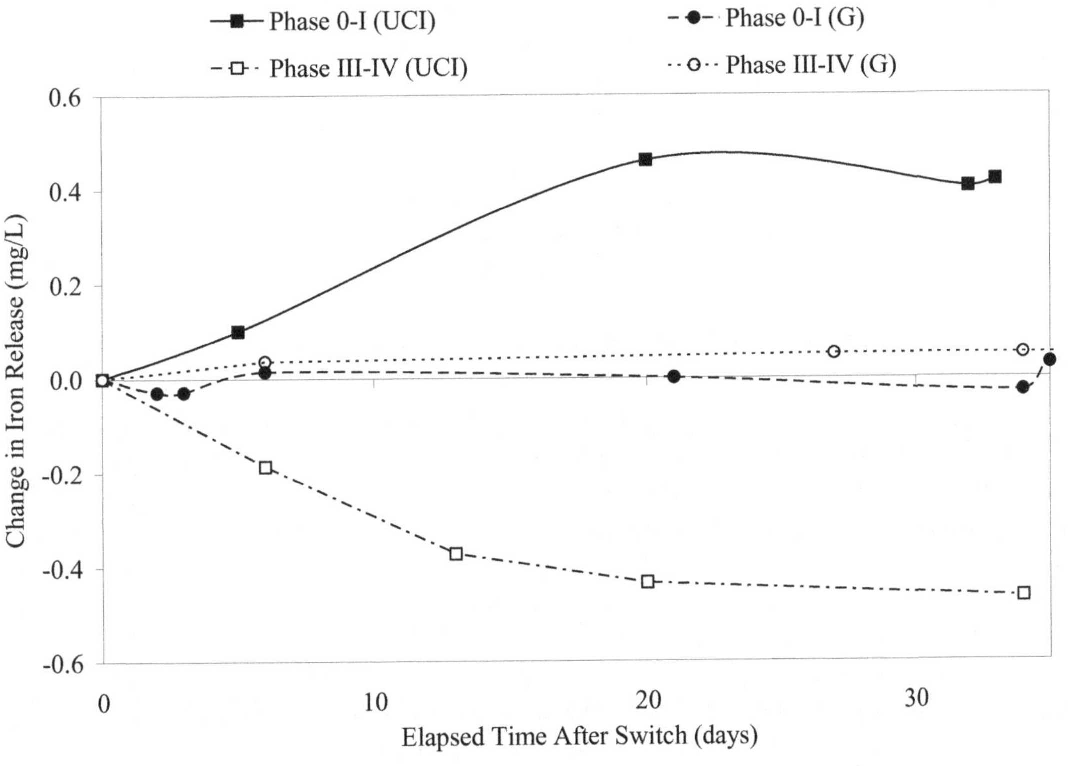

Figure 4.71 Typical response of change in total iron release with time elapsed from the change in water quality (G and UCI)

Data available for phase transitions from Phase 0-I, I-II, II-III and III-IV for PDS 15 and PDS 18 (UCI and G iron pipes) was used for the development of the transitional iron release model. Iron time series data immediately following each phase change were examined to define the process characteristics. The typical iron release response for both G and UCI is shown in Figure 4.71. It can be seen from Figure 4.71 that the transient response for UCI was completed at approximately 20 days. The trend for G is however not clear due to the relatively low iron release associated with the particular PDS. Therefore the entire phase data (90 days) was used in developing the transient response of G pipe.

The change in input blend could be represented as a step change in the input water quality. It is noted that coincident with the change in blend, the PDS were flushed with the new blend source, resulting in an immediate replacement of all of the PDS contents with the new source water. The transient response (i.e., the change in iron concentration as a function of time after the blend change) exhibited a pattern consistent with a first-order step response, as defined in equation (4.69). In this equation, the effluent iron concentration Fe_t represents the change in iron concentration exiting the reach of pipe relative to the concentration entering the reach of pipe. If the water quality entering the reach of pipe has minimal iron content (typical of finished SW or RO), the reported value for Fe_t would approximately equal the iron concentration in the output from the reach of pipe. For other conditions where the concentration of iron in the water entering the reach of pipe is not negligible (typical of GW), the iron concentration in the feed water must be added to Fe_t to approximate distribution system concentrations.

$$Fe_t = \Delta Fe_1 + (\Delta Fe_2 - \Delta Fe_1)\times\left(1 - e^{\left(-\theta^{(T-20)}\,k\,t\right)}\right) \tag{4.69}$$

where Fe_t = effluent total iron released after t days from time of phase change (mg/L)

ΔFe_1 = steady-state total iron before change (mg/L)

ΔFe_2 = effluent total iron at steady-state after the phase change (mg/L)

θ = iron temperature correction factor

k = iron rate constant (day^{-1})

t = time after change (days)

The time series data were evaluated using a non-linear least squares regression to estimate numerical values for the rate constant (k) and temperature correction factor (θ) for both pipe materials. The resulting estimates of the rate constant are provided in Table 4.30. Statistically determined values of the rate constant (k) were 0.116 and 0.068 day^{-1} for the UCI and G PDS, respectively. The respective temperature correction factors (theta) were 1.02 and 1.07. The R^2 for the regression was 0.97 and 0.79 for the UCI and G pipes respectively. Paired t-test for difference between the actual and predicted total iron released using 10% of the data that was not used in model development. The t-statistic for the difference in actual and predicted total iron released from the UCI PDS was 0.8 and less than the critical t statistic (3.18) of a two tailed t-test at 4 degrees of freedom and a 95 % confidence interval. The actual t-statistic for the G PDS was 1.32, which was less than the critical t statistic (2.36) for a two tailed t-test at 7 degrees of freedom and 95 % confidence interval. Hence, no statistical difference was found between the actual and predicted total released iron from UCI or G PDS. It is noted that operation throughout these transition periods corresponded to laminar flow conditions. Similar data for turbulent flow transitions in water quality is not available.

Table 4.30
UCI and G pipes rate constants (k) for each phase change

Pipe Material	k (day^{-1})	θ
UCI	0.116	1.03
G	0.068	1.07

Considering a rate constant of 0.116 day^{-1} it would take approximately 26 days (20°C) to reach a new steady state following the change in water quality. This duration is confirmed by examining Figure 4.71. To clearly visualize the first order response following the change in water quality a three dimensional plot describing driving force, change in iron release concentration in UCI PDS concentration following change in water quality and time elapsed is shown in Figure 4.72.

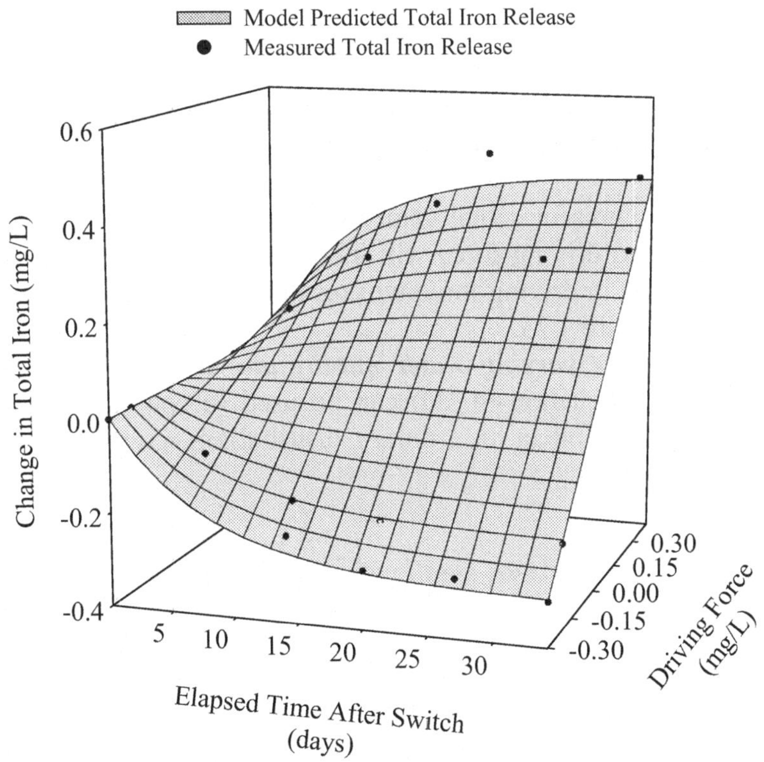

Figure 4.72 Three dimensional plot of change in total iron release, driving force and time elapsed following change in water quality in UCI pipe

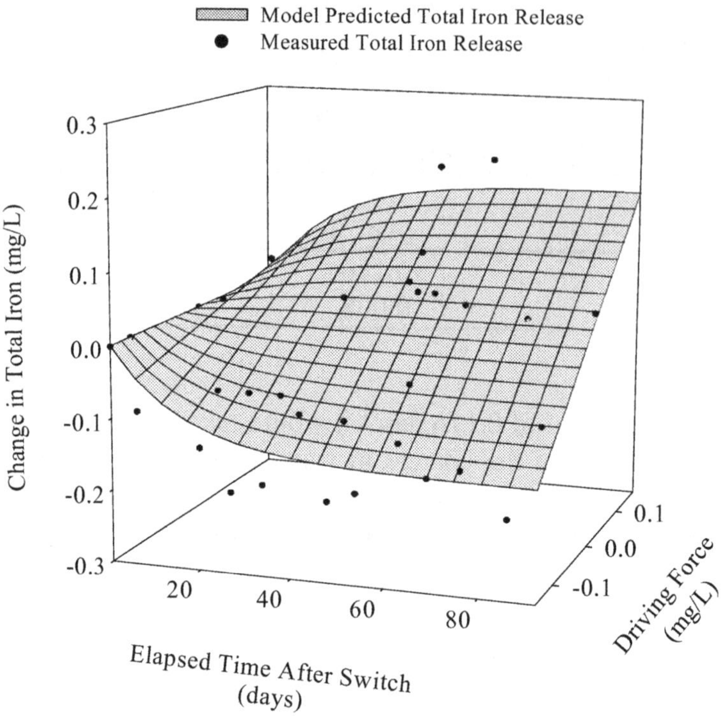

Figure 4.73 Three dimensional plot of change in total iron release, driving force and time elapsed following change in water quality in G pipe.

The dots in the plot represent actual values while the surface shows the regression line. The plot reveals that increased driving force will increase the total iron concentration reaching the consumer while a decrease in driving force will result in a lower total iron release. A similar plot is shown for G pipe in Figure 4.73.

Transient release under varying water quality and hydraulic conditions. The flux model developed previously (equation (4.21)) can be used to enable simulation of total iron release during the transition period in an arbitrary distribution system for any arbitrary water quality and hydraulic condition.

$$Fe_t = \Delta Fe_{ss,1} + \left(\Delta Fe_{ss,2} - \Delta Fe_{ss,1}\right) \times \left(1 - e^{\left(-\theta^{(T-20)}\, k\, t\right)}\right) \qquad (4.70)$$

where Fe_t = effluent total iron released after t days (mg/L)
$\Delta Fe_{ss,1}$ = steady-state iron concentration before change (mg/L)
$\Delta Fe_{ss,2}$ = steady-state iron after change (mg/L)

Water quality and flux models were developed previously that have the ability to describe the steady-state total iron release due to change in water quality and hydraulic characteristics of the G (equations (4.4) and (4.21)) and UCI (equations (4.3) and (4.21)) distribution systems. To confirm the applicability of the steady-state iron release model $\Delta Fe_{ss,1}$ and $\Delta Fe_{ss,2}$ were predicted and used in equation (4.70) for the single material UCI and G PDS. A typical plot of predicted versus actual change in total iron with time is shown in Figure 4.74.

Figure 4.74 shows the ability of the transient model to closely predict the total iron concentrations during the transition period by applying the steady-state water quality and flux models to predict the driving force. The t-statistic for the UCI PDS was 1.57, which was less than the critical t statistic of 2.08 for a two tailed t-test at 15 degrees of freedom and 95 % confidence interval. The t-statistic for the G PDS was 1.11, which was less than the critical t statistic of 2.01 for a two tailed t-test at 49 degrees of freedom and 95 % confidence interval. The results indicate the transition model accurately describes the total iron release from UCI and G pipes.

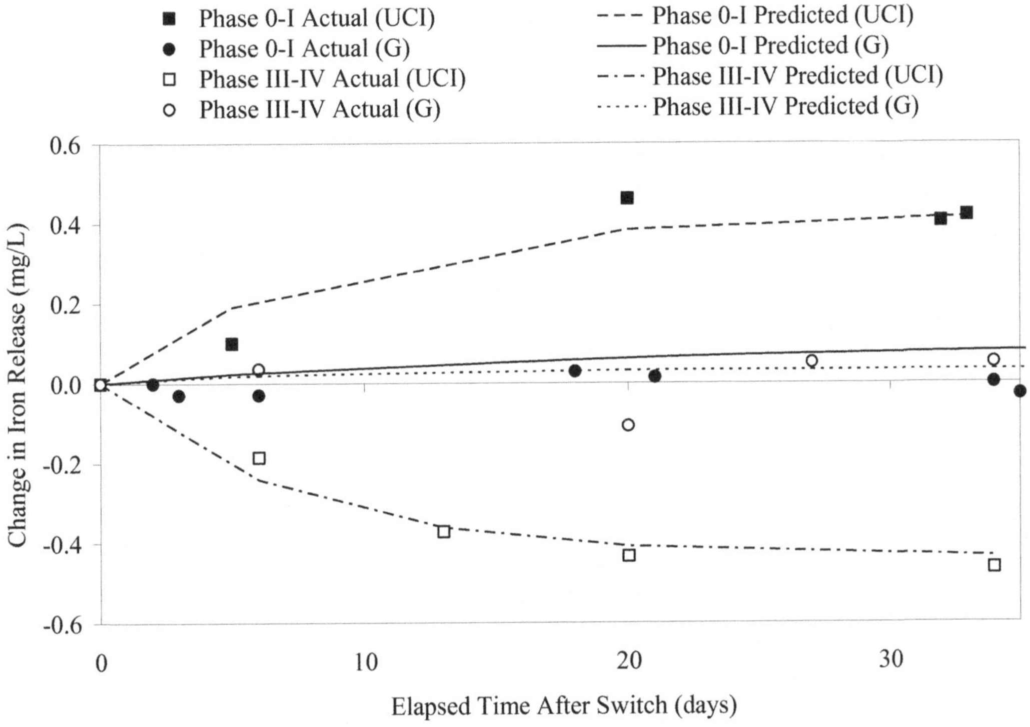

Figure 4.74 Actual versus predicted change in total iron with time elapsed from the change in water quality

Transient Total Iron Release Simulations for Different Pipe Geometries and Flow Conditions

The steady-state water quality and flux models developed previously can be used to enable simulation of total iron release during the transition period in an arbitrary distribution system for any arbitrary water quality and hydraulic condition. Completion of the calculations requires knowledge of the pipe material, pipe geometry (length and diameter), velocity (or flow rate), and water chemistry. The methodology of determining the steady-state iron release at any given water quality and hydraulic conditions is similar to the one outlined in the steady-state iron release section. The following simulations were performed for a 1 mile long UCI pipe while changing from 100% GW to 100% RO and 100% SW. The source water quality used for simulations is shown in Table 4.31. For brevity only UCI pipes are discussed. G pipes follow a similar trend but at far lower total iron release than UCI pipes. Therefore, UCI pipes can be considered the worst case scenario for iron release considerations.

Impact under laminar flow conditions. The simulations were performed for a 6 inch diameter, 1 mile length UCI pipe with flow velocity of 0.007 ft/sec (HRT = 8.73 days, Reynolds

Number = 350, Rate constant = 0.116 day^{-1}). Three cases were simulated, representing a change in input water quality from 100% GW to the three sources described in Table 4.31 (SW, RO, or a blend of 60% GW, 30% SW, and 10% RO). The results of the simulations are shown in Figure 4.75.

Table 4.31
Source water quality for simulations

Parameter	100% GW	100% SW	100% RO	Blend
Alkalinity (mg/L as $CaCO_3$)	225	50	50	155
Chloride (mg/L)	15	10	50	17
Sodium (mg/L)	10	15	30	13.5
Sulfate (mg/L)	10	180	30	63
Temperature (°C)	20	20	20	20
SiO_2 (mg/L)	14	10	4.5	11.9
DO (mg/L)	7.5	7.5	7.5	7.5

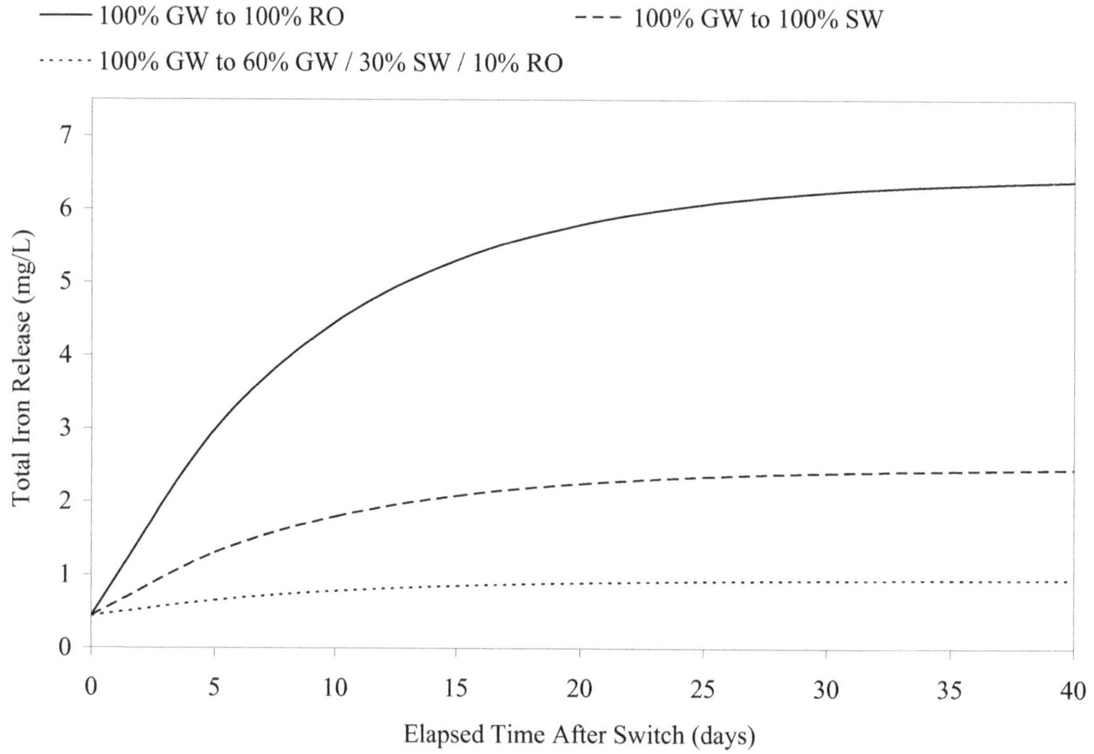

Figure 4.75 Transient simulation of blend change from GW to other sources/blends under laminar flow conditions, UCI pipe

As expected from the impacts of water chemistry, greater increases in total iron release were realized after switching from 100% GW to 100% RO than switching from 100% GW to 100% SW or 100% GW to 60% GW-30%SW-10%RO blend.

Impact under turbulent flow conditions. The simulations were also performed for an identical UCI pipe (6 inch diameter) with water flowing at 0.15 ft/sec velocity (HRT=9.8 hours). The results of the simulations are provided in Figure 4.76.

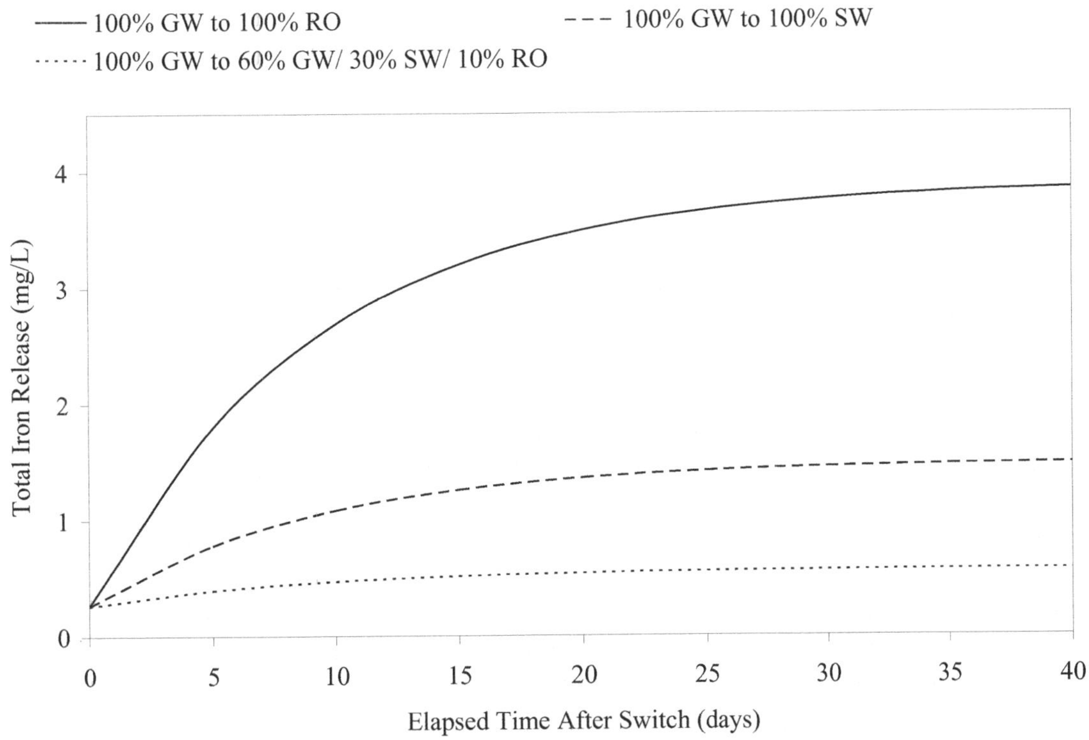

Figure 4.76 Transient simulation of blend change from GW to other sources/blends under turbulent flow conditions, UCI pipe

A comparison of predicted total iron concentration under laminar (Figure 4.75) and turbulent (Figure 4.76) conditions reveal that the potential for elevated iron concentrations is greater under laminar flow conditions. Laminar conditions usually exist in low-flow and dead-end regions of the distribution systems. This observation is consistent with the experience that red water conditions are characteristic of dead-ends in the distribution system. This observation is also consistent with the steady-state analysis reported previously.

Practical Implications of Iron Transition

The preceding discussion of transition data indicates that the dynamic response for a change in water quality will occur very gradually. Small values of the rate constant suggest that any alteration in water quality must be sustained for an extended period of time in order to express the change in concentration that would be predicted by the steady-state model. As noted previously, a period of three rate constants ($k\ t = 3$) is widely accepted as the practical duration required to reach a new steady state after a disturbance. The time duration to achieve steady state is therefore estimated to be approximately 26 days for the UCI system and almost 44 days for the G system. These durations are confirmed by examination of the raw data as plotted in Figure 4.71. Using the average estimates of the rate constant for the two pipe materials reported in Table 4.30, the portion of the maximum increase in iron concentration can be calculated for various exposure durations to an altered water quality. The portion of the steady-state iron increase is defined by equation (4.71) and can be visualized in Figure 4.77.

$$\frac{Fe_t - \Delta Fe_1}{\Delta Fe_2 - \Delta Fe_1} = 1 - e^{\left(-\theta^{T-20}\, k\, t\right)} \tag{4.71}$$

As indicated in Figure 4.77, a disturbance in water quality with one day duration would cause only a slight change in iron concentration. For the cast iron pipe, with a rate constant of 0.116 day^{-1}, the one-day duration would result in a fractional response of 11% of the steady-state change in iron concentration. For example, if the anticipated steady-state increase in iron concentration associated with a blend change is 1.5 mg/L, but the duration of the blend change is only one day, then the actual experienced increase in iron concentration would be 0.16 mg/L (=0.11 x 1.5 mg/L). For the G pipe (rate constant = 0.068 day^{-1}), the corresponding increase in iron concentration after one day would be only 7% of the steady-state change. Thus these dynamics suggest that unless a change in water quality is sustained for many days, the associated impairment in iron concentration would be much smaller than predicted by the steady-state models discussed previously.

This conclusion is supported by the data obtained from the high frequency line (PDS14). A plot of iron release data for the first year of operation is provided in Figure 4.78. The data indicate a relatively constant iron value for the high frequency PDS, despite rapid changes in the input blend. For comparison, the corresponding data for the three principal sources are also provided in Figure 4.78. The response of the high frequency PDS is centered about a value that would be expected for a blend of the three principal blends.

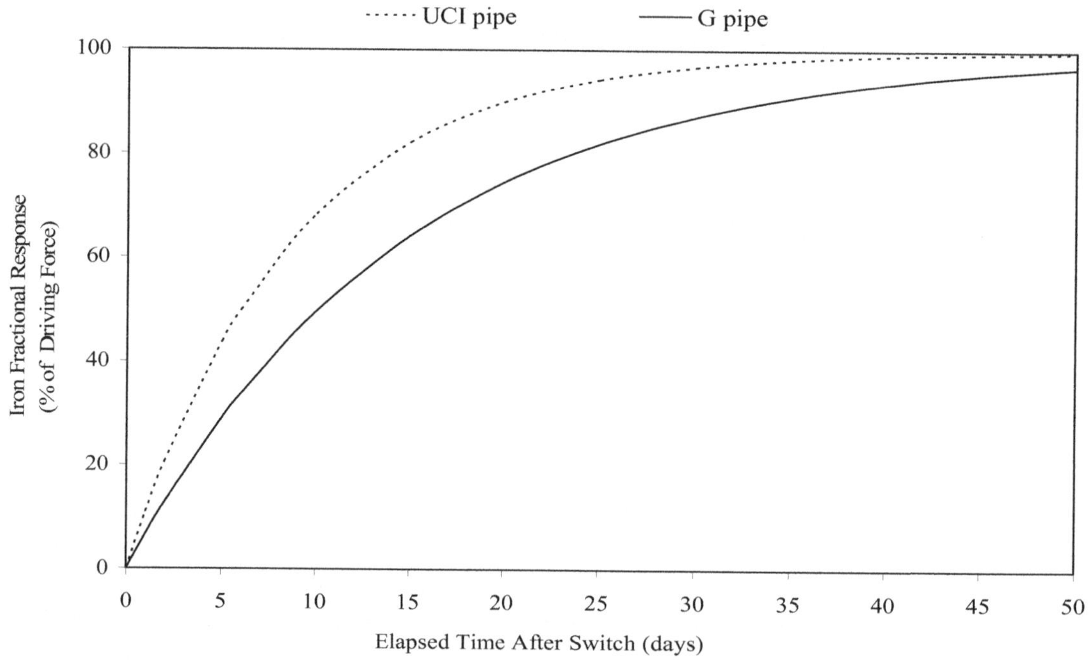

Figure 4.77 Percentage change in total iron release with time elapsed from the change in water quality for UCI and G pipes

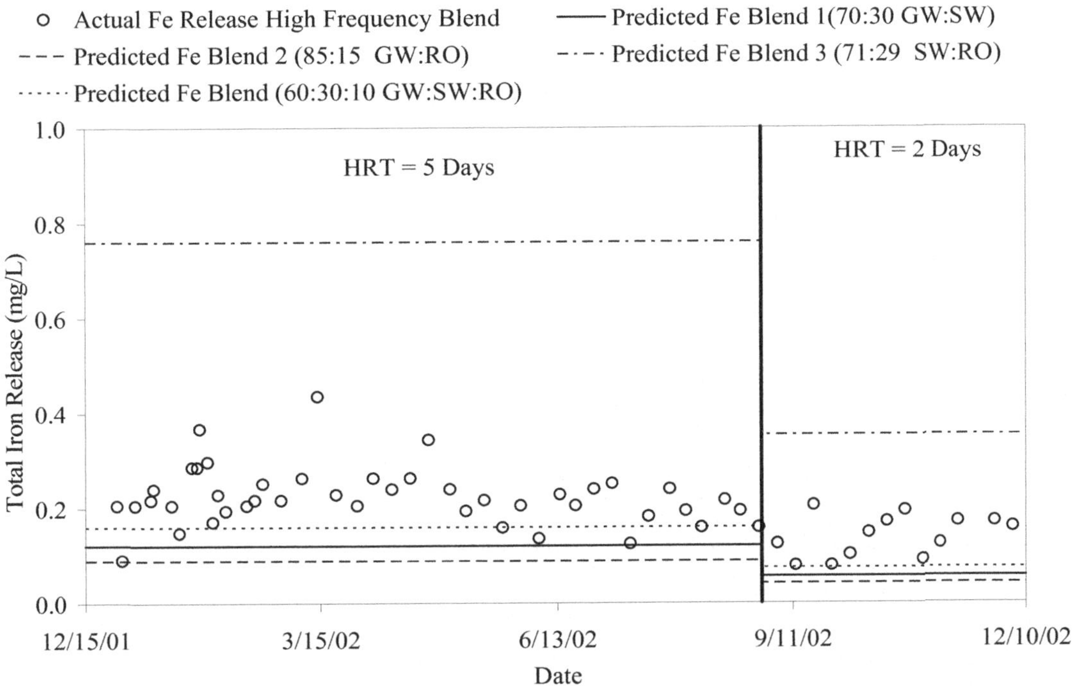

Figure 4.78 Observed iron release from GW, SW, RO water and high frequency blend PDS (12/15/01 to 12/10/02)

Iron release simulations were carried out on the data available from 12/15/01 to 2/23/02 (Phase I). The average source water quality was used for each source and the blend water quality determined by mass balance on the blend percentages. Table 4.32 shows a typical blend change regime and the water qualities associated with the particular blend. Steady-state iron release model developed earlier was used to determine the driving factor for the transient release behavior. Figure 4.79 shows the corresponding transient iron release simulations for the period between 12/15/01 to 2/23/02.

Thus, it can be concluded that due to the higher frequency of the change in blends any adverse effects resulting from the change in water quality were mitigated. Statistical hypothesis testing of the actual and simulated transient iron release data was made using t-test between means. The actual and critical t statistics were 2.1 and equal, which means there was no practical difference between the actual and predicted total iron released in PDS 14 (the high frequency line) and the transition model accurately described total iron release during rapid blend change.

Table 4.32
Transient water quality and simulation for iron release in high blend frequency line (PDS 14)

Blend	Time (Days)	Alky (mg/L $CaCO_3$)	Cl (mg/L)	Na (mg/L)	SO_4^{2-} (mg/L)	DO (mg/L)	T (°C)	HRT (Days)	Steady-State Fe Prediction (mg/L)	Fe Actual (mg/L)	Transient Fe Predicted (mg/L)
2	0.5	186	34	21	22	8	22.0	5	0.09	0.22	0.22
1	0.5	163	30	21	91	8	22.0	5	0.12		0.21
2	0.5	186	34	21	22	8	22.0	5	0.09		0.21
3	0.5	62	60	39	175	9	22.0	5	0.68		0.23
2	0.5	186	34	21	22	8	22.0	5	0.09		0.22
3	0.5	62	60	39	175	9	22.0	5	0.68		0.25
3	1	62	60	39	175	9	22.0	5	0.68		0.30
1	0.5	163	30	21	91	8	22.0	5	0.12		0.29
1	1	163	30	21	91	8	22.0	5	0.12		0.27
1	1.5	163	30	21	91	8	22.0	5	0.12		0.24

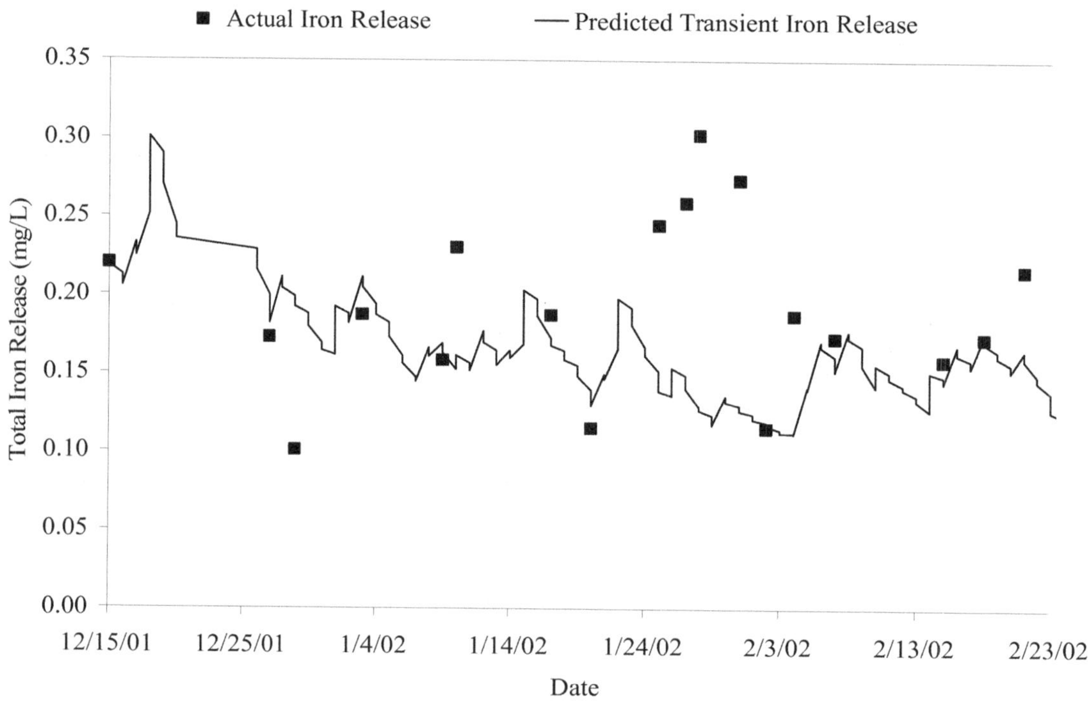

Figure 4.79 Transient iron release from high frequency line (PDS 14) from 12/15/01 to 2/23/02

Summary of Transient Total Iron Release in Aqueous Phase

Visual inspection of total iron release data indicated that the release versus time response were typical of an exponential relationship that was affected by water quality change.

Transient iron release model was developed which was consistent with the observed exponential relationship of observed release following phase changes in both the G and UCI pipes. The model developed was verified on 10% of data set aside for validation as well as for the steady-state model predicted driving forces.

The value of the rate constant at 20°C ($k_{20^\circ C}$) for transient total iron release model was found to be 0.116 day^{-1} and the value of θ was found to be 1.02 for the UCI pipe. This indicated that at least 26 days are needed for the total iron response in UCI pipes to reach a steady state following a change in water quality.

The transient response of the high blend frequency line PDS 14 was consistent with the low rate constant (k) that would require at least 26 days for steady-state total iron release.

Member governments that anticipate a change in water quality that impacts iron release should be aware that though the anticipated impact may take 26 days to stabilize, any remedial action taken by changing water quality will take an equal amount of time to have a noticeable effect.

The value of the rate constant at 20°C ($k_{20^\circ C}$) for transient total iron release model was found to be 0.068 day^{-1} and the value of θ was found to be 1.07 for the G pipe. This indicated that at least 44 days are needed for the total iron response in G pipes to reach a steady state following a change in water quality.

The effect of water quality on the transient response was consistent with the steady-state predicted driving forces. That is higher iron release at elevated chlorides, sulfates, sodium, DO, temperature and HRT and lower iron release due to elevated alkalinity. Thus GW and blends rich in GW (> 60% GW) would exhibit a comparatively lower iron release compared to RO and SW sources.

Simulations on the effect of laminar or turbulent flow conditions indicate higher potential iron concentrations at laminar conditions.

HPC Transition

HPC Transition Phases I thru IV

HPC transition was investigated throughout the normal phase changes, but the sampling intervals for monitoring HPCs were too far apart to provide any insight into HPC transition. HPCs, like AOC and BDOC were sampled at the beginning and end of each phase. A summary of HPC transition during phase changes through the first year is shown in Figure 4.80 with chloramine residual.

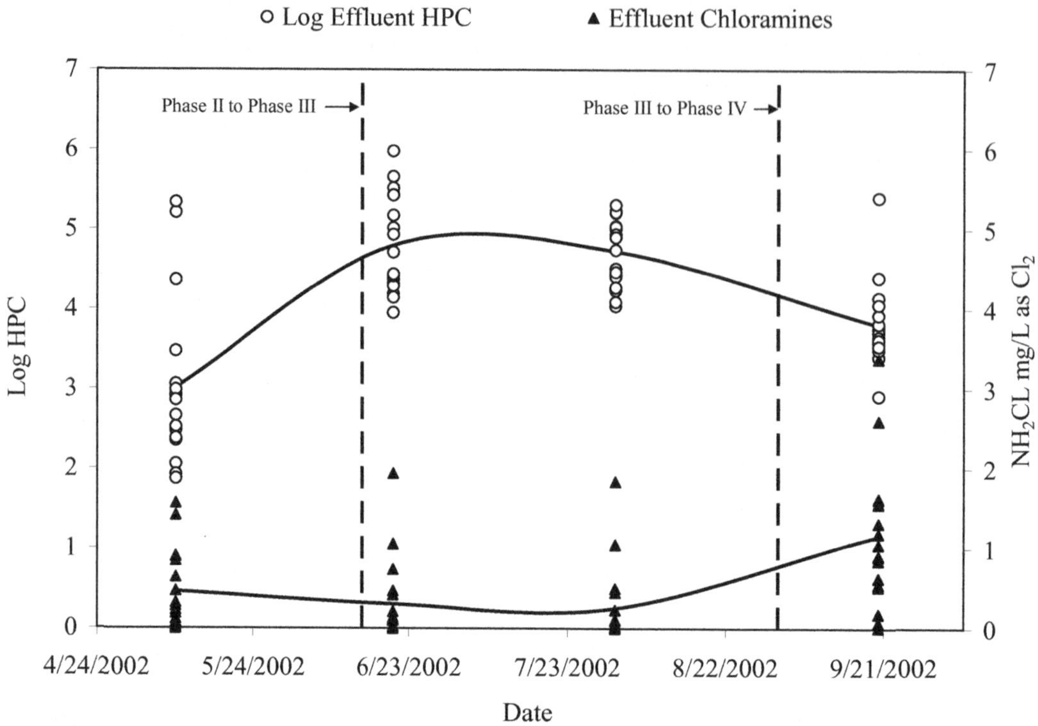

Figure 4.80 HPC and chloramine residuals through phase II to III and through phase III to IV

The average Log HPCs and NH_2Cl residuals for all PDS are shown in Figure 4.80 and show the inverse relationship between bacterial growth and residual in drinking water distribution systems. When residuals are relatively high based on the data in Figure 4.80, log HPCs are low, when residuals are relatively low (near zero), log HPCs are high. The effect can be grossly described by the exponential relationship shown in Figure 4.81. Clearly, transition effects on HPCs due to water quality change from phase to phase can not be assessed by the data collected during the first year of field investigation. However, a relationship between HPCs and residual is shown and stresses the need for residual maintenance in drinking water distribution systems.

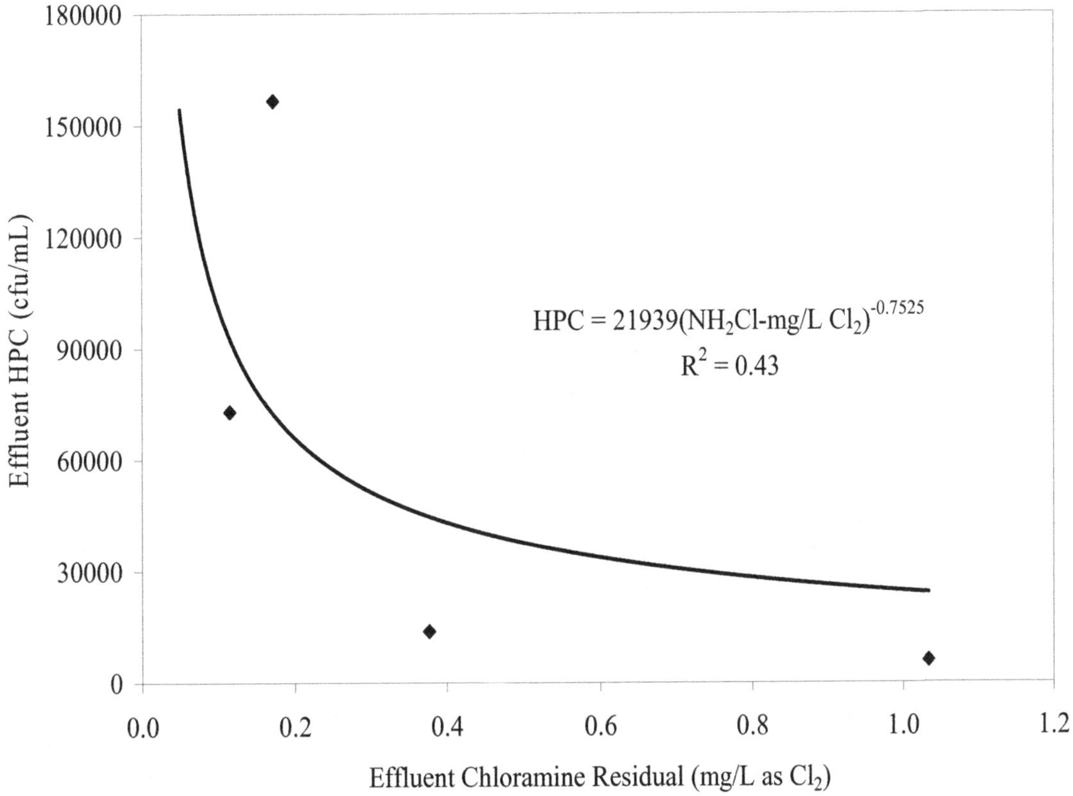

Figure 4.81 PDS average HPCs versus PDS average chloramine residual for the first year of field operation

High Frequency HPC Transition

At the end of phase VI, a secondary investigation on the effects of water quality change (phase change) on HPC transition was conducted. At the end of phase VI, the blends were changed in PDS 1 through 8 as shown in Table 4.33 and the high HPC sampling frequency experiment during transition was conducted. The blends were changed by switching the feed lines among PDSs.

Table 4.33
High frequency testing for HPC change during phase transition

	Date of Change	Source Water Change	Disinfectant	HPC sampling
PDS 1	6/27/2004	100 % G1 to 100 % S1	Cl_2	3-4/wk
PDS 2	6/27/2004	100 % S1 to 100 % G1	Cl_2	3-4/wk
PDS 3	6/27/2004	100 % RO to 63%G1/27%S1/11%RO	Cl_2	3-4/wk
PDS 4	6/27/2004	63%G1/27%S1/11%RO to 100 % RO	Cl_2	3-4/wk
PDS 5	6/27/2004	100 % G1 to 100 % S1	NH_2Cl	3-4/wk
PDS 6	6/27/2004	100 % S1 to 100 % G1	NH_2Cl	3-4/wk
PDS 7	6/27/2004	100 % RO to 63%G1/27%S1/11%RO	NH_2Cl	3-4/wk
PDS 8	6/27/2004	63%G1/27%S1/11%RO to 100 % RO	NH_2Cl	3-4/wk

HPC transition over time. Log HPCs versus time through the high sampling frequency tests are shown in Figure 4.82 and Figure 4.83 for free chlorine and chloramine residuals respectively. The figures are shown side by side to facilitate comparison of HPCs in Cl_2 and NH_2Cl environments. The transition date (water quality change) is shown by a vertical dotted line in Figure 4.82 and Figure 4.83. The change in steady state water quality was implemented on June 27, 2002 in PDS 1-8. The HRT in PDS 1-8 was two days, but as shown in Figure 4.82 and Figure 4.83 almost five days is required before steady state water quality is achieved, which would correspond to the time when HPCs release was maximized. Hence alkalinity change following the steady state water quality switch in PDS 1-8 indicates the HPC and alkalinity changes were similar over time, which further supports the correlation of steady state water quality change and HPC release in the PDS. However, it is possible the HPC transition could have been affected by the physical switching of lines among PDS 1-8. The removal of the flexible feed lines from the various peristaltic pumps controlling finished water flow rates from the selected finished water storage tanks may have resulted in disruption of biofilm that discharged a HPC spike to the PDSs. Such action could cause HPC peaks in all PDSs irrespective of water quality. The two causal factors (bacterial release from biofilms due to chemical change in the bulk water and physical disruption of films during a change of feed lines) should be investigated further.

The sharp rise in HPCs after the transition data was irrespective of PDS residual concentration and form as the same magnitude of HPC change occurred in PDS 1-8 regardless of residual or concentration. However the HPCs in the chloraminated PDSs (5 through 8) are one to 1.5 logs higher than the HPCs in the free chlorine PDSs (4 through 8) before the water quality change. Although the HPC population was significantly higher in the chloraminated PDS relative to the free chlorine PDS, the maximum HPC concentration achieved in the presence of either residual were approximately equal, meaning nearly the same mass of HPCs were released in the presence of either residual. Since the original HPC concentrations in the free chlorine PDSs were lower than in the chloraminated PDSs, more HPCs would have to be released from the free chlorine PDSs to equal the HPCs in the chloraminated PDSs. However, since both the chlorine and chloraminated PDSs approach six log HPCs, this difference is practically insignificant. Finally, there is some indication of a downward HPC trend in most of the PDSs.

Clearly, a change in steady state water quality or something occurring in the same time period and the steady state water quality change caused a significant release of HPCs in all PDSs involved in the high frequency HPC sampling investigation. However, it can not be dismissed that the HPC change may have been caused by switching the feed lines among PDSs.

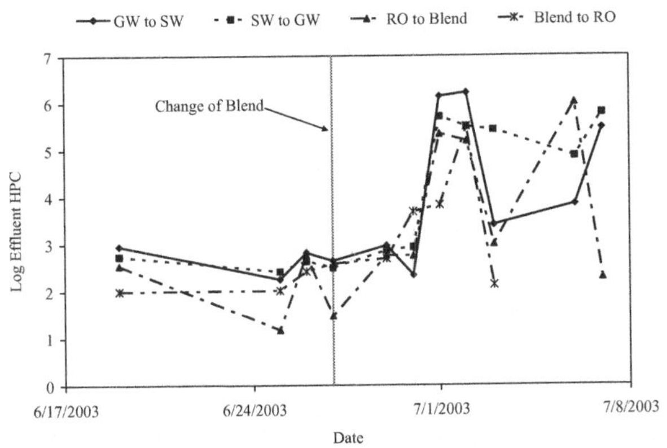

Figure 4.82 PDS effluent Log HPC by source switch as a function of time, for free chlorine residual

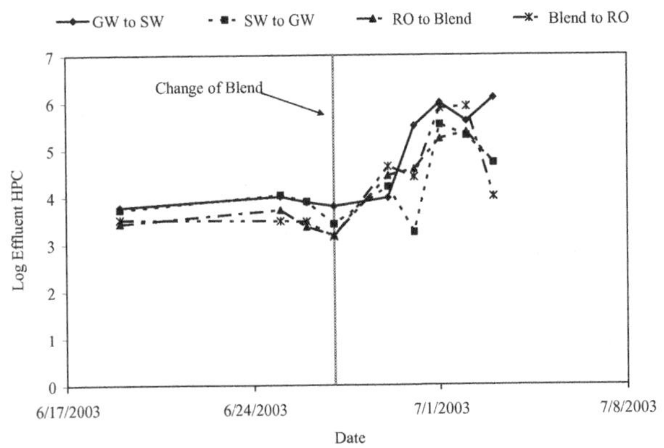

Figure 4.83 PDS effluent Log HPC by source switch as a function of time, for chloramine residual

HPC transition modeling. As shown in Figure 4.82 and Figure 4.83, the HPC transition peaks about four to five days after the water quality switch. PDS effluent stream alkalinity was monitored to determine mass transport through the PDS and was modeled using equation (4.72). Driving forces and rate constants are required for equation (4.72) and were determined by direct observation of alkalinity before and after the water quality change and non-linear regression respectively. The rate constant was significant at a 0.0001 confidence interval. Predicted and actual alkalinity versus time following the water quality change in the PDS is shown in Figure 4.84. R^2 was 0.95 for the mixing model. The significance of the PDS mixing model is that it accurately describe flow distribution through the PDS and that the reciprocal of the rate constant, ($k = 0.72$ days^{-1}) would be the time constant (1.40 days). The time constant represents the time for 62 % of the mass entering the PDS to pass through the PDS. Ninety-five percent of the mass entering the PDS will pass through the PDS in approximately 4.2 days, which corresponds to the point of maximum HPC release in PDS 1 through 8.

$$Alk_t = Alk_0 + \Delta Alk_{ss} \times \left(1 - e^{(kt/)}\right) \quad (4.72)$$

where, Alk_t = Effluent alkalinity after switch (mg/L $CaCO_3$)
Alk_0 = Steady state alkalinity before switch (mg/L $CaCO_3$)
Alk_0 = Steady state alkalinity after switch (mg/L $CaCO_3$)
ΔAlk_{ss} = $Alk_1 - Alk_0$ (mg/L $CaCO_3$)

k	=	First order time constant ($days^{-1}$)
t	=	Elapsed time after water quality switch (days)

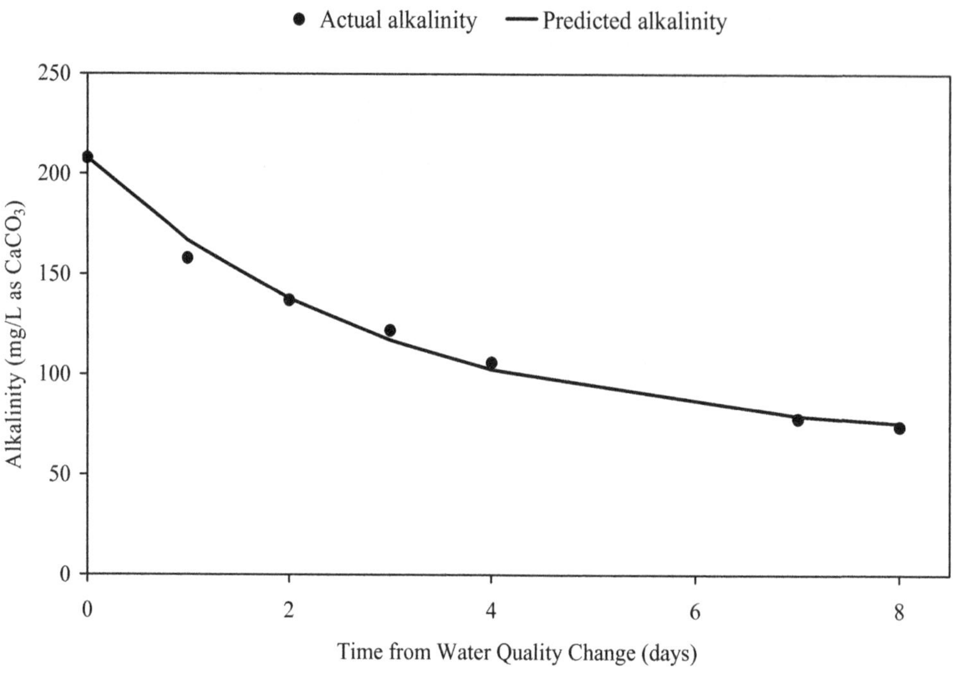

Figure 4.84 Alkalinity transition during HPC transition investigation for PDS-1

Equation (4.72) describes alkalinity transition during PDS water quality changes and it does provide a quantitative means of determining when transition is complete in the PDS effluent. However, it does not peak and then decrease as shown in Figure 4.80 and Figure 4.81 for HPC variation during the HPC transition experiment. The HPC transition data was regressed against time and in the second order function shown in equation (4.73) to develop a model that would predict HPC release and duration. Although a composite model for each PDS was statistically significant, the R^2 for the composite model was 0.00, which indicates the HPC release could not be accurately described for all PDSs by a general composite model. As shown in Table 4.34, the R^2 are acceptable when the model is applied to each PDS separately. The HPC transition model does predict the HPCs will return to their original values in approximately ten days and peak at approximately five days for all PDSs. Predicted and actual Log HPC for free chlorine and chloramines from a blended to a RO finished water are shown in Figure 4.85 and Figure 4.86 for PDS 4 and 8 respectively.

Table 4.34
Statistical parameters for HPC transition model

Term	PDS-1	PDS-2	PDS-3	PDS-4	PDS-5	PDS-6	PDS-7	PDS-8	Composite
$Log(HPC_0)$	2.67	2.58	2.10	2.30	3.90	3.79	3.62	3.60	3.09
A	1.64	1.69	1.96	2.99	2.12	0.68	1.75	2.49	1.59
$k_{hpc}(day^{-1})$	0.252	0.201	0.286	0.584	0.370	0.182	0.417	0.518	0.885
p-Value $Log(HPC_0)$	0.00	0.00	0.01	0.00	0.00	0.00	0.00	0.00	0.00
p-Value for A	0.12	0.00	0.12	0.22	0.00	0.41	0.07	0.17	0.00
p-Value for k	0.05	0.00	0.04	0.01	0.00	0.61	0.03	0.08	0.00
R^2	0.50	0.88	0.48	0.53	0.97	0.51	0.79	0.62	0.00
Conf. Interval	0.06	0.00	0.07	0.10	0.00	0.12	0.00	0.00	0.00

$$Log(HPC) = Log(HPC_0) + A \times time \times exp(-k_{hpc}t) \quad (4.73)$$

where $Log(HPC_0)$ = Initial Log(HPC)
A = Coefficient
time = Time (days)
k_{hpc} = Reaction rate ($days^{-1}$)

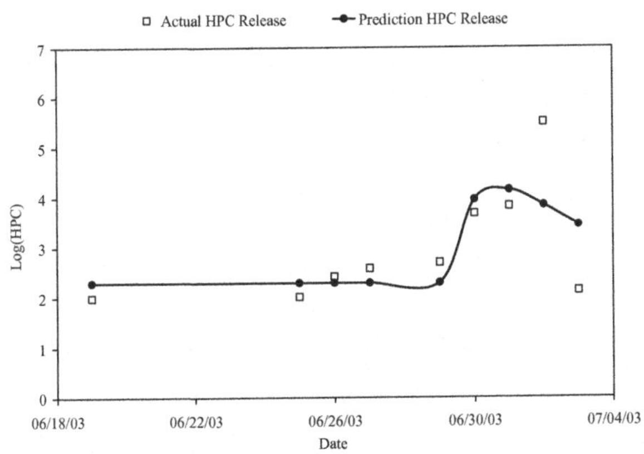

Figure 4.85 HPC response to change from blend to RO (free chlorine)

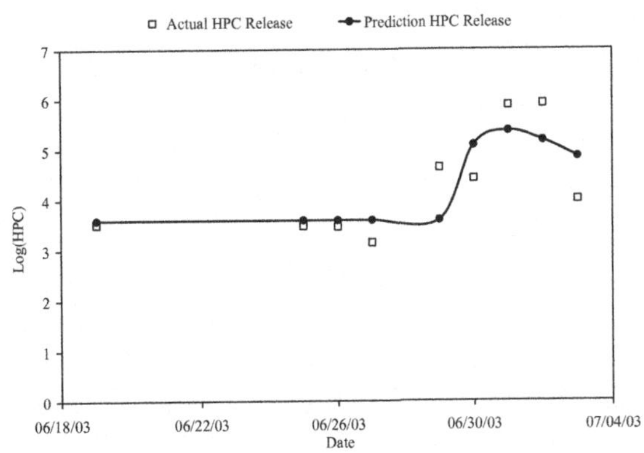

Figure 4.86 HPC response to change from blend to RO (chloramines)

Relating water chemistry to the magnitude of the peak HPC observed during the transition would have provided a useful tool in estimating the effect of blending on transient HPC release. An attempt was made to model the driving force (A) as a function of water chemistry parameters, including alkalinity, chlorides and sulfates. No significant relationship was found between the water chemistry and the driving force (A). The effect of water chemistry on transient HPC release was inconclusive.

HPC variations through PDS 1 through 4 were described as a function of temperature, time and chlorine residual for free chlorine as shown in equation (4.74). HPC variations through PDS 5 though 8 were described in a similar manner as shown in equation (4.75) for chloramine residuals. Predicted and actual HPCs from both of these models was within a 95 % confidence interval. R^2 for these models were 0.52 and 0.47 respectively. Both the models indicate HPCs increase with time and temperature, and decrease with increasing residual, which is what the actual data showed. Hence the models are descriptive of the observed HPCs during the HPC transition experiments.

$$Log\,(HPC) = \frac{1.19^{(T\text{-}20)} \times 1.12^{t}}{1.51 \times Cl_2^{\,0.14}} \tag{4.74}$$

where HPC = cfu/ml
T = temperature (°C)
t = time (days)
Cl_2 = free chlorine residual (mg/L as Cl_2)

$$Log\,(HPC) = \frac{1.06^{(T\text{-}20)} \times 1.08^{t}}{0.42 \times NH_2Cl^{\,0.04}} \tag{4.75}$$

where NH_2Cl = chloramine residual (mg/L as Cl_2)

A three dimension graph of actual and predicted Log HPCs normalized to 20°C versus time and residual is shown in Figure 4.87. Log HPCs were normalized to a common temperature in order to remove the effects of temperature.

The lower plane in Figure 4.87 is the plane of predicted Log HPCs for free chlorine residuals for varying time and residual. The clear circles are the actual HPCs for free chlorine residuals at associated time and residuals. The higher plane in Figure 4.87 is the plane of predicted Log HPCs for chloramine residuals for varying time and residual. The dark circles are the actual HPCs for chloramine residuals at associated time and residuals. Both planes increase sharply at low residuals, which emphasize the importance of residual to microorganism growth in drinking water distribution systems. The chloramine plane is significantly above the free chlorine plane at all points except where both residuals approach zero. Free chlorine more effectively retarded HPC proliferation in the PDSs.

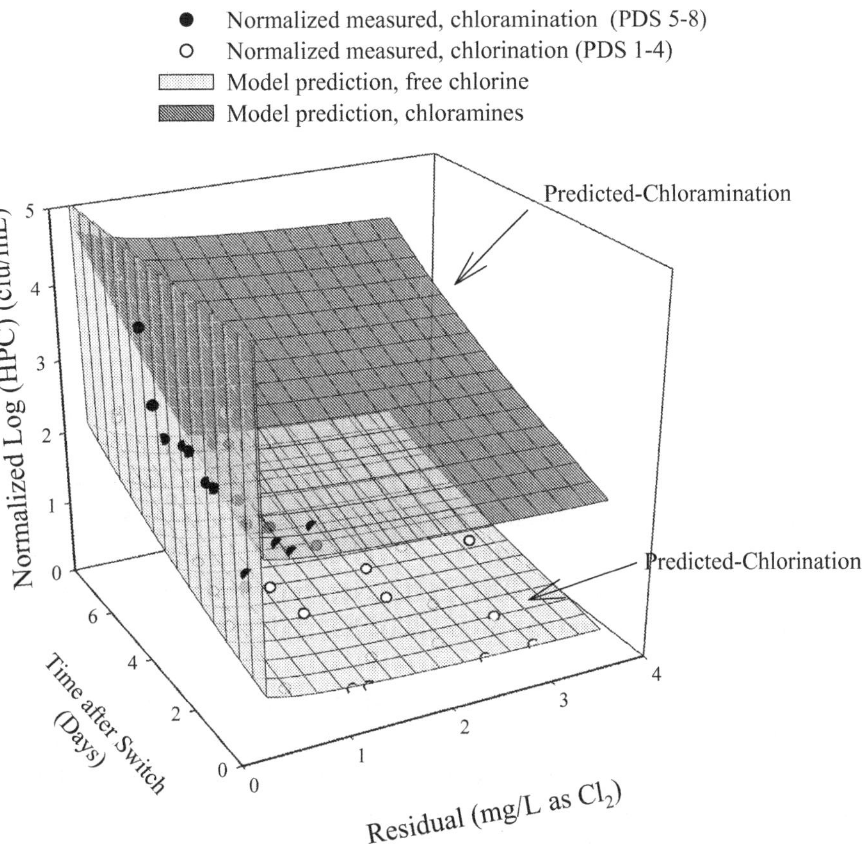

Figure 4.87 Actual and predicted HPCs versus residual and time following steady state water quality change

HPC Transition Summary

More HPC sampling was needed to investigate HPC transition during steady state water quality change in the first year of field study investigation. However, the HPC data clearly showed HPC populations were directly related to residual for chloramine or chlorine, and that increasing residual retarded HPC growth.

A high frequency HPC experiment was conducted in the field at the end of phase VI. Water quality blends were switched among PDSs 1 through 8 such that four identical blends were used between PDSs 1 through 4 for free chlorine, and PDSs 5 through 8 for chloramines. HPC samples were collected seven days before and seven days after the blend change. The high frequency HPC experiment showed a rapid increase in HPCs in all PDSs corresponding to the change in steady state water quality.

Modeling HPC transition from the time of steady state water quality change with a second order time model found that maximum HPCs were achieved approximately five days after the water quality change and that HPCs would return to the HPC concentrations prior to the water quality change in approximately ten days after the change. The time to reach maximum HPCs was related to a mixing model using alkalinity as a surrogate that confirmed alkalinity transition was complete in approximately eight days.

HPC transition was also modeled as a function of residual, time and temperature by non-linear regression. The actual and predicted HPCs showed that HPC populations decreased as residual increased; decreased as temperature decreased and increased with time. The actual and predicted data for free chlorine and chloramine residuals demonstrated that exponential growth occurred when either residual approached zero. The average Log (HPC) in the PDSs receiving free chlorine before the water switch was 2.48. The average Log (HPC) in the PDSs receiving chloramines before the water quality switch was 3.75, which was approximately twenty times more than the HPC populations in the free chlorine PDSs.

RESIDUAL MODELING

Free chlorine and chloramine residual in distribution systems decay with time due to reactions with the pipe wall and with organic and inorganic compounds in solution. Free chlorine and chloramine decay was found to vary depending on the pipe material as shown in Figure 4.88. The rate of free chlorine and chloramine dissipation for PVC and LCI is similar and significantly slower than for cast iron and G. PVC and lined iron are considered inert materials with a very small residual demand. The iron-based pipes showed a rapid decline in residual concentration due to the reaction of free chlorine and chloramine with the pipe wall and materials attached to the pipe surface.

Figure 4.88 shows the difference in chlorine residual between the influent and effluent in the single material lines after a residence time of 5 days. The influent residual for all the lines was 4 mg/L as Cl_2. UCI and G pipes lost all the residual while PVC and LCI maintained most of their residual.

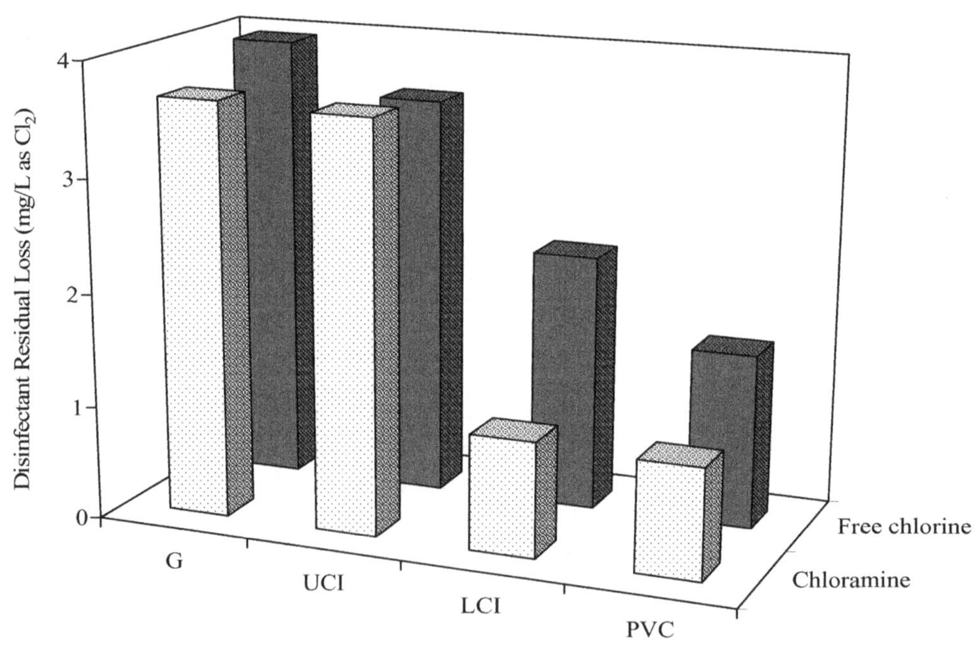

Figure 4.88 Disinfectant residual dissipation by pipe material

Free chlorine and chloramine residual dissipation was also found to depend on the water quality characteristics of the water. Figure 4.89 shows the loss of chlorine residual between influent and effluent observed for the different finished waters used during the project. The finished waters had a wide range of water quality characteristics as shown in Table 3.9. From Figure 4.89, the membrane treated GW (RO and G4), which were the waters with the highest relative quality experienced the least residual loss while the conventionally treated water (G1) lost the most residual.

Figure 4.90 shows the variation in chloramine residual demand through three PDSs in relation with the water temperature. The loss of residual is between 1 and 2 mg/L as Cl_2 from January to March of 2002 when the average water temperature was around 20°C, and increases to 3-4 mg/L as Cl_2 from May to June of 2002 when the temperature rose to 27°C. The same trend was observed for all the PDSs and PDS 1, 3, and 5 are presented as an example. Water temperature is denoted by the solid black symbol and is linked to the secondary y axis located on the right hand side of the figure. Chloramine demand is signaled by the empty symbol and can be read on the y axis located on the left hand side.

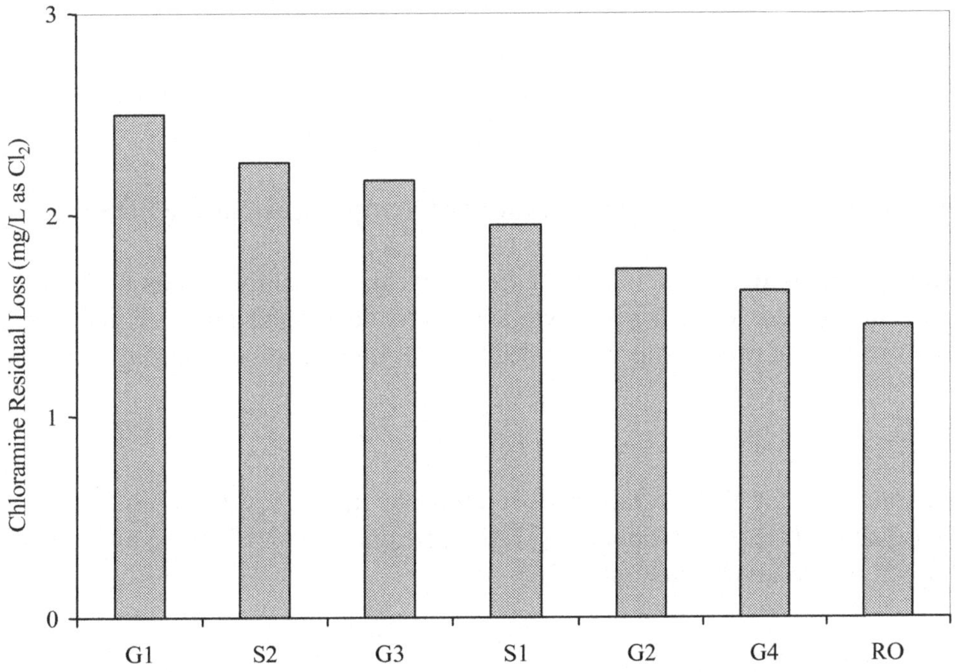

Figure 4.89 Variation in chloramine demand by type of finished water

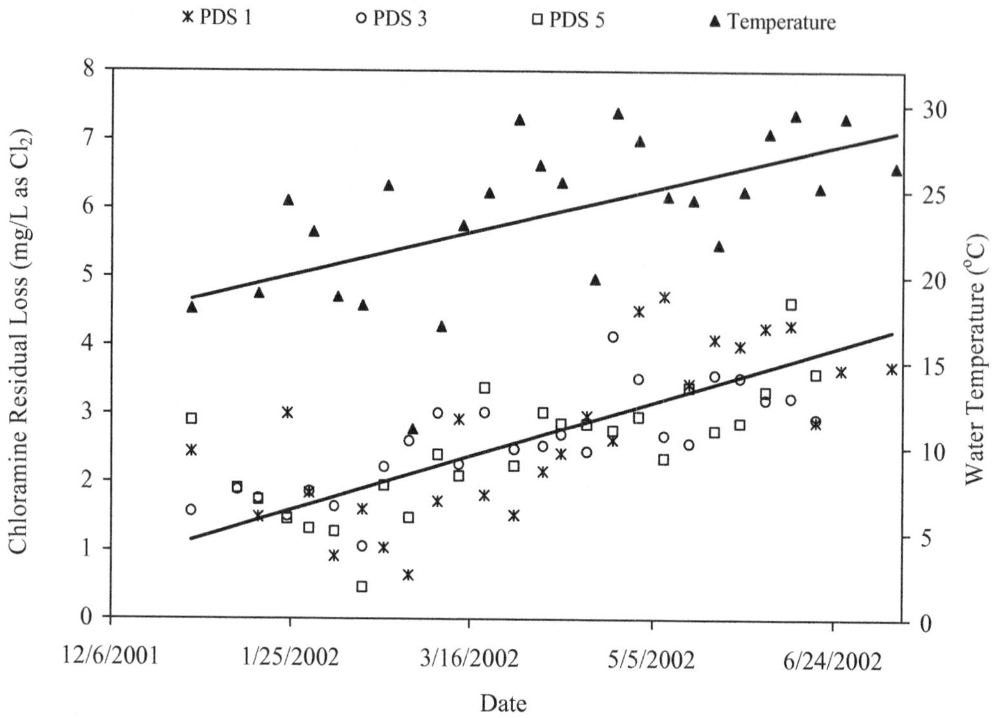

Figure 4.90 Variation in chloramine demand with temperature in the PDS lines

Residual models were developed for free chlorine and chloramines in the PDSs from the data collected during the field study. Models were developed and verified for disinfectant residuals as a function of time, temperature, PDS water quality, pipe material and residence time.

Overall Decay

An expression for the overall rate of chlorine decay inside a pipe can developed by substituting Equation (2.11) and Equation (2.19) in Equation (2.10) as shown in the Literature Review Chapter, which gives equation (4.76):

$$\left(\frac{dC}{dt}\right)_{Total} = \left[-\left(K_b \times UV254 + \frac{K_W}{D_p}\right) \times A^{(Temp-20)}\right] \times C \qquad (4.76)$$

Integration of equation (4.76) yields equation (4.77), which is the expression needed to predict chlorine residual with time in pipes given temperature and UV-254. K_b and K_w will vary by pipe material.

$$Cl_{2(t)} = Cl_{2(0)} \times \exp\left[-\left(K_b \times UV254 + \frac{K_W}{D_p}\right) \times A^{(Temp-20)}\right] \times time \qquad (4.77)$$

where K_b = bulk decay constant (cm/hr)

K_W = wall decay constant (in/hr)

$Cl_{2(t)}$	=	chlorine concentration at time t (mg/L as Cl_2)
$Cl_{2(0)}$	=	initial chlorine concentration (mg/L as Cl_2)
A	=	temperature correction coefficient
T	=	temperature in degrees C
t	=	time expressed in hours
D_p	=	pipe diameter (inches)

Free Chlorine Model

Free chlorine decayed at different rates depending on pipe material. Lower rates of chlorine consumption occurred in PVC and LCI pipes while chlorine concentration declined rapidly in the iron-based pipes. Decline of free chlorine residual increased in PDSs irrespective of material as time, UV-254 (organic content) and temperature increased.

Free chlorine decay was found to follow first-order kinetics with respect to chlorine concentration. The general formula of the model is shown in equation (4.77). In this model, the rate of chlorine decay is expressed using two constants: one for the bulk liquid reactions, K_b, and the other for pipe wall reactions, K_W. The bulk rate constant is dependent on UV-254, temperature and time and the wall rate constant is dependent on temperature, pipe diameter and time. Both kinetic constants were developed by non-linear regression. The model coefficients and R^2 for the free chlorine model are shown in Table 4.35.

Table 4.35
Free chlorine model coefficients and R^2

Material	Parameter	Value	R^2
PVC	K_b	0.048	0.97
	K_W	0.007	
	A	1.15	
Lined Cast Iron	K_b	0.084	0.93
	K_W	0.007	
	A	1.18	
Unlined Cast Iron	K_b	0.172	0.97
	K_W	0.063	
	A	1.13	
Galvanized	K_b	0.614	0.93
	K_W	0.054	
	A	1.04	

Effect of Pipe Material

The model coefficients show that chlorine decay is highly affected by the pipe material. In general, the rate of chlorine dissipation for PVC and LCI is slower than that for UCI and G. As shown in Table 4.35, the value of K_b for PVC and lined-iron are 0.048 and 0.084 cm/h, which indicates that there is little effect of these materials on bulk rate decay. For UCI pipes the value of K_b is 0.17 cm/h, which is four times larger than PVC and two times larger than LCI. The value of K_b for G is 0.61 cm/h which is approximately ten times larger than that of PVC and LCI. The larger values of K_b for iron-based pipes may be explained by the reactions of chlorine with materials that dissolved pipe walls into the water increasing the chlorine demand of the bulk liquid.

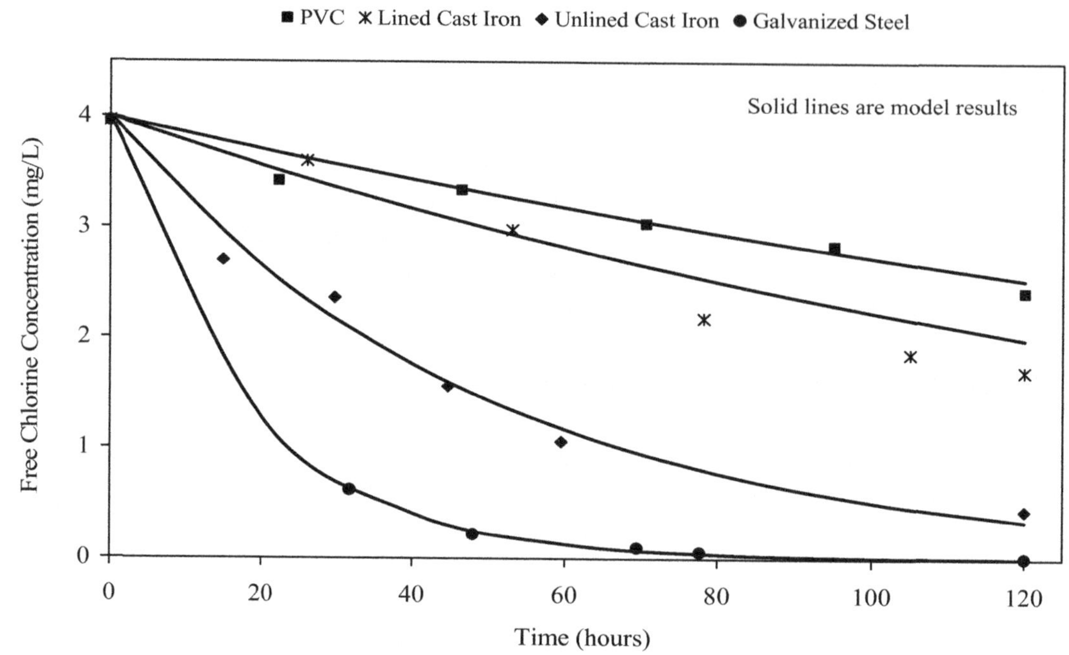

Figure 4.91 Effect of pipe material on free chlorine decay decay

K_W is almost identical for PVC and lined-iron pipes (0.007 in/hr), which indicates that both materials react with free chlorine in a similar way. Based on the K_W values it can be said that PVC and LCI are fairly inert materials and have a very low chlorine demand. On the other hand, UCI and G have a K_W value that is ten times larger than PVC and LCI pipes. This may be explained by the reactivity of chlorine with the ferrous material of the pipe or the presence of additional biological activity on the unlined metallic pipes. The corrosion products on UCI and G pipe provide greater surface area for biological growth than the PVC and LCI pipes. Figure 4.91 shows the residual dissipation profile for different material pipes at 20°C and 1.5 mg/L NPDOC. All the pipes were 6 inches in diameter with the exception of G pipe which had a diameter of 2 inches.

PVC and LCI pipes lost chlorine residual at a slower rate than cast iron and G pipes. After 5 days, PVC and LCI pipes maintained most of the chlorine residual while cast iron and G lost almost all residual.

Effect of Water Quality

Temperature and the organic content of the water also affect the rate of chlorine dissipation. The rate of chlorine decay increases with an increase in temperature. The change in rate is incorporated with an expression similar to the Arrhenius equation and is shown in equation (4.78).

$$K_T = K_{20} \times A^{(T-20)} \quad (4.78)$$

The temperature correction factor, A, were between 1.04 (for G) and 1.18 (for LCI). These values mean that the rate of chlorine decay doubles for each 5°C increment of temperature for LCI pipes and each 17°C for G pipes. The effect of temperature on chlorine dissipation can make it difficult to maintain a residual in the distribution system in summertime conditions.

Figure 4.92 shows the effect of temperature on free chlorine decay on water with an organic content of 1 mg/L as NPDOC and pipes with a diameter of 6 inches. As shown in Figure 4.92, at 10°C the chlorine residual will be present in significant amounts for all the pipe materials even after 5 days. However, for temperatures above 18°C in G and 24°C in UCI pipes no residual will be present after 5 days. Even though the decay rate increases appreciably with temperature, the PVC and LCI pipes will be able to maintain their free chlorine residuals above the minimal regulatory limit of 0.2 mg/L as Cl_2. This increase in chlorine dissipation rates with an increase in temperature may be due to the increased rate of: bacterial growth or other distribution system phenomena at higher temperatures as well as the increase on the rate of chlorine reaction with water at higher temperatures.

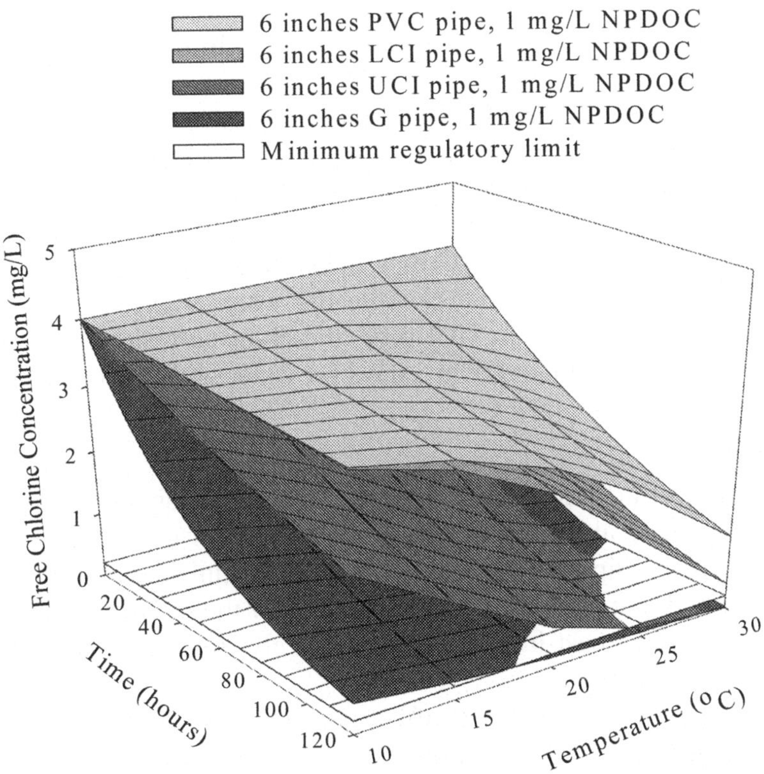

Figure 4.92 Effect of temperature on free chlorine decay for different materials

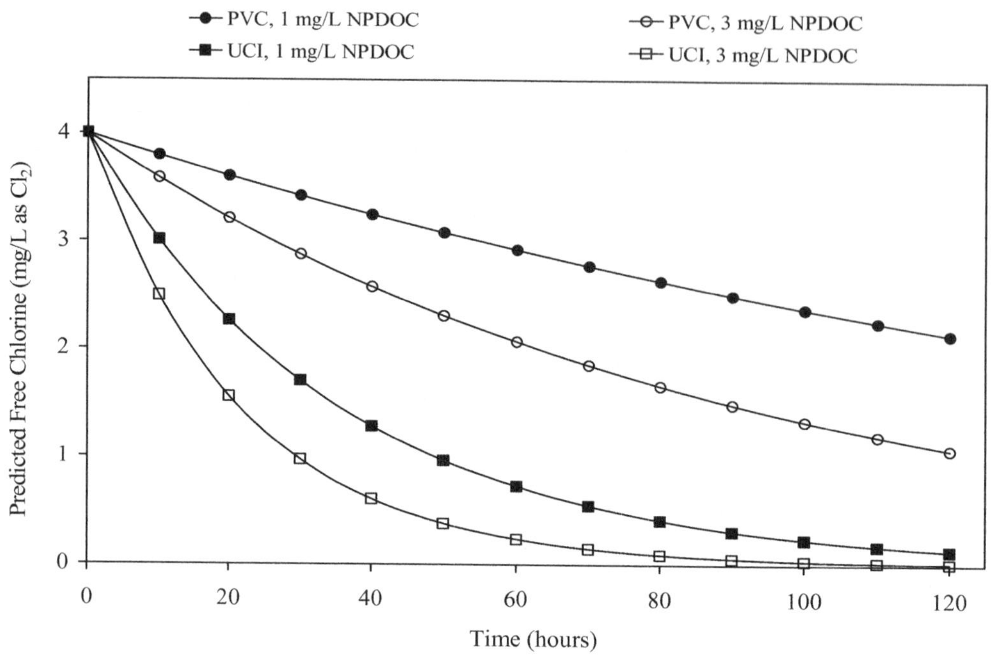

Figure 4.93 Effect of organic content on predicted free chlorine decay

Organic content also has an effect on chlorine decay. Figure 4.93 shows the change in chlorine decay rate for PVC and unlined cast-iron pipes at NPDOC concentrations of 1 mg/L and 3 mg/L. The plots were generated for pipes with a diameter of 6 inches and a temperature of 25°C.

As seen from Figure 4.93, in PVC pipe the water with 1 mg/L NPDOC will have a free chlorine residual of 2 mg/L after 5 days, while the water with 3 mg/L NPDOC will have a chlorine residual of only 1 mg/L. In UCI pipe, the chlorine residual will be lost in 4 days in water with 1 mg/l NPDOC and in less than 3 days for water with 3 mg/L NPDOC.

Effect of Pipe Diameter

The pipe diameter affects the rate at which chlorine dissipates in the distribution system. The rate of chlorine decay at the pipe wall is dependent on the surface area to volume ratio. When the diameter of the pipe is increased, that ratio diminishes and the relative effect of the wall decay is reduced. Figure 4.94 and Figure 4.95 show the effect of pipe diameter on chlorine decay for PVC and unlined cast-iron pipes, respectively, for water at 25°C temperature and a NPDOC concentration of 1 mg/L.

Since the change in diameter will affect only the wall component of the rate decay, the relative effect of changing the diameter of the pipe varies depending on the type of pipe. PVC is a relatively inert material with low chlorine demand at the wall and so the increase in pipe diameter has only a small effect on the rate of decay. Figure 4.94 shows that after 60 hours, water running through a PVC pipe with a diameter of 6 inches will have a chlorine residual of 2.9 mg/L as Cl_2 while the same water running through a 48 inch pipe will have a residual of 3.3 mg/L as Cl_2. In contrast, pipes made of materials with larger demand at the wall, such as unlined cast-iron, will have a more drastic change. Figure 4.95 shows that after 60 hours, water running through a 6 inches UCI pipe will have a chlorine residual of 0.7 mg/L as Cl_2 while the same water running through a 48 inches pipe of the same material will have a residual of 1.9 mg/L as Cl_2.

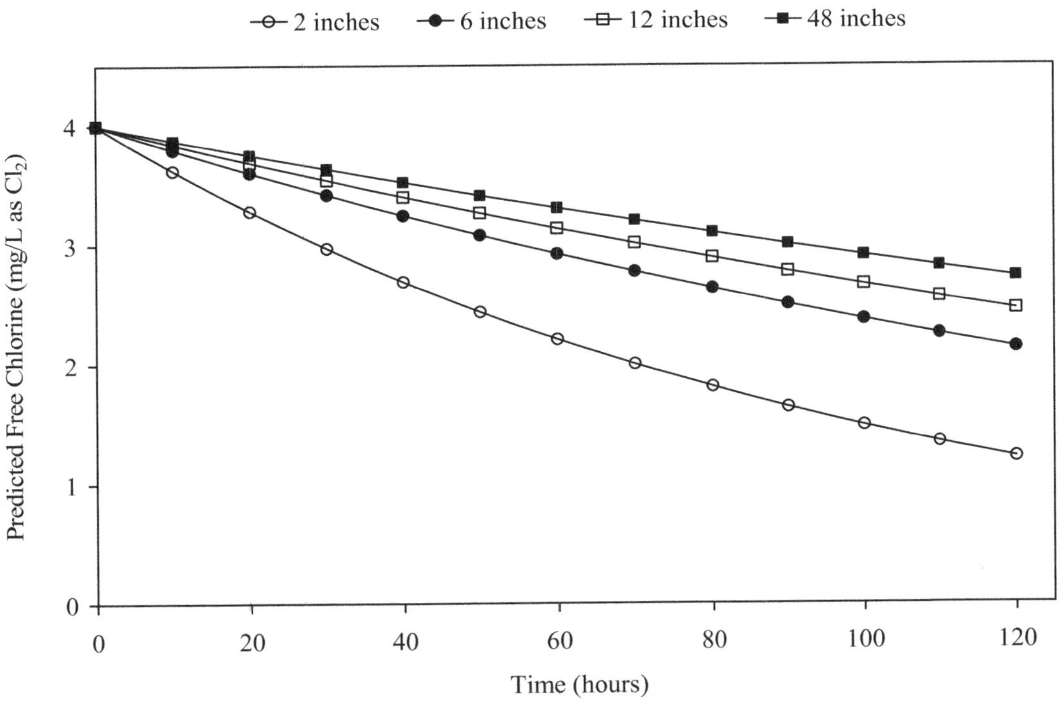

Figure 4.94 Effect of pipe diameter on predicted free chlorine decay for PVC pipes

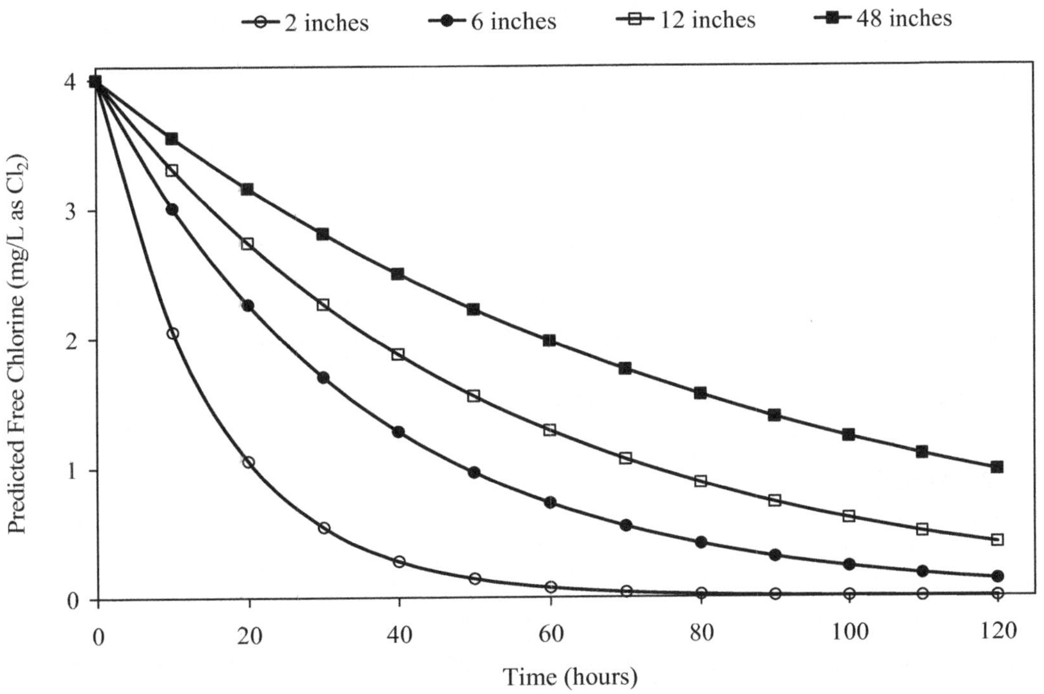

Figure 4.95 Effect of pipe diameter on predicted free chlorine decay for unlined cast-iron pipes

Free Chlorine Model Verification

The model for free chlorine was verified to the 95% confidence interval using independent data. These data was collected from the same pilot distribution system but was not used for the model development. The actual independent free chlorine residual versus the predicted free chlorine residuals is shown in Figure 4.96. The solid line at 45 degrees represents perfect match between predicted and actual values. All the points are located very close to the line which indicates that the model is useful to predict free chlorine residual in distribution systems.

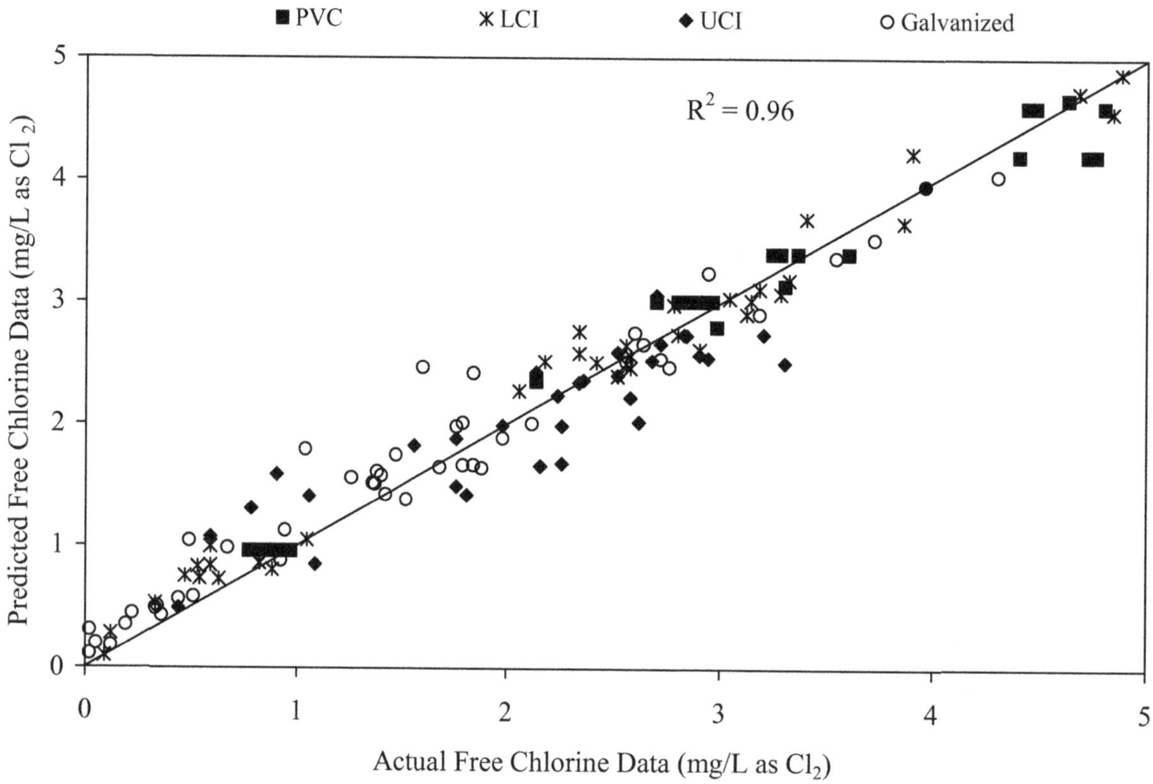

Figure 4.96 Free chlorine model verification

Chloramine Model

The same model shown in equation (4.77) was used to describe either free chlorine or chloramine residuals. The model parameters varied by pipe material and residual, and are shown in Table 4.36 for chloramines.

Similar to free chlorine, the rate of chloramine decay is expressed using two constants: one for the bulk liquid reactions, K_b, and the other for pipe wall reactions, K_W. The bulk rate constant is dependent on UV-254, temperature and time and the wall rate constant is dependent on temperature, pipe diameter, and time. The values of the kinetic constants were found to be dependent on the characteristics of the pipe material. Table 4.36 shows the model coefficients for the chloramine model and the values of R^2 obtained for each pipe material.

Table 4.36
Chloramine model coefficients and R^2

Material	Parameter	Value	R^2
PVC	K_b	0.020	0.90
	K_W	0.007	
	A	1.16	
Lined Cast Iron	K_b	0.017	0.90
	K_W	0.011	
	A	1.16	
Unlined Cast Iron	K_b	0.147	0.87
	K_W	0.084	
	A	1.11	
Galvanized	K_b	0.319	0.87
	K_W	0.073	
	A	1.12	

Effect of Pipe Material

As seen in Table 4.36, K_b for PVC and LCI are the lowest (0.02 and 0.018 cm/h), which indicates that these pipe materials do not influence the bulk decay rate for chloramine. In contrast, K_b for UCI (0.16 cm/h) and G (0.32 cm/h) are about 8 and 16 times larger. This difference may be explained by the increased bulk chlorine demand due to reactions of chloramine with materials that dissolve into water from the metallic pipe walls.

K_W is similar for PVC and LCI pipes (0.007 and 0.011 in/hr), which indicates that both materials react in a similar way with chloramine. PVC and LCI are considered fairly inert materials and have a minimum chloramine demand. UCI and G pipes, on the other hand, have a K_W that is 10 times larger than PVC and LCI pipes. This may be explained by the reactivity of chloramine with the ferrous material of the pipe or additional biological activity on the unlined metallic pipes. Previous studies (Vikesland and Valentine, 2002) have shown that chloramine reacts with iron at the pipe wall to form ammonia and chlorides as shown in equation (4.79).

$$2Fe(II)_{surf} + NH_2Cl + H^+ \rightarrow 2Fe(III) + NH_3 + Cl^- \tag{4.79}$$

The general profile observed for chloramine dissipation for the different types of pipe is shown in Figure 4.97. The curves were obtained using the models for water at 20°C and with an organic content of 1.5 mg/L as carbon. All pipes were 6 inches in diameter with the exception of G which had a diameter of 2 inches.

The lowest rate of chlorine dissipation was observed for PVC and LCI pipes while chloramine residual was rapidly lost in UCI and G pipes. PVC and lined-iron maintained 75 % of the initial chloramine residual after 5 days, while unlined cast-iron and G lost all residual in the same amount of time.

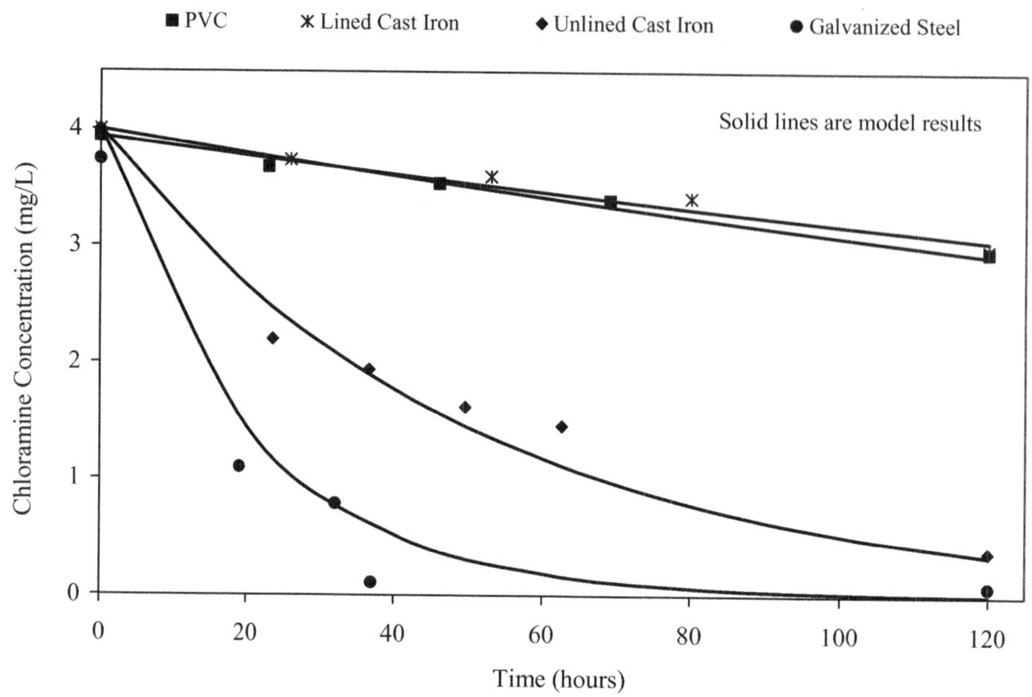

Figure 4.97 Effect of pipe material in chloramine decay

Effect of Water Quality

The rate of chloramine decay is also affected by other factors such as temperature, organic content of the water and the diameter of the pipe. In a similar manner to free chlorine, the variation on chloramine dissipation with temperature was modeled using the Arrhenius equation. The temperature correction factor obtained for the model range from 1.11 for unlined cast-iron to 1.16 for PVC and LCI which means that the decay constants will double for every 5 to 7°C increment in temperature. Figure 4.98 shows the effect of temperature on chloramine decay for water with 1 mg/L of organic content as NPDOC and an initial chloramine concentration of 4 mg/L as Cl_2 in pipes with a diameter of 6 inches.

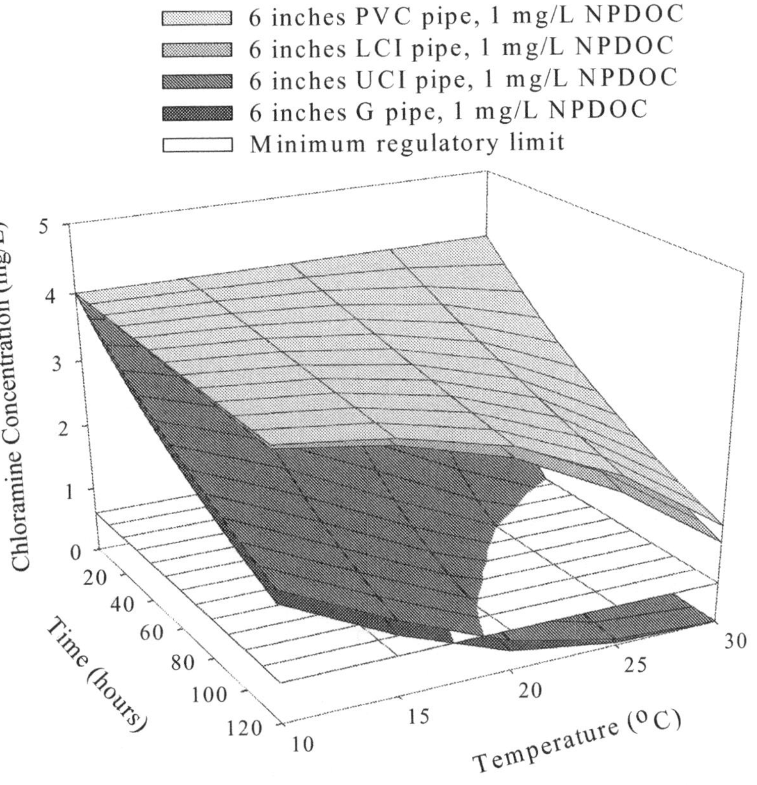

Figure 4.98 Effect of temperature on chloramine decay

PVC and LCI pipes maintain a chloramine residual of 3.5 mg/L even after 5 days at a temperature of 10°C, but the predicted residual quickly depletes to around 1 mg/L when the temperature increases to 30°C. For temperatures above 18°C, the model predicts that UCI and G pipes will not carry the minimum chloramine residual of 0.6 mg/L required by regulations for more than 1 day.

The organic content of the water also has an influence on the rate of chloramine dissipation. Figure 4.99 shows the change in chloramine decay rate for 6-inch PVC and UCI pipes at NPDOC concentrations of 1mg/L and 3 mg/L and a temperature of 25°C.

The rate of chloramine decay increases with an increase in the organic content of water. For the PVC pipe, the water with 1 mg/L of NPDOC will have a residual of 2.5 mg/L as Cl_2 after 5 days. In contrast, the water with 3 mg/L of organic will have a residual of 1.9 mg/L as Cl_2. In UCI pipe the residual will be completely lost after two days in both waters; after 40 hours, the chloramine residual will be 1.1 mg/L as Cl_2 in the water with 1 mg/L of NPDOC and 0.6 mg/L as Cl_2 in water with 3 mg/L of NPDOC.

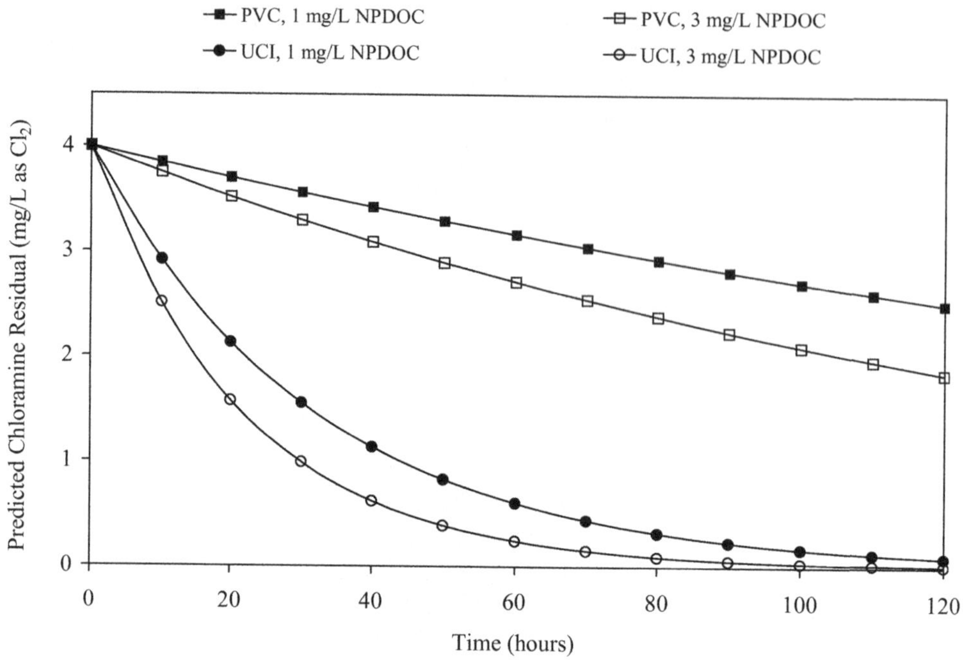

Figure 4.99 Effect of organic content on predicted chloramines decay

Effect of Pipe Diameter

The dissipation rate of chloramine in a distribution system is also affected by the pipe diameter. Smaller pipe diameter results in a faster rate of chloramine loss. Figure 4.100 and Figure 4.101 show the chloramine decay in PVC and unlined cast-iron pipes for different diameters.

Since PVC is a material with low chloramine demand at the wall, the increase of pipe diameter has only a small effect on the rate of decay. Figure 4.100 shows that after 60 hours, water running through a 6-inch PVC pipe will have a chloramine residual of 3.2 mg/L as Cl_2 whereas the same water running through a 48 inches pipe will have a chloramine residual of 3.6 mg/L as Cl_2. In contrast, pipes made of materials with larger demand at the wall, such as UCI, will have a more drastic change. Figure 4.101 shows that after 60 hours, water running through a 6-inch UCI pipe will have a chloramine residual of 0.6 mg/L as Cl_2 whereas the same water running through a UCI pipe with a diameter of 48 inches will have a chloramine residual of 2.1 mg/L as Cl_2.

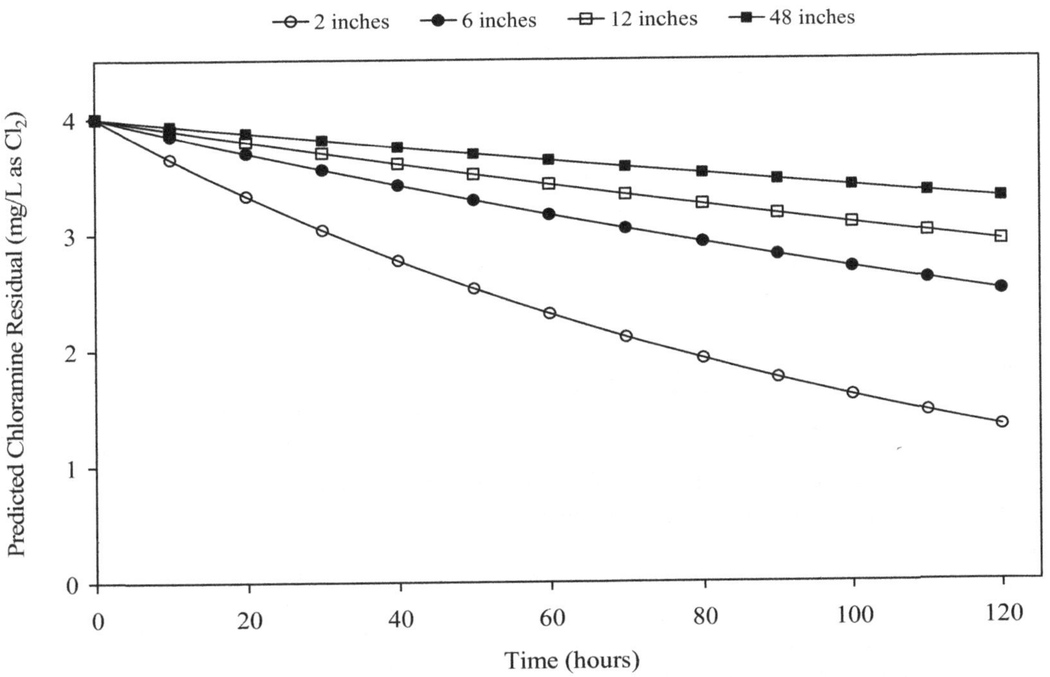

Figure 4.100 Effect of pipe diameter on predicted chloramine decay for PVC pipes

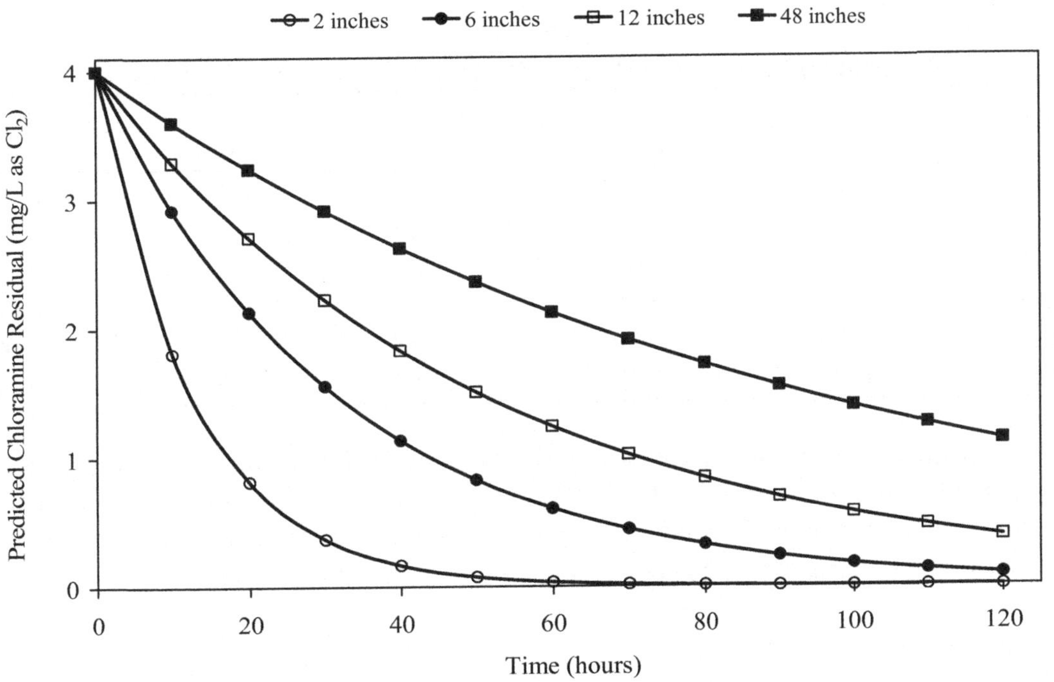

Figure 4.101 Effect of pipe diameter on predicted chloramine decay for cast-iron pipes

Effect of Flow Velocity

The effect of flow velocity on chloramine decay was also evaluated for G and UCI pipe material only and was done at the same time that the effect of flow velocity on iron release was evaluated. These experiments were done at the end of P V, and consequently were not done for free chlorine nor PVC and LCI pipe material. The purpose of the experiments on residual was to investigate the effect of turbulence on chloramine decay.

The rate of chloramine decay was determined for the UCI and G pipes under different flow regimes. The velocity of water flowing through the distribution system was regulated to obtain Reynolds numbers between 5 and 9,000. Table 4.37 shows the change of the kinetic decay constants with the Reynolds number for G pipe with a diameter of 2 inches.

Table 4.37 shows that as the Reynolds number increases, the mass transfer rate of chlorine between the bulk and the pipe wall, K_F, increases producing a higher rate of chlorine decay at the pipe wall, K_W. The equations that defined the relationship between K_W, K_F and K_{W1} as shown in equation (4.80) were given in the literature review and are equations (2.13), (2.15), (2.16) and (2.19).

$$K_W = -\frac{4K_{W1}K_F}{(K_{W1}+K_F)} \tag{4.80}$$

Table 4.37
Variation of kinetic constants with Reynolds number for G pipe

Reynolds Number	HRT (days)	K (hr^{-1})	K_F (in/hr)	K_W (in/hr)	K_{W1} (in/hr)
5	4	0.052	1466	0.073	0.018
13	2	0.052	1466	0.073	0.018
391	0.06	0.090	1467	0.149	0.037
2279	0.01	0.167	1023	0.331	0.083
2934	0.008	0.197	1278	0.393	0.098
5025	0.005	0.435	2053	0.869	0.217

The overall rate of chloramine decay for G shows a linear increase with the Reynolds number under laminar and transitional flow conditions. Concerns for safe operation of the pilot distribution system prevented the study of Reynolds numbers larger than 5000 in the G pipe. The relation is presented in Figure 4.102 and can be expressed as equation (4.81).

$$K = 7 \times 10^{-5} R_e + 0.0419 \tag{4.81}$$

Table 4.38 presents the values of overall kinetic decay constant, K, composite pipe wall kinetic decay constant, K_W, and pipe wall reaction rate constant, K_{W1} obtained for different Reynolds numbers for cast iron pipe with a diameter of 6 inches.

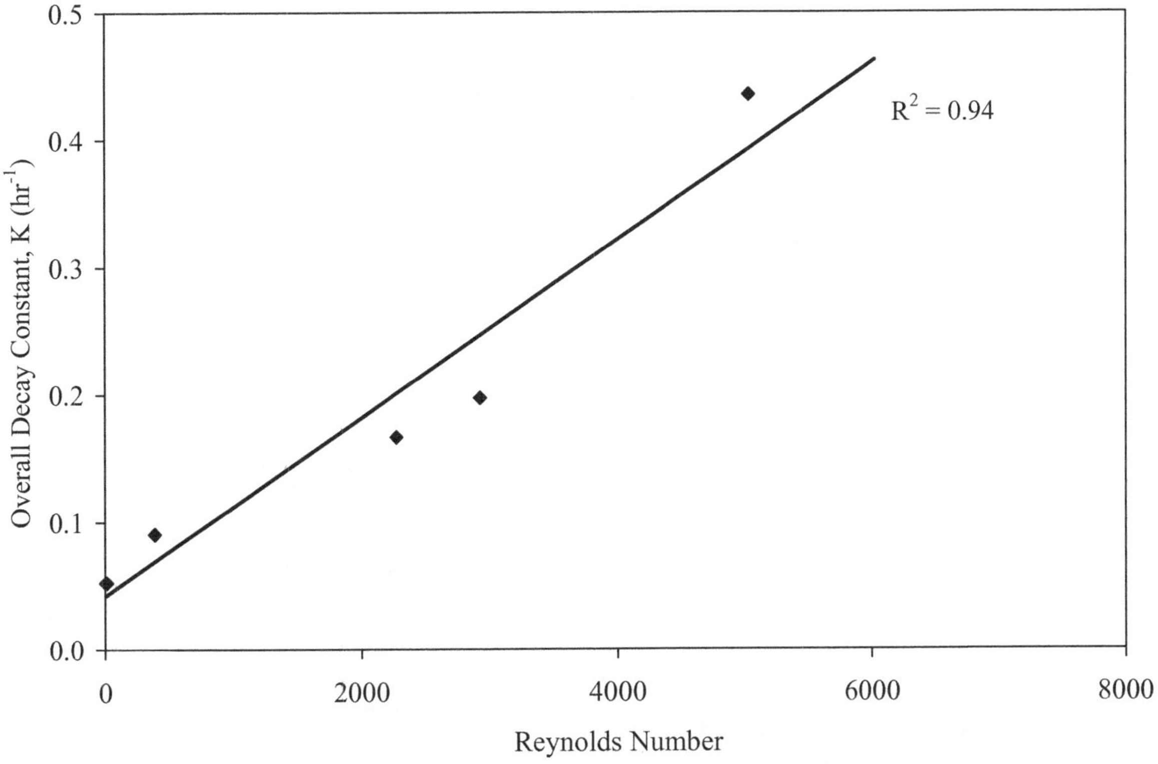

Figure 4.102 Change of overall chloramine decay constant K with Reynolds number for G pipe.

Table 4.38
Variation of kinetic constants with Reynolds number for cast-iron pipe

Reynolds Number	HRT (days)	K (hr^{-1})	K_F (in/h)	K_W (in/hr)	K_{W1} (in/hr)
11	5	0.02	489	0.08	0.005
27	2	0.02	489	0.08	0.005
264	0.2	0.02	489	0.06	0.004
834	0.06	0.04	489	0.19	0.010
1321	0.04	0.03	489	0.16	0.008
2643	0.02	0.18	389	1.06	0.044
4185	0.01	0.11	583	0.63	0.026
6527	0.008	0.04	861	0.22	0.010
8682	0.006	0.06	1107	0.37	0.016

The change of the overall chloramine decay rate constant, K with the Reynolds number for UCI pipes is presented in Figure 4.103. K increased linearly with Reynolds number up to a value of approximately 3000, corresponding to the increase in the mass transfer rate of chloramine towards the pipe wall. For fully turbulent flow conditions (Re > 4000) the effect of the mass transfer rate diminishes and the rate of chloramine decay stabilizes and this behavior can be represented by the equation (4.82).

$$K = 0.01 \times R_e^{(0.2209)} \tag{4.82}$$

The results showed that turbulence affected chloramine residual in G and UCI pipes. However not enough data was collected to incorporate turbulence into the bulk and wall reaction constants for the chloramine residual model because of lack of variation in temperature, pipe material and TOC (UV254).

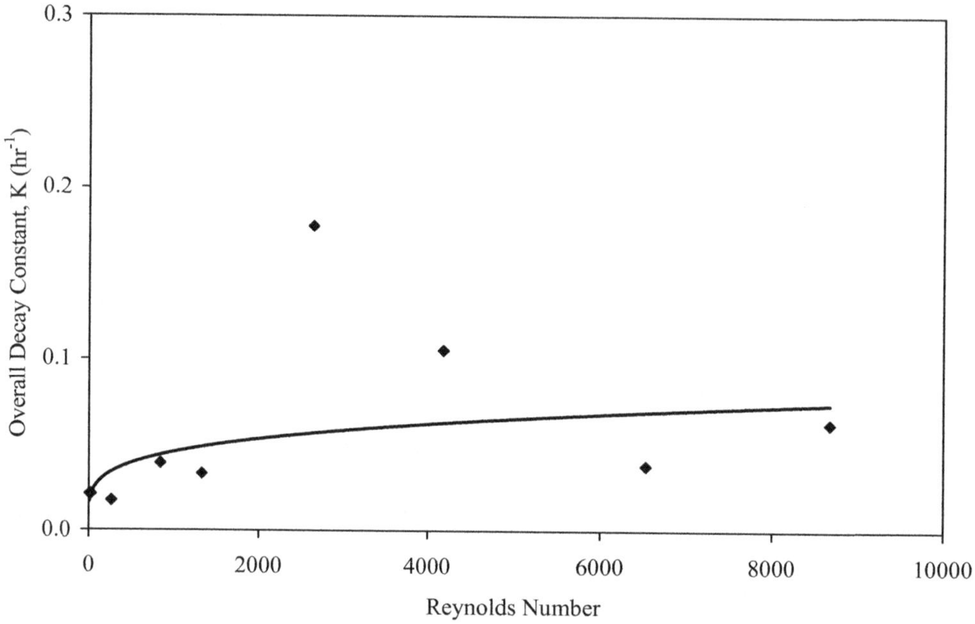

Figure 4.103 Change of overall chloramine decay constant K with Reynolds number for cast-iron pipe

Chloramine Model Verification

The model for chloramine was verified using independent data to the 95% confidence interval. That data was collected from the same pilot distribution system but was set aside and not used for the model development process. The comparison between actual and predicted chloramine residual for the different pipe materials is presented in Figure 4.104. The solid line at 45 degrees represents the perfect prediction of the actual values of chloramine residual. All the points are located very close to the line which indicates that the model developed is useful to predict chloramine residual in distribution systems.

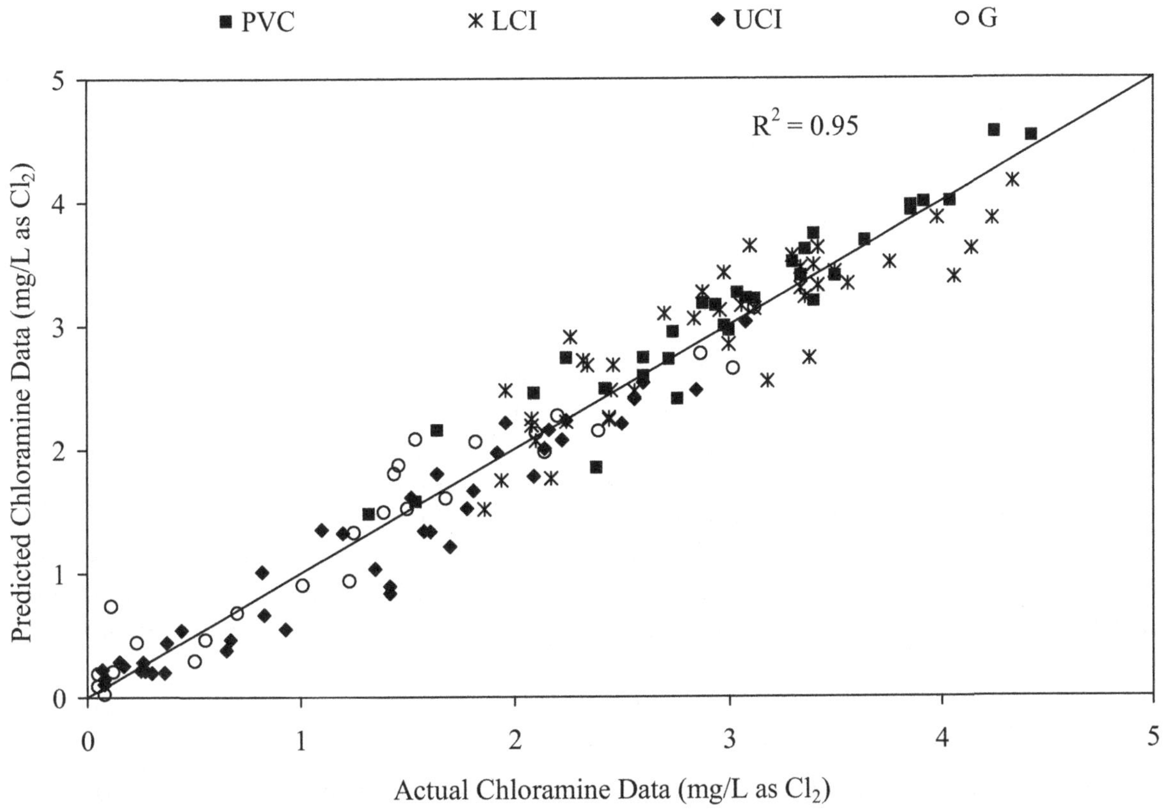

Figure 4.104 Chloramine model verification

Summary

The rate of decay of free chlorine and chloramines depended on pipe material, temperature and water quality. Lower rates of free chlorine and chloramines decay occurred in PVC and LCI pipes while free chlorine and chloramine decayed more rapidly in unlined iron-based pipes (UCI and G).

Separate empirical models were developed to predict free chlorine and chloramine residual in distribution system as a function of time, water quality, pipe material and pipe geometry. Free chlorine and chloramine residual was found to decay following first-order kinetics with respect to chlorine concentration. Decline of free chlorine and chloramine residual increased in PDSs irrespective of material as time, UV-254 (organic content) and temperature increased.

In the models, the rate of residual decay was expressed using two constants: one for the bulk liquid reactions, K_b, and the other for pipe wall reactions, K_W. The bulk rate constant is dependent on UV-254, temperature and time and the wall rate constant is dependent on temperature, pipe diameter and time. Both kinetic constants were developed by non-linear regression. The model coefficients and R^2 for the free chlorine and chloramine models are shown in Table 4.35 and Table 4.36 respectively.

The bulk water reaction rates (K_b) for PVC and LCI are lower than the bulk water reaction rates (K_b) for UCI and G, which indicates that the unlined metal materials from the UCI and G pipe walls influence the bulk decay rate and that the lined materials from the PVC and

LCI pipe walls do not influence the bulk decay rate. The wall reaction rates (K_W) are similar for PVC and LCI pipes, which indicates that both materials react in a similar way with free chlorine and chloramine residual. UCI and G pipes, on the other hand, have K_W that are between 7 and 10 times larger than PVC and LCI pipes. This may be explained by the reactivity of chlorine with the ferrous material of the pipe or additional biological activity on the unlined metallic pipes.

Temperature and the organic content of the water also affect the rate of free chlorine and chloramine dissipation. The increase of the rate of residual decay with the increase in temperature was modeled using an expression similar to the Arrhenius equation. The effect of temperature on chlorine dissipation can make it difficult to maintain a residual in the distribution system in summertime conditions. The rate of free chlorine and chloramine decay is also dependent on the organic content of the water. Chlorine residual is maintained longer in waters with low organics than in waters with high concentration of organics. No relation between pH and the rate of chlorine decay was observed due to the small range of values utilized (7.9 to 8.3).

The pipe diameter affects the rate at which free chlorine and chloramine dissipate in the distribution system. The rate of residual decay at the pipe wall is dependent on the wetted surface area to volume ratio. When the diameter of the pipe is increased, that ratio diminishes and the relative effect of the wall decay is reduced. The effect of the pipe diameter is more significant for metallic pipes since they have larger K_W than PVC and LCI.

The following steps could be taken to increase the time free chlorine and chloramine residuals are maintained in a distribution system:

- UCI and G are not the best kind of pipe since they result in high rates of residual dissipation relative to lined or PVC pipe. This could result in complete loss of residual in some parts of the distribution system, especially when temperature is elevated and water age is high.
- Frequency of monitoring free chlorine and chloramine residuals in distribution systems should be based on location, season and pipe material. Monitoring should increase as water temperature increases. Locations in the distributuion system that have high residence times such as dead zones or have G or UCI pipes warrant increased monitoring and residual maintance.
- The models for free chlorine and chloramine residual could be used to predict residual concentrations at different points within the distribution system and can determine the need for locations where rechlorination, flushing or other means of residual maintenance should be implemented.

Practical Application of Free Chlorine and Chloramines Models – User Manual

In order to apply the models for free chlorine and chloramines in a distribution system, knowledge of temperature and organic content of the water, pipe diameter and material, as well as the hydraulic residence time (HRT) is necessary. The HRT for a determined reach of pipe could be calculated by dividing the pipe length by the average water velocity through the pipe as shown in equation (4.83).

$$HRT = \frac{Pipe\ Length\,(ft)}{Average\ \text{Water Velocity}\left(\text{ft}/_{\text{sec}}\right)} \tag{4.83}$$

The flow velocity in a full size distribution system could be estimated by assuming a velocity of 1 ft/sec from 6 AM to 9 AM, noon to 3 PM and 6 PM to 9 PM, and zero all other times. Hence the equivalent daily average velocity would be 0.33 ft/sec. As is shown in Table 4.39, the equivalent length varies from 75.6 to 0.27 miles in an actual DS for HRTs of 14 to 0.05 days.

Table 4.39
Equivalent pipe length and HRT for actual distribution systems

Assumed HRT		Water Velocity	Equivalent Actual DS Pipe Length	
(days)	(hours)	(ft/sec)	(ft)	(miles)
14	336	0.33	399168	75.6
7	168	0.33	199584	37.8
5	120	0.33	142560	27
2	48	0.33	57024	10.8
1	24	0.33	28512	5.4
0.5	12	0.33	14256	2.7
0.25	6	0.33	7128	1.35
0.1	2.4	0.33	2851	0.54
0.05	1.2	0.33	1426	0.27

It is unlikely that 75.6 miles of pipe would be encounter in a distribution system, however these equivalent lengths give the reader a reference point to better understand the findings of the current report.

Predictions of free chlorine or chloramine residual can be made using the respective model as follows:

Examples

1) Determine the effluent free chlorine and chloramines residual for a 1000 foot length of 2-inch galvanized pipe for 100%GW, and a blend of 60%/30%/10% (GW/SW/RO), at a temperature of 28°C and an influent residual concentration of 4 mg/L. Repeat for 1000 foot of 6-inch UCI pipe.

Based on Table 3.9, the corresponding values of UV254 for 100%GW, and the 60/30/10 blend are 0.0700 cm^{-1}, and 0.0557 cm^{-1} respectively.

<u>*Galvanized pipe:*</u>

In the case of free chlorine, the residual model for galvanized pipes takes the form:

$$Cl_{2-t} = Cl_{2-0} \exp\left[-\left(0.614 * UV254 + \frac{0.054}{Dp}\right) * 1.04^{(T-20)} * time\right]$$

a) For 100%GW:

UV254 = 0.070
Time = 1000ft/0.33 ft/sec = 3000 sec = 0.83 hours
Dp = 2 inches

T = 28°C

The model predicts:

$$Cl_{2-t} = 4*\exp\left[-\left(0.614*0.070+\frac{0.054}{2}\right)*1.04^{(28-20)}*0.83\right]$$

$$Cl_{2-t} = 3.69 \text{ mg/L as } Cl_2$$

b) For the blend of 60%/30%/10% (GW/SW/RO) : UV254= 0.0557

The model predicts:

$$Cl_{2-t} = 4*\exp\left[-\left(0.614*0.0557+\frac{0.054}{2}\right)*1.04^{(28-20)}*0.83\right]$$

$$Cl_{2-t} = 3.73 \text{ mg/L as } Cl_2$$

In the case of chloramines, the residual model for galvanized pipes takes the form:

$$Cl_{2-t} = Cl_{2-0}\exp\left[-\left(0.319*UV254+\frac{0.073}{Dp}\right)*1.12^{(T-20)}*time\right]$$

a) For 100%GW: UV254= 0.070

The model predicts:

$$Cl_{2-t} = 4*\exp\left[-\left(0.319*0.070+\frac{0.073}{2}\right)*1.12^{(28-20)}*0.83\right]$$

$$Cl_{2-t} = 3.54 \text{ mg/L as } Cl_2$$

b) For the blend of 60%/30%/10% (GW/SW/RO) : UV254= 0.0557

The model predicts:

$$Cl_{2-t} = 4*\exp\left[-\left(0.319*0.0557+\frac{0.073}{2}\right)*1.12^{(28-20)}*0.83\right]$$

$$Cl_{2-t} = 3.54 \text{ mg/L as } Cl_2$$

Unlined cast iron pipe:

In the case of free chlorine, the residual model for UCI pipes takes the form:

$$Cl_{2-t} = Cl_{2-0}\exp\left[-\left(0.172*UV254+\frac{0.063}{Dp}\right)*1.13^{(T-20)}*time\right]$$

a) For 100%GW: UV254= 0.070

Time = 1000ft/0.33 ft/sec = 3000 sec = 0.83 hours

Dp = 6 inches

T = 28°C

The model predicts:

$$Cl_{2-t} = 4*\exp\left[-\left(0.172*0.070+\frac{0.063}{6}\right)*1.13^{(28-20)}*0.83\right]$$

$$Cl_{2-t} = 3.81 \text{ mg/L as } \mathrm{Cl_2}$$

b) For the blend of 60%/30%/10% (GW/SW/RO) : UV254= 0.0557
The model predicts:

$$Cl_{2-t} = 4 * \exp\left[-\left(0.172 * 0.0557 + \frac{0.063}{6}\right) * 1.13^{(28-20)} * 0.83\right]$$

$$Cl_{2-t} = 3.83 \text{ mg/L as } \mathrm{Cl_2}$$

In the case of chloramines, the residual model for UCI pipes takes the form:

$$Cl_{2-t} = Cl_{2-0} \exp\left[-\left(0.147 * UV254 + \frac{0.084}{Dp}\right) * 1.11^{(T-20)} * time\right]$$

a) For 100%GW: UV254= 0.070
The model predicts:

$$Cl_{2-t} = 4 * \exp\left[-\left(0.147 * 0.070 + \frac{0.084}{6}\right) * 1.11^{(28-20)} * 0.83\right]$$

$$Cl_{2-t} = 3.82 \text{ mg/L as } \mathrm{Cl_2}$$

b) For the blend of 60%/30%/10% (GW/SW/RO) : UV254= 0.0557
The model predicts:

$$Cl_{2-t} = 4 * \exp\left[-\left(0.147 * 0.0557 + \frac{0.084}{6}\right) * 1.11^{(28-20)} * 0.83\right]$$

$$Cl_{2-t} = 3.83 \text{ mg/L as } \mathrm{Cl_2}$$

2) Determine the chlorine residual for all pipe links in the hypothetical subdivision layout shown below (Figure 4.105) assuming a chloramine residual of 4.0 mg/L as Cl_2 at the reservoir (R). The hydraulic parameters (velocities, flows, and head loss) were determined using a standard software package. The water quality was assumed to match the blend of groundwater (60%), surface water (30%), and desalination (10%) described in previous problems.

Assume the main distribution line to be 24-inch PVC pipe (P35, P30, P22, P34) and the remaining piping is galvanized. Repeat the analysis assuming all the non-PVC pipe is unlined cast iron.

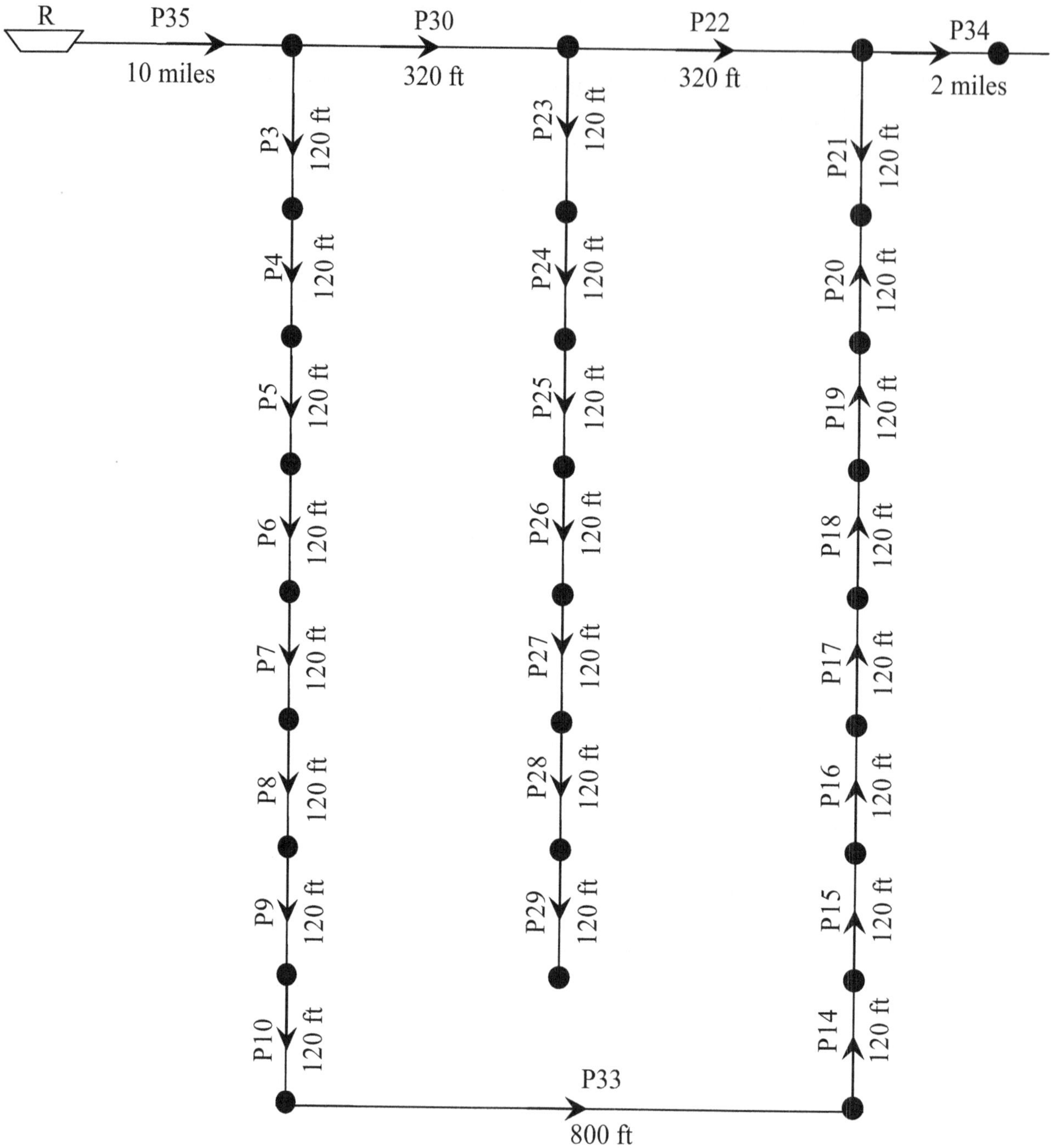

Figure 4.105 Hypothetical subdivision layout

The results are presented in Table 4.40 and Table 4.41 for galvanized steel and unlined cast iron respectively.

Table 4.40
Free chlorine and chloramine residual predictions for hypothetical subdivision using PVC and galvanized pipes

End of Pipe	Pipe Material	Length ft	Diameter Inches	Velocity Ft/sec	NH_2Cl residual	Cl_2 Residual	HRT hrs	Σ HRT hrs
R	-	-	-	-	4.00	4.00	0	0
P35	PVC	52,800	24	1.0	3.74	3.50	14.8	14.8
P30	PVC	320	24	1.0	3.73	3.49	0.09	14.9
P22	PVC	320	24	1.0	3.73	3.49	0.09	15.0
P34	PVC	10,560	24	1.0	3.68	3.40	2.98	18.0
P3	G	120	4	0.1913	3.68	3.46	0.17	15.0
P4	G	120	4	0.1789	3.62	3.42	0.19	15.2
P5	G	120	4	0.1665	3.55	3.37	0.20	15.4
P6	G	120	4	0.1540	3.49	3.32	0.22	15.6
P7	G	120	4	0.1416	3.41	3.27	0.24	15.8
P8	G	120	4	0.1292	3.34	3.22	0.26	16.1
P9	G	120	4	0.1168	3.25	3.16	0.29	16.3
P10	G	120	4	0.1044	3.16	3.09	0.32	16.7
P33	G	800	4	0.0920	2.55	2.64	2.42	19.1
P14	G	120	4	0.0796	2.45	2.57	0.42	19.5
P15	G	120	4	0.0672	2.35	2.49	0.50	20.0
P16	G	120	4	0.0548	2.22	2.39	0.61	20.6
P17	G	120	4	0.0423	2.07	2.27	0.79	21.4
P18	G	120	4	0.0299	1.88	2.11	1.11	22.5
P19	G	120	4	0.0175	1.58	1.87	1.90	24.4
P20	G	120	4	0.0051	0.88	1.22	6.54	**30.9**
P21	G	120	4	0.0073	2.48	2.59	4.57	19.5
P23	G	120	2	0.3475	3.69	3.47	0.10	15.0
P24	G	120	2	0.2979	3.63	3.43	0.11	15.1
P25	G	120	2	0.2482	3.57	3.40	0.13	15.2
P26	G	120	2	0.1986	3.49	3.35	0.17	15.4
P27	G	120	2	0.1489	3.38	3.29	0.22	15.6
P28	G	120	2	0.0993	3.23	3.20	0.34	16.0
P29	G	120	2	0.0496	2.96	2.93	0.67	16.6

Table 4.41
Free chlorine and chloramine residual predictions for hypothetical subdivision using PVC and unlined cast iron pipes

End of Pipe	Pipe Material	Length ft	Diameter Inches	Velocity Ft/sec	NH_2Cl residual	Cl_2 Residual	HRT hrs	Σ HRT hrs
R	-	-	-	-	4.00	4.00	0	0
P35	PVC	52,800	24	0	3.74	3.50	14.8	14.8
P30	PVC	320	24	1.0	3.73	3.49	0.09	14.9
P22	PVC	320	24	1.0	3.73	3.49	0.09	15.0
P34	PVC	10,560	24	1.0	3.68	3.40	2.98	18.0
P3	UCI	120	4	0.1913	3.69	3.46	0.17	15.0
P4	UCI	120	4	0.1789	3.65	3.41	0.19	15.2
P5	UCI	120	4	0.1665	3.60	3.37	0.20	15.4
P6	UCI	120	4	0.1540	3.55	3.32	0.22	15.6
P7	UCI	120	4	0.1416	3.49	3.27	0.24	15.8
P8	UCI	120	4	0.1292	3.43	3.21	0.26	16.1
P9	UCI	120	4	0.1168	3.37	3.15	0.29	16.3
P10	UCI	120	4	0.1044	3.29	3.08	0.32	16.7
P33	UCI	800	4	0.0920	2.80	2.62	2.42	19.1
P14	UCI	120	4	0.0796	2.72	2.55	0.42	19.5
P15	UCI	120	4	0.0672	2.63	2.46	0.50	20.0
P16	UCI	120	4	0.0548	2.53	2.36	0.61	20.6
P17	UCI	120	4	0.0423	2.40	2.24	0.79	21.4
P18	UCI	120	4	0.0299	2.22	2.08	1.11	22.5
P19	UCI	120	4	0.0175	1.96	1.83	1.90	24.4
P20	UCI	120	4	0.0051	1.26	1.18	6.54	**30.9**
P21	UCI	120	4	0.0073	2.75	2.57	4.57	19.5
P23	UCI	120	2	0.3475	3.69	3.46	0.10	15.0
P24	UCI	120	2	0.2979	3.65	3.42	0.11	15.1
P25	UCI	120	2	0.2482	3.59	3.37	0.13	15.2
P26	UCI	120	2	0.1986	3.52	3.31	0.17	15.4
P27	UCI	120	2	0.1489	3.43	3.23	0.22	15.6
P28	UCI	120	2	0.0993	3.30	3.11	0.34	16.0
P29	UCI	120	2	0.0496	3.05	2.89	0.67	16.6

As shown in Table 4.40 and Table 4.41, the water in the 24-inch PVC pipe will maintain a stronger residual after 18 hours when chloramines are used (3.68 mg/L as Cl_2) compared with free chlorine (3.40 mg/L as Cl_2).

When the pipe loop material is galvanized steel (Table 4.40), the predicted residual will be higher for free chlorine (1.22 mg/L as Cl_2) than for chloramines (0.88 mg/L as Cl_2) after a cumulative HRT of 30.9 hours. When the pipe loop material is UCI (Table 4.41), the predicted residual will be greater for chloramines (1.26 mg/L as Cl_2) than for free chlorine (1.18 mg/L as Cl_2) after 30.9 hours.

It is of interest to note that the presence of a "dead zone" or region of low velocity occurs in the pipe loop in the vicinity of pipe links 20-21. The corresponding cumulative HRT of the water to reach that point (30.9 hours) is greater than the cumulative HRT in the linear pipe segment corresponding to pipe links 23 to 29 (16.6 hours). The lowest free chlorine and chloramine residual would be found in the vicinity of the called "dead zone" with residuals in the range of 0.9 to 1.3 mg/L as Cl_2 depending on the pipe material and disinfectant type.

Summary Prediction Tables

In this section several tables are presented that show predictions of free chlorine and chloramine residuals with time for different water sources, water temperatures, pipe diameters and pipe materials.

The data in Table 4.42 and Table 4.43 show the summary of predicted free chlorine and chloramines residual respectively in sections of 2-inch diameter galvanized steel pipe for different HRTs. The predictions were made for 100% groundwater, 100% surface water and a blend of 60%/30%/10% (GW/SW/RO), at temperatures of 20°C and 28°C.

The results show that at the velocity of 0.33 ft/sec the water would have flowed through 0.25 miles and 0.5 miles of pipe after 1.1 and 2.2 hours. In addition, 1.8, 10.8, and 27 miles of pipe are needed to obtain HRTs of 8, 48 and 120 hours respectively. After 1, 2, and 8 hours at 20°C, the predicted residual are similar for free chlorine and chloramines for each type of water source with chloramines performing slightly better than free chlorine.

When the water temperature increases to 28°C, greater free chlorine concentrations are predicted irrespective of the water source. Free chlorine residuals in the range of 3.2-3.4 mg/L as Cl_2 and 1.9-2.3 mg/L as Cl_2 are predicted after 2.2 and 8 hours respectively, compared with chloramine residuals in the range of 2.9-3.1 mg/L as Cl_2 and 1.3-1.5 mg/L as Cl_2 for the same HRTs.

The results for 48 and 120 hours are not discussed because is very unrealistic to expect the existence of 10.8 and 27 miles section of 2-inch diameter galvanized pipe in an actual distribution system.

Table 4.42
Free chlorine residual predictions for 2 inches galvanized steel pipes for GW, SW, and blend in a distribution system

Pipe Diameter (in)	Flow Velocity (ft/sec)	HRT (Hr)	Temperature (°C)	Equivalent Distance (miles)	Free Chlorine Residual (mg/L as Cl_2)		
					100% GW	100% SW	60/30/10% Blend
2	0.33	1.1	20	0.25	3.70	3.78	3.74
2	0.33	2.2	20	0.5	3.43	3.58	3.50
2	0.33	8	20	1.8	2.29	2.67	2.45
2	0.33	48	20	10.8	0.14	0.36	0.21
2	0.33	120	20	27.0	0.00	0.01	0.00
2	0.33	1.1	28	0.25	3.60	3.71	3.65
2	0.33	2.2	28	0.5	3.24	3.44	3.33
2	0.33	8	28	1.8	1.86	2.31	2.05
2	0.33	48	28	10.8	0.04	0.15	0.07
2	0.33	120	28	27.0	0.00	0.00	0.00

Table 4.43
Chloramine residual predictions for 2 inches galvanized steel pipes for GW, SW, and blend in a distribution system

Pipe Diameter (in)	Flow Velocity (ft/sec)	HRT (Hr)	Temperature (°C)	Equivalent Distance (miles)	Chloramine Residual (mg/L as Cl_2)		
					100% GW	100% SW	60/30/10% Blend
2	0.33	1.1	20	0.25	3.75	3.79	3.77
2	0.33	2.2	20	0.5	3.51	3.59	3.55
2	0.33	8	20	1.8	2.50	2.71	2.59
2	0.33	48	20	10.8	0.24	0.39	0.30
2	0.33	120	20	27.0	0.00	0.01	0.01
2	0.33	1.1	28	0.25	3.41	3.50	3.45
2	0.33	2.2	28	0.5	2.90	3.07	2.98
2	0.33	8	28	1.8	1.25	1.53	1.37
2	0.33	48	28	10.8	0.00	0.01	0.01
2	0.33	120	28	27.0	0.00	0.00	0.00

Free chlorine and chloramines predictions for unlined cast iron pipes of 6 and 24 inches in diameter are presented in Table 4.44 and Table 4.45. Table 4.44 shows free chlorine residual predictions for 100% GW, 100% SW, and the 60%/30%/10% blend at temperatures of 20°C and 28°C. Table 4.45 shows the chloramine residual predictions under the same conditions.

Comparison of Table 4.44 and Table 4.45 shows that free chlorine and chloramine behave similarly in UCI pipes. Both disinfectants can maintain a residual greater than 1.2 mg/L as Cl_2 for at least 10.8 miles (48 hours) at 20°C in pipes with a diameter of 6 inches. When the temperature increases to 28°C, all the waters would maintain a residual between 0.3 and 0.4 mg/L as Cl_2. For 24-inch UCI pipes and a water temperature of 20°C, all the waters will show a residual greater than 0.7 mg/L as Cl_2 after flowing through 27 miles of pipe (5-day HRT). For a temperature of 28°C, all the waters disinfected with chloramines will carry a residual 0.1-0.2 mg/L greater than the respective water source using free chlorine. After 10.8 miles (48 hours), chloraminated GW will present a residual of 0.87 mg/L as Cl_2 compared with 0.62 mg/L as Cl_2 for free chlorinated GW.

Table 4.44
Free chlorine residual predictions for 6 and 24 inches UCI pipes for GW, SW, and blend in a distribution system

Pipe Diameter (in)	Flow Velocity (ft/sec)	HRT (Hr)	Temperature (°C)	Equivalent Distance (miles)	Free Chlorine Residual (mg/L as Cl_2)		
					100% GW	100% SW	60/30/10% Blend
6	0.33	8	20	1.8	3.34	3.49	3.41
6	0.33	48	20	10.8	1.36	1.77	1.53
6	0.33	120	20	27.0	0.27	0.52	0.36
24	0.33	8	20	1.8	3.56	3.72	3.63
24	0.33	48	20	10.8	1.98	2.58	2.23
24	0.33	120	20	27.0	0.69	1.33	0.92
6	0.33	8	28	1.8	2.48	2.78	2.61
6	0.33	48	28	10.8	0.23	0.45	0.31
6	0.33	120	28	27.0	0.00	0.02	0.01
24	0.33	8	28	1.8	2.93	3.29	3.09
24	0.33	48	28	10.8	0.62	1.24	0.84
24	0.33	120	28	27.0	0.04	0.22	0.08

Table 4.45
Chloramine residual predictions for 6 and 24 inches UCI pipes for GW, SW, and blend in a distribution system

Pipe Diameter (in)	Flow Velocity (ft/sec)	HRT (Hr)	Temperature (°C)	Equivalent Distance (miles)	Chloramine Residual (mg/L as Cl_2)		
					100% GW	100% SW	60/30/10% Blend
6	0.33	8	20	1.8	3.29	3.42	3.35
6	0.33	48	20	10.8	1.25	1.56	1.38
6	0.33	120	20	27.0	0.22	0.38	0.28
24	0.33	8	20	1.8	3.58	3.72	3.64
24	0.33	48	20	10.8	2.06	2.59	2.28
24	0.33	120	20	27.0	0.76	1.34	0.98
6	0.33	8	28	1.8	2.56	2.79	2.66
6	0.33	48	28	10.8	0.27	0.46	0.34
6	0.33	120	28	27.0	0.00	0.02	0.01
24	0.33	8	28	1.8	3.10	3.38	3.22
24	0.33	48	28	10.8	0.87	1.46	1.10
24	0.33	120	28	27.0	0.09	0.32	0.16

Table 4.46 and Table 4.47 show the predictions of free chlorine and chloramine residual for lined cast iron pipes conducting 100% GW, 100% SW and a 60%/30%/10% blend of GW, SW, and RO. Predictions were made for pipes with diameters of 6 and 24 inches and water temperatures of 20°C and 28°C.

Analysis of the tables shows that chloramines would be maintained longer in LCI pipes than free chlorine regardless of the water source and water temperature for pipe diameters greater than 6 inches. At 20°C, waters using chloramines will have residuals of 2.8-3.0 mg/L as Cl_2 and 3.3-3.5 mg/L as Cl_2 for 6 and 24 inches diameter pipes respectively after 27 miles (120 hours), compared with 1.7-2.4 mg/L as Cl_2 and 1.9-2.6 mg/L as Cl_2 for free chlorine. When the temperature increases to 28°C, waters using chloramines will have residuals of 1.2-1.5 mg/L as Cl_2 and 2.1-2.6 mg/L as Cl_2 for 6 and 24 inches diameter pipes respectively after 27 miles (120 hours), compared with 0.2-0.6 mg/L as Cl_2 and 0.3-0.8 mg/L as Cl_2 for free chlorine.

Table 4.46
Free chlorine residual predictions for 6 and 24 inches LCI pipes for GW, SW, and blend in a distribution system

Pipe Diameter (in)	Flow Velocity (ft/sec)	HRT (Hr)	Temperature (°C)	Equivalent Distance (miles)	Free Chlorine Residual (mg/L as Cl_2)		
					100% GW	100% SW	60/30/10% Blend
6	0.33	8	20	1.8	3.78	3.86	3.82
6	0.33	48	20	10.8	2.85	3.24	3.02
6	0.33	120	20	27.0	1.72	2.37	1.98
24	0.33	8	20	1.8	3.81	3.89	3.84
24	0.33	48	20	10.8	2.97	3.38	3.15
24	0.33	120	20	27.0	1.91	2.63	2.20
6	0.33	8	28	1.8	3.24	3.51	3.36
6	0.33	48	28	10.8	1.12	1.82	1.39
6	0.33	120	28	27.0	0.17	0.56	0.29
24	0.33	8	28	1.8	3.32	3.60	3.44
24	0.33	48	28	10.8	1.31	2.13	1.63
24	0.33	120	28	27.0	0.25	0.83	0.42

Table 4.47
Chloramine residual predictions for 6 and 24 inches LCI pipes for GW, SW, and blend in a distribution system

Pipe Diameter (in)	Flow Velocity (ft/sec)	HRT (Hr)	Temperature (°C)	Equivalent Distance (miles)	Chloramine Residual (mg/L as Cl_2)		
					100% GW	100% SW	60/30/10% Blend
6	0.33	8	20	1.8	3.90	3.92	3.91
6	0.33	48	20	10.8	3.46	3.55	3.50
6	0.33	120	20	27.0	2.78	2.97	2.87
24	0.33	8	20	1.8	3.95	3.96	3.96
24	0.33	48	20	10.8	3.70	3.79	3.74
24	0.33	120	20	27.0	3.28	3.50	3.38
6	0.33	8	28	1.8	3.70	3.75	3.72
6	0.33	48	28	10.8	2.49	2.71	2.58
6	0.33	120	28	27.0	1.22	1.51	1.34
24	0.33	8	28	1.8	3.83	3.89	3.86
24	0.33	48	28	10.8	3.09	3.36	3.21
24	0.33	120	28	27.0	2.09	2.59	2.30

Table 4.48 and Table 4.49 present the predictions for free chlorine and chloramines respectively, in PVC pipes with diameters of 6 and 24 inches. The predictions were calculated for water temperatures of 20°C and 28°C.

Table 4.48 and Table 4.49 show that under the same conditions, the chloramine residual would be maintained longer in PVC pipes than free chlorine. At 20°C, waters using chloramines will have residuals of 2.9-3.2 mg/L as Cl_2 and 3.3-3.5 mg/L as Cl_2 for 6 and 24 inches diameter pipes respectively after 27 miles (120 hours), compared with 2.3-2.8 mg/L as Cl_2 and 2.6-3.1 mg/L as Cl_2 for free chlorine. If the water temperature rises to 28°C, waters using chloramines will have residuals of 1.5-1.9 mg/L as Cl_2 and 2.1-2.6 mg/L as Cl_2 for 6 and 24 inches diameter pipes respectively after 27 miles (120 hours), compared with 0.8-1.3 mg/L as Cl_2 and 1.1-1.8 mg/L as Cl_2 for free chlorine.

Table 4.48
Free chlorine residual predictions for 6 and 24 inches PVC pipes for GW, SW, and blend in a distribution system

Pipe Diameter (in)	Flow Velocity (ft/sec)	HRT (Hr)	Temperature (°C)	Equivalent Distance (miles)	Free Chlorine Residual (mg/L as Cl_2)		
					100% GW	100% SW	60/30/10% Blend
6	0.33	8	20	1.8	3.86	3.91	3.88
6	0.33	48	20	10.8	3.22	3.47	3.33
6	0.33	120	20	27.0	2.32	2.79	2.52
24	0.33	8	20	1.8	3.88	3.93	3.91
24	0.33	48	20	10.8	3.36	3.61	3.47
24	0.33	120	20	27.0	2.58	3.10	2.80
6	0.33	8	28	1.8	3.58	3.72	3.64
6	0.33	48	28	10.8	2.06	2.58	2.28
6	0.33	120	28	27.0	0.76	1.33	0.98
24	0.33	8	28	1.8	3.66	3.80	3.72
24	0.33	48	28	10.8	2.34	2.93	2.59
24	0.33	120	28	27.0	1.05	1.84	1.35

Table 4.49
Chloramine residual predictions for 6 and 24 inches PVC pipes for GW, SW, and blend in a distribution system

Pipe Diameter (in)	Flow Velocity (ft/sec)	HRT (Hr)	Temperature (°C)	Equivalent Distance (miles)	Chloramine Residual (mg/L as Cl_2)		
					100% GW	100% SW	60/30/10% Blend
6	0.33	8	20	1.8	3.92	3.94	3.93
6	0.33	48	20	10.8	3.54	3.65	3.59
6	0.33	120	20	27.0	2.94	3.17	3.04
24	0.33	8	20	1.8	3.95	3.97	3.96
24	0.33	48	20	10.8	3.69	3.80	3.74
24	0.33	120	20	27.0	3.27	3.53	3.38
6	0.33	8	28	1.8	3.74	3.80	3.77
6	0.33	48	28	10.8	2.67	2.95	2.79
6	0.33	120	28	27.0	1.46	1.87	1.63
24	0.33	8	28	1.8	3.83	3.89	3.86
24	0.33	48	28	10.8	3.07	3.39	3.21
24	0.33	120	28	27.0	2.06	2.64	2.30

BIOLOGICAL STABILITY

Biostability is a relatively new concept to drinking water that involves interaction between the bulk water and biofilms in the distribution systems, both of which are affected by several factors. Biostability is a concept which addresses the overall tendency of the water to promote or suppress microbial proliferation, and can be viewed as an assessment of overall distribution system quality with respect to microorganisms. It pertains to the proliferation of microorganisms in the water distribution system and does not address the ecology of opportunistic or other pathogens or coliforms. Biostability describes aggregate proliferation and does not address the fate of specific subpopulations. As a result biological instability may or may not favor the proliferation of coliforms or a specific pathogen(s) depending on a host of additional factors.

Organic carbon is considered to be the limiting nutrient with respect to heterotrophic biostability in most drinking water distribution systems (LeChevallier et al. 1991). The organic carbon that drives heterotrophic growth is referred to as biodegradable organic matter (BOM). BOM is typically quantified using the assimilable organic carbon (AOC) or the biodegradable organic carbon (BDOC) assays. AOC has been correlated with HPC proliferation in the literature, although this type of correlation is less established for BDOC (Van der Kooij 1992, LeChevallier et al. 1987).

In the presence of chlorine residual, LeChevallier et al. (1995) suggested that AOC levels of less than 100 μg/L were necessary for biological stability in drinking water distribution systems. His statements were based on the occurrence of total coliforms, which are regulated by the total coliform rule. BOM levels are only one of many factors that determine heterotrophic instability (LeChevallier et al. 1995). Biological stability implies no more than an acceptable concentration of microorganisms in a drinking water distribution system; lack of biological stability implies excessive microorganisms in the same.

Bulk Biostability

Bulk biostability was assessed by HPC population as no coliforms were ever observed in any PDS influent or effluent samples. A general assessment of biostability without regard to temperature, pipe materials or water quality is shown in Figure 4.106. Biostability as assessed by HPC population is shown to be dependent on PDS effluent residual. The data set represented in Figure 4.106 is from the first year of operation. HPC population dramatically increases when the PDS effluent residual is less than approximately 0.1 mg/L as Cl_2 and appears to enter a transition zone of visually apparent exponentially increasing growth at 0.2 to 0.3 mg/L chloramine as Cl_2.

A rational argument can be made that the HPCs corresponding to those residual concentrations defined the transition zone to and the zone of biological instability. Another method of defining the transition zone to and the zone of biological instability can be the HPC goals of 500 or 1,000 cfu/mL and 10,000 cfu/mL, respectively. The beginning of the transition zone to biological instability would be defined by a regulatory definition and the beginning of the zone would be arbitrarily defined as an order of magnitude higher.

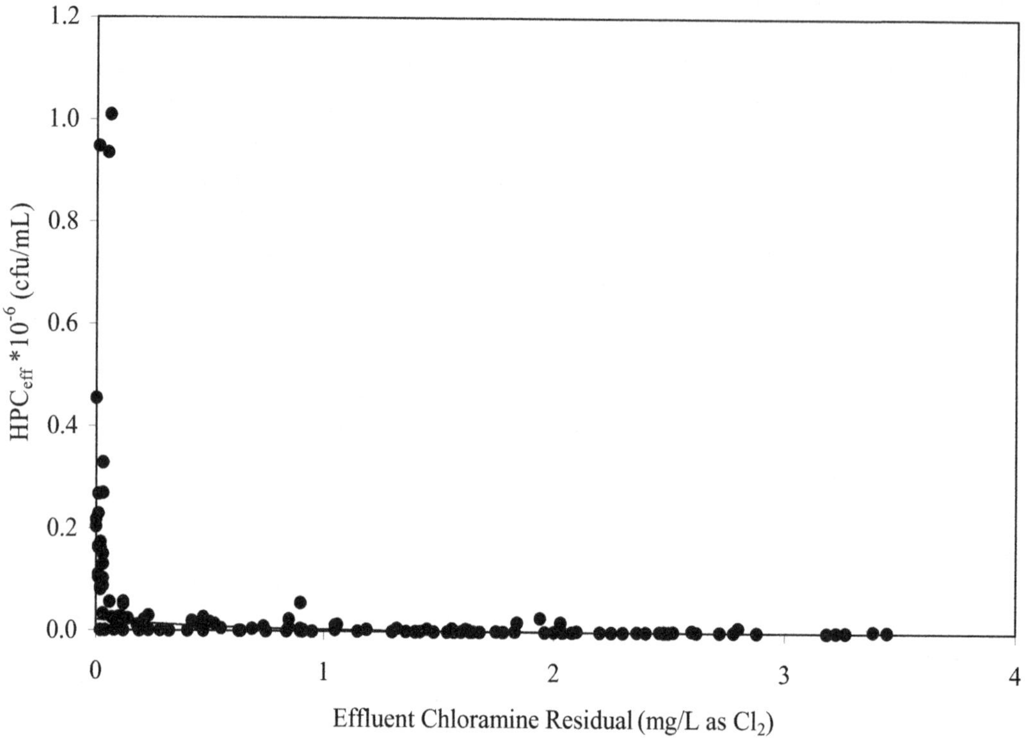

Figure 4.106 Effect of effluent chloramine residual (T $Cl_{2\text{-}eff}$) on effluent HPC

Although the HPC axis shown in Figure 4.106 is too compressed to allow the associated residuals to be defined, the HPC relationship with HPCs shown in Figure 4.106 clearly shows a zone of unchecked microbiological growth corresponding to total residual, which can easily be defined as biological instability. HPC populations corresponding to the transition zone and beginning of biological stability will be defined later in this section.

HPC Modeling

Relationships of HPCs with AOC, temperature and residual were investigated by visual inspection of graphs and statistical analyses of the data in the bulk and biofilm. Bulk biostability was assessed by relating HPC proliferation to water quality parameters by non-linear regression. Pipe material was represented by dummy variables in the regression analysis with H being assigned to the hybrid lines. Independent water quality variables included in the initial regression were AOC, residual, temperature, DOC or UV254, calcium, sulfates, silica, chlorides and alkalinity. With the exception of AOC and the saturation constant, independent variables that were not within the 95 % confidence level (CL) were eliminated from the data set and the regression was repeated until only independent variable that were within the 95 % CL remained. When more than one independent variable was outside the 95 % CL, the variable most outside the 95 % CL (highest p value) was eliminated and the regression was repeated. The final regression equation relating Log HPC_{AOC} to water quality is shown in equation (4.84), which is designated as Log HPC_{AOC} to clarify this model from other Log HPC models. The selected model is a non-linear model, which includes exponential water quality terms with exponents whose magnitude and sign are determined by regression. The residual term included a retardant term (saturation constant) plus the actual residual in the denominator, which maintains convergence at zero residual.

$$Log(HPC)_{AOC} = (2.6488H + 2.6477UCI + 3.3410LCI + 3.0633PVC + 2.8364G) \times \frac{AOC_{\text{inf}}^{0.0398}}{(0.0177 + TCl_{2-eff})^{0.1310}} \qquad (4.84)$$

where Log $(HPC)_{AOC}$ = Log HPCs (cfu/mL)
AOCinf = Influent AOC
T = Temperature (°C)
H = Hybrid PDS (0,1)
UCI = UCI PDS (0,1)
LCI = LCI PDS (0,1)
G = G PDS (0,1)
$TCl_{2\text{-eff}}$ = Effluent Chloramine Residual (mg/L as Cl_2)

As shown in Table 4.50, all dummy coefficients for material were significant; however, AOC_{eff} ($p = 0.27$) and the saturation constant ($p = 0.53$) were not significant at the 95 % confidence level. The dummy variable coefficient for PVC is larger than the dummy variable coefficient for G, which could be incorrectly interpreted as the PVC environment was more conducive to HPC growth than the G environment, which is not the case. The relatively high PVC dummy variable coefficient was caused by the inclusion of residual in the model and the higher residual in the PVC environment relative to the G environment. If the same residuals could be maintained in the G and PVC environments, then the PVC HPCs would be predicted to exceed the G HPCs, which is not unreasonable as leaching zinc may be toxic to HPCs.

Table 4.50
Statistically determined coefficients and parameters for Log HPC model with AOC

	Coefficient	Std. Error	t	P
Saturation Constant	0.0177	0.0283	0.6254	0.5334
AOC_{inf}	0.0398	0.0358	1.1100	0.2702
Residual	0.1310	0.0319	4.1083	<0.0001
Hybrid	2.6488	0.4076	6.4978	<0.0001
UCI	2.6477	0.4649	5.6947	<0.0001
LCI	3.3410	0.6360	5.2529	<0.0001
PVC	3.0633	0.5951	5.1474	<0.0001
G	2.8364	0.5314	5.3371	<0.0001

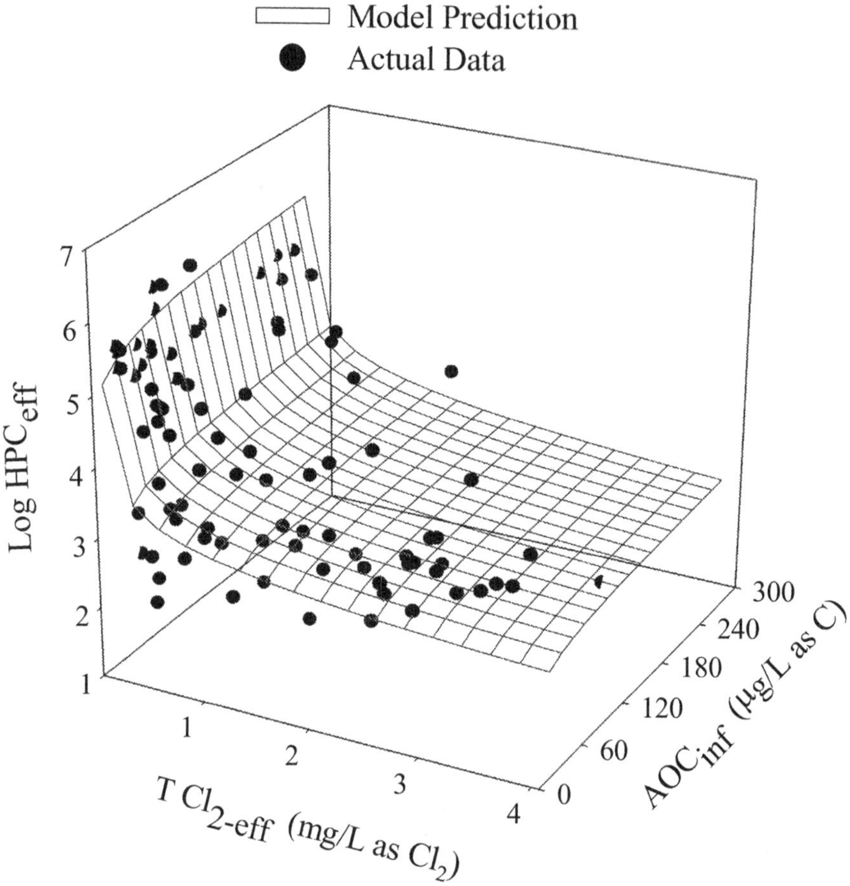

Figure 4.107 Predicted Log HPC versus effluent chloramine residual (T $Cl_{2\text{-}eff}$) and Log AOC for hybrid PDS

The predicted and actual Log HPCs versus residual and AOC are shown in Figure 4.107, which illustrate the impact of AOC and residual for this model. The coefficient of determination for this model is 0.51. As shown in Figure 4.107 and after inspection of equation (4.84), the predicted HPC population is approximately one order of magnitude less at a 1 mg/L residual than at a zero residual; whereas the predicted HPC population increases by approximately 0.2 of a log for an AOC variation of 10 to 1000 µg/L as C. Residual impacts the predicted HPC population significantly more than AOC. AOC did vary by process with the highest AOC measured at approximately 300 µg/L as C. The highest average AOC was 110 µg/L as C, which was in conventionally treated GW (G1). The lowest average AOC was 43 µg/L as C, which was in RO. Using these AOC values in equation (4.84), results were 500 cfu/mL higher for predicted HPCs in the G1 source than in the RO source. This difference would be significant at low HPC populations; however the predicted HPC populations in both sources were over 5000 cfu/mL. Consequently, the predicted difference in HPC populations due to the low and high AOCs at the average residual is about 10% of the HPC population, which is not significant at these levels. At zero residual, this difference increases to approximately 9,000 cfu/mL. Both populations were in excess of 40,000 cfu/mL, but the high and low AOC difference increases the HPC populations by approximately 20%. Finally, it must be remembered that AOC was statistically insignificant and was used to illustrate the predicted impact on HPCs. The projections show that the impact on HPC populations increased significantly as residual decreased. AOC was thoroughly investigated but was found to be statistically insignificant to HPC in the PDSs when residual and pipe material and temperature were considered.

Another empirical model for Log HPCs was developed that only considered statistically significant terms. The model was developed in the same manner as the model shown in equation (4.84) except that only statistically significant terms were retained. Independent water quality variables included in the initial regression were AOC, residual, temperature, DOC or UV254, calcium, sulfates, silica, chlorides and alkalinity. Dummy variables were used to represent pipe material. This log HPC model is shown in equation (4.85). All pipe materials, temperature and effluent residual were statistically significant with respect to log HPC. AOC, DOC or any other inorganic parameters were not significant. There were at times significant numbers of HPC in the feed standpipe for the PDS. Hence, a similar model for ΔHPC (HPC_{in} - HPC_{out}) was developed and found to have the same form, significant parameters and approximately the same magnitude of coefficients and exponents shown in equation (4.85). Consequently, equation (4.85) was adequate to describe HPC in the PDS effluent.

$$LogHPC = \left(2.6407H + 2.5964UCI + 3.3322LCI + 2.9237PVC + 2.9111G\right) \times \frac{1.029^{(Temp-20)}}{(0.0310 + T\,Cl_{2-eff})^{0.11476}} \quad (4.85)$$

The coefficients and statistical parameters associated with equation (4.85) are shown in Table 4.51. All terms are statistically significant except the saturation constant, which was retained in the model to maintain convergence at zero residual and illustrate the effects of zero residual. The coefficient of determination (R^2) was 0.57 for predicted and actual Log HPC using equation (4.85).

A three dimensional plot of predicted and actual Log HPC versus chloramine residual and temperature using equation (4.85) is shown in Figure 4.108. The plane or net is the grouping of predicted Log HPCs; while the clear circles are the actual data points. The dark curves within the net are curves for constant Log HPC 3 and 4; whereas the curves on the bottom plane are the same Log HPC 3 and 4 curves projected on the chloramine residual and temperature plane of Figure 4.108.

Table 4.51
Statistical coefficients and parameters for Log HPC model without AOC

	Coefficient	Std. Error	t	P
Constant	0.0310	0.0358	0.8654	0.3886
$TCl_{2\text{-eff}}$	0.1476	0.0345	4.2764	<0.0001
θ	1.0290	0.0087	117.7563	<0.0001
H	2.6407	0.1436	18.3883	<0.0001
UCI	2.5964	0.2460	10.5529	<0.0001
LCI	3.3322	0.3666	9.0902	<0.0001
PVC	2.9273	0.3401	8.6064	<0.0001
G	2.9111	0.2758	10.5535	<0.0001

Biostability is defined in terms of HPC populations in this work. Transition to biological instability was defined as beginning at 3 Log HPC and biological instability was defined as beginning at 4 Log HPC. Those points in Figure 4.108 in terms of temperature and total residual as mg/L as Cl_2 are for (a) 15°C at 0.06 mg/L for 3 Log HPC and greater than 4 mg/L for 4 Log HPC; for (b) 25°C at 1.1 mg/L for 3 Log HPC and 0.06 mg/L for 4 Log HPC; and for (c) 35°C three Log HPC is not possible from 0 to 4 mg/L and at 1.1 mg/L for 4 Log HPC. Log HPC increased from 3 to 4 when temperature was increased from 25°C to 35°C at 1.1 mg/L residual; hence a ten-degree temperature rise increases the HPC population such that biological stability is classified or located at entering the zone of transition at 25 °C to entering the zone of biological instability at 35 °C.

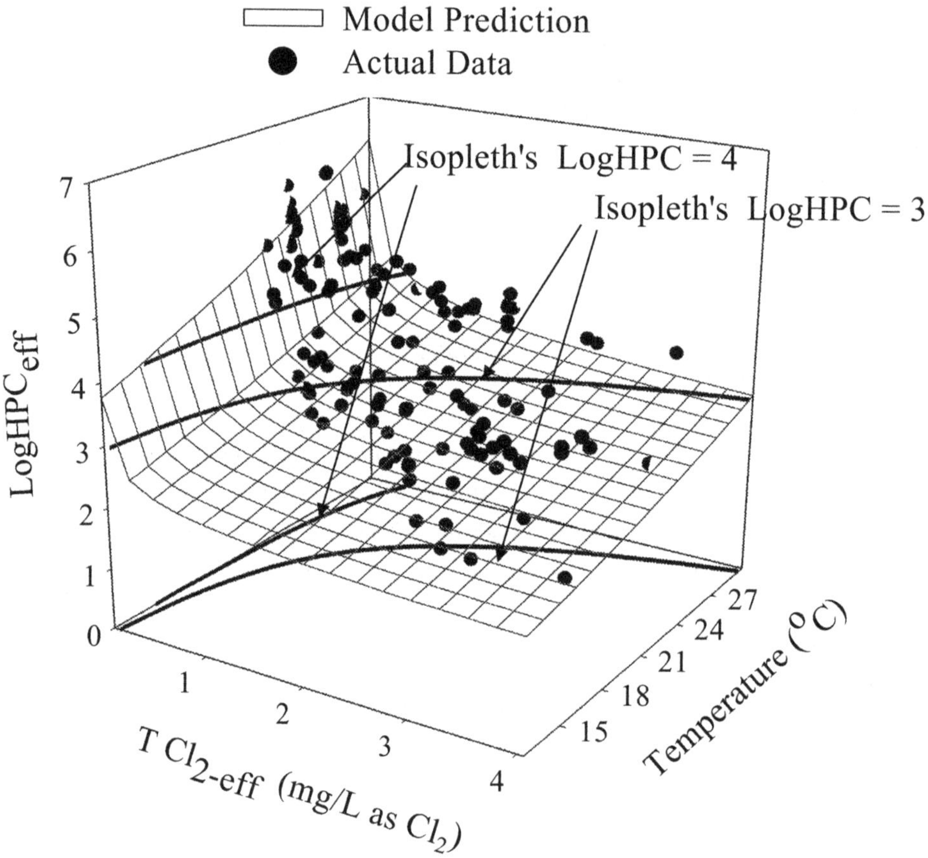

Figure 4.108 Predicted and actual Log HPCs versus chloramine residual (T $Cl_{2\text{-eff}}$) and temperature for the hybrid PDSs

Given that the Log HPC relationship shown in equation (4.85) is non-linear, it is difficult to state whether temperature or residual has the greatest impact on HPCs. Both impacts are significant, but the impact of residual is more significant that the impact of temperature as residual approaches zero; and the impact of temperature is most significant on HPCs at residuals greater than 1 mg/L as Cl_2. Given that utilities can do nothing to control temperature in a distribution system, residual is the only practically significant parameter on HPCs and requires more active management during high temperature periods.

Finally the Log HPC model shown in equation (4.85) can be proactively utilized to anticipate biological instability by incorporating the chlorine residual model shown in equation (4.77) into equation (4.85). This was done for the five day and two day HRT data collected through the project and shown in Figure 4.109 as a plot of predicted versus actual Log HPC. The correlation of determination was 0.53 and the predicted data was not significantly different from the actual data as determined by a paired t test at the 95 % CL.

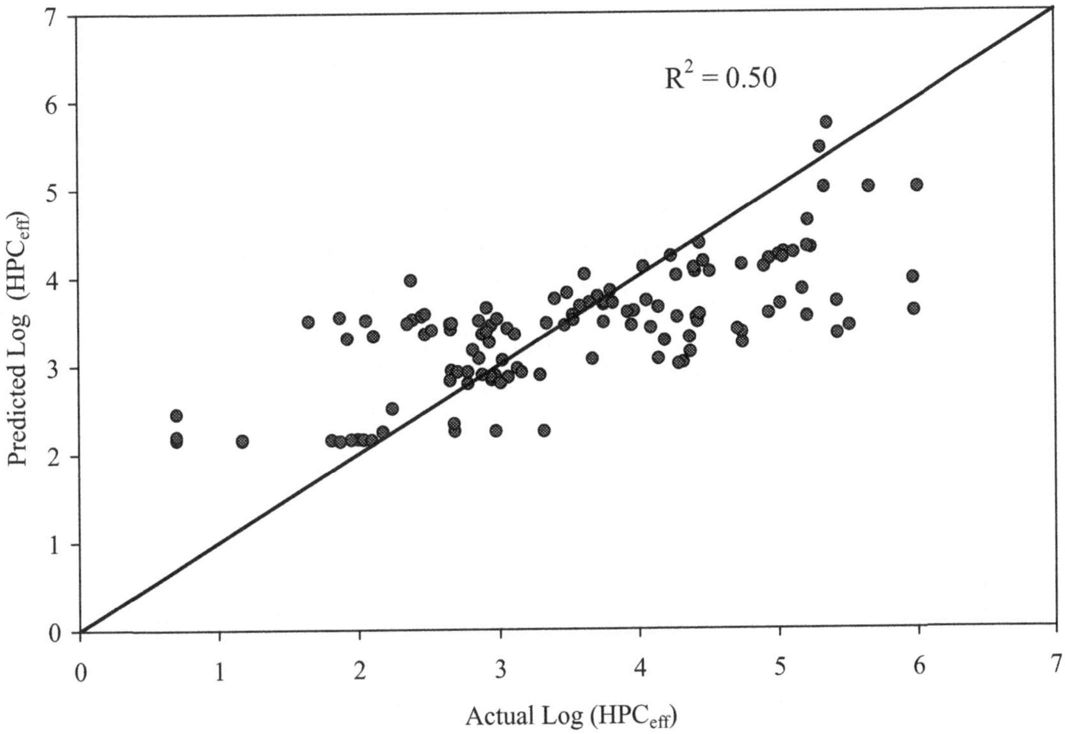

Figure 4.109 Actual versus predicted effluent Log HPC using chloramine residual predicted by chloramine residual model

Biofilm Modeling

Biofilms are always present, can never practically be removed and represent reservoirs for microorganisms in drinking water distribution systems. Hence, utilities have to successfully co-exist with biofilms. Biofilms function in part like any other microorganism population in that biofilms somewhat like microorganisms in a batch process go through acclimation, growth, steady state and death processes. Consequently, distribution system biofilms should be kept at a steady state existence in equilibrium with a strong residual. In such an environment, the bulk residual would be able to deactivate almost all of the microorganisms as they are released from the biofilm. Residuals enter biofilms by diffusion and dissipate via oxidation reactions with viable organisms and non-viable material that can function as an electron donor. Strong residuals can reduce "or beat back" biofilms somewhat but can not eliminate biofilms because the non-viable growth products of microorganism build a solid barrier that protects them from residuals.

If a residual is lost, biofilm growth, like microorganism growth in the bulk liquid, will increase and a non-desirable environment will develop. Note microorganism growth can occur in either the water or the biofilm in the PDS or in a distributuion system. Here, the water in the PDS is referred to as bulk water and the biofilm is called biofilm. Utilities have no other practical choice in America for successful biofilm control than maintenance of a strong residual in drinking water distribution systems.

The data set used to model PEPA consisted of average PEPAs for each coupon in the fourteen hybrid PDS over the four quarterly phases during the first year of operation. Hence, 224 PEPA observations were included in the data set. As shown in Figure 4.110, Log PEPA for all materials were significantly higher for the two-day HRT observations than for the five-day HRT observations. However, |ΔAOC| is significantly higher at a five day HRT than at a two day HRT as shown in Figure 4.111. Since |ΔAOC| is a bulk parameter and could be related to bulk HPCs, the relationship shown in Figure 4.111 could suggest that bulk HPCs would vary directly with |ΔAOC| and indirectly with biofilm HPCs. Such a relationship suggests an equilibrium approach can be used to describe the relationship between bulk and biofilm HPCs.

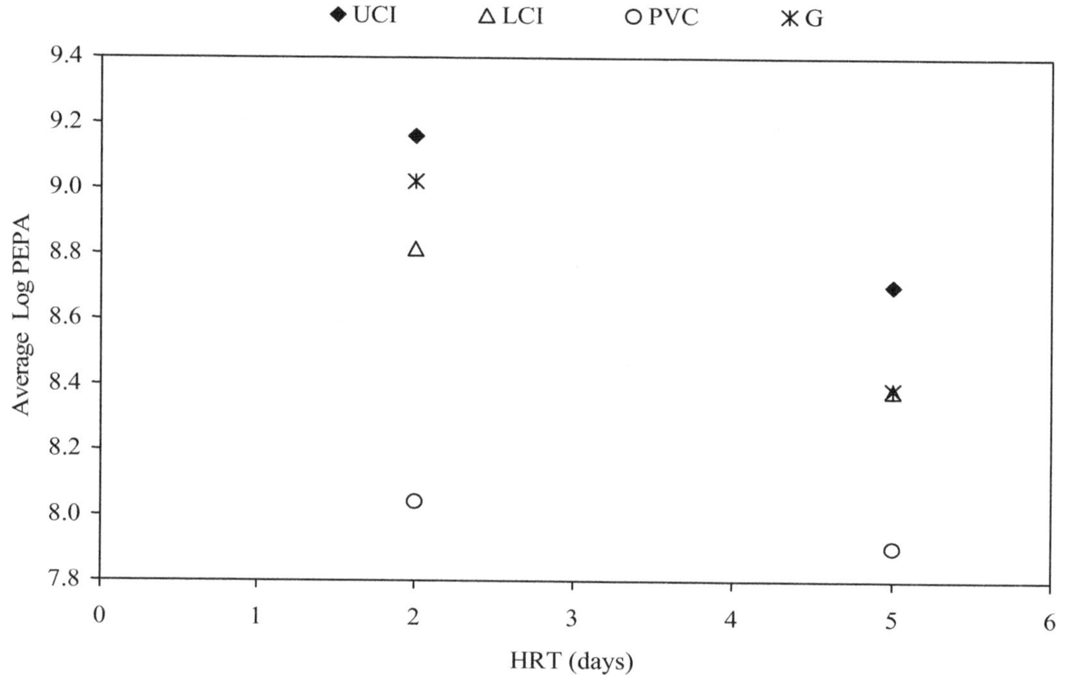

Figure 4.110 Average Log PEPA versus HRT for different materials

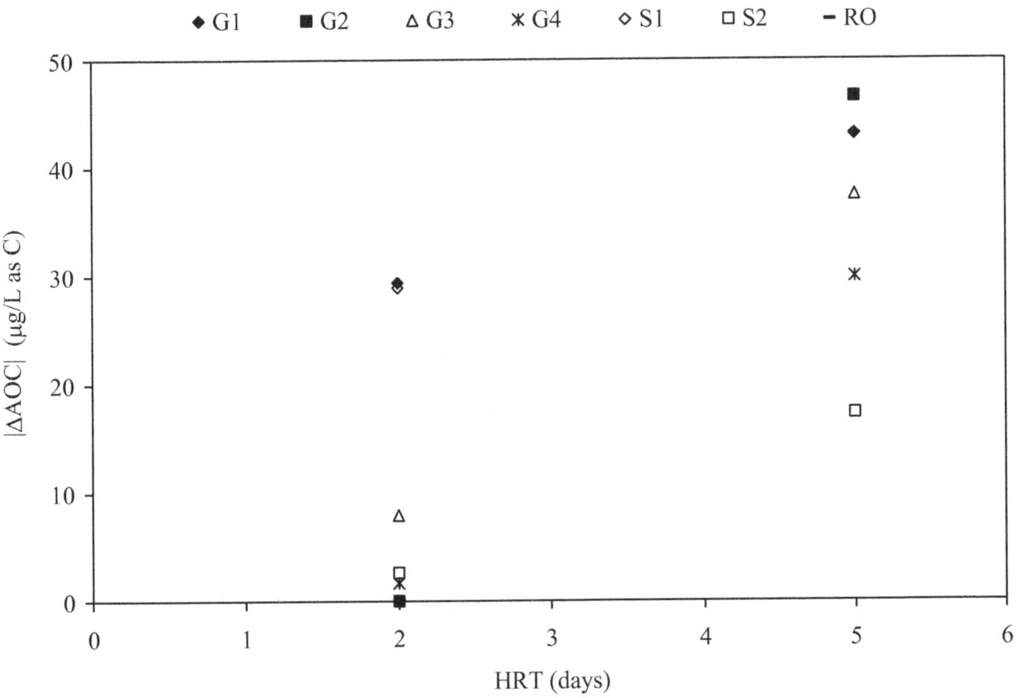

Figure 4.111 |ΔAOC| versus 5 days and 2 days HRT for different water sources

An empirical model for Log PEPA was developed in the same manner as previous models in that HPCs, AOC, residual, temperature, DOC or UV254, calcium, sulfates, silica, chlorides and alkalinity were the independent variables that were regressed against Log PEPA; and dummy variables were used to represent pipe material. The empirical Log PEPA model is shown in equation (4.86) and shows that all material terms, temperature, residual, Log HPC and HRT were statistically significant at the 95 % CL. The statistically determined coefficients, exponents and associated parameters are shown in Table 4.52. The coefficient of determination, R^2, for the Log PEPA model was 0.87.

$$Log\ PEPA = (8.6966UCI + 8.3791LCI + 7.7651PVC + 8.3993G) \times \frac{1.0232^{(Temp-20)}}{T\ Cl_{2-eff}^{\ 0.0112}(LogHPC)^{0.0292}\ HRT^{0.0271}} \tag{4.86}$$

Further review of equation (4.86) reveals the impact of material, HRT, HPC and residual on PEPA. PEPA increases as temperature increases and decreases as residual increases, which is intuitively rational. However, as found by model development as shown in equation (4.86), PEPA increases as HPCs and HRT decrease, which is not intuitively rational at first glance. Prior studies have established that the bulk liquid HPCs can both attach to the biofilm, or alternately HPCs can detach or slough from the biofilm and enter the bulk liquid. It is also well established that, while in the biofilm, bacteria are protected by, and held in the biofilm by, exocellular polymers (ECP). ECP helps limit the diffusion of bulk liquid residual into the depths of the biofilm (Geldreich 1996). Our data was taken after a period of time that was intended to be sufficient for establishing a pseudo-equilibrium for attachment and detachment of HPCs at the biofilm-bulk liquid interface. For PDSs where bulk liquid HPCs were low the inverse relationship of bulk and biofilm HPC implies that biofilm detachment/sloughing was low under these low bulk HPC/high bulk residual environments. The low detachment rate might have been due to an increase in ECP, which is a common biofilm response to oxidants/anti-microbial agents (Sutherland 1999). In addition higher ECP levels might have increased the rate of attachment resulting in increased biofilm. The inverse relationship of HRT to biofilm levels again could be due to increased ECP levels at the low HRT/high residual environment.

Table 4.52
Statistical coefficients and exponents for Log PEPA model

	Coefficient	Std. Error	t	P
θ	1.0232	0.0013	796.8906	<0.0001
T $Cl_{2\text{-eff}}$	0.0112	0.0024	-4.6295	<0.0001
Log HPC	0.0292	0.0103	-2.8377	0.0050
HRT	0.0271	0.0069	-3.9388	0.0001
UCI	8.6966	0.1329	65.4293	<0.0001
LCI	8.3791	0.1302	64.3680	<0.0001
PVC	7.7651	0.1204	64.4815	<0.0001
G	8.3993	0.1279	65.6646	<0.0001

Predicted versus normalized actual log PEPA is shown in Figure 4.112 and Figure 4.113 using equation (4.86). Since there are four independent variables in equation (4.86), two different figures were required to represent the effects of the independent variables.

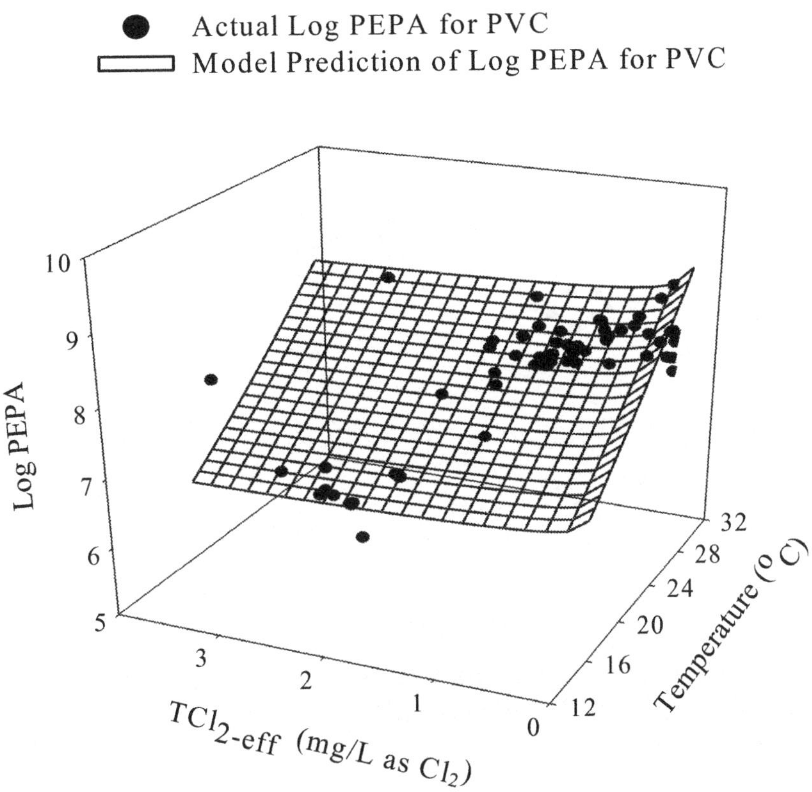

Figure 4.112 Predicted and normalized Log PEPA to 2 days HRT and 4 Log HPC versus temperature and effluent chloramine residual (T $Cl_{2\text{-eff}}$)

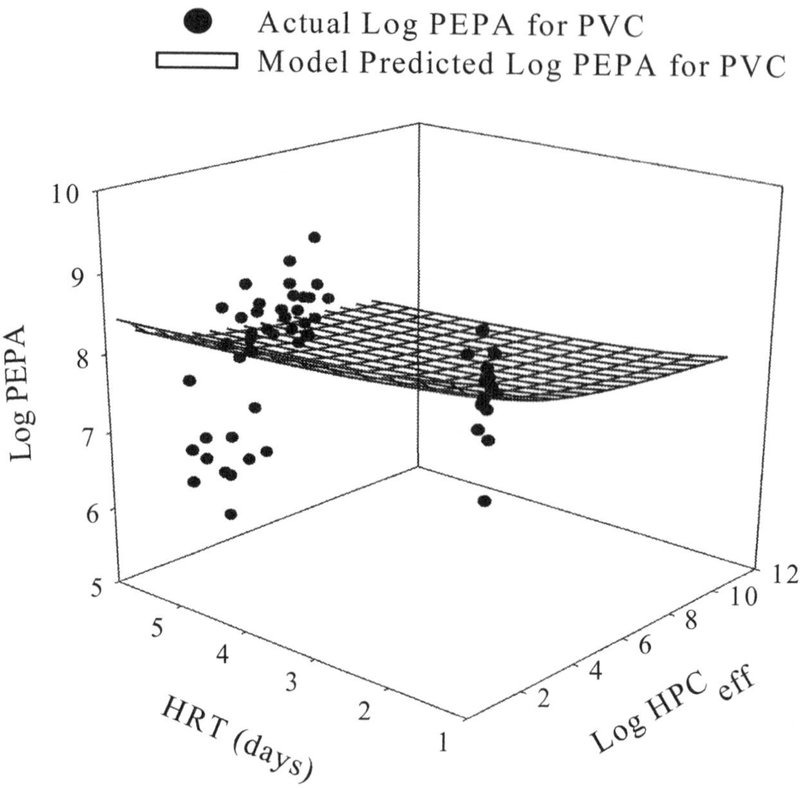

Figure 4.113 Predicted versus actual normalized PEPA to 2 mg/L chloramine and 25°C for PVC pipe material at varying HRT and Log HPC_{eff}

Log PEPA was normalized by holding Log HPC and HRT constant at 4 and 2 days respectively, and then using equation (4.86) to calculate Log PEPA for variable temperature and total residual as shown in Figure 4.112. Log PEPA was normalized by holding total residual and temperature constant at 2 mg/L as Cl_2 and 25°C, and then using equation (4.86) to calculate Log PEPA for varying Log HPC and HRT as shown in Figure 4.113.

As shown in Figure 4.112, chloramine residual has little impact on the magnitude of estimated Log PEPA; however, temperature has a significant impact on Log PEPA. The biofilm density hardly changes from 7 logs beginning at near zero residual to a total residual of 4 mg/L as Cl_2; whereas, the biofilms density increases from approximately 7 logs to more than 9 logs from 15°C to 30°C. Although statistically significant, the impact of HRT and Log HPC_{eff} is practically not significant relative to temperature as shown in Figure 4.113. Predicted PEPA varies from about 8.4 to 8.8 logs from 2 days to 7 days HRT; whereas, varying Log HPC in the effluent from 7 to 1 increases PEPA by approximately one-half of a log.

Other Correlations

Although bulk and biofilm biostability are being defined with regard to HPCs, there are some relationships between other water quality parameters that have significance to biological activity in drinking water distribution systems. Such parameters are AOC, BFHPCs and DO.

An empirical model for $|\Delta AOC|$ was developed in the same manner as previous models in that HPCs, AOC, residual, temperature, DOC or UV254, calcium, sulfates, silica, chlorides and alkalinity were the independent variables that were regressed against Log PEPA; and dummy variables were used to represent pipe material. The empirical $|\Delta AOC|$ model is shown in equation (4.87) and shows that only AOC_{inf} and temperature were statistically significant at the 95 % CL.

$$|\Delta AOC| = AOC_{inf}^{0.73} \times 1.1033^{(T-20)} \qquad (4.87)$$

where $|\Delta AOC|$ = $|AOC_{inf} - AOC_{eff}|$ (µg/L as C)

Figure 4.114 shows model-predicted data versus actual data for $|\Delta AOC|$ as a function of temperature and AOC_{inf}. The statistically determined coefficients, exponents and associated parameters are shown in Table 4.53. The coefficient of determination, R^2, for the $|\Delta AOC|$ model was 0.57. The statistical model shown in equation (4.87) is reminiscent of a first order reaction that is adjusted for temperature, which is commonly used to describe BOD reduction in microbiological systems.

Table 4.53

Statistical summary for $|\Delta AOC|$ as function of temperature and AOC_{inf}

	Coefficient	Std. Error	t	P
AOC_{inf}	0.7283	0.0390	18.6944	<0.0001
θ	1.1033	0.0314	35.1661	<0.0001

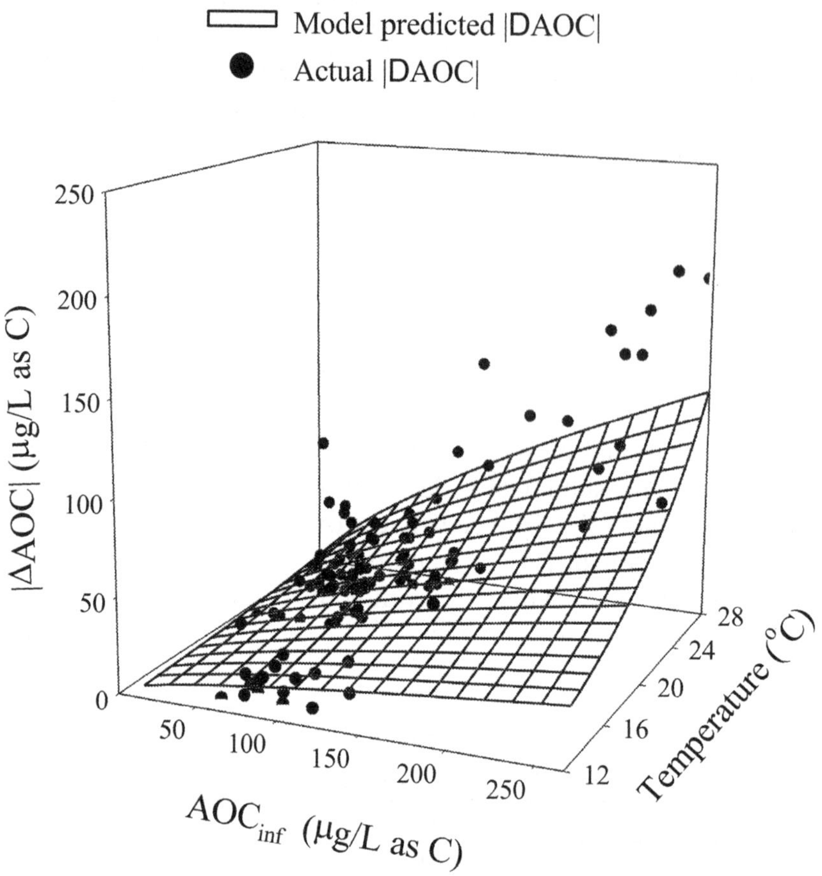

Figure 4.114 |ΔAOC| as a function of temperature and AOC_{inf}

Specifically there was no relationship between AOC and residual. This is really not surprising as residual exists in mg/L units and AOC exists in μg/L units. AOC has to overcome three orders of magnitude to affect residual stoichiometrically. This is noted here, because AOC is a rather new parameter to drinking water and its impact on residual is typically recognized as none.

The ratio of disinfectant concentration to substrate concentration has been used as the biocide to food ratio in nitrification studies (Harrington et al 2002). The Biocide:Food ((total residual - mg/L as Cl_2):(AOC - μg/L as C)) or the B/F ratio was defined, for the present study, as the ratio of residual and influent AOC. As shown in Figure 4.115, |ΔAOC|s of less than 25 μg/L as C were obtained for B/F ratio greater than 50, which generally corresponded to lower HPC populations and biostable conditions. Such relationships infer biological stability can be obtained by residual maintenance or AOC reduction.

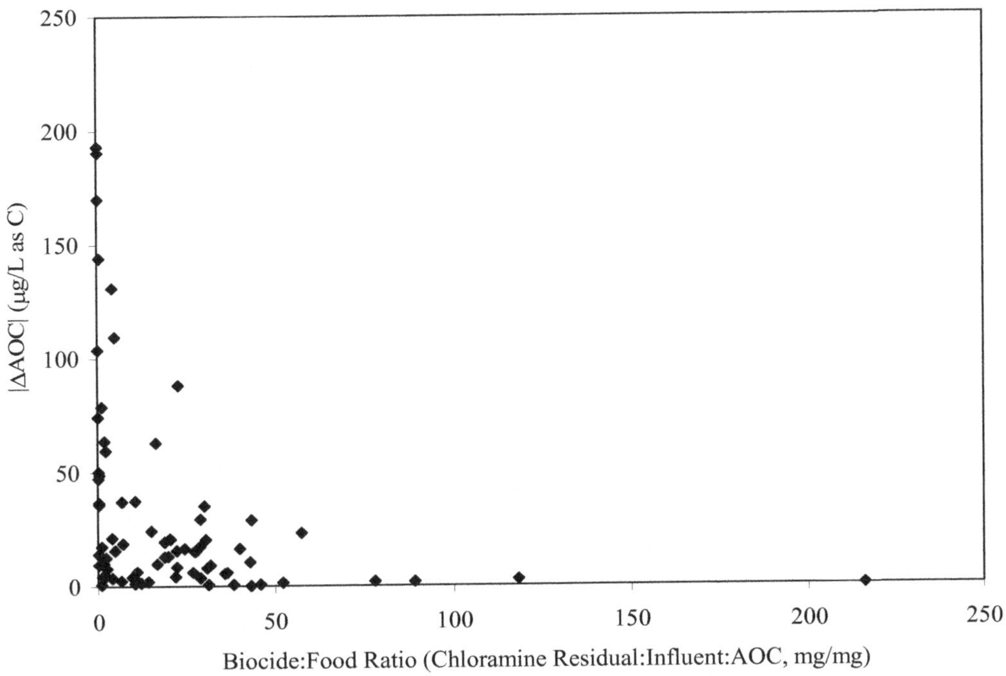

Figure 4.115 |ΔAOC| as a function of biocide:food ratio

A strong correlation was obtained between effluent DO and chloramine residual (Figure 4.116). Both residual and DO are excellent electron acceptors that are involved in wall reactions (e.g. corrosion) with unlined metals. Autotrophic growth (i.e. nitrification) also depletes DO. Low DOs were observed both when nitrification was occurring as well as when residual was low without nitrification. Unfortunately, the strong correlation between DO and residual kept both from being significant in all regression analyses as they were clearly confounding. However as opposed to residual, DO depletion is a symptom and not a cause of biological instability.

The correlation between Log PEPA and Log BFHPC is shown in Figure 4.117. As expected there is a linearly positive correlation between Log PEPA and Log BFHPC. Both Log PEPA and Log BFHPC are higher in the biofilms associated with the unlined metal coupons (UCI and G) than in the coupons associated with the non-metal surfaces (PVC and LCI). PEPA is two to five orders of magnitude greater than BFHPC because PEPA measures both culturable and non-culturable bacteria by quantifying the enzymatic activity of heterotrophs.

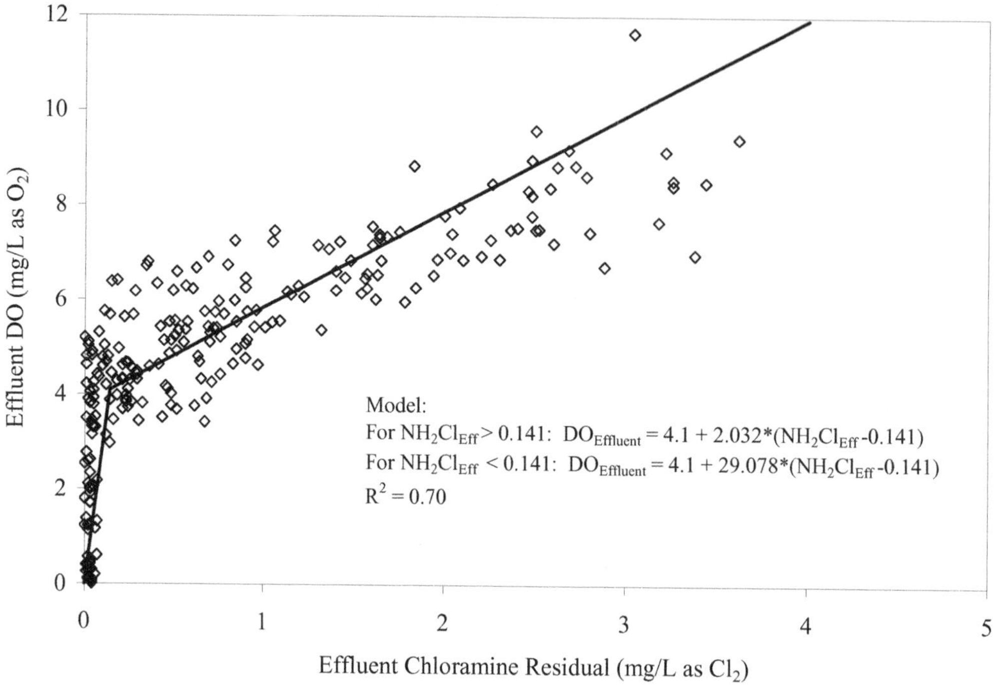

Figure 4.116 Correlation of effluent chloramine residual with effluent DO

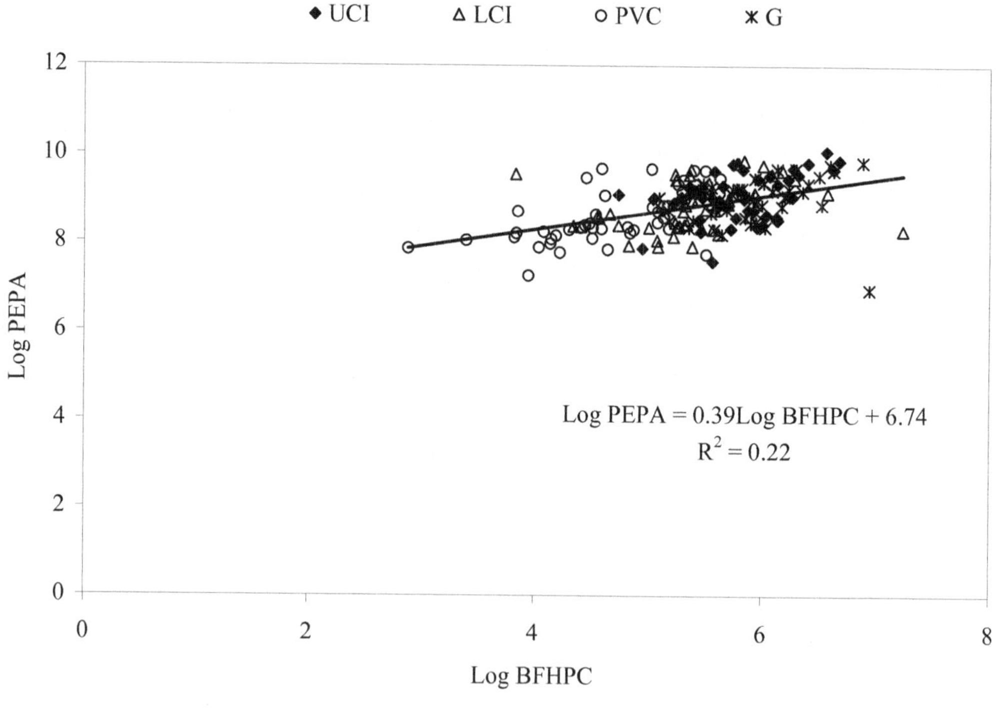

Figure 4.117 Log PEPA versus Log BFHPC by material

Biological Annular Reactor Experiments

As described in the Material and Methods section of this report, the BAR is a device designed to allow measuring and monitoring accumulation of film, whether microbial, organic or inorganic. Two experiments were conducted to compare observed trends during operation of the PDS and cradles and obtain kinetic information on biofilm growth. All sources waters were chloraminated at a target concentration of 4 mg/L Cl_2. Findings for each experiment are discussed in separate sections and a summary is provided.

Experiment One

A water produced by nanofiltration (G4), which is produced from a blend of RO, enhanced coagulated SW (S1) and ground water (G1), was compared to a water produced by enhanced coagulated, ozonated and carbon filtered SW (S1) in two BARs. Despite the fact that G4 seemed to be less biologically stable than S1 with regard to $|\Delta AOC|$ as shown in Figure 4.118, bulk liquid results were similar between the two BARs with respect to effluent HPC, nitrogen species levels (nitrates and nitrites), ΔDO and effluent chloramine residual as shown in Table 4.54 and Table 4.55, HPCs were fairly stable at 10^4 cfu/mL.

PEPA and BFHPC showed similar trends, and biofilm inventory in the BAR fed by S1 was consistently greater than in the BAR fed by G4 (Figure 4.119). Since bacterial material from the biofilm can be released into the distribution bulk liquid phase through sloughing, greater biofilm inventory implies a greater risk for release of viable cells in the bulk liquid and may lead to greater bacterial regrowth. A higher biofilm inventory may not necessarily be correlated to bioinstability. PEPA results were about 3.5 orders of magnitude greater than results from BFHPC, which can be explained by the fact that BFHPC measures only the viable, culturable, heterotrophic fraction of the biofilm, while PEPA measures the whole activity of the biofilm. It seems that biofilm inventory did not vary significantly (less than one order of magnitude) after primary colonization (between 9/19/02 and 10/03/02).

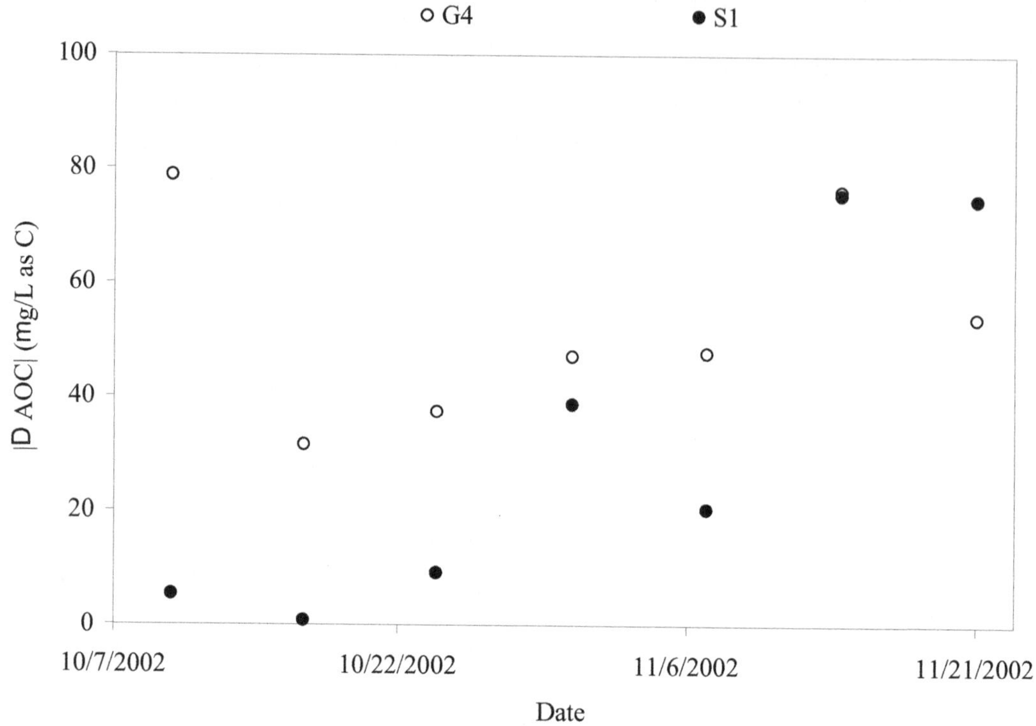

Figure 4.118 BAR experiment one: variation of |ΔAOC| with time

Table 4.54
Average water quality for BAR1(G4) – experiment one

Sample Port	DO mg/L as O_2	Residual mg/L as Cl_2	Nitrite mg/L as N	Nitrate mg/L as N	HPC cfu/mL	AOC µg/L as C
Feed	8.1	4.2	0.01	0.16	1.4E+03	47
Effluent	8.3	3.3	0.01	0.17	3.7E+04	75

Table 4.55
Average water quality for BAR2(S1) – experiment one

Sample Port	DO mg/L as O_2	Residual mg/L as Cl_2	Nitrite mg/L as N	Nitrate mg/L as N	HPC cfu/mL	AOC µg/L as C
Feed	8.5	4.4	0.01	0.25	1.5E+03	77
Effluent	8.1	3.4	0.01	0.42	1.5E+04	84

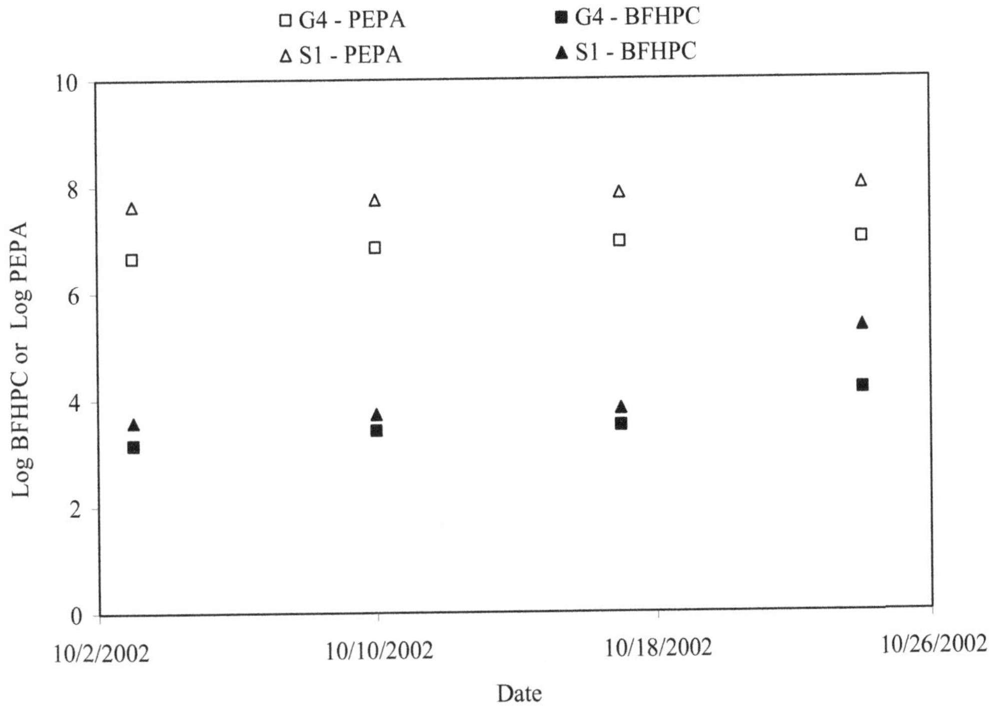

Figure 4.119 BAR experiment one: variation of Log BFHPC and Log PEPA with time

Experiment Two

Conventionally treated GW (G1) was compared to a water produced by enhanced coagulated, ozonated and carbon filtered SW (S1) in two BARs. No significant difference was observed for the two BARs receiving GW (G1) and SW (S1), for water quality (Table 4.56 and Table 4.57) or fixed biomass (Figure 4.120). Effluent HPC, nitrogen species levels (nitrates and nitrites), ΔDO and effluent chloramine residual exerted similar influences on both BARs. Slight but not significant increases of biofilm inventory were observed between 2/6/03 and 3/13/03. The increased biofilm inventory may be related to a 4.4°C rise in temperature during this period. Figure 4.120 and the other data imply that the source water had little effect on log BFHPC or log PEPA values.

Table 4.56
Average water quality for BAR1(G1) – experiment two

Sample Port	DO mg/L as O_2	Residual mg/L as Cl_2	Nitrite mg/L as N	Nitrate mg/L as N	HPC cfu/mL	AOC µg/L as C
Feed	7.0	3.8	0.00	0.01	2.9E+03	35
Effluent	7.6	3.0	0.00	0.02	3.0E+04	107

Table 4.57
Average water quality for BAR2(S1) – experiment two

Sample Port	DO mg/L as O_2	Residual mg/L as Cl_2	Nitrite mg/L as N	Nitrate mg/L as N	HPC cfu/mL	AOC µg/L as C
Feed	8.2	4.1	0.00	0.11	7.0E+03	103
Effluent	7.4	4.0	0.00	0.11	7.4E+04	294

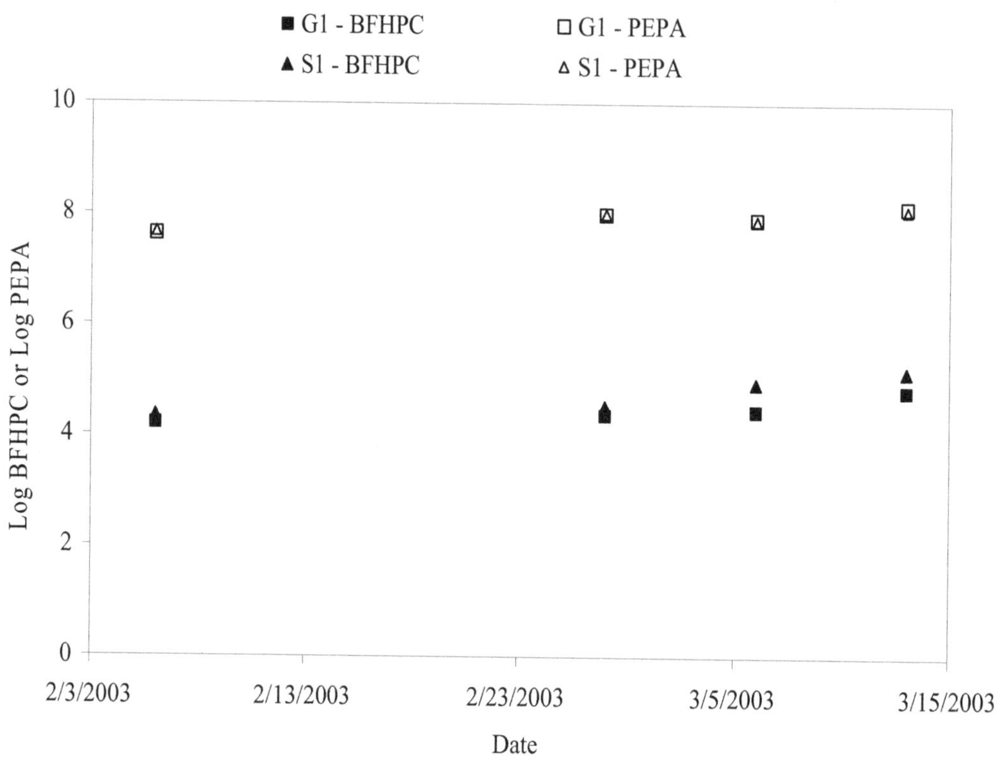

Figure 4.120 BAR experiment two: variation of Log BFHPC and Log PEPA with time

Process Biostability

Although this was a distribution system project, the project scheme allowed the assessment of process biostability. The assessment of process biostability was done by comparing finished water bulk stream measures of AOCs, BDOCs and HPCs and finished water biofilm measures of PEPA relative to process and pipe material.

Finished Water AOC

Average finished water AOCs for the first year of the filed study are shown in Figure 4.121 with the standard error as a bar and whisker graph. Conventionally treated ground water (G1) had the highest average finished water AOC (110 µg/L as C) whereas RO finished water had the lowest average finished water AOC (43 µg/L as C). As RO was simulated by filtering the same GW source as was used for G1, there is little doubt that RO membrane filtration removed AOC from the GW source. The average finished water AOC for G4 and S2 was 57 and 55 µg/L as C, respectively. The G4 process consisted of nanofiltration of a blended source of coagulated SW, conventionally treated ground water and RO filtered ground water. The S2 process consisted of enhanced ferric sulfate coagulation, conventional filtration of SW followed by nanofiltration. The same Osmonics nanofilter was used in the G4 and S2 processes. Although inferences concerning G4 are difficult because it was a blend, there is little doubt that nanofiltration removed additional AOC relative to the S1 process as the same SW source was used for S1 and S2. The S1 process consisted of enhanced ferric sulfate coagulation, ozonation (O_3) and filtration by biologically activated carbon (BAC) with a minimum of a five minute empty bed contact time (EBCT). G2 and G3 were softened waters. G2 utilized the same raw GW source as G1; whereas, G3 utilized the same blended source as G4. The average AOC difference between G1 (110 µg/L as C) and G2 (89 µg/L as C) is slight considering the precision of the AOC test, which is 80 % to 120 %. AOC data was lost for the beginning of P 1 and P IV, hence 84 AOC measurements were taken during the first year of operation. The blended softening process (G3) produced the highest average AOC of any finished water.

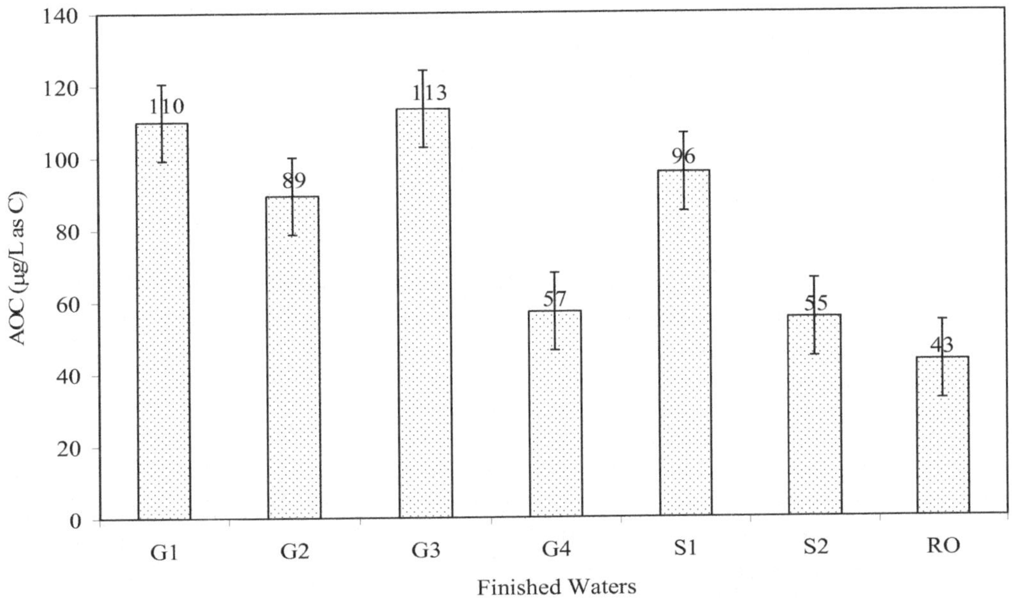

Figure 4.121 Average and standard error of finished water AOC versus source

The AOC of every finished water was compared in one versus one grouping using a paired t test. The P-values for the t tests that compared differences in finished water AOCs are shown in Table 4.58 by process. The paired t-test determines if there is a statistical difference in means of two compared series of values at a given CL. For engineering applications, this confidence level is typically selected to be 95 %. A confidence level of 95 % corresponds to an alpha value equal to 0.05. The value for alpha represents the probability of incorrectly concluding that the means are not equal when they are in fact equal. If the calculated P-value using the actual data is less than the specified value for alpha, then there is a statistically significant difference in the means at a confidence level equal to one minus alpha. Alternately, for a calculated P-value equal to P, it may be stated with a confidence level equal to 100 – 100 x

P percent that the means are significantly different. For example, if the calculated P-value equals 0.18, then it could be concluded with a confidence level of 82 % that the means are significantly different. The statement would be expected to be incorrect 18 times out of 100. The data in Table 4.58 has different fonts, which denotes the 95 % CL, the 90 % CL and the 80 % CL which correspond to alpha values of 0.05, 0.10, and 0.20, respectively. The statistical data was presented in this manner so that a most meaningful finished water AOC interpretation was possible.

Table 4.58
Alpha values from process paired t test for determining finished water AOC differences

	Process						
	G1	G2	G3	G4	S1	S2	RO
G1	-	0.4852	0.9029	0.0782	0.6269	0.0690	***0.0285***
G2	0.4852	-	0.4131	0.2751	0.8310	0.2498	*0.1234*
G3	0.9029	0.4131	-	0.0608	0.5436	0.0534	***0.0215***
G4	0.0782	0.2751	0.0608	-	*0.1942*	0.9512	0.6412
S1	0.6269	0.8310	0.5436	*0.1942*	-	*0.1747*	0.0815
S2	0.0690	0.2498	0.0534	0.9512	*0.1747*	-	0.6854
RO	***0.0285***	*0.1234*	***0.0215***	0.6412	0.0815	0.6854	-

Italicized, underlined and bold alpha values are less than 0.05
Underlined alpha values are greater than 0.05 and less than 0.10
Italicized alpha values are greater than 0.10 and less than 0.20

As shown in Table 4.58, the alpha values for the comparison of G3 and RO, and the comparison of G1 and RO are less than 0.05. This means that average AOC produced by G1 (110 μg/L as C) and G3 (110 μg/L as C) were statistically higher than the average AOC produced by the RO process (43 μg/L as C) at the 95 % CL. The average AOC of G1 was also statistically higher than the average AOC produced by the G4 (57 μg/L as C) and the S2 (55 μg/L as C) processes at the 90 % CL. Following this assessment with a review of the G4 and S2 rows provides a complete assessment of nanofiltration removal of AOC. G4 and S2 AOCs are statistically lower than G3 AOCs at the 90 % CL, and different from S1 AOC at the 80 % CL, which indicates that nanofiltration with this membrane removed AOC that enhanced coagulation and softening did not. Finally, although the average finished water AOC from the RO process was approximately 30 % lower than the average AOCs from the nanofiltered surface and ground waters, there was no statistical difference between RO and nanofiltered AOCs (G4 and S2).

The average BDOCs for the different finished waters are shown in Figure 4.122. The BDOC order by process is different that the AOC order by process. However, G1 and RO produced the highest and lowest average finished water AOCs and highest and lowest average finished water BDOCs. The G3 BDOC is relatively much lower than the G1 and G2 BDOCs. The S1 BDOC is lower than the S2 BDOC.

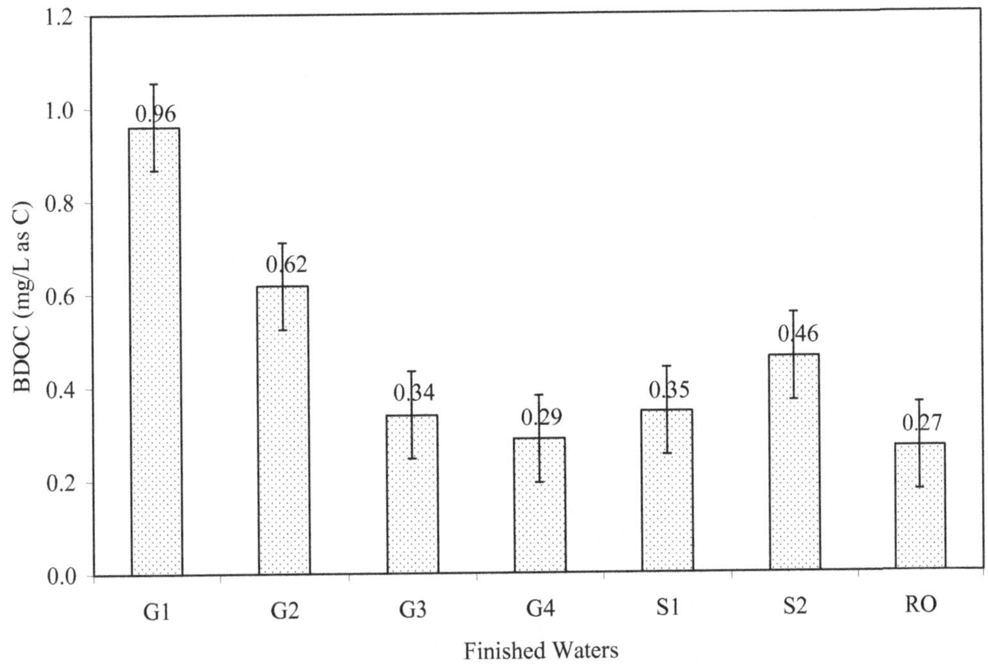

Figure 4.122 Average and standard error of finished water BDOC versus source

The alpha values for the BDOC comparison by process are shown in Table 4.59. The G1 row Table 4.59 shows that the G1 BDOC is statistically higher at the bracketed % CL than the BDOC associated with the G3 (90 %), G4 (95 %), S1 (90 %), S2 (80 %) and RO (95 %) processes.

Table 4.59

Alpha values from process paired t test for determining finished water BDOC differences

	Process						
	G1	G2	G3	G4	S1	S2	RO
G1	-	0.2786	0.0568	***0.0402***	*0.0596*	*0.1207*	***0.0352***
G2	0.2786	-	0.3765	0.2948	0.3892	0.6185	0.2673
G3	0.0568	0.3765	-	0.8659	0.9810	0.6946	0.8157
G4	***0.0402***	0.2948	0.8659	-	0.8472	0.5754	0.9488
S1	*0.0596*	0.3892	0.9810	0.8472	-	0.7122	0.7973
S2	*0.1207*	0.6185	0.6946	0.5754	0.7122	-	0.5329
RO	***0.0352***	0.2673	0.8157	0.9488	0.7973	0.5329	-

Italicized, underlined and bold alpha values are less than 0.05
Underlined alpha values are greater than 0.05 and less than 0.10
Italicized alpha values are greater than 0.10 and less than 0.20

The average Log HPCs produced by each finished water process are shown in Figure 4.123. As with average process AOCs and BDOCs, the highest and lowest average Log HPCs were produced by the G1 and RO processes, respectively. The differences in Log HPCs among the remaining processes are slight, especially when the normal variation of HPC measurements is considered. The P values for the paired Log HPC t tests are shown in Table 4.60 and shows that the Log HPCs from G1 are statistically higher that the Log HPCs from every other process at the 95 % CL, except for G2, which is statistically different at the 94.96 % CL (for practical purposes equal to a 95% CL). However, those are the only Log HPC differences that can be ascertained from Table 4.60.

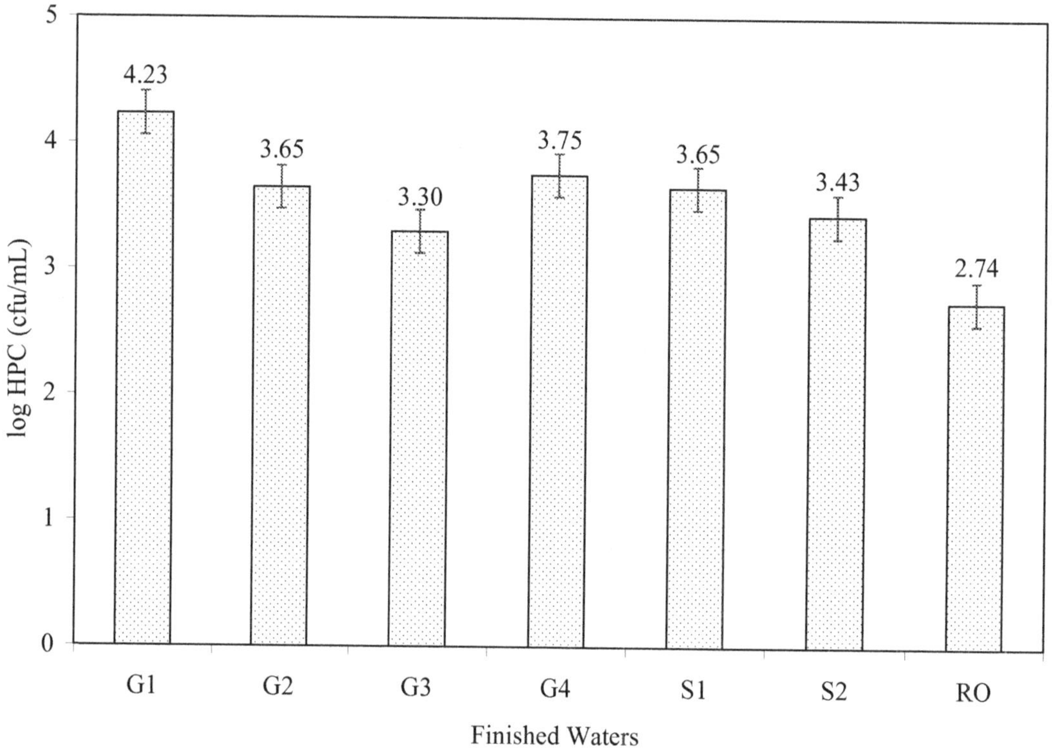

Figure 4.123 Average and standard error of finished water Log HPC versus source

Table 4.60
Alpha values from process paired t test for determining HPC differences

	Process						
	G1	G2	G3	G4	S1	S2	RO
G1	-	0.0504	***0.0148***	***0.0496***	***0.0316***	***0.0152***	***0.0060***
G2	0.0504	-	0.6942	0.8435	0.9974	0.7690	0.5175
G3	***0.0148***	0.6942	-	0.5192	0.6607	0.9022	0.7980
G4	***0.0496***	0.8435	0.5192	-	0.8281	0.5837	0.3472
S1	***0.0316***	0.9974	0.6607	0.8281	-	0.7400	0.4675
S2	***0.0152***	0.7690	0.9022	0.5837	0.7400	-	0.6914
RO	***0.0060***	0.5175	0.7980	0.3472	0.4675	0.6914	-

Italicized, underlined and bold alpha values are less than 0.05
Underlined alpha values are greater than 0.05 and less than 0.10
Italicized alpha values are greater than 0.10 and less than 0.20

The average Log PEPAs are shown by process in Figure 4.124. The highest average Log PEPAs were produced on coupons receiving G1 and RO finished waters, which is surprising relative to RO finished water. The lowest average Log PEPAs were produced on coupons receiving nanofiltered surface and blended waters, which makes the average Log PEPA from the RO environment seem out of sequence. The sequence of ground water average Log PEPAs is rational assuming that increasing degrees of treatment would reduce the Log PEPA potential accordingly as G1 Log PEPA> G2 Log PEPA> G3 Log PEPA> G4 Log PEPA. The sequence of SW average Log PEPAs is also rational based on degree of treatment as S1 Log PEPA>S2 Log PEPA. There are over 200 PEPA observations considered in Figure 4.124, which eliminates the possibility of an anomaly for the RO average Log PEPA observation.

The alpha values for the t test for average Log PEPAs by process is shown in Table 4.61. Average G1 Log PEPA is statistically higher than the average Log PEPA at the bracketed % CL for G4 (80%), S1 (80%) and S2 (90%). Average G2 Log PEPA is statistically higher than average S2 Log PEPA at the 90 % CL. Average G4 and S2 Log PEPA are statistically higher than the average RO Log PEPA at the 80 % CL.

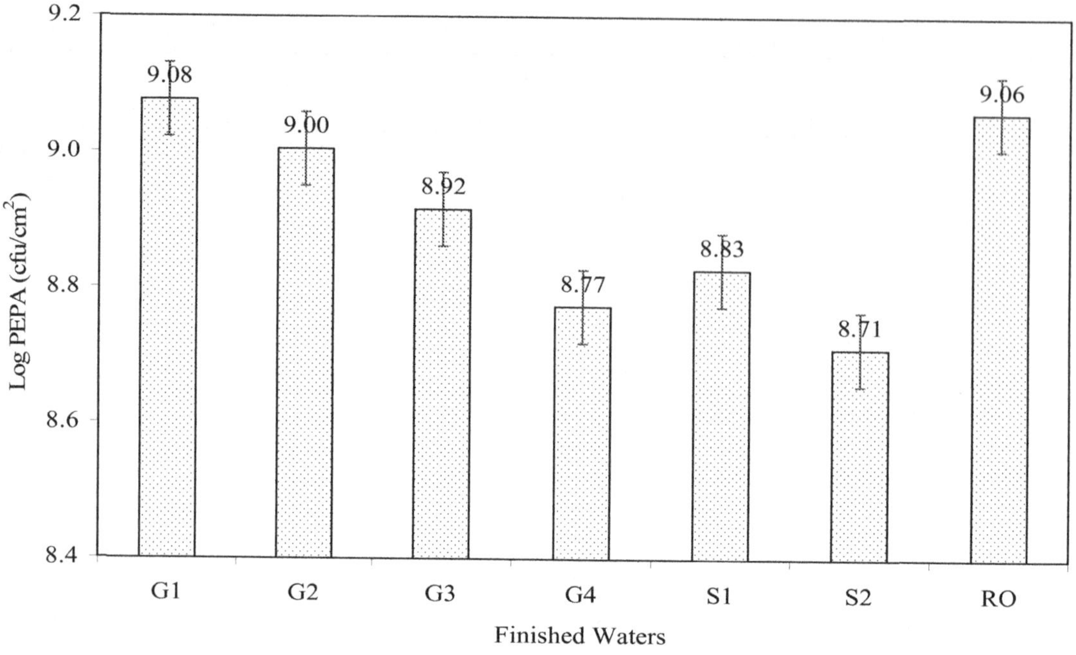

Figure 4.124 Average and standard error of finished water Log PEPA versus source

The average Log PEPAs by coupon materials are shown in Figure 4.125. The lined materials (PVC and LCI) produced the lowest average Log PEPAs. The unlined metal materials (UCI and G) produced the highest average Log PEPAs. Although there is less than an order of magnitude between the average Log PEPAs for any of the materials, a considerable amount of cells are gained in drinking water distribution systems biofilms going from 10^8 to 10^9 cell/cm^2.

Table 4.61
Alpha values from process paired t test for determining finished water PEPA differences

	Process						
	G1	G2	G3	G4	S1	S2	RO
G1	-	0.6328	0.2723	*0.1189*	*0.1751*	<u>0.0783</u>	0.9034
G2	0.6328	-	0.5337	0.2769	0.3776	*0.1967*	0.7214
G3	0.2723	0.5337	-	0.6404	0.7941	0.5014	0.3284
G4	*0.1189*	0.2769	0.6404	-	0.8366	0.8370	*0.1499*
S1	*0.1751*	0.3776	0.7941	0.8366	-	0.6804	0.2164
S2	<u>0.0783</u>	*0.1967*	0.5014	0.8370	0.6804	-	*0.1006*
RO	0.9034	0.7214	0.3284	*0.1499*	0.2164	*0.1006*	-

Italicized, underlined and bold alpha values are less than 0.05
Underlined alpha values are greater than 0.05 and less than 0.10
Italicized alpha values are greater than 0.10 and less than 0.20

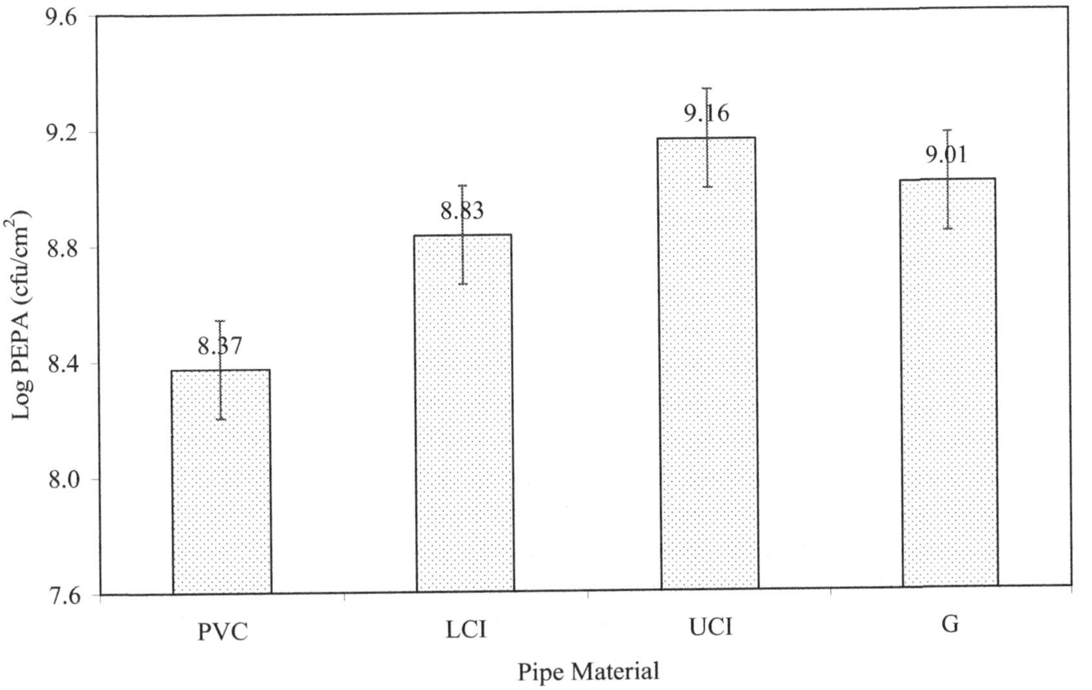

Figure 4.125 Average and standard error of Log PEPA versus material

The magnitude of difference in the average Log PEPA by material can be seen in the paired t test for difference shown in Table 4.62. The average Log PEPA for every material is statistically significant at either the 95, 90 or 80 % CL except the G average Log PEPA that is statistically higher than the UCI average Log PEPA at the 79.17 % CL.

Table 4.62
Alpha values from process paired t test for determining material PEPA differences

	Pipe Material			
	PVC	LCI	UCI	G
PVC	-	0.0905	***<0.0001***	***0.0037***
LCI	0.0905	-	***0.0046***	0.2083
UCI	***<0.0001***	***0.0046***	-	*0.1065*
G	***0.0037***	0.2083	*0.1065*	-

Italicized, underlined and bold alpha values are less than 0.05
Underlined alpha values are greater than 0.05 and less than 0.10
Italicized alpha values are greater than 0.10 and less than 0.20

Table 4.63
Process ranking by relative position based on lowest biological parameter

Points for Position	HPC	AOC	BDOC	PEPA	Rank Order	Weight
1	RO	RO	RO	G4	RO	9
2	G3	S2	G4	S2		
3	S2	G4	G3	S1	G4, S2, S1	12, 12, 12
4	G2	G2	S1	G3		
5	S1	S1	S2	G2	G3	16
6	G4	G1	G2	RO	G2	19
7	G1	G3	G1	G1	G1	27

Although arbitrary, some quantitative assessment of the water treatment processes can be made in regard to biological quality of the finished water. As shown in Table 4.63, the processes were ranked by relative position among processes based on lowest biological parameter. For example, RO received 1 point for have the lowest average Log HPC, 1 point for having the lowest average AOC, 1 point for having the lowest average BDOC and 6 points for having the second highest average BDOC, which summed to 9 as shown in the weight column of Table 4.63. The data in Table 4.63 indicates that biological water quality in the finished water increases as degree of treatment increases.

Summary Heterotrophic Biostability

Bulk Liquid

- A biologically unstable state in drinking water distribution systems has been defined in the literature as when AOC concentrations exceed 100 µg/L as C in the presence of chlorine residual. This definition was based on the occurrence of total coliforms in drinking water samples, which corresponded to AOCs greater than 100 µg/L as C in the presence of a residual. Based on this definition and as no total coliforms were ever found in any sample taken in this work, all systems were always biologically stable. However, that is difficult to accept when six logs of HPCs were measured in several samples. Bulk liquid heterotrophic biostability was assessed in this work by HPC populations. From three to four log HPCs were considered to the beginning of the transition to biological instability and above four log HPCs was considered biologically unstable. Selection of three and four logs HPC as the beginning of the transition and biological instability zones was correlated to a chloramine residual of approximately 0.1-0.2 and 0.5 mg/L as Cl_2. As these residuals were both below the regulatory minimum of 0.6 mg/L as Cl_2, biological stability was almost always all about residual and was maintained as long as residual regulatory requirements were met.
- An empirical model relating effluent Log HPC to pipe material, AOC and residual was developed. All terms were significant but AOC was only significant at the 73 % CL. Residual impacted Log HPC significantly more than AOC for the anticipated

ranges of residual and AOC in drinking water. Predicted Log HPC increased significantly as residual approached zero.

- A second empirical model was developed that related Log HPC to pipe material and temperature. Log HPC increased as temperature increased and residual decreased. The impact of residual is more significant than the impact of temperature on Log HPC as residual approaches zero, but the impact of temperature is most significant at residuals greater than 1 mg/L on Log HPC. Given utilities can do nothing to control temperature in a distribution system, residual is the only practically significant parameter for controlling HPCs and requires more active management during high temperature periods.

Biofilms

- A non-linear model was developed for biofilm density (PEPA) as a function of pipe material, temperature, total residual, Log HPC and HRT. More than 200 observations of PEPA were included in the data set that was regressed to develop the model. PEPA increased with temperature and decreased with total residual, Log HPC and HRT. The variation of PEPA with Log HPC indicated that cells in the bulk and biofilm were in pseudo-equilibrium and proliferated more in the environment that was most suitable for their growth. When residuals were high bulk proliferation was suppressed, and biofilm detachment was probably indirectly suppressed by increased ECP or a change in ECP composition as has been observed in other studies (Sutherland 1999). When residuals were low, bulk proliferation was observed and the accompanying decrease in dissolved oxygen probably led to a reduced depth of DO diffusion into the biofilm, resulting in increased sloughing. However these mechanisms are only speculative since neither ECP nor DO in the biofilm were quantified in this study, but these hypotheses are consistent with prior literature (Sutherland 1999, Geldreich 1996). Log HPCs varied directly with time, which then resulted in the negative relationship with biofilm cell density (PEPA). Temperature impacted PEPA significantly more than HRT, Log HPC_{eff} and residual, all of which had relatively little impact on PEPA. The model predicts biofilms will always be active for environments common to drinking water distribution systems, which is consistent with the hypothesis that biofilms act as reservoirs for microorganisms and are a constant source of microorganisms to the bulk water under normal conditions. As American utilities can not control biofilms with extensive and costly AOC reduction, American utilities must maintain a strong residual to maintain biological integrity and stability in drinking water distribution systems.
- An empirical model was developed for $|\Delta AOC|$ that was a function of AOC_{inf} and temperature. The model was similar to a first order model in that the amount of AOC change was dependent on the amount of AOC present, and was typical of models that describe BOD consumption by microorganisms.
- A Biocide:Food ratio ((chloramine residual - mg/L as Cl_2):(AOC - µg/L as C)) was developed and correlated to $|\Delta AOC|$. The graphical correlation showed that when the ratio was high, $|\Delta AOC|$ was very low: and when the ratio was low, $|\Delta AOC|$ was very

high. This relationship and observation indicates that biological stability can be achieved by residual maintenance or AOC reduction.

- DO and total residual varied in the same manner such that they were confounding and could not be used together for development of regression models; however, monitoring DO in a drinking water distribution system can be used as an indicator of autotrophic and heterotrophic bioinstability since it is consumed by both types of organisms in addition to being consumed by abiotic reactions that consume oxidants and thus correspond to pipe environments that also consume residual. DO monitoring is relatively inexpensive and could be used by locating a DO probe in the distribution system which would then send the data back to the plant (e.g. remote sensing, SCADA). If DO levels decrease significantly the utility could then measure residual, HPC, TC, and nitrate and nitrite to determine what sort of event is occurring.

Biological Annular Reactors (BARs)

- BAR experiments were conducted that compared biostability of a blended nanofiltered water (G4) with a SW treated by enhanced coagulation, ozonation and biological activated carbon filtration (S1). The $|\Delta AOC|$ and biofilm inventory in the BAR containing the nanofiltered blend (G4) was greater than the $|\Delta AOC|$ and biofilm inventory in the BAR containing the enhanced coagulated SW (S1). There were no differences observed between BARs with respect to bulk liquid heterotrophic biostability (HPCs).
- BAR experiments were conducted that compared biostability of a conventionally treated GW (G1) with a SW treated by enhanced coagulation, ozonation and biological activated carbon filtration (S1). No differences were observed between BARs with respect to bulk HPCs or biofilm inventories (BFHPC and PEPA).

Process Comparison

- Finished water biostability of conventionally treated ground water (G1), softened ground water (G2), enhanced coagulated, ozonated and BAC filtered SW (S1), RO membrane filtered ground water (RO), enhanced coagulated-filtered and nanofiltered SW (S2), softened blend of S1, G1 and RO (G3) and a nanofiltered blend of S1, G1 and RO (G4) was assessed by a visual and statistical comparison of average AOC, BDOC, Log HPC and biofilm (Log PEPA). All bulk water biostability parameters (AOC, BDOC, and Log HPC) were lower in RO finished water and higher in conventionally treated ground water. Average finished water AOC indicated nanofiltration reduced AOC relative to coagulation or softening; however this relationship was not observed for BDOC or Log HPC. Average process biofilm density (PEPA) of treated ground waters with the exception of RO was ordered as G1>G2>G3>G4. The average process biofilm density of treated SWs was ordered as S1>S2. The order of biofilm densities of treated surface and ground waters excepting RO indicated that biostability increased as the degree of treatment increased with nanofiltered finished waters being the most biostable. Average process biofilm density (PEPA) was the highest in RO finished water, which was seemingly out of sequence.

NITRIFICATION

There are five subsections in the nitrification section of the Results and Discussions Chapter, which describe stoichiometry associated with nitrification, a nitrification episode, a controlled nitrification investigation, correlations of biological parameters that were common to the nitrification episode and controlled study and a nitrification model for drinking water distribution systems.

Nitrification Stoichiometry

Biological nitrification in drinking water systems proceeds sequentially in two steps as shown in equations (4.88) and (4.89). A reduced form of nitrogen, which is typically ammonia in chloraminated drinking water distribution systems, must be available for the bacteria (nitrifiers) to biochemically oxidize ammonia to nitrites, and nitrites to nitrates. Ammonia is the product of chloramine degradation as shown in equation (4.90) and is the primary nitrogen source and Nitrosomonas bacteria are considered as the primary ammonia oxidizing bacteria (AOB). Nitrosomonas bacteria are responsible for the conversion of ammonia to nitrite and as shown in equation (4.88). Nitrobacter is traditionally the major nitrite oxidizing bacteria (NOB) as shown in equation (4.89). Molecular studies now imply that Nitrospira may be more important in nitrite oxidation along with the more easily cultured Nitrobacter. Nitrites can also react chemically with chloramines as shown in equation (4.91).

$$55NH_4^+ + 76O_2 + 109HCO_3^- \rightarrow C_5H_7O_2N + 54NO_2^- + 57H_2O + 104H_2CO_3 \qquad (4.88)$$

$$400NO_2^- + NH_4^+ + 4H_2CO_3 + HCO_3^- + 195O_2 \rightarrow C_5H_7O_2N + 3H_2O + 400NO_3^- \qquad (4.89)$$

$$2NH_2Cl + 2H_2O \rightarrow 2NH_3 + 2HCl + O_2 \qquad (4.90)$$

These reactions indicate that biological nitrification requires a high oxygen input, consumes alkalinity, which decreases pH due to the generation of protons.

$$NH_2Cl + NO_2^- + H_2O \rightarrow NO_3^l + NH_4^+ + Cl^- \qquad (4.91)$$

During biological nitrification, ammonia (NH_4^+) combines with oxygen and bicarbonate to produce bacterial cell mass, nitrite (NO_2^-), water and carbonic acid. Ammonia, carbonic acid, bicarbonate and nitrite are required in the succeeding reaction to form additional cell mass and nitrates. The stoichiometric change of water quality parameters involved in the biochemical nitrification reaction is shown in Table 4.64.

Table 4.64
Stoichiometric mass change due to nitrification (mg/L)

	mg/L Mass Change Due to Nitrification Reactions				
AOB NH_4	NH_4	O_2	HCO_3	NO_2	H_2CO_3
	1 ↓	3.45 ↓	8.63 ↓	0.99 ↑	8.37 ↑
NOB NO_2	NO_2	NH_4	H_2CO_3	HCO_3	NO_3
	1 ↓	0.0025 ↓	0.044 ↓	0.011 ↓	1 ↑
Abiotic NH_4	NH_2Cl	NO_2	NO_3	NH_4	
	1 ↓	0.2 ↓	0.2 ↑	0.2 ↑	

Assuming there is no ammonia in the source water and ammonia is only added to form chloramines, then there are only two sources of ammonia in the distribution system. Ammonia is available from (1) excess ammonia used in the formation of chloramines and (2) the dissipation of chloramines in the distribution system. Based on the simplified formation reaction for chloramines shown in equation (4.92), very little excess ammonia will remain when a 5/1 chlorine to ammonia-nitrogen mass ratio (Cl_2/NH_3-N) is used for chloramine formation.

$$NH_3 + Cl_2 \rightarrow NH_2Cl + HCl \tag{4.92}$$

Assuming most utilities would form 4 mg/L NH_2Cl as Cl_2, Cl_2/NH_3-N ratios of 4/1 and 3/1 would result in remaining or excess NH_3-N of approximately 0.2 and 0.5 mg/L NH_3-N. However the degradation reaction with water or anything as shown in equation (4.93) will result in approximately 1 mg/L NH_3-N. Hence, any use of chloramines provides NH_3-N for nitrification. Low chloramines residuals and difficulty maintaining residuals are associated with nitrification episodes.

$$2NH_2Cl + 2H_2O \rightarrow NH_3 + HCl + O_2 \tag{4.93}$$

Nitrates will be biologically reduced following biological nitrification if anaerobic conditions are developed in the distribution systems. Although denitrification rarely occurs in drinking water distribution systems, such conditions are highly undesirable because of associated taste, odors, black water and other adverse water quality effects. A denitrification reaction assuming acetate as a carbon source is shown in equation (4.94).

$$5CH_3COOH + 8NO_3^- \rightarrow 3N_2 + 5CO_2 + 7H_2O + 6OH^- \tag{4.94}$$

The purpose of showing equations (4.88) through (4.94) is to explain what water quality parameters can be theoretically associated with biochemical reactions starting with ammonia in the distribution system. Decreasing ammonia, alkalinity, pH and DO, and increasing nitrite, and nitrate are theoretical indicators of biological nitrification in distribution systems. However continued DO consumption will produce anaerobic conditions and denitrification will occur, which stops nitrification and results in nitrogen loss, which indicates adverse water quality conditions in the distribution system. The growth of any aerobic bacteria will consume DO and is retarded by a strong residual. Hence, the maintenance of strong residual as noted in the biostability section is essential to reducing nitrification. Decreasing ammonia, alkalinity, pH and DO, and increasing nitrite, and nitrate are indicators of nitrification. Of these indicators, pH and DO are mostly easily monitored in the field, and of those, DO is the best indicator of biological activity in distribution systems as DO is measured linearly and pH is measured on a log scale.

Nitrification Episode

The initial investigation was conducted as a response to a nitrification episode that occurred from May 2002 to December 2002 and investigated nitrification with respect to changing water quality in a distribution system as a part of the first year plan of study. Nitrification was initially unplanned but did occur after 6 months of operation during the first year of the field study. The primary reason nitrification occurred in the PDSs was that either bulk liquid residual was lost, or that nitrification was initiated in some portion of the biofilm that was protected from residual but was still aerobic (ammonia would also be present due to the presence of heterotrophs and the deamination of proteins/amino acids during normal metabolism or due to natural cell death/lysis). Either scenario would result in the reaction of nitrite with residual, which would then produce ammonia and low residual levels, leading to more nitrification and the well known nitrite-chloramine vicious cycle. Excess ammonia was also added during chloramination before the nitrification episode, and could have significantly contributed to the availability of ammonia for nitrification as does chloramine degradation. Once the residual was lost, the general variation of water quality parameters associated with nitrification was predictable from the stoichiometry associated with biological nitrification. The following text describes the water quality variation during the unplanned nitrification episode by pipe material.

General Observations

Temperature was a major independent variable affecting nitrification. The average influent and effluent chlorine residual and temperature for all the 18 PDSs during the one-year operation (12/06/01 to 12/23/02) is presented in Figure 4.126. The chloramine concentration in the finished waters was targeted slightly above 4.0 mg/L as Cl_2. The temperature range from 12/06/01 to 06/13/02 and from 06/14/02 to 09/12/02 was 7-31°C and 22-32°C respectively.

Before March 2002, the average effluent residual was around 2.0 mg/L at temperatures of 10 to 20°C. From March to May 2002, the average effluent residual was above 1.0 mg/L but began to decrease as temperature increased from 20°C to 30°C. Although the residual in the finished water storage tanks was above 4.0 mg/L as Cl_2, the effluent residual of the PDS was lost from May 2002. From 06/14/02, the chlorine residual in all finished waters was maintained above 4.5 mg/L as Cl_2, but no significant change in effluent chlorine residual was observed. After the reduction of HRT from 5 to 2 days on 08/30/02, significant effluent residual was detected in all PDSs. Hence increasing the chloramine residual slightly in the finished water storage tanks prior to entering the PDS was ineffective. Reducing the HRT from five to two days, was highly effective in restoring residual in the PDS effluent. However there was essentially no residual in many of the PDSs during the summer and a favorable environment was created for microbiological growth in the PDSs, which included biological nitrification.

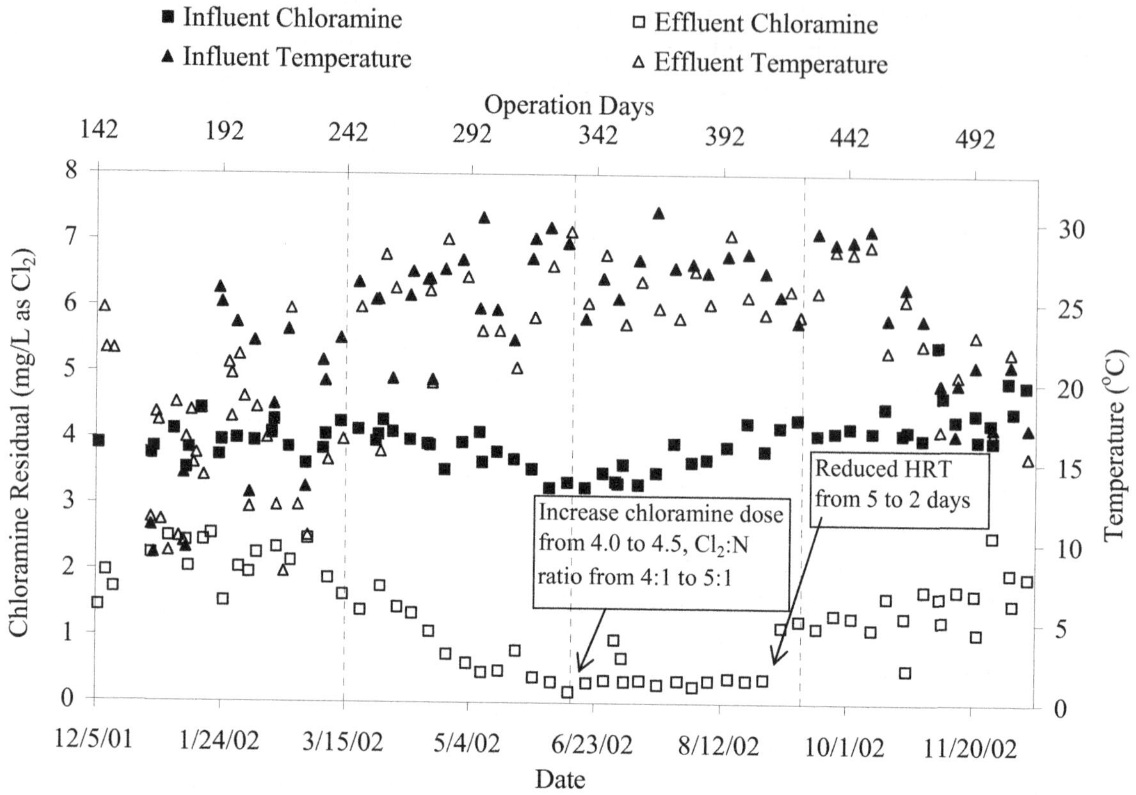

Figure 4.126 The average chloramine residual and temperature for all the 18 PDSs

Water Quality Profiles during Nitrification

The water quality profiles were dependent on pipe material, which affected chloramine residual. During the episode in phase 3 and beginning of phase 4, residuals were lost in a number of hybrid lines, and in the pure material UCI and G PDSs. Residual was not lost in PVC and LCI PDSs. Hence water quality profiles from phase IV (9/13/02 to 12/15/02) are presented in the hybrid, UCI and G PDSs. The hybrid profiles show general trends. The UCI and G profiles show water quality profiles in PDSs which in the UCI typified maximum nitrification and in the G the most rapid residual loss. No effluent residual was present in the UCI and G PDS during the nitrification episode. The G PDS experienced the longest exposure to a residual free environment and developed a microbiological environment that did not significantly nitrify, since it was also anaerobic. Most of the other PDSs experienced nitrification. The least nitrification was experienced in PDSs which maintained the strongest residuals (the lined CI and the PVC PDS). These two pure material lines had a maximum nitrite value of 0.10 and 0.07 mg/L-N (with averages much lower than that). These values are above 50 ppb which is generally the level at which utilities begin to suspect the beginnings of a nitrification event. However in the unlined CI PDS the high value was 3.52 mg/L-N nitrite and almost half the observations were between 2 and 3 mg/L-N.

As noted, the primary objective of the investigation into the unplanned nitrification episode was the correlation of major water quality variables to nitrification. The variation of nitrogen species in the PDSs was investigated in regard to this objective. The hybrid PDS nitrogen profiles were averaged and used to illustrate water quality change as shown in Figure 4.127. Profiles for ammonia-nitrogen (NH_3-N, as mg/L-N), nitrite-nitrogen (NO_2-N, as mg/L N), NO_X-N, (sum of nitrite and nitrate, as mg/L-N) and monochloramine-nitrogen (NH_2Cl-N, as mg/L N) are shown in Figure 4.127. The monochloramine-nitrogen (as mg/L-N) was determined stoichiometrically as one fifth of the chloramine residual (mg/L as Cl_2). The reported NO_X-N concentration is the sum of the actual nitrite and nitrate concentrations.

The chlorine residual decreased through the pipe due to chlorine demand. The chlorine demand arises from the reactions of chorine with organic matter, pipe materials, corrosion products and biofilms. The chloramine residual curve in Figure 4.127 is convex-up because the chlorine demand of the pipe in the hybrid PDSs increased with sequence of material. The order of chlorine demand was PVC < LCI < UCI < G, which produced a convex-up chlorine residual curve through the PDSs.

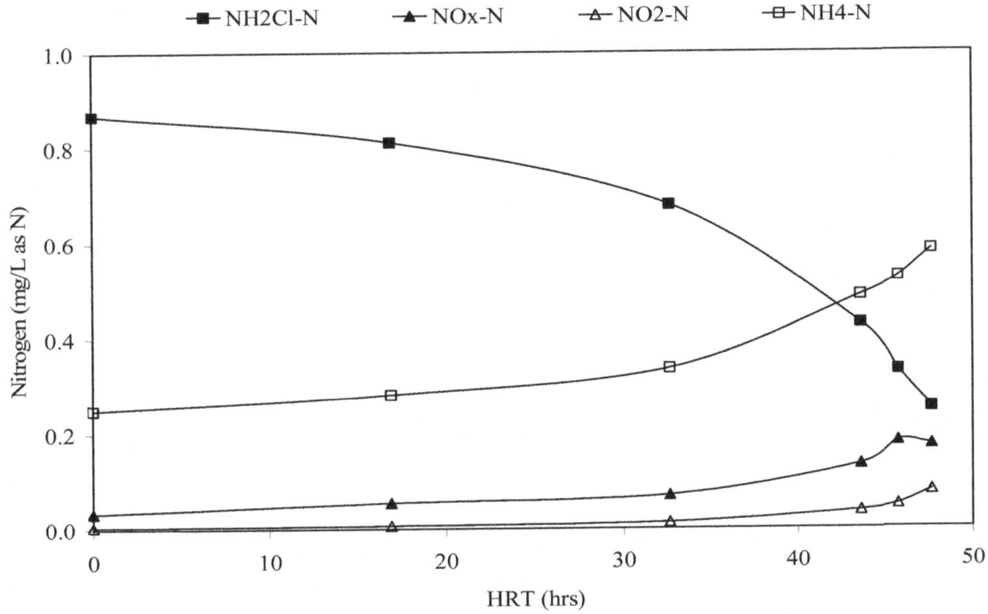

Figure 4.127 The average nitrogen profiles in the hybrid PDSs

The release of ammonia can be clearly observed from the NH_2Cl -N and NH_3-N curves shown in Figure 4.127. The loss of NH_2Cl -N is relatively close to the gain of NH_3-N. There were no significant nitrites or nitrates in the PDS influent stream, which indicates there was no nitrification and very little loss of residual in the finished water tanks as 0.8 mg/L of NH_2Cl-N corresponds to 4 mg/L of chloramines at a 5:1 Cl_2:NH_3 ratio. However there was approximately 0.25 mg/L of free NH_3-N entering the PDS, which came from degradation of chloramines in the standpipe. The detection of average nitrite and nitrate in the PDSs effluent readily indicates the occurrence of nitrification within the PDSs.

The nitrogen profiles in PDS 15 (UCI) were different from that of the hybrid PDSs (Figure 4.128), in that ammonia curve increased and then decreased after 33 hrs. Initially, ammonia release from chloramine degradation exceeded ammonia consumption from nitrification. However, this relationship was reversed after 33 hours, and ammonia consumption from nitrification exceeded ammonia release from chloramine degradation. Nitrification of ammonia did not exceed ammonia release in the hybrid PDSs as illustrated in Figure 4.127, presumably because of the reduced exposure to UCI pipe. Nitrification in the UCI PDS was more extensive than in the hybrid PDSs. As noted in the Biostability section, this was presumably because the biofilm environment in the UCI pipe was more receptive to bacteria.

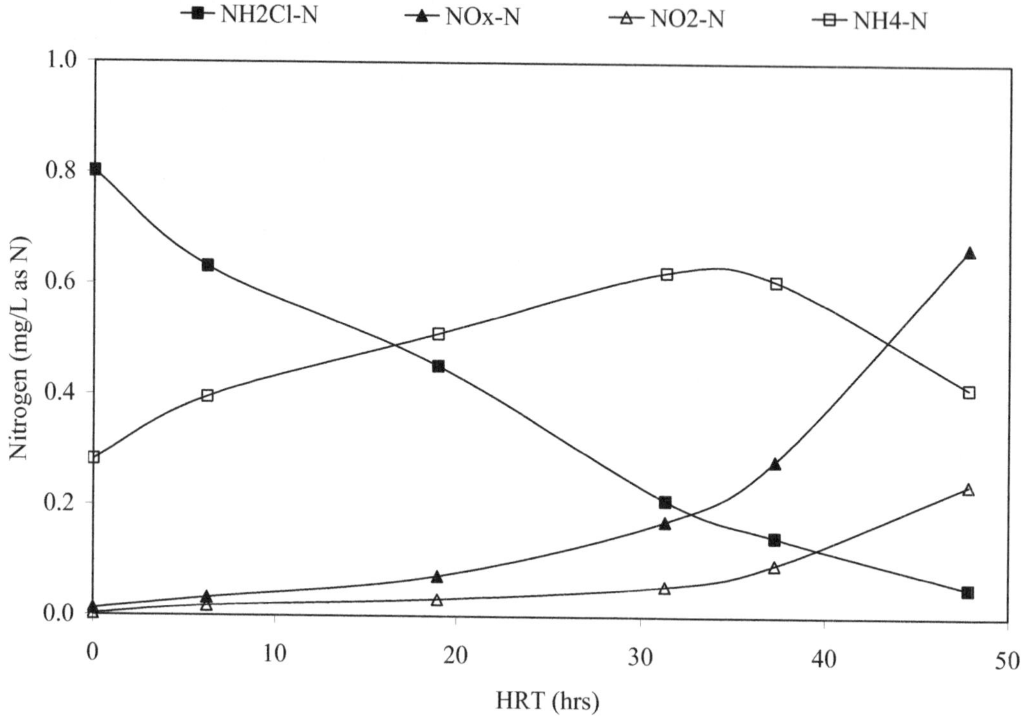

Figure 4.128 The average nitrogen profiles in PDS 15 (UCI) at normal operation

Figure 4.129 illustrates the nitrogen profiles in PDS 18 (G) during the nitrification event that started in the summer of 2002 and continued into the fall (phases III and IV). G was the most reactive material. Residual was lost in the G PDS (PDS 18) at about 15 hrs, and DO was depleted at about 30 hrs. It was no surprise that nitrification did not occur under the anaerobic conditions in PDS 18 since nitrifiers are obligate aerobes. The slight rise in nitrites and nitrates at 52 hours was probably due to the diffusion of oxygen in the effluent standpipe, which was the effluent sampling location for PDS 18. AOBs were only measured for effluent and biofilm samples and thus no profiles such as this were obtained for that parameter.

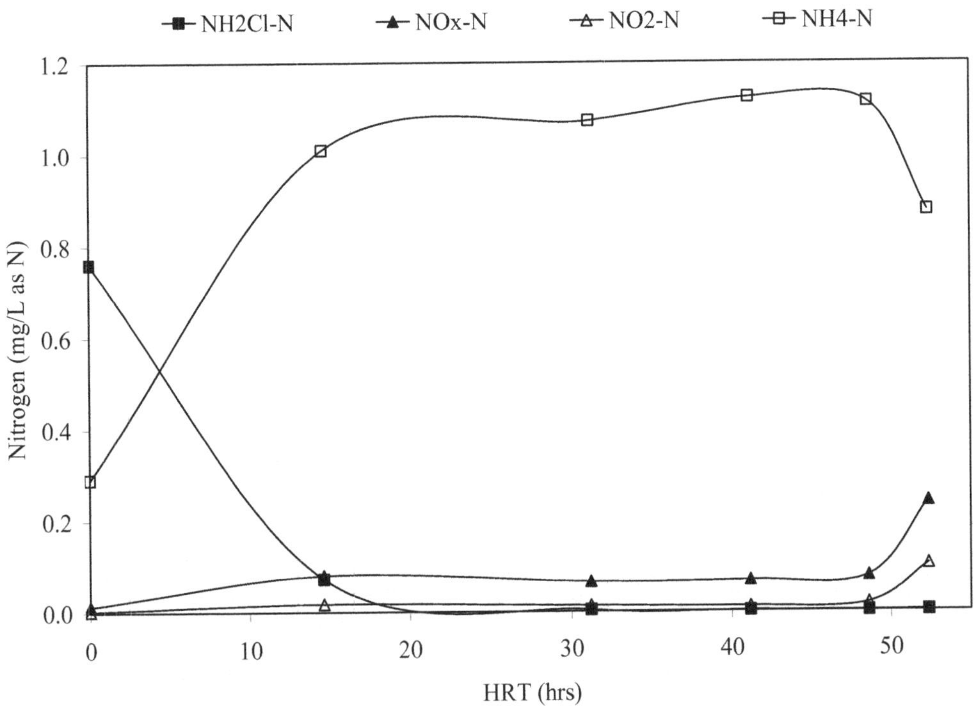

Figure 4.129 The average nitrogen profiles in PDS 18 (G) at normal operation.

Nitrogen balance. The total nitrogen balance for samples taken from the internal ports in PDS5 is shown in Table 4.65. The nitrogen balance was based on the following: (1) The NH_2Cl -N was calculated by dividing total monochloramine concentration by 5, (2) free chlorine concentration was disregarded and not included in the mass balance, (3) NO_X-N included NO_3-N and NO_2-N, and (4) Total nitrogen was calculated adding NH_4-N, NO_X-N and NH_2Cl -N together. The nitrogen mass balances by date and port as shown in Table 4.65 are within 15 %, and show that nitrogen species were balanced satisfactorily on the dates. Although nitrogen balances are excellent research tools for determining residual loss, ammonia loss and nitrification in distribution systems, utilities are able to identify nitrification episodes much more rapidly with nitrite and nitrate analysis. Mass balance analysis could be of use to further understand and quantify nitrification events.

Table 4.65
Nitrogen balance for PDS 5

Date	10/10/2002	10/25/2002	11/08/2002	11/22/2002
Feed	1.0	1.3	1.2	1.2
Port 4	0.9	1.3	1.3	1.3
Port 5	0.9	1.3	1.3	1.1
Port 6	0.7	1.2	1.3	1.1
Port 7	0.8	1.3	1.3	1.1
Effluent	0.9	1.2	1.3	1.1

DO. The average DO profiles for the PDSs are presented in Figure 4.130. The G PDS consumed more DO than any other PDS and became near anaerobic at 35 hrs. The PVC and LCI PDS consumed very little DO, so the DO profile was essentially flat. The hybrid PDSs had modest DO consumption. DO was always consumed in the PDSs, but significantly increased during nitrification or excessive bacterial activity. The order of DO consumption by material was the same as the order of chlorine demand, which was PVC < LCI < UCI < G. DO consumption increased during nitrification because of bacterial activity that was associated with nitrification. The DO profiles illustrate better maintenance of PDS water quality when inert materials (PVC, LCI) relative to reactive materials (UCI, G) are used to transport drinking water. Decreasing DO alone does not indicate nitrification but it does suggest the onset of bacterial driven water quality problems in distribution systems and is an excellent monitoring parameter for distribution system water quality.

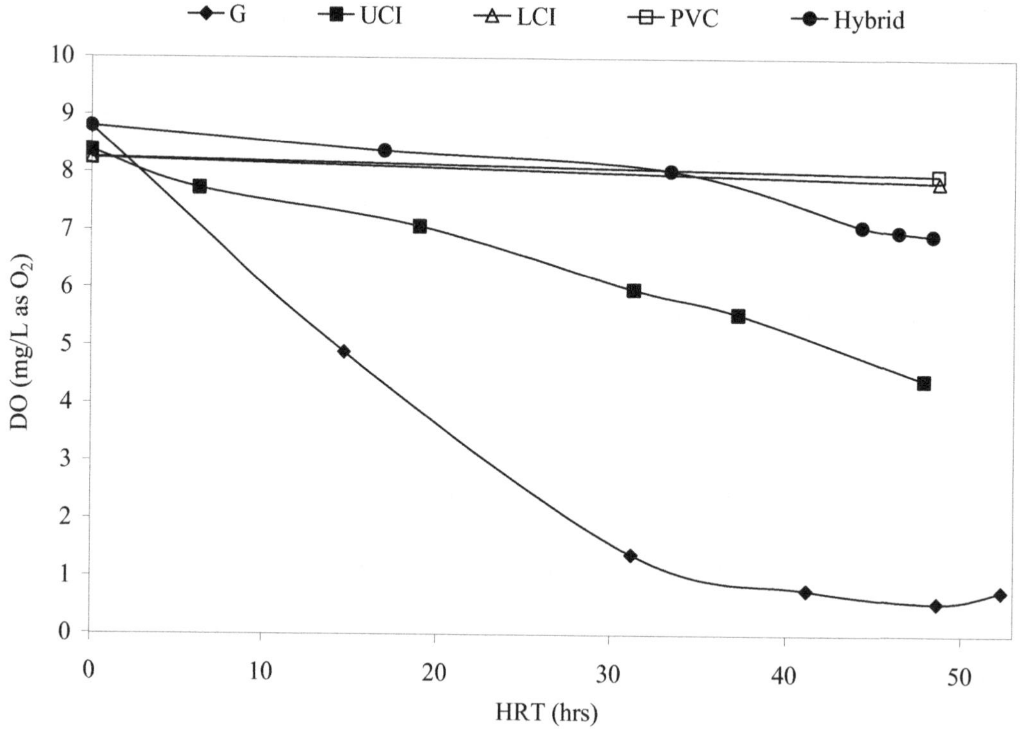

Figure 4.130 The average DO profiles through the PDSs

Alkalinity. Theoretically, alkalinity should be consumed during nitrification, however, no significant alkalinity consumption was observed. The average alkalinity profile through the UCI and Hybrid PDSs under nitrification is presented in Figure 4.131. As stated by Odell (1996) the change in alkalinity due to nitrification may be not significant enough to be detected analytically, or the change may be buffered by the carbonate system in the water or calcium carbonate solids on the internal wall of the pipes. These observations indicate that nitrification can not be monitored by monitoring alkalinity changes in distribution systems.

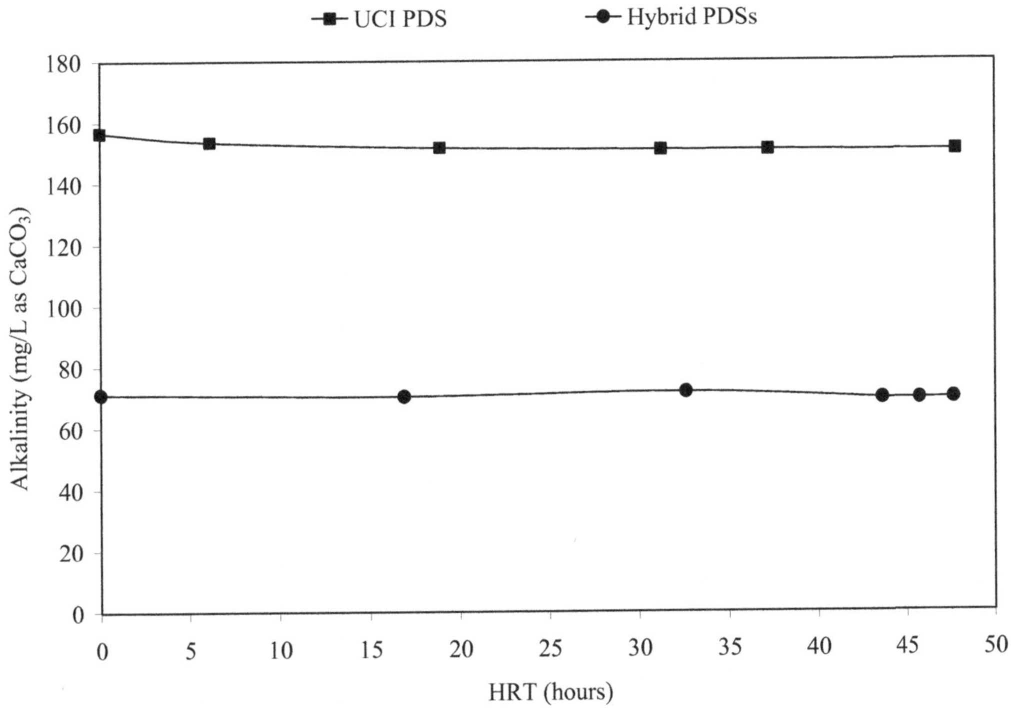

Figure 4.131 The average alkalinity profile through the PDSs

pH. The pH profile through the UCI and Hybrid PDSs under nitrification is illustrated in Figure 4.132. pH decreased 0.3 pH units (from 8.0 to 7.7) during nitrification, which was expected as protons are a byproduct of nitrification. Decreasing pH was observed at 30 hrs for hybrid PDSs and at 10 hrs for UCI PDS, which essentially coincided with the initiation of nitrification in the PDSs. pH did change as expected during nitrification but a pH change of 0.1 to 0.3 pH units alone would be difficult to attribute to nitrification, especially when utilities distribute waters with pHs above pH_s. Decreasing distribution system pH is best viewed as a secondary indicator of nitrification.

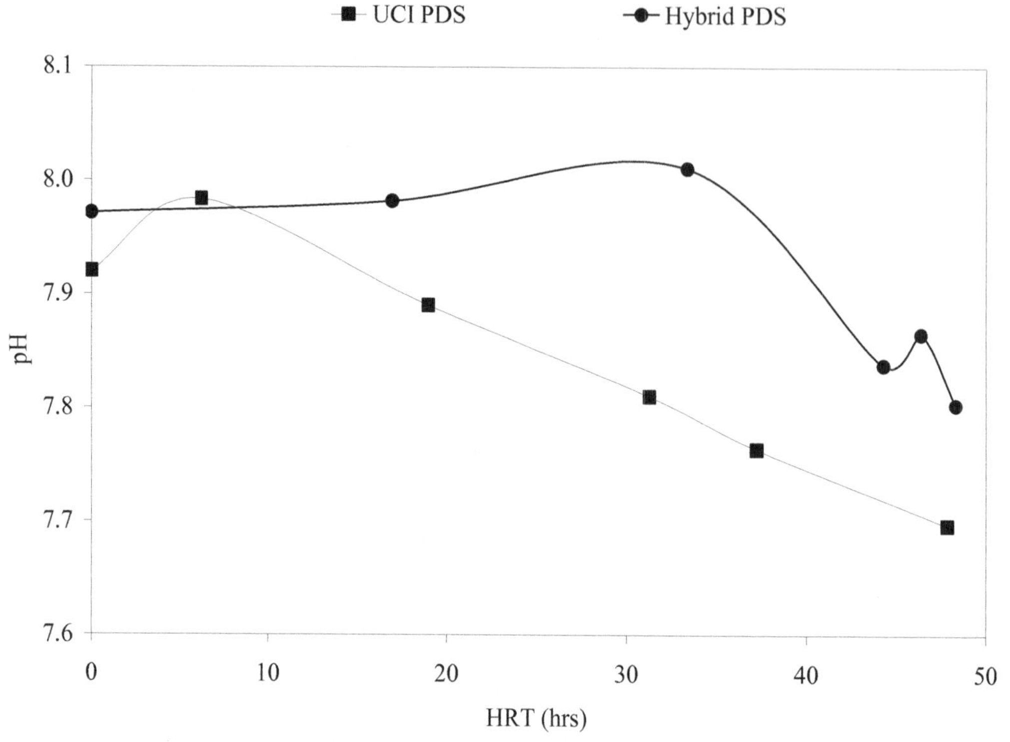

Figure 4.132 The average pH profile through the PDSs

Comparison of actual and theoretical water quality variations. To explain the water quality changes seen in PDS 15, biological nitrification was considered to be the primary source of nitrites and nitrates and a comparison was made between actual and theoretical water quality variations due to nitrification. Theoretical water quality changes based on biological reactions were calculated and compared to observed water quality changes in PDS 15 in Table 4.66. The comparison illustrates the difference between theoretical and actual water quality changes. Actual nitrite and nitrate concentrations were used to determine the expected theoretical change in HCO_3, pH, O_2, and NH_3-N as shown in Table 4.66. To determine the expected theoretical water quality at a specific HRT the theoretical parameter change was added (production) or subtracted (consumption) to the feed stream water quality. An additional adjustment was made for ammonia-N. Assuming all nitrogen released from monochloramine degradation is available as ammonia-N, released ammonia was added to the theoretical ammonia calculation.

Table 4.66
PDS 15 theoretical and actual water quality during nitrification episode (mg/L)

HRT	ID	NH_2Cl-N	NH_3-N	O_2	HCO_3 as CaCO3	NO_2-N	NO_3-N	pH	ΣN
0	Actual	0.85	0.26	7.8	154	0	0.00	7.71	1.11
	Actual	0.66	0.37	7.2	154	0.01	0.01	7.87	1.05
6	Theoretical		0.43	7.7	153.9			7.69	
	Difference		0.06	0.5	-0.1			-0.18	
	Actual	0.36	0.53	6.0	152	0.02	0.02	7.71	0.93
19	Theoretical		0.71	7.6	153.7			7.68	
	Difference		0.18	1.6	1.7			-0.03	
	Actual	0.16	0.58	5.1	152	0.05	0.08	7.65	0.87
31	Theoretical		0.82	7.3	153.1			7.64	
	Difference		0.24	2.2	1.1			-0.01	
	Actual	0.10	0.55	4.6	150	0.08	0.23	7.64	0.96
37	Theoretical		0.69	6.5	151.8			7.57	
	Difference		0.14	1.9	1.8			-0.07	
	Actual	0.03	0.31	3.3	152	0.27	0.60	7.61	1.21
48	Theoretical		0.19	4.3	147.7			7.39	
	Difference		-0.12	1.0	-4.3			-0.22	

The theoretical and actual values are not equal, however some observations can be made. Actual and theoretical ammonia increase as chloramines degrade for the initial 31 hrs, and decrease after 31 hrs as the rate of nitrite and nitrate production becomes greater than the rate of monochloramine degradation and corresponding ammonia release. Hence ammonia is an intermediate product during nitrification when chloramines are used for residual maintenance.

There is increasing oxygen demand throughout the PDS for the each observation to 31 hrs, which is probably due to utilization of DO by aerobic microorganisms. The difference between the actual and expected DO concentrations also increases during this period. More oxygen is consumed than required by the nitrification reactions, which is expected as other aerobic bacterial species are active. After 31 hrs the differences decrease, indicating actual oxygen demand is not increasing at the same rate biological nitrification increases. The conversion of nitrites to nitrates by monochloramines would contribute to this effect by reducing the actual oxygen demand. The production of 0.60 and 0.27 nitrates and nitrites should reduce bicarbonate alkalinity from 154 to 148 as mg/L as $CaCO_3$ and pH from 7.71 to 7.39 according to theory. Actual alkalinity and pH decreased but less than what was expected and was less consistent than expected. Such variations in alkalinity and pH could be the result of alkalinity addition from calcium carbonate solids in the PDS.

Nitrogen was summed in the last column and represents the sum of monochloramine, ammonia, nitrite and nitrate nitrogen for each increment through PDS 15. The N balance for each HRT is reasonable ranging from 0.87 to 1.21 mg/L N. Influent and effluent total nitrogen were 1.11 mg/L and 1.21 mg/L N. The data in Table 4.66 shows the variation of NH_2Cl-N, NH_3-N, DO and pH with NO_2-N and NO_3-N formation generally followed the variations predicted by microbiological nitrification reactions as shown in equations (4.88) and (4.89). The variation of alkalinity during nitrification was less than predicted by microbiological nitrification.

Source Water Effect

Nitrification was not specific to any source water in the PDSs nor was any source water protected from nitrification. TKNs for PDSs 1 thru 18 from March to June 2002 are shown in Figure 4.133. TKNs were the primary method of measuring ammonia and organic nitrogen. There was no change in influent or effluent TKN in any of the PDSs in March 2002. There was a decrease in effluent TKNs in 16 of the 18 PDSs in June 2002, which was due to the conversion of ammonia to nitrites and nitrates by nitrification. The finished waters were either unblended or blended combinations of the seven sources and as shown in Figure 4.133, there was no difference among source waters.

The percentage of TKN loss was directly proportioned by percentage of source water in each PDS and then plotted versus blend composition, which is Figure 4.134. If TKN in the PDS influent was the same in every PDS and experienced the same degree of nitrification regardless of blend composition, then the slope of a regression line for TKN loss versus blend percentage would be 0.01. As shown in Figure 4.134, slope is 0.01, which is the equivalent of a straight line with exact proportioning by % blend and shows TKN loss increases linearly with blend percentage. Although there is some variation among the data, Figure 4.134 indicates no difference in nitrification among sources. The reason for the lack of difference among processes for nitrification is the ammonia for nitrification came from the degradation of the initial chloramine concentration which was the same for all waters. Nitrification was not associated with the water quality of the processes.

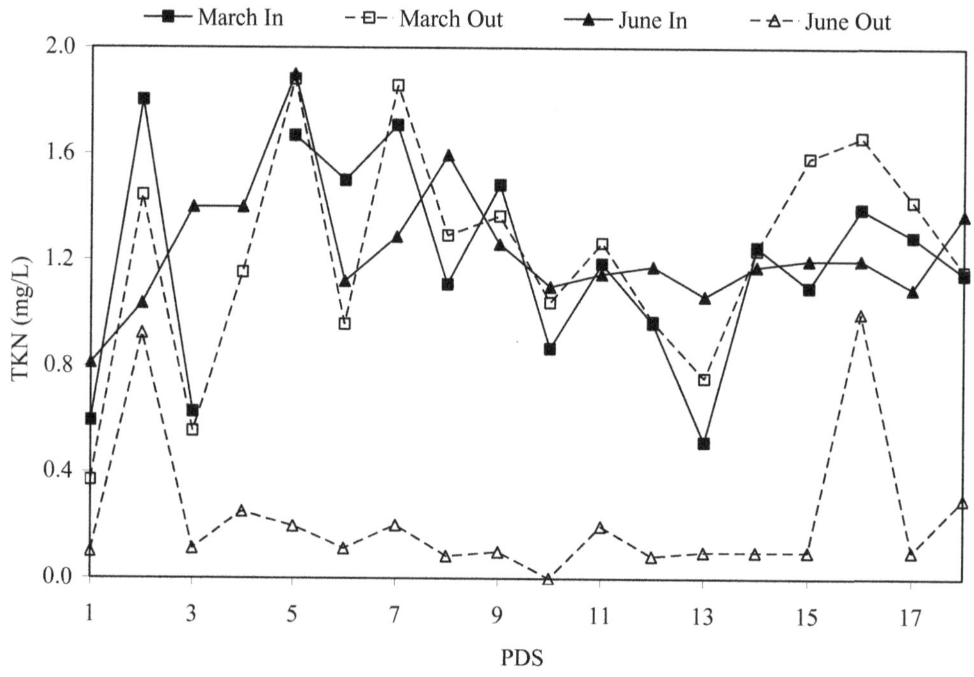

Figure 4.133 PDS TKN before and after nitrification episode

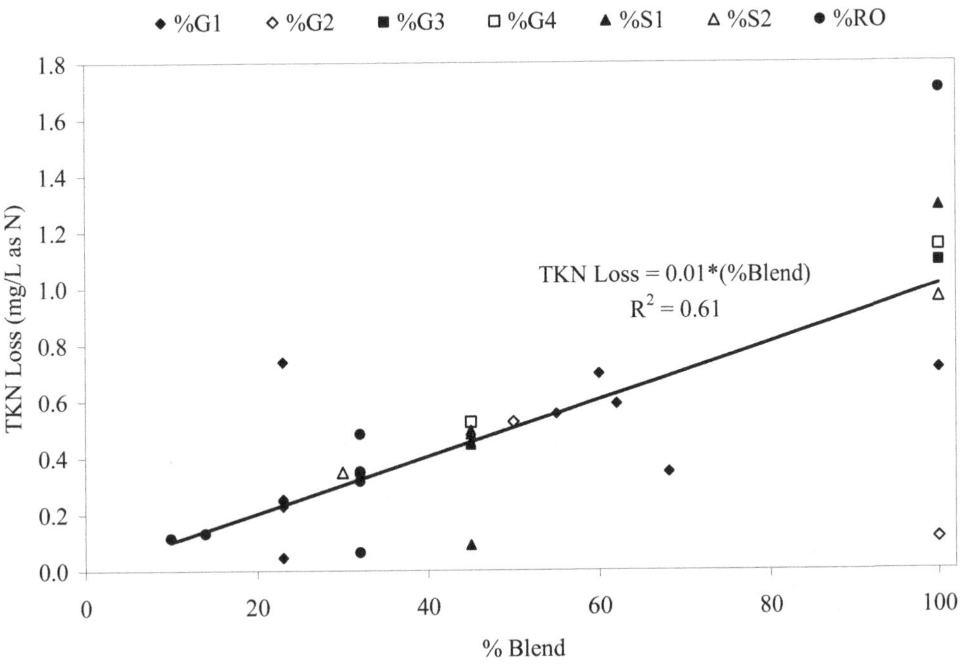

Figure 4.134 Linearly proportioned TKN loss by source percentage in PDS finished water during nitrification episode

Controlled Nitrification Investigation

The controlled nitrification study occurred during phase VI from May through June, 2004. The effects of ammonia-N, Cl_2:NH_3 ratio, flushing on nitrification with respect to pipe material and time were investigated in the nitrification study.

Nitrite Nitrate Correlation

Nitrification can be monitored by measurement of nitrites. As shown in Figure 4.135, a strong correlation existed between effluent nitrite and NO_X-N. AOBs produce nitrite from ammonia. Nitrite was biologically oxidized by NOBs or chemically oxidized by chloramines to nitrate. The NO_X-N concentrations were approximately 137 % of nitrite concentrations, which meant nitrate concentrations were 37% of nitrite concentrations. Because of the strong correlation between nitrite and NO_X-N, only nitrite is used for the discussion of nitrification. The relationship between nitrite and NO_X-N was similar for the four pure material PDSs and the hybrid PDSs. Nitrites are the first product and provide a more proactive signal of biological nitrification, are directly correlated to nitrate production from nitrification, are easily monitored in the field at low concentrations and are regulated at 1 mg/L NO_2-N. Monitoring nitrites as an indicator of nitrification was more practical and effective in this field study and would be a better monitoring parameter than nitrates for utilities seeking to control nitrification in drinking water distribution systems.

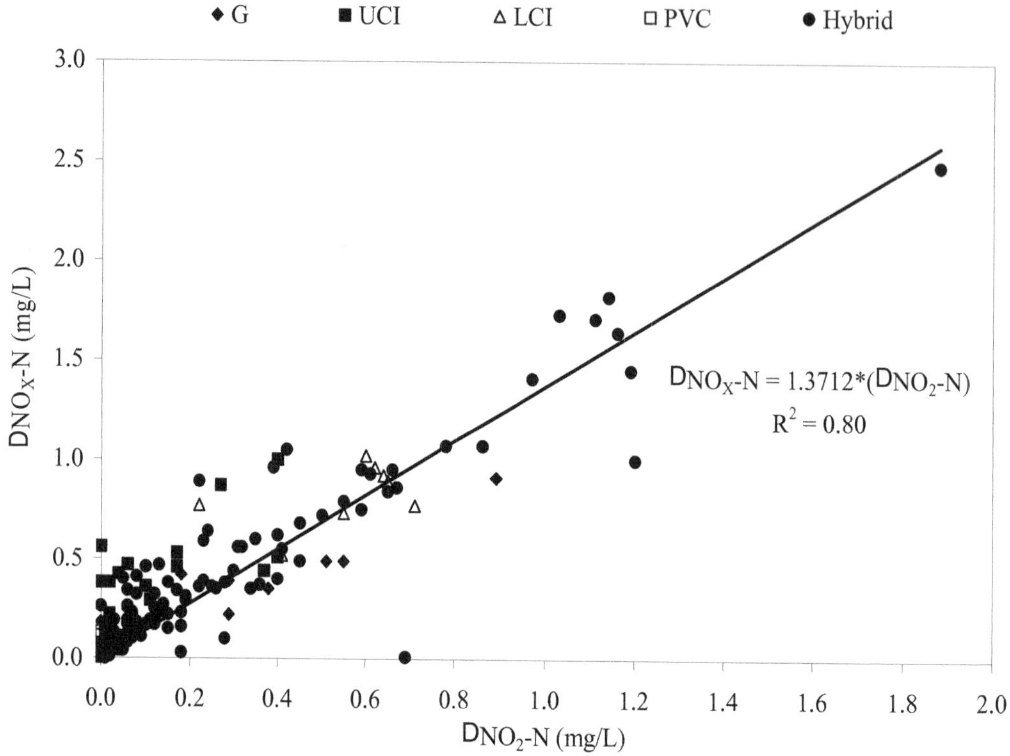

Figure 4.135 Relationship of nitrite production and NO_X-N production

Ammonia-N

The correlation between influent ammonia-N and effluent nitrite is illustrated in Figure 4.136. Higher ammonia in the influent resulted in higher nitrite in the effluent, because there was more available food for AOBs that could be converted to nitrite and nitrate. Clearly, nitrification is reduced when ammonia-N is minimized in the finished water. Ammonia-N can not be eliminated in drinking water distribution systems when chloramines are used because chloramines degrade to ammonia and chlorides. This work shows increasing ammonia-N increases nitrification, and thus is undesirable in the distribution system.

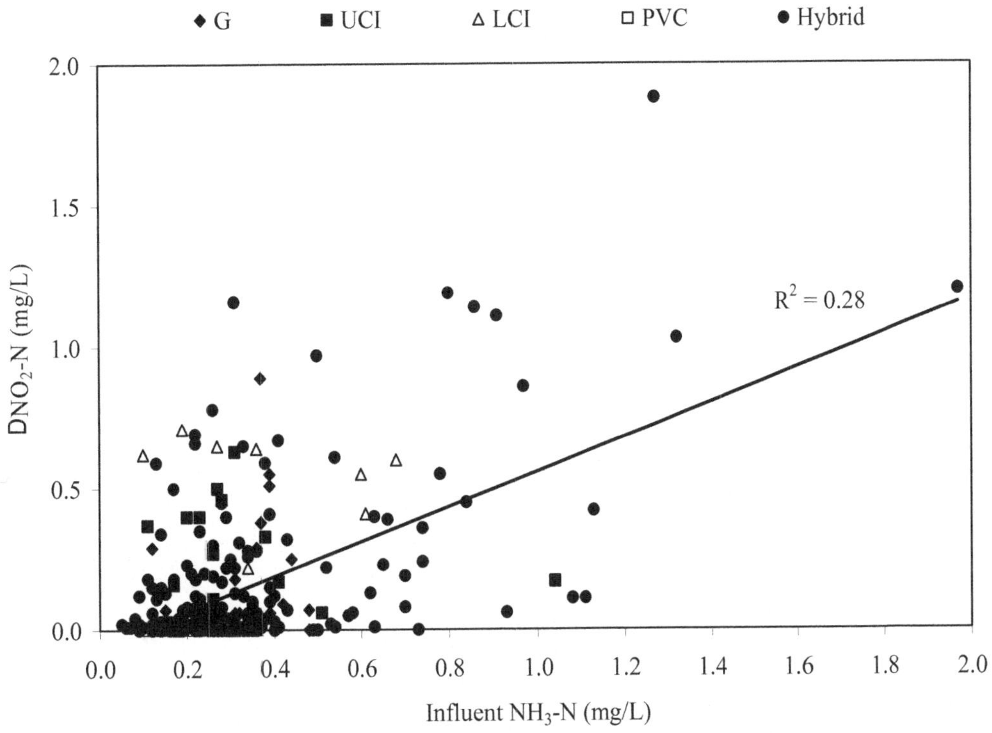

Figure 4.136 Relationship of influent ammonia and NO_2-N production

The relationship of AOB concentration and the influent ammonia in the bulk water by water sources is presented Figure 4.137. The AOB concentration was greatest in the blended water, which was followed by S1, RO and G1 finished waters. RO and G1 sources produced relatively low AOB growth through the PDSs. The data in Figure 4.137 provides evidence of AOB dependence on influent ammonia. As shown in the early part of this section, the relationship of AOB to influent ammonia as shown from the field investigation is similar to the relationship of effluent nitrite to Cl_2:NH_3 ratios and influent ammonia. The statistical trend is correct but there is significant dispersion from the trend. Such observations are expected in field investigations of phenomena that are affected by several factors and shown in a two dimensional representation.

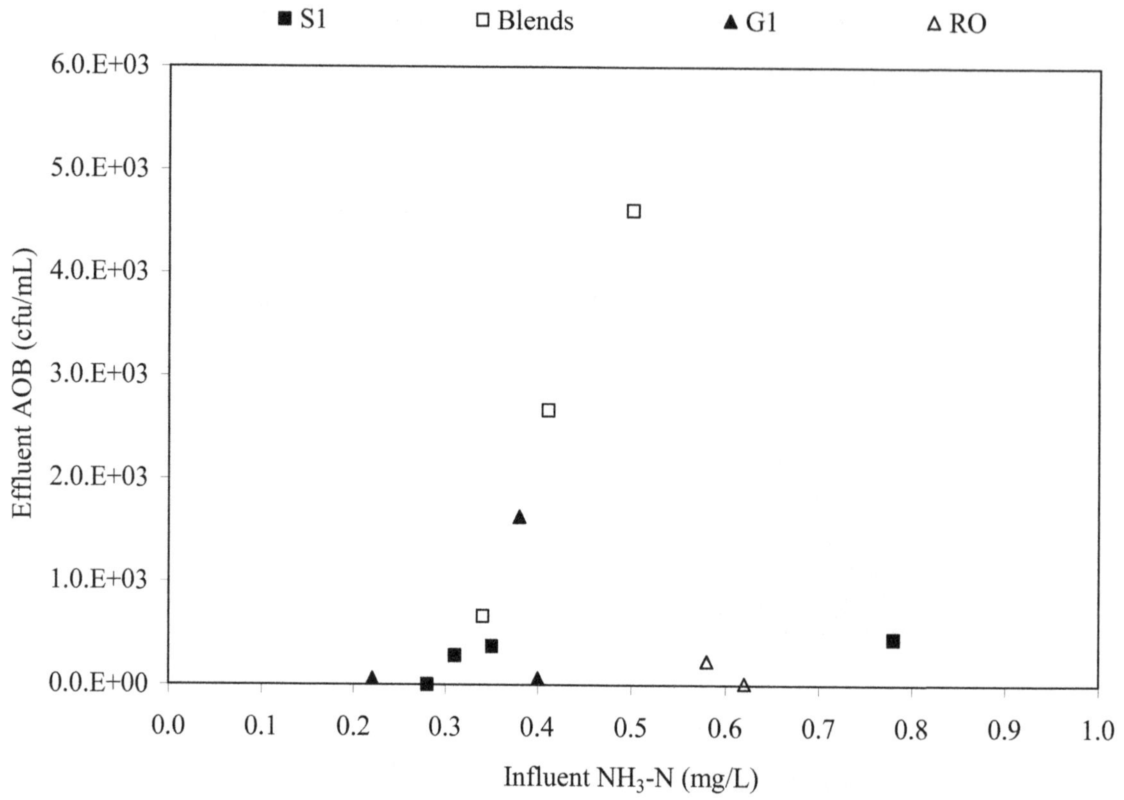

Figure 4.137 Relationship of AOB concentration and influent ammonia in the bulk water

The relationship of ammonia with biofilm AOB concentration is illustrated in Figure 4.138. There was no clear correlation between AOB growth in the biofilm and the ammonia concentration in the bulk. The highest AOB growth occurred on the G material when the ammonia concentration was above 0.5 mg/L as N. G coupons for biofilm AOB measurements were located in all the hybrid PDSs where there was nitrification. The G coupons should not be confused with PDS 18 where nitrification was not observed.

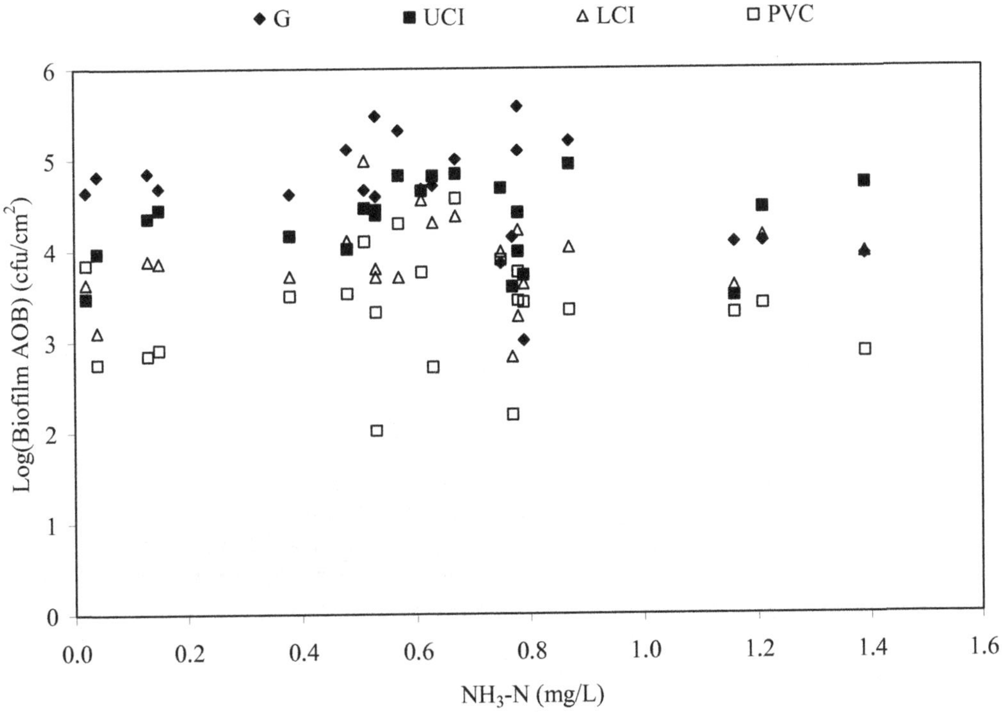

Figure 4.138 Relationship of biofilm AOB concentration and cradle ammonia

Chlorine to Ammonia Ratio

The relationship between $Cl_2:NH_3$ ratio and nitrite is presented in Figure 4.139. Higher nitrite was associated with lower $Cl_2:NH_3$ ratio. Lower $Cl_2:NH_3$ ratio means higher ammonia-N or lower chlorine residual or both. The former contributes more food to the system and the latter means less capacity to inhibit the nitrifiers growth. In support of this, the nitrification potential was found to be negatively related to $Cl_2:NH_3$ ratio. As with the relationship of AOB in the bulk and ammonia in the influent stream, the data in Figure 4.139 indicates there is a definite relationship between ΔNO_2-N and influent $Cl_2:NH_3$ ratio. Using a $Cl_2:NH_3$ ratio of 5:1 will practically eliminate ammonia-N (<0.05 mg/L NH_3-N) in the finished water entering the distribution. Using a higher ratio will produce di- and tri-chloramines, which can produce adverse taste and odor effects. Using a lower ratio will produce ammonia-N. This work clearly shows minimizing ammonia-N in the finished water before entry in the distribution system minimizes nitrification. A 5:1 $Cl_2:NH_3$ ratio was shown theoretically and in the field to minimize nitrification.

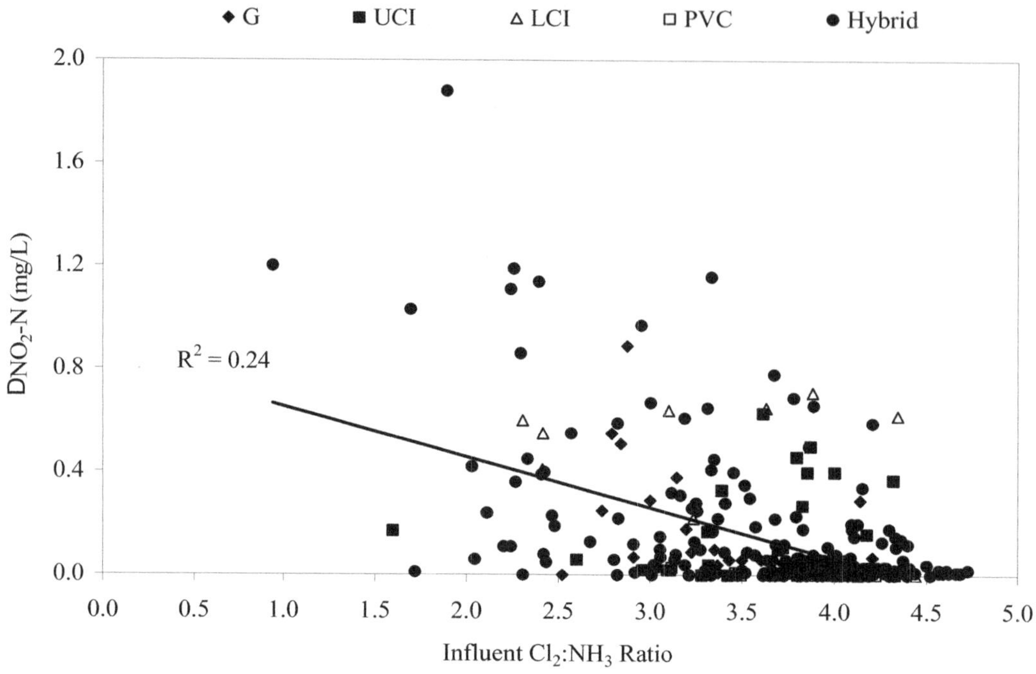

Figure 4.139 Relationship of nitrite and influent Cl_2:NH_3 ratio

Ammonia-N should be avoided in the plant and during chloramine boosts in the field, which means free chlorine should be used to boost chloramine residuals at a 5:1 Cl_2:NH_3 ratio in the distribution system if ammonia-N is to be minimized. This requires ammonia-N must be accurately measured in the field before free chlorine is applied, and the application of free chlorine must be at a 5:1 Cl_2:NH_3 ratio.

Flush Frequency and Flush Strength

Distribution system flushing is a viable strategy for controlling nitrification. Benefits from flushing are restoration of residual, removal of bulk microorganisms and removal of biofilms. Disadvantages of flushing with chloraminated water include water loss, labor loss and providing ammonia sources which can refuel nitrification. PDSs 9, 10, and 14 received the same blended water that was used to investigate flushing. Regular flushing was maintained for PDS 14 and was changed for PDSs 9 and 10. Before flushing, finished waters were pumped to the flush water storage tank, and the chloramine concentrations were adjusted as necessary.

Chloramination was restored on 05/02/03 in PDS 9 and 10 following the incubation period with no chloramine. From 05/02/03 to 06/06/03, the PDSs were flushed every 1.5 weeks with three pipe volumes at 1 ft/sec for a total of three flushes. From 06/06/03 to 07/04/03, the PDSs were flushed weekly with three pipe volumes at 1 ft/sec for a total of four flushes. From 05/16/03 to 06/27/03, PDS 14 was flushed biweekly with three pipe volumes at 1 ft/sec for a total of four flushes. The PDS HRT during the flush investigation was 2 days. The sampling events were always conducted before flushing. When the sampling and flushing occurred on the same day, the samples were actually the consequences of the previous flushing.

Isopleths relating residual loss, ammonia loss, nitrite production and nitrate production after flushing in and out of PDS 9, 10 and 14 are shown in Figure 4.140 for flush frequency and residual strength. The isopleth for residual loss during flush is shown in the upper right hand corner of Figure 4.140. The residual loss ranges from 2.9 to 3.3 mg/L as Cl_2 isopleths and show that PDS residual loss increased as chloramine concentration in the flush decreased and as flush frequency increased. Hence flushing more frequently and with a higher strength chloramine residual decreased the residual loss during normal operation, which indicates the PDS demand was renewable and adversely affected by flush frequency and strength. The residual loss increased with decreasing flush strength (chloramine concentration), and increased with flush frequency.

The ammonia loss is a direct measure of nitrification and is observed to vary from -0.3 to 0.1 mg/L as N. The negative number indicates the rate of chloramine degradation exceeded the rate of ammonia consumption. When the flush strength was 10 mg/L as Cl_2, the ammonia-N loss was about -0.4 mg/L, which means the ammonia-N in the effluent was 0.4 mg/L higher than in the influent. The higher ammonia in the effluent came from the degradation of chloramines, and had not been consumed to nitrite or nitrate by nitrification. The ammonia-N loss at 4 mg/L as Cl_2 flush strength was about 0.2 mg/L, which indicated that the all the ammonia released from chloramines and the ammonia-N had been converted to nitrite or nitrate. PDS ammonia-N loss decreased as the strength of the flush decreased and was unaffected by flush frequency for a 1 to 2 week flushing frequency.

The nitrite isopleth shows that the nitrite production varied from 0.2 to 0.4 mg/L. Nitrite produced in the PDSs increased as flush strength decreased and flush frequency increased, which is similar to the residual variations with flush concentration and frequency. The nitrate isopleth shows that nitrates produced in the PDSs increased as flush concentration decreased but were not affected by a 1 to 2 week variation in flush frequency. The nitrate-N production varied from 0.2 to 0.8 mg/L.

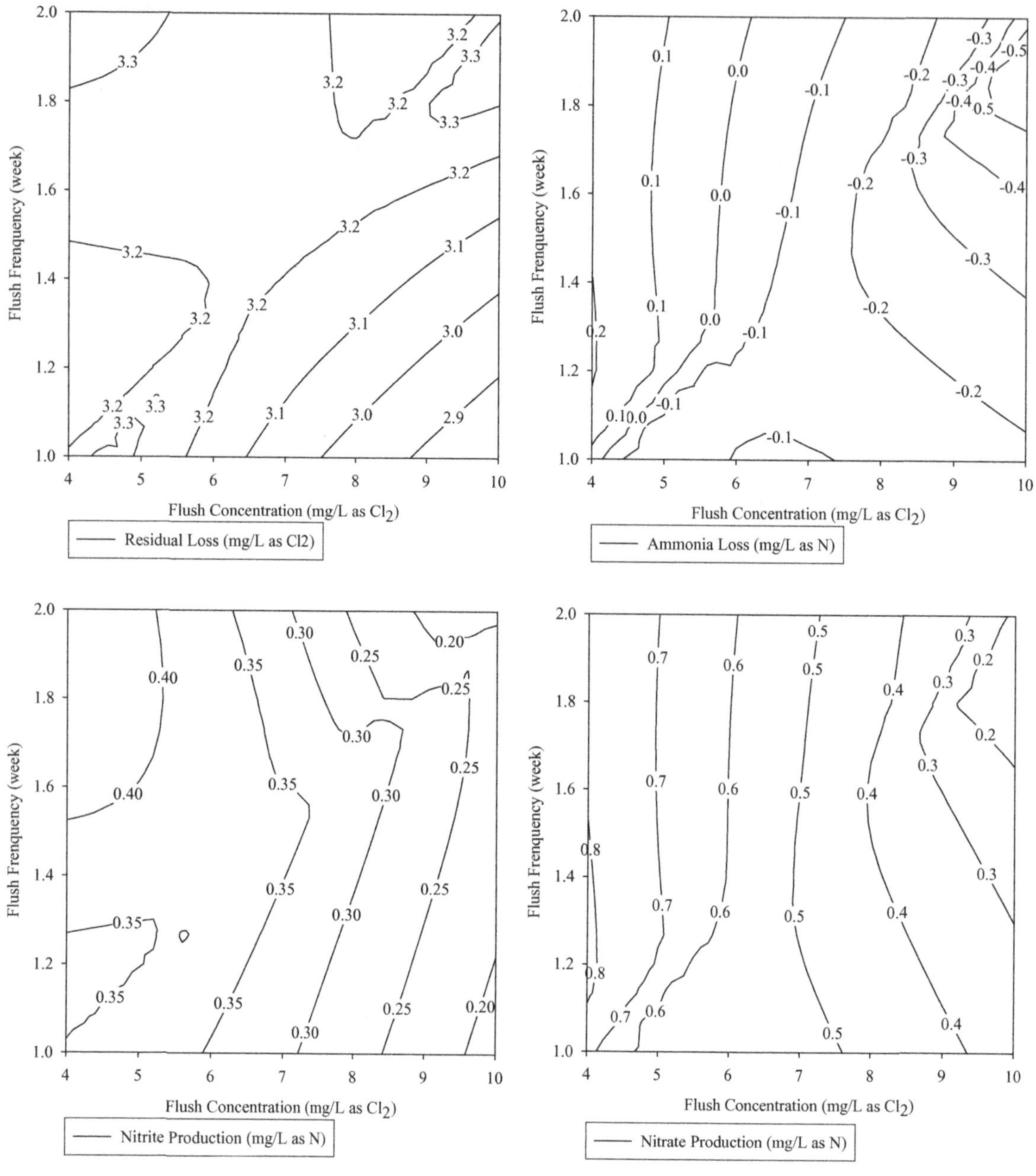

Figure 4.140 The effect of flush chloramine residual concentration and frequency on residual loss, ammonia loss, nitrite production, and nitrate production

Utilities may not be able to vary flush strength when flushing is implemented. In this work, the PDS were flushed every week during the first three phases and every two weeks in all studies thereafter. Our flushes were limited by quantity and normal operation utilizes very slow PDS velocities. This data shows nitrification is adversely affected by increasing flush frequency and flush strength, which should have the same effect in actual operation. However utilities should maintain records that assess flushing effectiveness. These records would record flush frequency, flush strength, temperature, residual change during flushing, residual change during regular operation, ammonia and nitrite change during regular operation so that flushing can be proactively implemented following flushing assessment.

When the flush strength was 4.0 mg/L as Cl_2, the residual loss was 3.2 mg/L as Cl_2, which was equivalent to about 0.6 mg/L as chloramines-N, the ammonia-N loss was 0.2 mg/L, and the nitrate-N production was 0.8 mg/L, which is equal to the sum of chloramine-N loss and ammonia-N loss. This observation reconfirms the previous observation of nitrogen balance in the PDSs without denitrification as shown in Table 4.65.

Denitrification

Figure 4.141 illustrates the nitrogen profiles in PDS 15 (UCI) during nitrification experimental study. The residual was lost at about 20 hrs. The nitrification was strong in this PDS as indicated by the presence of NOx-N at about 8 hrs. The nitrite and NO_X-N reached the highest concentration at about 30 hrs. The DO was depleted at about 30 hrs. After 30 hrs, the PDS was anaerobic, and the nitrite and NO_X-N decreased by 0.1 mg/L, which indicated denitrification. This reconfirms the fact that DO is an important parameter that the utilities should monitor. At certain locations of the distribution systems, the lack of DO may limit the occurrence of nitrification before nitrification starts, but an undesirable environment is created in the distribution system, which includes denitrification, adverse taste, adverse odor, black water and uncontrolled microbiological growth. Hence low nitrites or nitrates, without paired observations of DO and residual and associated trends with time means very little. It is possible for any utility to monitor nitrite, DO and residual at various locations in the distribution system, but accurate correlation to time will be difficult without a functional computer model that provides time by locations where samples are taken in the distribution system.

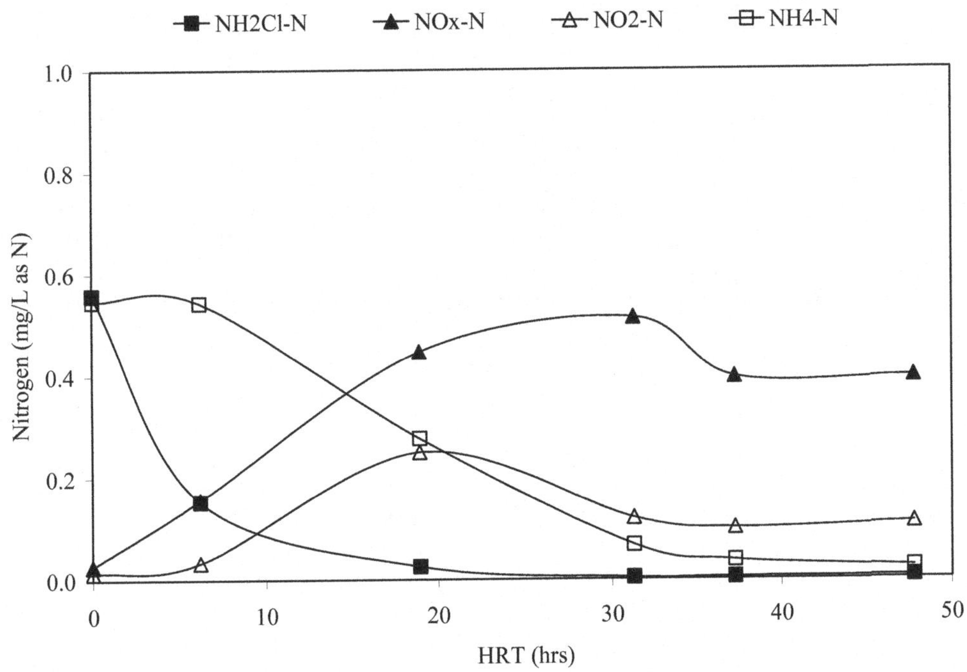

Figure 4.141 The average nitrogen profiles in PDS 15 (UCI) at nitrification experimental study

Free Chlorination

A one week free chlorine burn was instigated from 12/15/02 to 12/23/02 using a free chlorine residual of 5.0 mg/L Cl_2. Ammonia does not exist in the presence of free chlorine and was not present to fuel nitrification. As shown in Table 4.67, ammonia, nitrite or nitrate was not found in the PDS effluents. Without ammonia, nitrification does not occur.

Chloramination use was restored on 12/23/2002. As shown in Figure 4.142, nitrite concentrations were not found in PDS 5, 6, 7 and 8 effluents until after May 1, 2003. Even on June 26, 2003, the highest nitrite concentration detected was 0.2 mg/L as N. Nitrification was clearly eliminated by the free chlorine burn and nitrite did not appear in the PDS effluent until the end of May 2003, which corresponds to a 15 to 30°C temperature increase and three to four months of no detectable nitrification.

Table 4.67
Water quality during free chlorination

Parameter	Date	Sample	PDS 11	PDS 12	PDS 13	PDS 14
Free Cl_2	12/17/02	Feed	5.0	5.8	5.0	5.1
(mg/L as Cl_2)	12/18/02	Effluent	2.0	2.5	2.4	2.6
	12/19/02	Feed	6.0	5.6	6.3	6.1
	12/20/02	Effluent	2.5	2.7	2.5	2.3
NO_2-N	12/18/02	Feed	0.0	0.0	0.0	0.0
(mg/L as N)	12/19/02	Effluent	0.0	0.0	0.0	0.0
	12/20/02	Effluent	0.0	0.0	0.0	0.0
HPC	12/16/02	Feed	5.0	5.0	5.0	160
(cfu/mL)	12/16/02	Effluent	390	140	200	150
	12/19/02	Feed	5.0	5.0	5.0	35
	12/19/02	Effluent	250	650	150	200

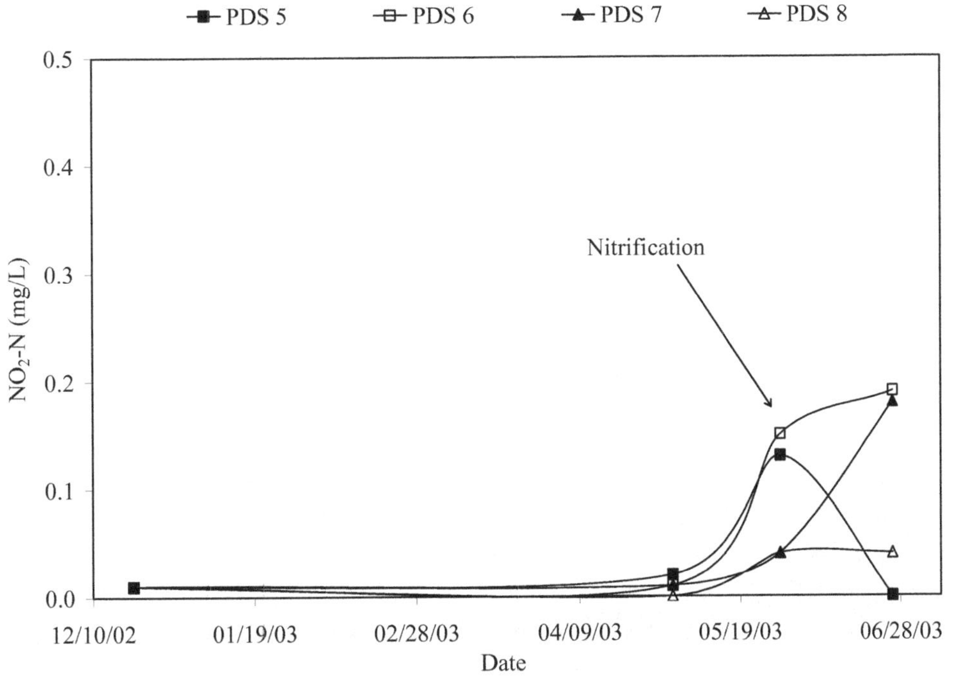

Figure 4.142 NO_2-N trend after free chlorination (PDS 5-8)

General Nitrification Correlations

Several parameters were monitored during the nitrification episode and controlled nitrification investigation that were more effectively presented as a combined data set. Those parameters primarily relate to nitrite production and are chloramine residual, AOBs in the bulk and the biofilm, HPCs in the bulk and biofilm, temperature.

Bulk Stream Correlations

The impact of effluent chloramine residual on the effluent nitrite is presented in Figure 4.143. It is obvious that nitrification occurred when residual was lost. The effluent residual threshold for nitrification was about 1.0 mg/L. Effluent nitrite was lower and there were no signs of nitrification when the effluent residual was above 1.0 mg/L. The direct effect of effluent residual on nitrification was more significant than the effect of influent residual. The PVC PDS had no nitrification even when the residual was lost, the residual in G PDS was always lost, but nitrification was low due to DO limitation. Nitrification in LCI PDS only occurred during the nitrification experimental study, which indicates pipes lined with relatively inert materials significantly reduced biological activity and are highly desirable in drinking water distribution systems relative to unlined metal pipes.

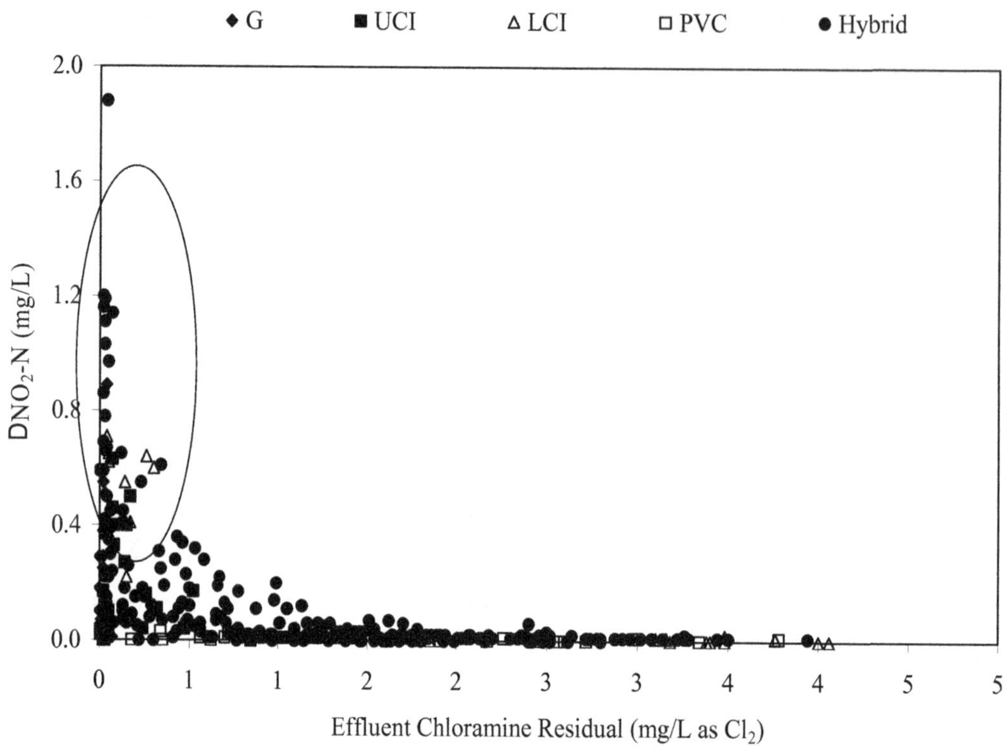

Figure 4.143 Relationship of nitrite production and effluent chloramine residual

The residual of chloramine acted as a biocide, and as illustrated in Figure 4.144, AOBs were negatively correlated to residual. The AOB concentrations increased exponentially as chloramines approached zero. When residuals were above 0.6 mg/L as Cl_2, the AOB growth was significantly inhibited. This is very similar to the exponential relationship between effluent HPCs and effluent chloramine residual shown in the biostability section. Higher effluent residuals certainly reduced if not eliminated nitrification. As shown in Figure 4.144, the highest AOBs occurred in sources that had the lowest effluent residuals.

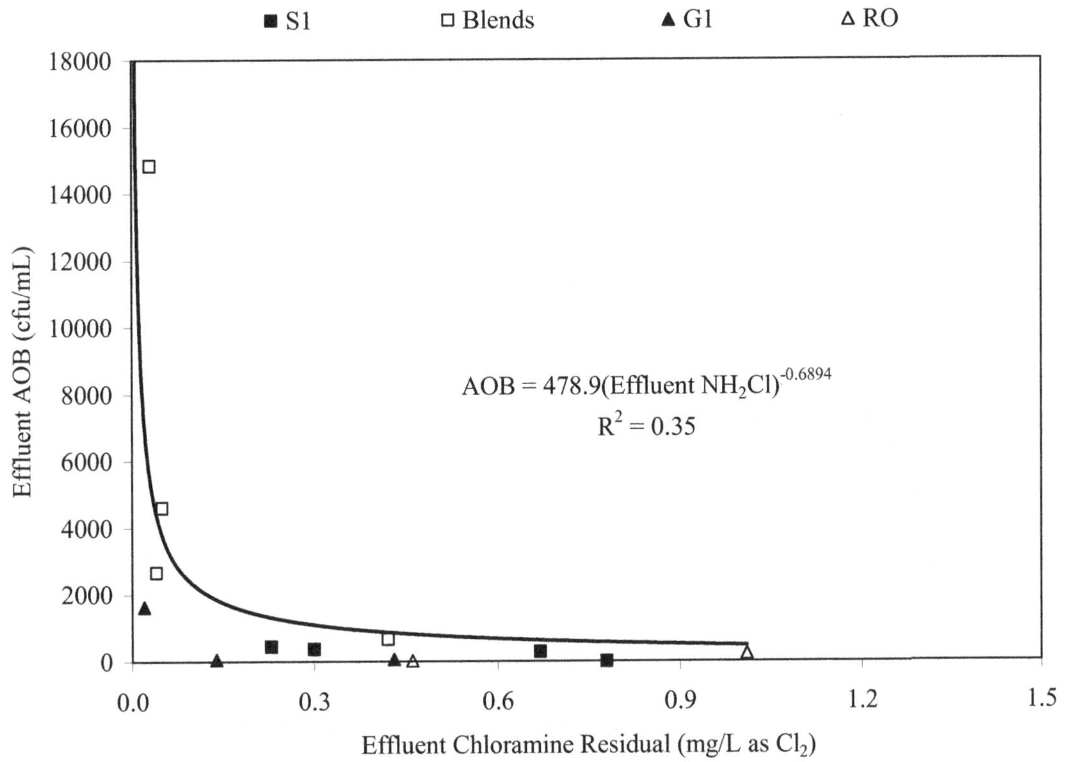

Figure 4.144 Relationship of effluent chloramine residual and AOB in the bulk water

Figure 4.145 provides the correlation between DO consumption and nitrite. The growth of nitrifiers consumes oxygen. Stoichiometrically 3.16 mg/L of oxygen is consumed per 1 mg/L of nitrite nitrogen produced from ammonia, and an additional 1.11 mg/L of oxygen will be consumed per 1 mg/L of nitrate nitrogen produced from nitrite. However, DO may also be consumed by heterotrophic bacterial growth, and corrosion process, but from a stoichiometry point of view, AOB metabolism can be a major factor during nitrification, and can also lead to increased heterotrophic activity and DO consumption. The circled data points in Figure 4.145 were mainly from G PDS and UCI PDS. The former did not show significant nitrification but had significant DO consumption, which was attributed to other than nitrifying microorganism activity and to denitrification during the nitrification experimental study respectively.

Figure 4.146 presents the relationship between DO consumption and AOB concentration, which as expected shows as AOBs grew more DO was consumed. Oxygen consumption is required by all aerobic microorganisms and is a good indicator of bacterial activity in a drinking water distribution system.

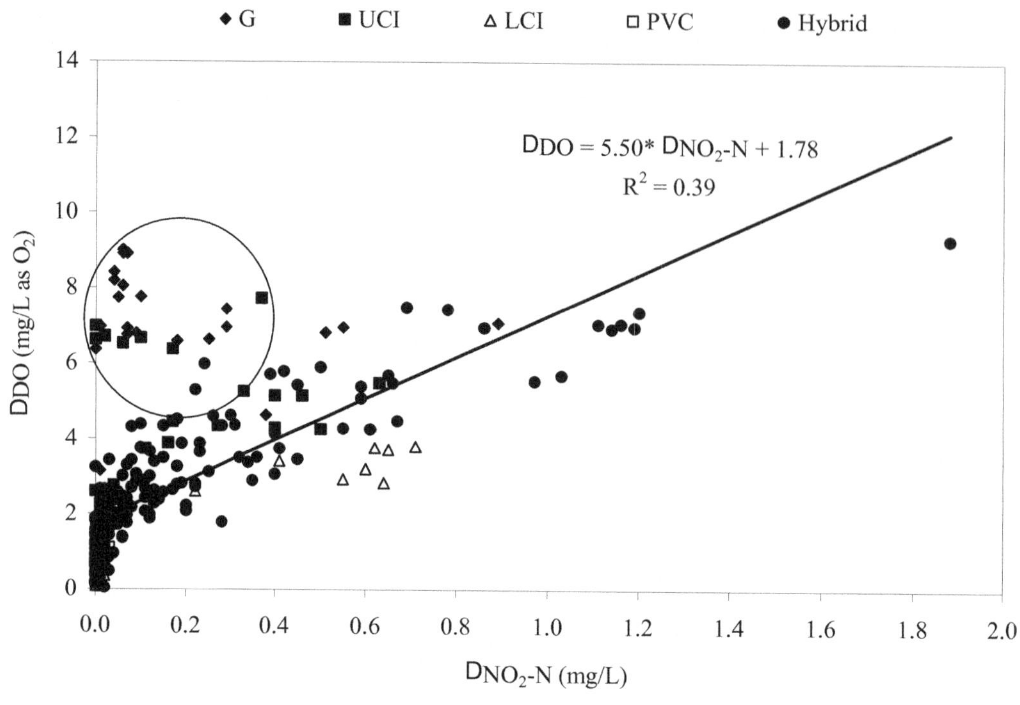

Figure 4.145 Relationship of DO consumption and NO_2-N production

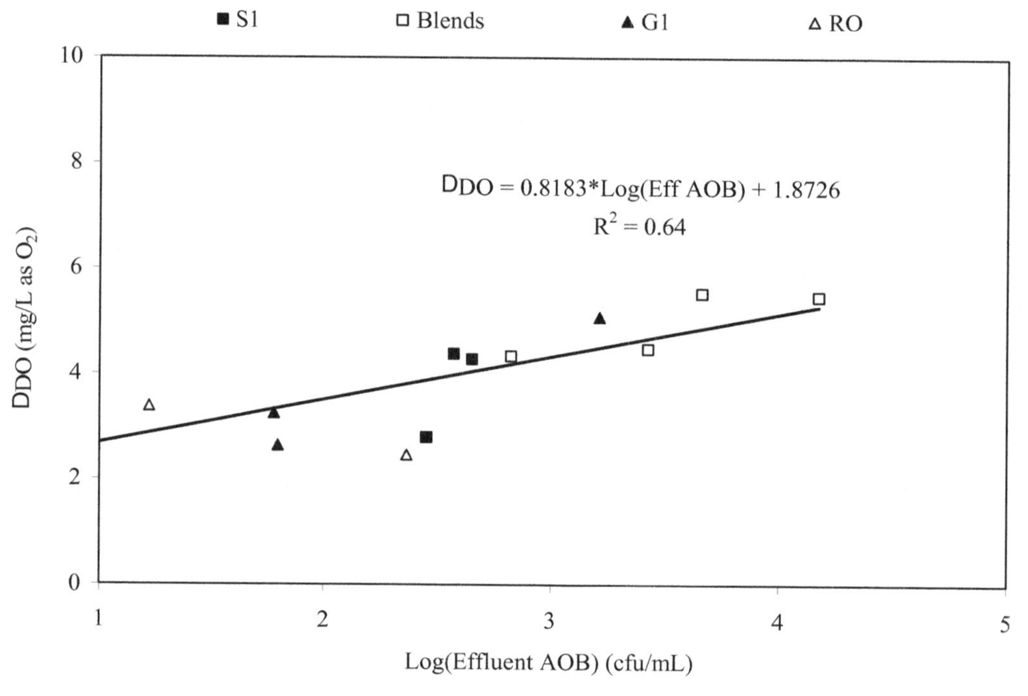

Figure 4.146 Relationship DO consumption and AOB concentration in the bulk water

Figure 4.147 illustrates the influence of temperature on nitrite. The data were from 09/16/02 to 12/14/02. During this period, water chemistry did not vary and provided an unbiased observation of temperature effects on nitrification. Nitrite correlated well with temperature. Hence, nitrification is more likely to occur under warmer water conditions and is likely to proceed more rapidly with increased temperatures.

Figure 4.148 presents the influence of temperature on effluent AOB concentration. The data were from the controlled nitrification study. AOB growth was significantly enhanced when the temperature was above 30°C; however the variation in AOB data indicates that other factors such as pipe material influenced AOB growth and activity. Also the presence of oxygen is a pre-requisite for AOB activity, being the only known electron acceptor for AOB metabolism. This may explain the lack of nitrification in PDS 18 (100% G material).

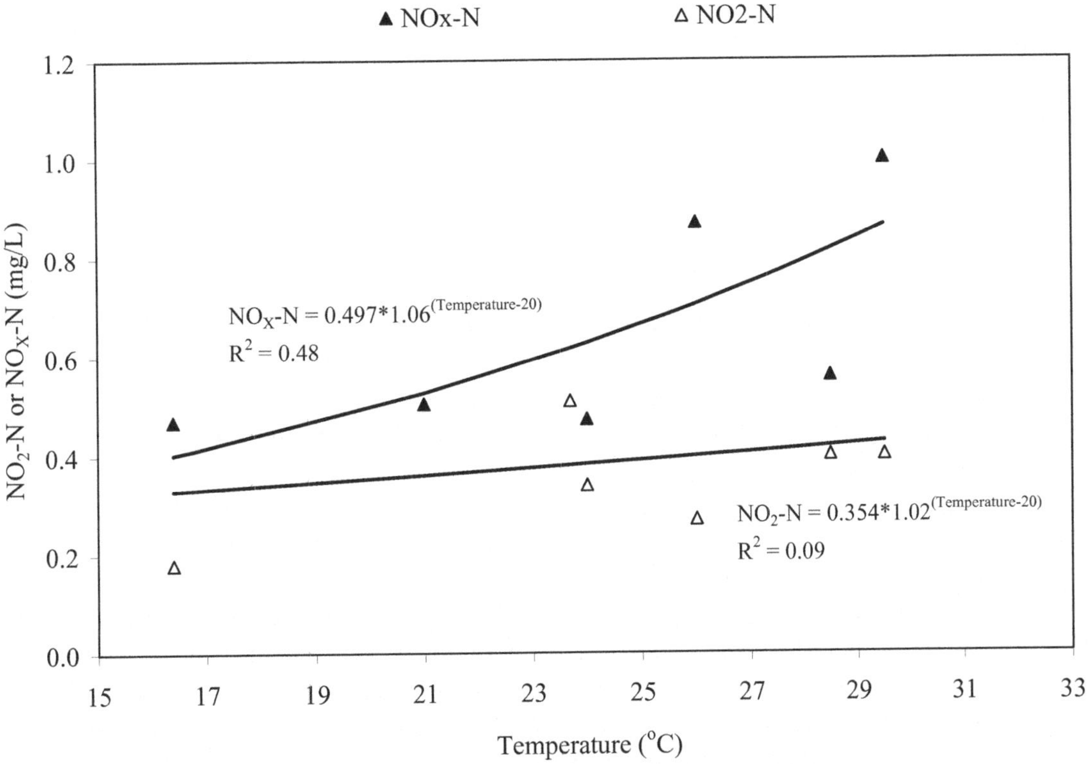

Figure 4.147 Relationship of effluent nitrite and NO_X-N with temperature in the bulk water

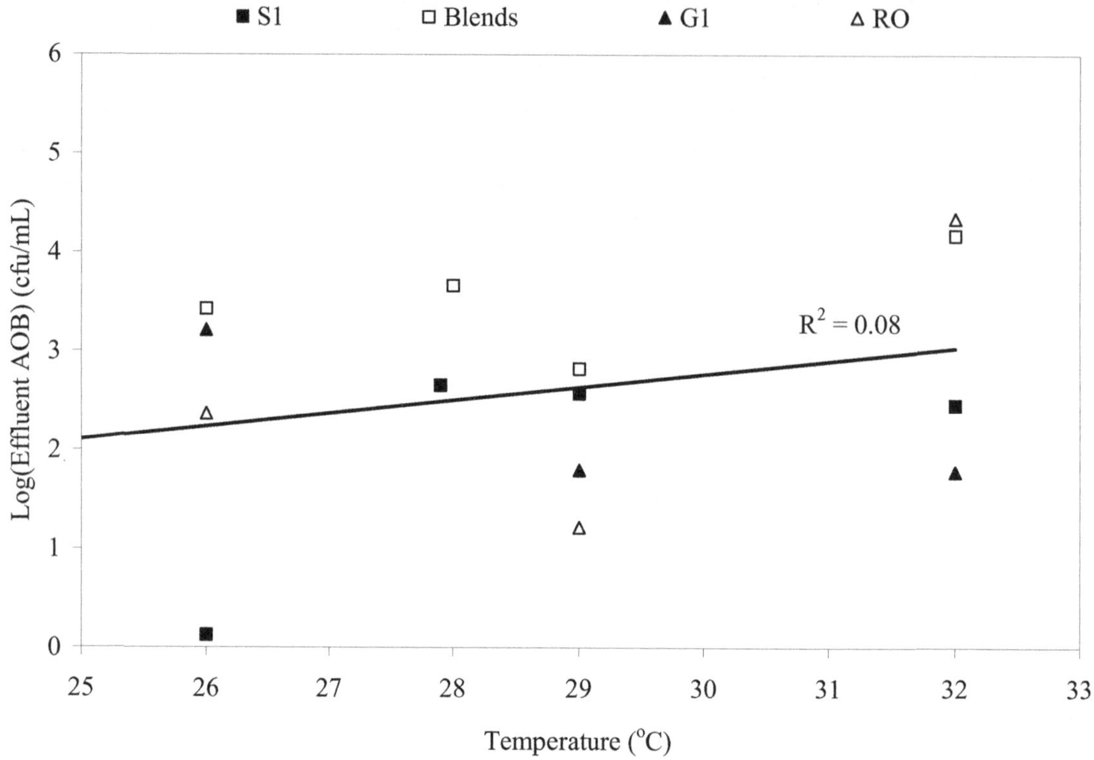

Figure 4.148 Relationship of temperature and effluent AOB in the bulk water

Figure 4.149 illustrates the relationship between nitrite production and AOBs using data from PDSs 11-14 during the controlled nitrification study. AOBs increased as nitrite increased, which is expected during biological nitrification. The blended water (63%G1/27%S1/11%RO) had the highest nitrite production and AOB proliferation. RO finished water had the least nitrite production, AOB proliferation and the lowest AOC.

The relationship of HPC with NO_2-N production is illustrated in Figure 4.150, and the relationship of HPC and AOB proliferation is illustrated in Figure 4.151. The two figures show that HPC growth was positively correlated with AOB proliferation, but the correlation of HPC with nitrite was tenuous at best. It can be seen that HPC proliferation was favored by AOB growth and by loss of residual during nitrification.

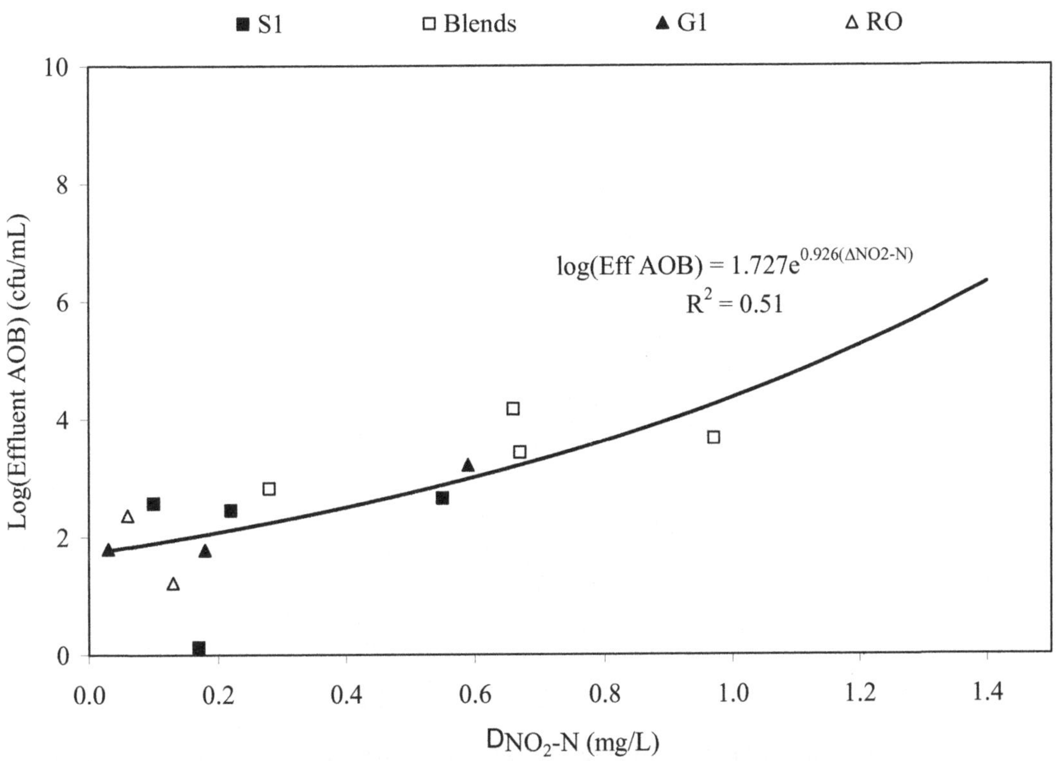

Figure 4.149 Relationship of NO_2-N and AOB in the bulk water

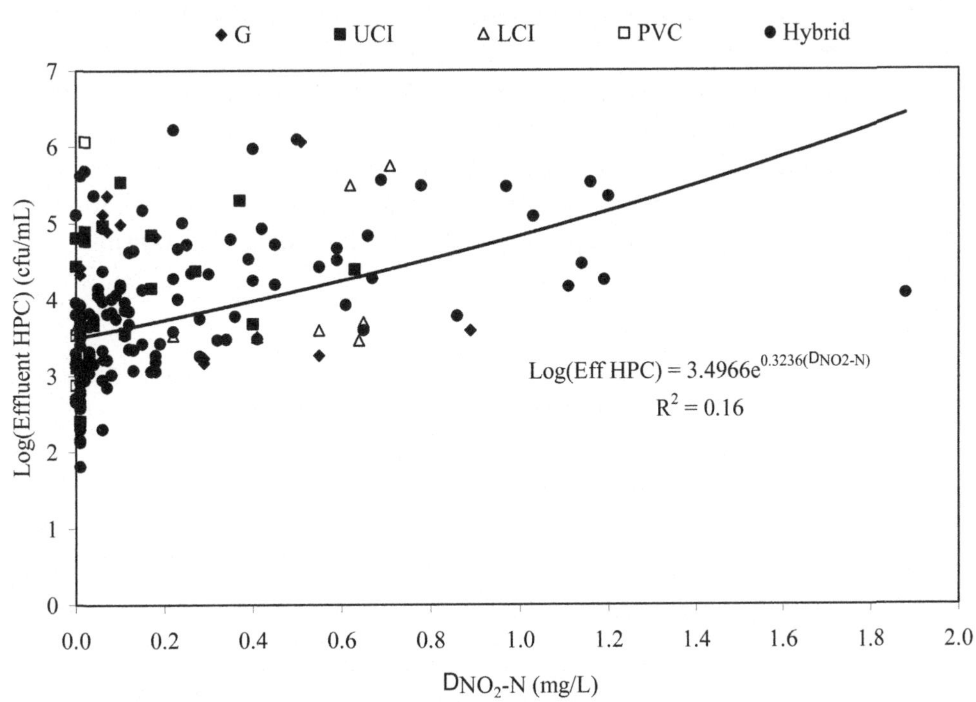

Figure 4.150 Relationship of NO_2-N production and HPCs in the bulk water

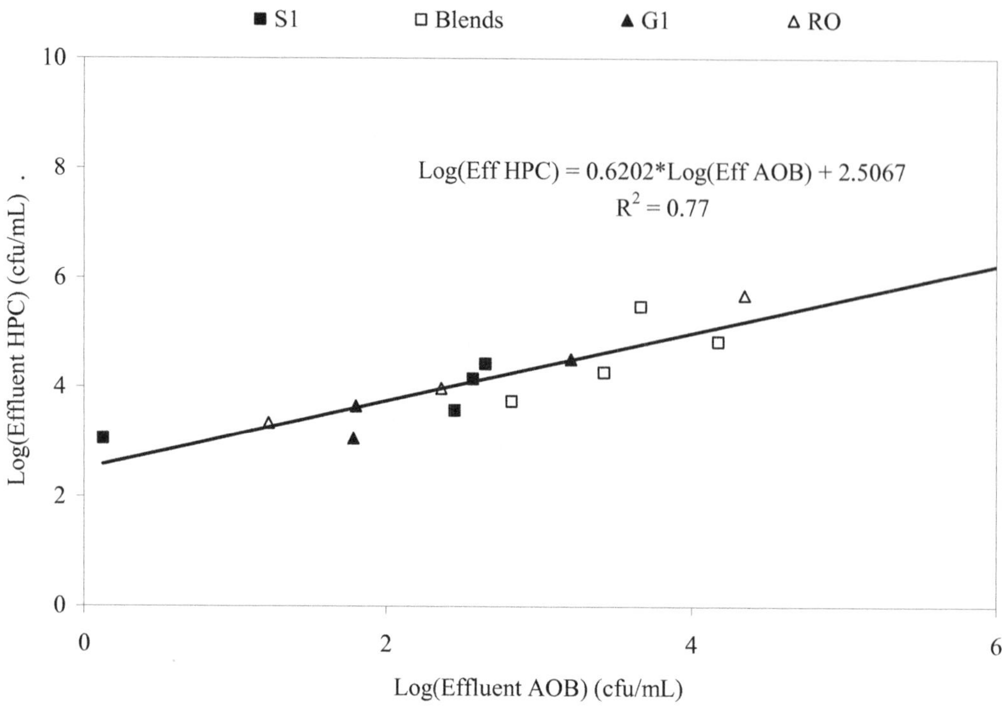

Figure 4.151 Relationship of HPC and AOB in the bulk water

The relationship of AOB proliferation and AOC is presented in Figure 4.152. The AOC levels for S1 SW was 30 - 160 µg/L, blends 28 - 120 µg/L, G1 40 - 100 µg/L and RO 4 - 26 µg/L. The AOC level sequence is S1>Blends>G1>RO. The figure shows the AOB proliferation is independent of the organic carbon concentration in the water. Judging from the literature this could be expected since all nitrifiers (both ammonia and nitrite oxidizers) are obligate chemolithotrophs (Holt et al 1994), and Nitrosomonas are frequently used as model autotrophic organisms since they are obligate autotrophs (Wood, Aurikko and Kelly 2004).

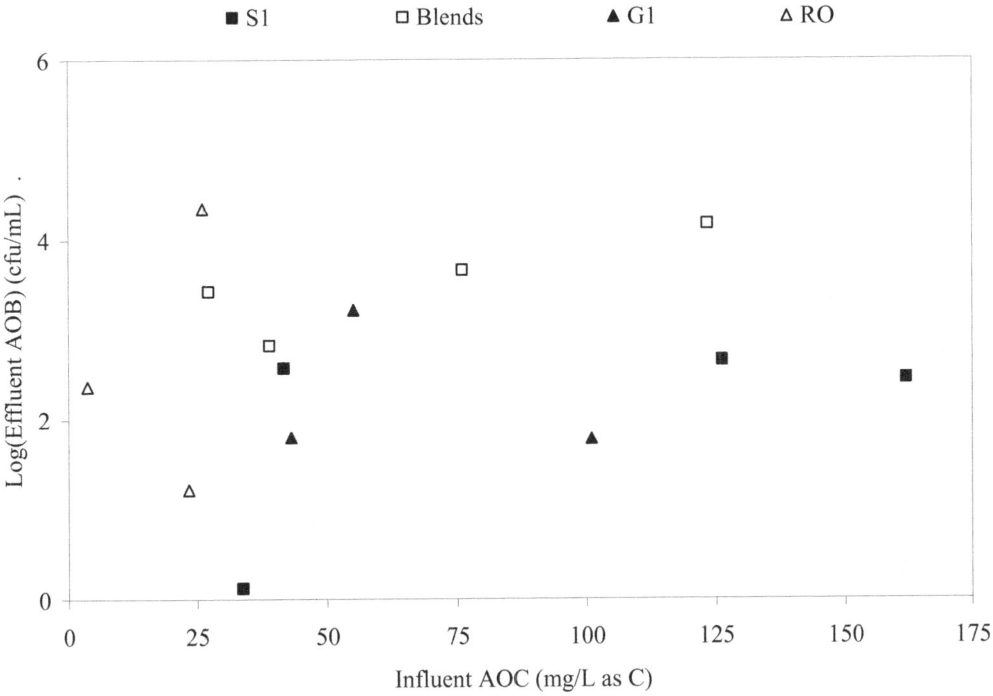

Figure 4.152 Relationship of AOC and AOB in the bulk water

Biofilm Correlations

The relationship of biofilm AOB density and residual in the bulk water is presented in Figure 4.153. These relationships are varied, but AOB concentrations in the biofilm appear to vary by PDS material and then by residual. There was higher AOB proliferation in the biofilm for unlined metal (UCI and G) coupons than in lined (LCI and PVC) coupons. There was little nitrification in the lined PDSs, hence there is no correlation to residual, but a general trend is apparent for biofilm AOBs in the unlined metal coupons. Although AOBs exist in the bulk and biofilm some generalizations can be made about the contribution of each to nitrification. Assume the AOB concentration in the biofilm is 10^5 cfu/cm^2 and in the bulk is 10^4 cfu/mL; which corresponds to the approximate maximum concentrations of AOB in the bulk and biofilm that were observed. The AOB mass in a one foot section of a 2-inch diameter pipe in the bulk would be approximately twice the AOB mass in the biofilm. The AOB mass in a one foot section of a 1-foot diameter pipe in the bulk would be approximately three times the AOB in the biofilm. Hence, nitrification could occur either in the bulk or the biofilm as the estimated difference between AOBs in the bulk or the biofilm is less than an order of magnitude. It is likely that biofilms harbor AOBs and continually release them to the bulk under normal operation. The bulk residual inactivates the AOBs and controls nitrification. If the residual is not strong enough to inactivate the bulk AOBs, nitrification increases until ammonia-N or DO diminishes, or a strong residual is restored. Utilities can do little to eliminate biofilms, but they can keep biofilm effects on the bulk in check with a strong residual.

The relationship of DO with biofilm AOB concentration is illustrated in Figure 4.154. The relationship is not a strong one but most of the values above 50,000 occurred when DO was greater than 5 mg/L. However limiting DO is not a viable strategy for preventing nitrification for utilities. Anaerobic conditions produce many adverse water quality conditions such as bad taste and odors and black water. Monitoring residual, nitrite, and nitrate are probably the most straightforward way of monitoring for nitrification events.

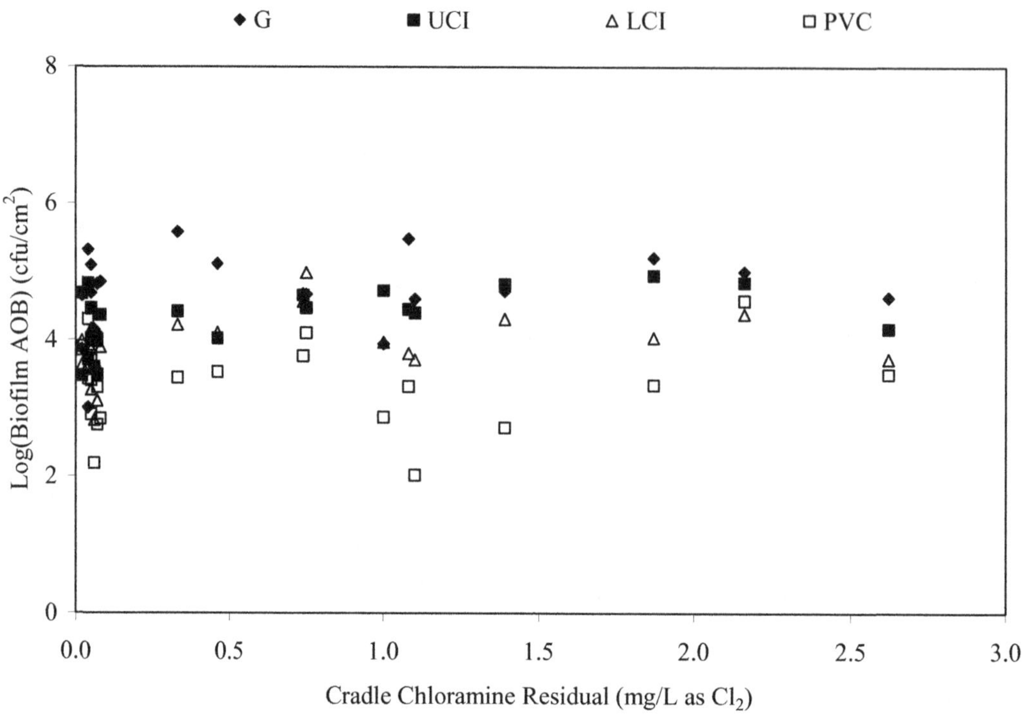

Figure 4.153 Relationship between effluent residual in the bulk water and AOB in the biofilm

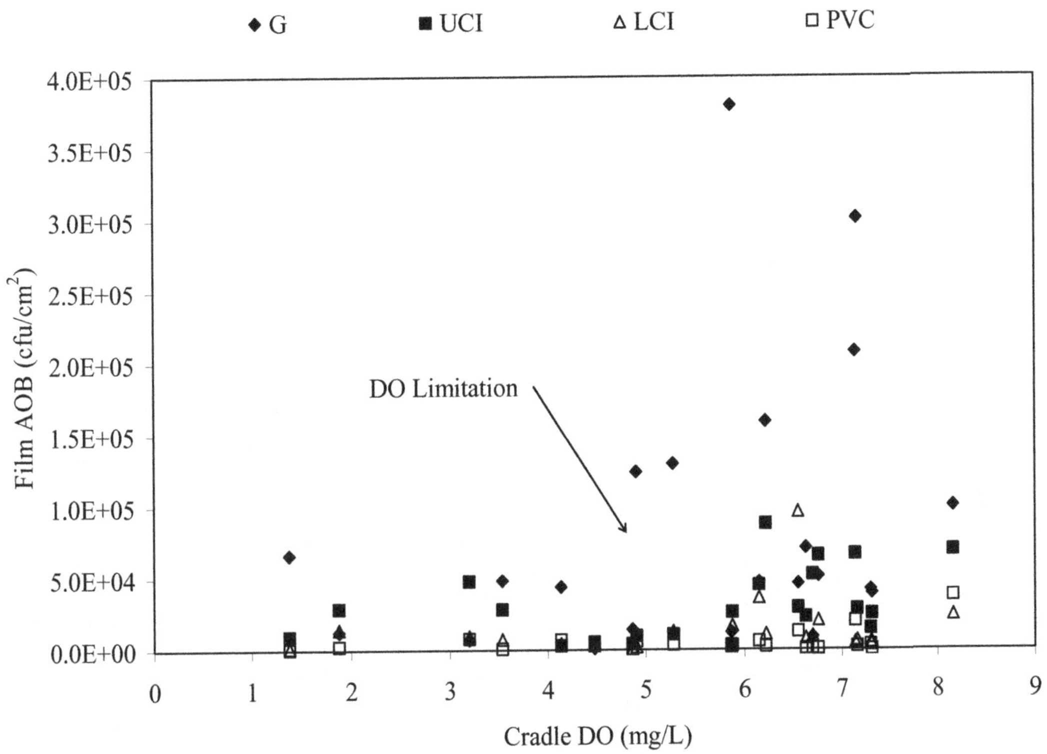

Figure 4.154 Relationship of biofilm AOB concentration and cradle DO

DO monitoring in the distribution system is potentially a useful tool to monitor distribution system water quality since on-line probes are inexpensive and reliable. Ultimately residual, nitrite, and nitrate are more informative, and there are probes to quantify these parameters. Nitrification can also be anticipated and located in the distribution system using DO profiles. Correlation of DO profiles by season and location can be proactively used to identify problematic distribution system locations. But the other 3 parameters would be preferred if costs and maintenance of probes is comparable.

Figure 4.155 illustrates the influence of temperature on biofilm AOB concentration. The highest AOB proliferation in the biofilm occurred at the highest temperature (30.5°C) on G material. But in general, there is no significant correlation between AOB concentration in the biofilm and the water temperature. This is likely because temperature was confounded by residual.

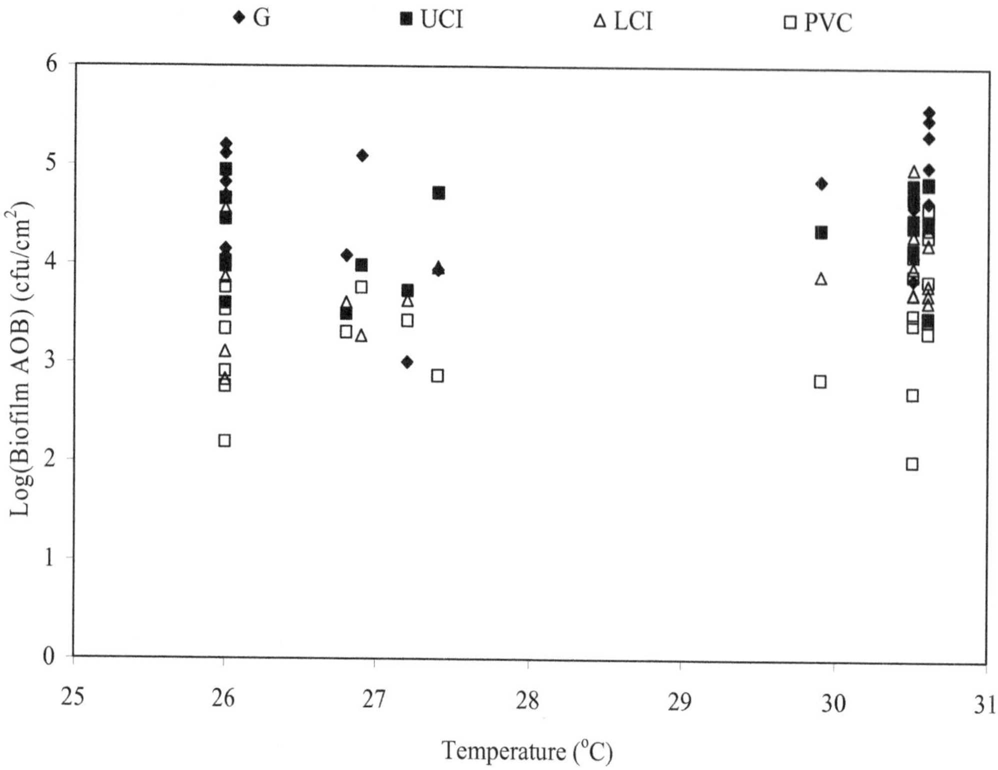

Figure 4.155 Relationship of temperature and AOB in the biofilm

Temperature is and should be monitored by utilities in distribution systems. Correlation of distribution system residuals, temperatures and bacterial activity is essential to reducing biologically oriented problems in the distribution system. As the rates for reactions typically increase with temperature, key parameters associated with nitrification (residual, DO, NH_3, HPCs, NO_2) should be more extensively monitored during high temperature conditions.

The relationship between AOB concentration in the biofilm and the nitrite production is illustrated in Figure 4.156. There is no significant correlation between AOB concentration in the biofilm and the nitrite production in the bulk. Four materials, namely UCI, LCI, PVC and G, were used for biofilm growth study. None of the biofilm AOB on individual materials showed any significant correlation with nitrite. Among these four materials, the G material had relative higher AOB growth than other materials, UCI material had the second highest AOB growth, and PVC had the least AOB growth. The sequence of AOB growth on the materials is: G>UCI>LCI>PVC.

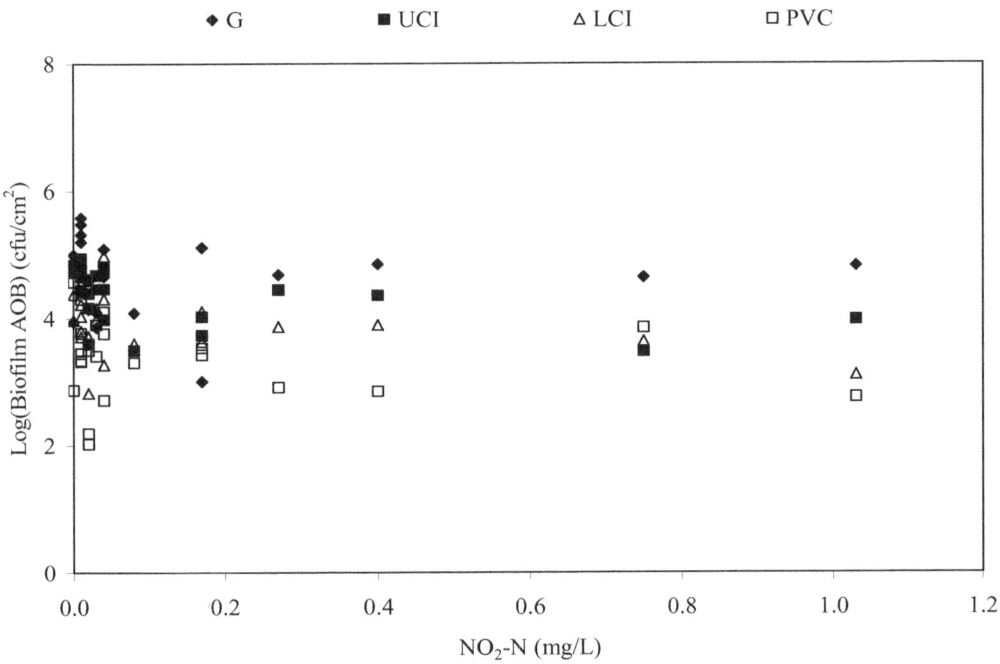

Figure 4.156 Relationship of NO_2-N production in the bulk water and AOB in the biofilm

The biofilm AOB did not correlate with bulk nitrite, ammonia, or residuals. This indicated that the effect of biocide on biofilm was different from the biocide effect on bulk water HPC. It may be that organisms living within the biofilm are protected from the residual since it may be consumed the deeper it diffuses into the biofilm.

The relationship between biofilm HPC on the four materials and NO_2-N concentration in the cradle is plotted in Figure 4.157. There was no significant trend between these two parameters, because heterotrophic bacteria were not direct players in nitrification

The relationship of HPC and AOB proliferation in the biofilm is presented in Figure 4.158. Similar to the bulk water, the correlation was significant. The highest AOB growth occurred on the G material, which was up to 3.8×10^5 cfu/cm^2, while the AOB density on the other materials was less than 10^5 cfu/cm^2. The heterotrophic growth on the G material was also the highest.

The relationship between biofilm PEPA and biofilm AOB is presented in Figure 4.159. There was correlation between the two parameters. Biofilm PEPA varied from 10^8 to 10^9 cells/cm^2, while AOB varied from 10^2 to 10^5 and was approximately 4 to 6 orders of magnitude lower than PEPA.

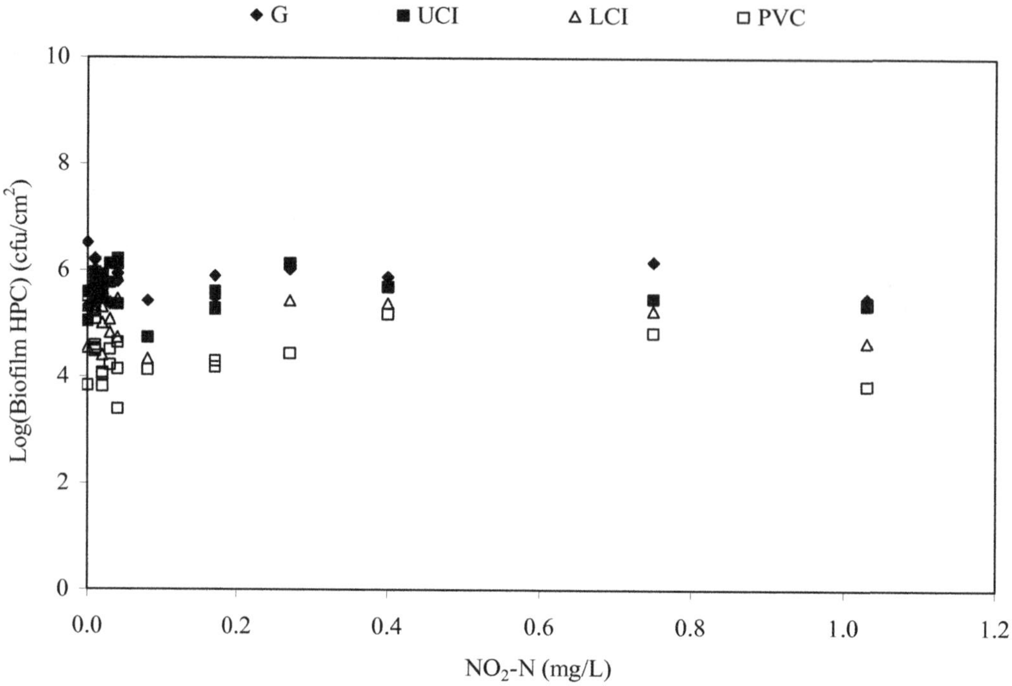

Figure 4.157 Relationship of NO_2-N and HPCs in the biofilm

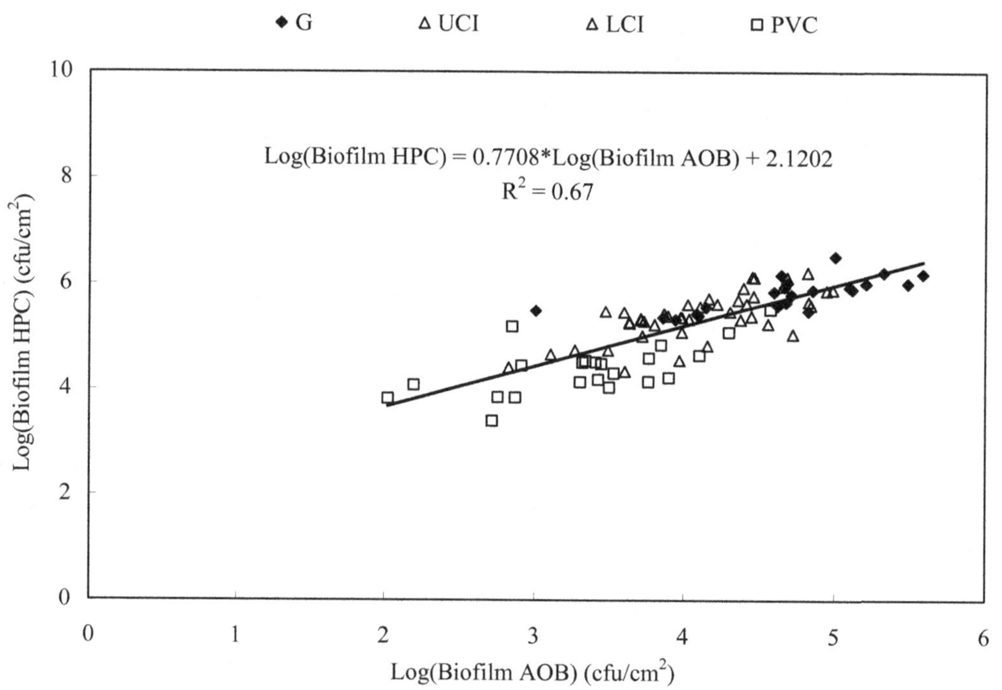

Figure 4.158 Relationship of HPC and AOB in the biofilm

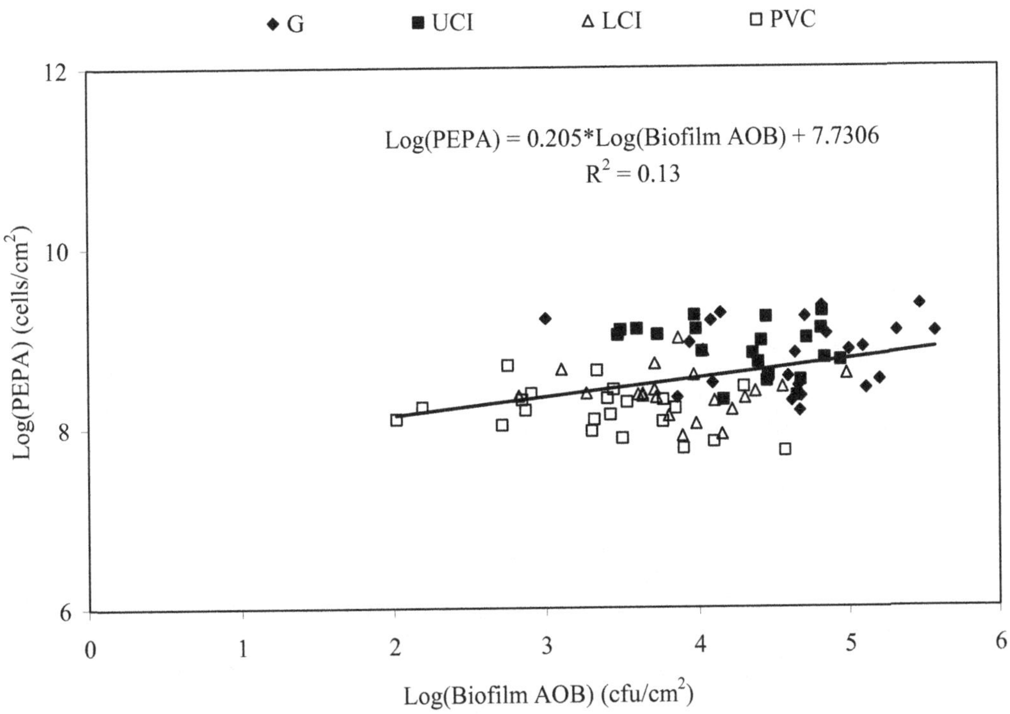

Figure 4.159 Relationship of PEPA and AOB in the biofilm

Although unlined coupons (UCI and G) have higher biofilm (PEPA) than lined coupons (LCI and PVC), there is not a linear trend relating PEPA and NO_2-N, as illustrated in Figure 4.160. However, the UCI and G PEPAs are generally higher than the UCI and PVC PEPAs for the same nitrite. PEPA measures the total bacterial inventory and since AOB are a small fraction of the total biofilm inventory a correlation with PEPA would be difficult to observe. Such relationships indicate that monitoring PEPA in a distribution system will be of little use to proactively or reactively control nitrification.

Figure 4.161 presents the relationship between the biofilm AOB proliferation and the influent AOC. Similar to the bulk water, the AOB proliferation in the biofilm was also independent of the organic carbon.

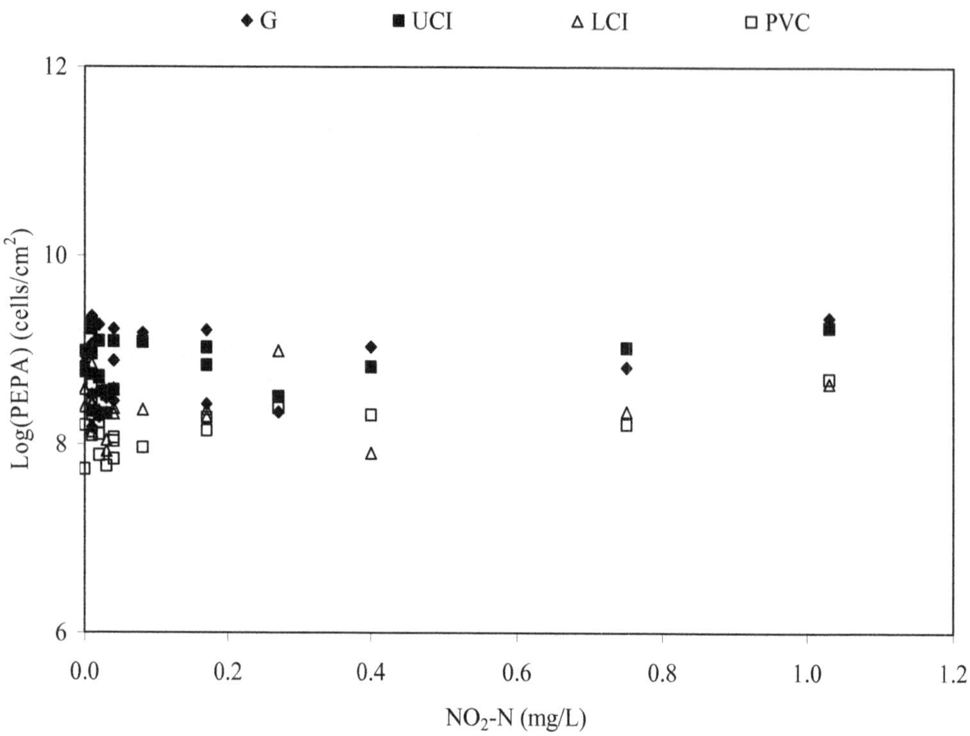

Figure 4.160 Relationship of NO_2-N and PEPA in the biofilm

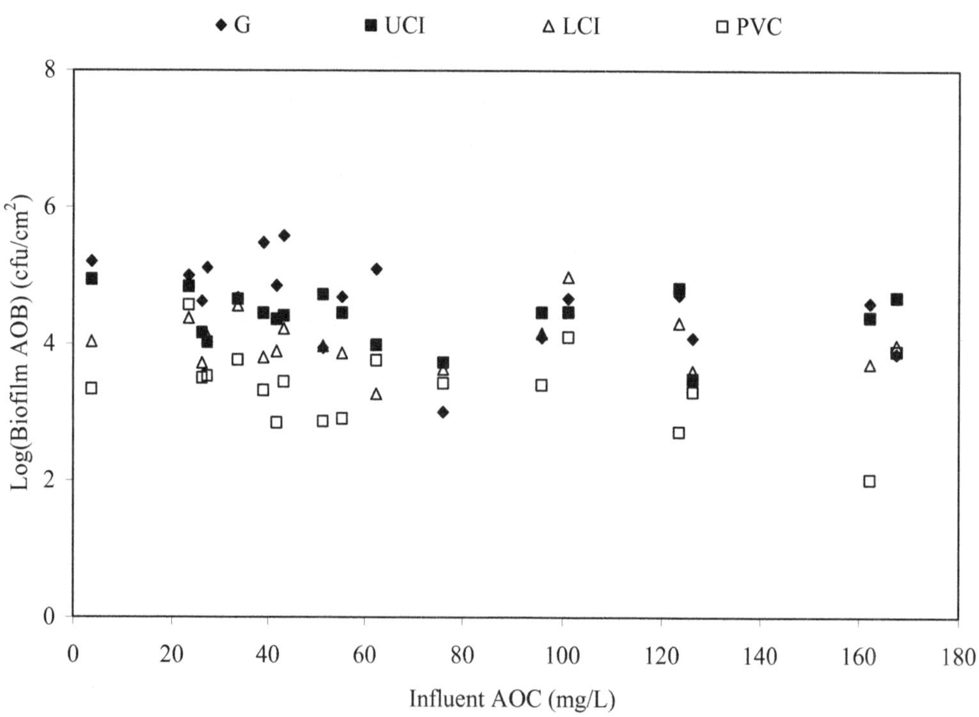

Figure 4.161 Relationship of AOC and AOB proliferation in the biofilm

Nitrification Modeling

Statistical Model

A statistical model was developed to determine the significance of independent water quality variables to nitrite formation in the PDSs. The data set from the nitrification episode and the controlled nitrification study was used for model development. The statistical assessment provides insight into mechanisms involving nitrification, which were used in development of the theoretical models and provided a means of predicting nitrification.

An Arrhenius relationship had been used previously to describe the effect of temperature effect on bulk nitrite and NO_X-N formation using data from the UCI PDS data that was collected during fall 2002 (09/16/2002 to 12/14/2002). All operating conditions were constant during this period except temperature. The relationships were shown previously in equations (4.95) and (4.96). Dummy variables were used to allow representation of all pipe material in development of the statistical models. All pipe material except PVC, ammonia-N, DO and chloramine residual were significant. No other inorganic, organic or biological parameters were significant. Due to denitrification during the controlled nitrification study, data from the UCI PDS was not included in the regression dataset. The statistical nitrification models are presented in equations (4.95) and (4.96).

$$NO_2\text{-}N = \frac{1.06^{(T-20)}\left(0.160 \times H + 0.186 \times UCI + 0.169 \times LCI + 0.037 \times G\right)}{(NH_3\text{-}N)^{-0.310}(DO)^{0.315}\left((NH_2Cl)^{0.722} + 0.188\right)} \quad (4.95)$$

$$NO_X\text{-}N = \frac{1.05^{(T-20)}\left(0.430 \times H + 0.776 \times UCI + 0.400 \times LCI + 0.103 \times G\right)}{(NH_3\text{-}N)^{-0.312}(DO)^{0.242}\left((NH_2Cl)^{0.585} + 0.295\right)} \quad (4.96)$$

where
NO_2-N = Nitrite N (mg/L as N)
NO_x-N = Nitrate + Nitrite N (mg/L as N)
T = Temperature (°C)
H = Hybrid PDS (0,1)
UCI = UCI PDS (0,1)
LCI = LCI PDS (0,1)
G = G PDS (0,1)
NH_3-N = Influent Ammonia N (mg/L as N)
DO = Effluent DO (mg/L as O_2)
NH_2Cl = Effluent Chloramine Residual (mg/L as Cl_2)

The models shown in equations (4.95) and (4.96) indicate that nitrification increases with increasing temperature and influent ammonia-N; and decreases with increasing effluent DO and residual. The negative relationship with effluent DO simply means that as more DO was consumed more nitrification occurred. If the DO term represented DO consumption instead of effluent concentration, the DO exponent would have been negative. The material coefficients was the largest for the UCI PDS, which indicates the UCI PDS environment was more favorable to nitrification than other materials. There was no material coefficient for PVC because the PVC PDS always maintained a strong residual, which prevented nitrification. Paired T-tests found no significant difference between the predicted and actual NO_2-N and NO_X-N, which are presented in Figure 4.162 and Figure 4.163.

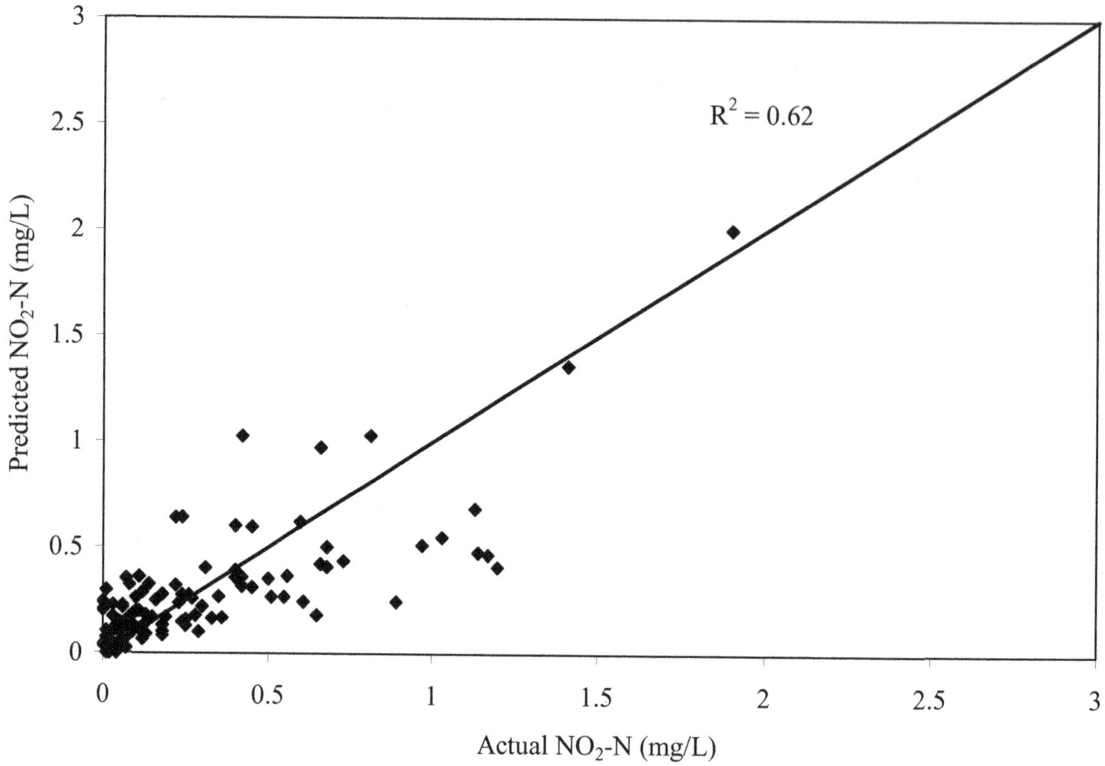

Figure 4.162 Statistically predicted NO_2-N versus actual NO_2-N

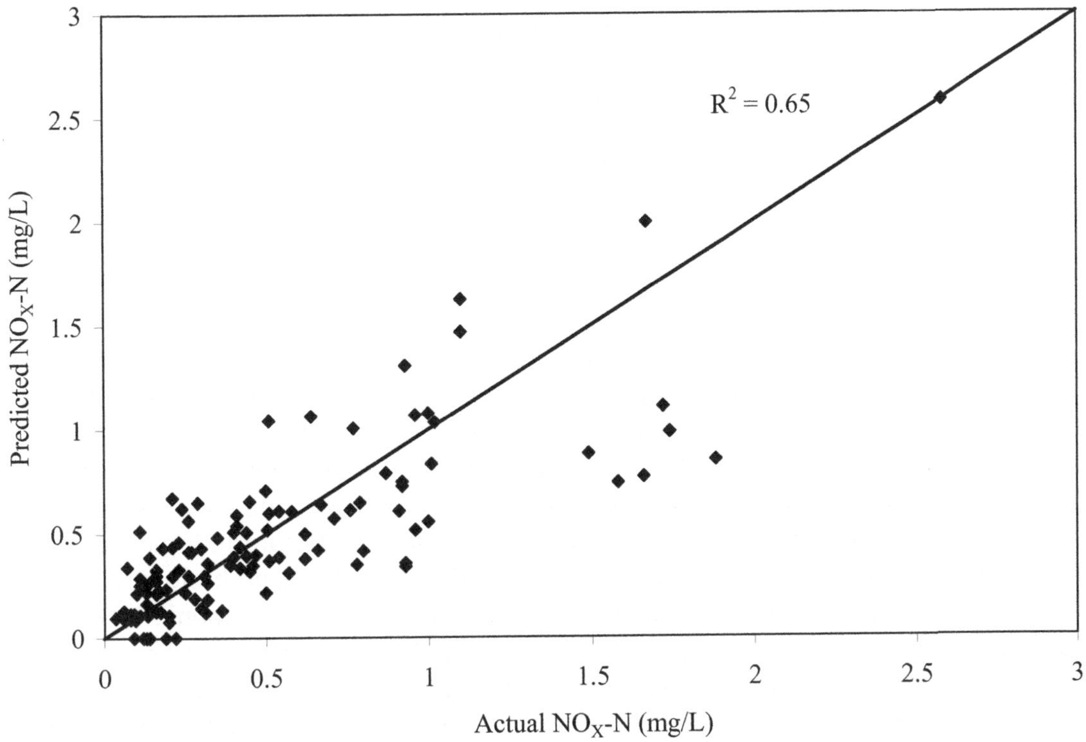

Figure 4.163 Statistically predicted NO_x-N versus NO_x-N

Monitoring

Recommended guidelines for monitoring biological stability and nitrification in a drinking water distribution system are shown in Table 4.68. The guidelines are distributed by location, temperature, pipe material and water quality. The intent of the matrix in Table 4.68 is to classify locations within the distribution system by difficulty of residual maintenance, and monitor those locations as frequently as necessary for residual maintenance.

There are three location groupings in Table 4.68. These groupings separate distribution system locations into near, average and extreme distances from points of residual addition. Near locations would be the same for all utilities, but medium and extreme locations would differ by utility. An extreme location in a densely populated or small plant service area may be five miles or less from the last point of residual addition. The same location for a geographically disperse service may be forty miles.

There are three temperature groups in Table 4.68. The effect of temperature on residual has been clearly identified. Monitoring frequencies must increase during warm weather conditions. Residual dissipates 20 to 30 % faster at 25°C than at 20°C and 20 to 30 % slower at 15°C than at 20°C. Hence, the monitoring frequencies need to be approximately twice as frequent in the extreme warm weather conditions as in the extreme cold weather conditions.

The effect of material has been clearly shown. There is more than an order of magnitude difference in the free chlorine and chloramine residual dissipation rates of LCI and PVC pipe relative to UCI and G pipes. Pipe material must be considered by utilities when monitoring frequencies are developed and those frequencies are specific to utilities.

The selection of recommended water quality parameters for monitoring was made after consideration of the data collected in this project. The recommended monitoring parameters are temperature, residual, DO, ammonia-N, HPCs, pH, nitrite-N and nitrate-N. All of these parameters except HPCs can be monitored in the field. HPCs were not mathematically significant to statistical or fundamental models developed to describe nitrification, but they were clearly associated with biostability and must be monitored. Inclusion of temperature, residual, ammonia-N and nitrite-N is obvious. Those parameters are all included in every model developed for nitrification. DO was statistically associated with nitrification and biological stability, but is a symptom not a cause of the same. DO at any one point will not benefit proactive actions in the distribution system, but a plot of DO concentration versus residence time and location will be extremely useful to identify trends and proactively mitigate nitrification or biological unstable conditions in the distribution system. pH is somewhat the same as DO in that trends are more useful that points, but pH is easily monitored with the same equipment that is used to monitor DO. Finally nitrate-N is included as a recommended monitoring parameter. There was no evidence collected in this work that suggested nitrification could not be adequately monitored by nitrite-N. Nitrates are a required regulatory monitoring parameter but the parameters shown in Table 4.68 are only for control of nitrification and biological stability and not for regulatory requirements. Conversion of all of the possible ammonia-N from 4 mg/L chloramines as Cl_2 to nitrite or nitrate would result in 0.8 mg/L of either NO_2-N or NO_3-N, which is 80 % of the 1 mg/L NO_2-N MCL or 8 % of the 10 mg/L NO_3-N MCL

Any monitoring plan for the distribution system must be practical and possible; and consider location, temperature and pipe material when frequencies when sampling points are defined. The information shown in Table 4.68 is a recommendation and should be adjusted by utilities to what is doable and meaningful.

Table 4.68
Recommended guidelines for monitoring for biological stability and nitrification

	Location Category 1											
	5-15°C				15-25°C				25-35°C			
	G	UCI	LCI	PVC	G	UCI	LCI	PVC	G	UCI	LCI	PVC
Temperature	Q	Q	Q	Q	BM	BM	BM	BM	BW	BW	M	M
Residual	Q	Q	Q	Q	BM	BM	BM	BM	BW	BW	M	M
DO	Q	Q	Q	Q	BM	BM	BM	BM	BW	BW	M	M
NH_3-N	Q	Q	Q	Q	BM	BM	BM	BM	BW	BW	M	M
HPC	Q	Q	Q	Q	BM	BM	BM	BM	BW	BW	M	M
pH	Q	Q	Q	Q	BM	BM	BM	BM	BW	BW	M	M
NO_2-N	Q	Q	Q	Q	BM	BM	BM	BM	BW	BW	M	M
NO_3-N	Q	Q	Q	Q	BM	BM	BM	BM	BW	BW	M	M
	Location Category 2											
	5-15°C				15-25°C				25-35°C			
	G	UCI	LCI	PVC	G	UCI	LCI	PVC	G	UCI	LCI	PVC
Temperature	BM	BM	Q	Q	BM	BM	BM	BM	W	W	M	M
Residual	BM	BM	Q	Q	BM	BM	BM	BM	W	W	M	M
DO	BM	BM	Q	Q	BM	BM	BM	BM	W	W	M	M
NH_3-N	BM	BM	Q	Q	BM	BM	BM	BM	W	W	M	M
HPC	BM	BM	Q	Q	BM	BM	BM	BM	W	W	M	M
pH	BM	BM	Q	Q	BM	BM	BM	BM	W	W	M	M
NO_2-N	BM	BM	Q	Q	BM	BM	BM	BM	W	W	M	M
NO_3-N	Q	Q	Q	Q	BM	BM	BM	BM	BW	BW	M	M
	Location Category 3											
	5-15°C				15-25°C				25-35°C			
	G	UCI	LCI	PVC	G	UCI	LCI	PVC	G	UCI	LCI	PVC
Temperature	M	M	BM	BM	BW	BW	BW	BW	3.5 D	3.5 D	BW	BW
Residual	M	M	BM	BM	BW	BW	BW	BW	3.5 D	3.5 D	BW	BW
DO	M	M	BM	BM	BW	BW	BW	BW	3.5 D	3.5 D	BW	BW
NH_3-N	M	M	BM	BM	BW	BW	BW	BW	3.5 D	3.5 D	BW	BW
HPC	M	M	BM	BM	BW	BW	BW	BW	3.5 D	3.5 D	BW	BW
pH	M	M	BM	BM	BW	BW	BW	BW	3.5 D	3.5 D	BW	BW
NO_2-N	M	M	BM	BM	BW	BW	BW	BW	3.5 D	3.5 D	BW	BW
NO3-N	M	M	BM	BM	BW	BW	BW	BW	3.5 D	3.5 D	BW	BW

Q=quarterly, BM=bimonthly, M=monthly, BW=every two weeks, W=weekly, D=Days

Summary

Obviously, utilities should monitor nitrates and nitrites in the distribution system. These observations showed that nitrite monitoring was an effective means of monitoring nitrification. Although there was no correlation between biofilms and nitrification, biofilms were correlated to HPCs, which were correlated to AOB. There is no more important water quality parameter to monitor in drinking water distribution systems than residual. Observation in this investigation clearly showed that nitrification was initiated once residual was lost. Additionally, once nitrification had begun, it could not be stopped simply by regaining strong chloramine residuals. Hence, these observations indicate that utilities should proactively anticipate where residuals will be lost in the distribution system in association with blend and seasonal water quality effects. Once anticipated, utilities should proactively take actions to prevent residual loss. Although utility personnel are aware of many problems, such actions are difficult to adequately achieve without detailed computer models describing distribution system residence times, flows, pressure, water balance and water quality. There are many hydraulic models that are utilized to monitor flow and pressure in drinking water distribution systems, and some of these use linear average methods to determine water quality. However there are few models, if any, implemented in utility distribution systems that predict chlorine residuals as a function of hydraulics, material, geometry and water chemistry. Such models could significantly benefit water quality in distribution systems.

Episode

- Biochemical nitrification is a complex microbiological process. Nitrites, nitrates, pH, DO and ammonia-N varied during the nitrification episode in a systematic manner with residence time in the pilot distribution systems.
- Biochemical nitrification during the nitrification episode was associated with loss of residual, pH and initially increasing ammonia-N. Nitrification is most likely to occur at elevated temperatures common to spring and summer.
- The loss of residual was strongly associated with pipe material. When the bulk water interfaces with relatively inert pipe material (such as the inner walls of PVC or LCI pipes) the residual loss is significantly reduced and nitrification can be mitigated. No nitrification was observed in the pure material PVC PDS because the pipe wall had practically no residual demand and all bacterial growth was held in check because of the strong residual. The growth of nitrifiers and other aerobic microorganisms was uncontrolled at higher temperatures in unlined metal pipes (G and UCI) because of total residual loss.
- During the nitrification episode, the ammonia N for either biological nitrification or eventual denitrification came from ammonia-N or chloramine-N associated with the decay of chloramines.
- DO can be a limiting factor for nitrifiers growth. However no evidence was gathered during the episode that limiting DO would benefit distribution system water quality. Whereas nitrifiers are aerobic and can not function as facultative or anaerobic bacteria, the resulting growth of anaerobic bacteria and the associated water quality problems would be worse than nitrification in most cases.
- No significant alkalinity variation was observed in the pilot distribution systems during biochemical nitrification. Monitoring alkalinity as an indicator of nitrification is this work was ineffective. An inorganic carbon source is required for biological nitrification, in the form of bicarbonate, other dissolved carbonate/carbonic acid species, or dissolved carbon dioxide. It is possible that alkalinity consumption was

not observed because consumed bicarbonate was replaced by dissolution of calcium carbonate solids on the pipe surface.

- Once nitrification occurred, increasing the influent chloramine residual from 4 to 4.5 mg/L as Cl_2 was ineffective for reducing nitrification as the additional 0.5 mg/L of residual was vastly inadequate to maintain residual. Residual maintenance throughout the distribution system is essential to prevent nitrification, but restoration of a chloramine residual cannot completely stop nitrification once it becomes established.
- Once nitrification occurred, increasing the Cl_2:N from 4:1 to 5:1 was ineffective for reducing nitrification. Since the chloramine residual was completely dissipated, increasing the Cl_2:NH_3 ratio from 4:1 to 5:1 reduced available NH_3-N from 1 to 0.8 mg/L , which had little effect on reducing nitrification.
- Implementing a free chlorine burn at 5 mg/L as Cl_2 for one week stopped nitrification. If nitrification occurs, free chlorine burns are a tool that can be used by a utlity to stop nitrification.
- Reducing the hydraulic detention time from 5 to 2 days was effective for reducing but did not eliminate nitrification. The reduced HRT restored the effluent chloramine residual, which did effectively reduce nitrification during the nitrification episode and can be a primary tool for controlling biochemical nitrification because of residual maintenance. Reducing distribution system residence time will have no effect on biological nitrification unless a weak chloramine residual (< 1 mg/L) is increased to a strong chloramine residual (>2 mg/L). Utilities can flush to restore residual, however other means of controlling nitrification are required given a poor distribution system design or reaches in the distribution system consisting of unlined metal pipes. Other means would be consideration of water age during distribution system design and construction, replacing unlined metal pipes in distributions, rerouting existing distribution systems to reduce residence time (dead ends) and free chlorine burns.

Controlled Study

- Primary water quality parameters that provide a systems approach to monitoring nitrification are temperature, residual, DO, HPCs, nitrite and nitrate in the bulk water.
  - The potential for nitrification increases as temperature increases because of decreased residual and microbiological growth. Nitrite production increased exponentially with temperature, such that nitrite production in the bulk water during summer (30°C) is almost five fold more than nitrite production in winter (15°C).
  - Nitrification was not observed until residual was completely lost. Bulk nitrites and bulk AOBs were negatively correlated to chloramine residual. Bulk AOBs were positively correlated to nitrites and bulk HPCs. The relationships among nitrites, AOBs and HPCs found in this work indicate nitrification was biologically driven in the bulk water stream. Maintenance of residuals above 1.0 mg/L as Cl_2 is critical for the prevention of nitrification.
  - Residual DO in the distribution decreased and was negatively correlated to nitrite production during nitrification. DO was also negatively correlated to HPCs.

Decreasing DO can occur with nitrification due to increased growth of non-nitrifying bacteria. However, there is no circumstance when DO should not be monitored and assessed for biologically or chemically driven adverse water quality effects in drinking water distribution systems.

  - HPCs vary significantly in drinking water distribution systems but were generally positively correlated to key water quality indicators of biological nitrification. HPCs typically increased with increasing AOBs and decreasing DO, which corresponded to increased nitrites during biological nitrification.

- Biofilms and nitrification control
  - Biofilms were correlated to pipe material in that unlined metal pipes (UCI and G) had higher PEPA, HPC and AOB cell densities than lined pipes (PVC and LCI), which means that biofilm densities increased as residual demand of the pipe wall or the biofilm increases and that biofilm densities varied with chloramine residual. Replacing distribution system pipes with high chlorine demand is the same as replacing distribution system pipes with high biofilm density potential and will reduce the potential for nitrification and enhance nitrification control. This is an acceptable but expensive method for biofilm control that is available to utilities.
  - Although nitrite production was correlated to biofilm AOBs and biofilm HPCs were correlated to biofilm AOBs, general mass balances of AOB in the bulk and biofilm indicated AOB in biofilms were less than AOB in the bulk, but that either location could provide AOB to cause nitrification. Correlations of biofilm AOB to bulk residual, biofilm HPCs to bulk nitrites, PEPA to AOB biofilm, PEPA to AOB bulk, and PEPA to nitrites were weak. Typically biofilm correlations to bulk nitrification were weak and did not provide any evidence that regular biofilm monitoring would benefit utility maintenance of drinking water distribution systems. Nevertheless, biofilms can serve as a reservoir for nitrifying bacteria that are available to inoculate the bulk liquid given the right conditions (available DO and ammonia, low residual). Control of biofilms is essential and can only be achieved by maintaining a strong residual in these waters.

- Nitrite was positively correlated to the sum of nitrites and nitrates or only nitrates, and can be used to effectively monitor the NO_x species in drinking water distribution systems. Nitrates were approximately thirty-seven percent of nitrite concentrations.

- Nitrification increased as ammonia-N increased. Ammonia-N in the finished water and in the distribution system should be minimized to reduce biological nitrification. Raw water contributions of ammonia-N should be assessed and specific unit operation for ammonia removal should be considered in the treatment process.

- Nitrification decreased as the Cl_2:NH_3 ratio decreased. A Cl_2:NH_3 ratio of 5:1 will produce very little ammonia-N. Cl_2:NH_3 ratios higher than 5:1 will produce dichloramines and trichloramines, which may be more bactericidal than monochloramines, which will produce taste and odors. Cl_2:NH_3 ratios lower than 5:1 will produce ammonia-N and increases the potential for nitrification. Low Cl_2:NH_3 ratios are seemingly unjustified given the control systems that are available for controlling chlorine and ammonia addition in water treatment.

- Nitrification was reduced as flushing frequencies and flush strength (chloramine concentration) increased. Increasing chloramine concentrations in the flush from 4 to

10 mg/L as Cl_2 reduced nitrification as measured by nitrite production more than increasing flush frequency from bimonthly to weekly events in this work. Flushing with a higher strength residual than is commonly used in the distribution system may not be available, whereas flushing at a higher frequency is available to utilities to control nitrification. However, all evidence gained in this work indicates free chlorine burns are necessary to stop biological nitrification in drinking water distribution systems once a significant episode has occured.

- Any utility using chloramines must aggressively manage distribution system water quality to control nitrification. High Cl_2:NH_3 ratios (5:1), residual maintenance, distribution system flushing and water quality monitoring are essential to nitrification control. Nitrification will be more difficult to control as water temperature increases in distribution systems using chloramines for residual maintenance.

Modeling

- A statistical model was developed from the entire nitrification data base that related NO_2-N and NO_X-N formation by pipe material to increasing temperature and influent ammonia-N, and decreasing effluent DO and chloramine residual. No other inorganic, organic or biological parameters were statistically significant to NO_2-N and NO_X-N formation.
- Guidelines for monitoring biological stability and nitrification in drinking water distribution systems were developed. The guidelines separated monitoring frequencies by location, temperature and parameters. The most intensely monitored conditions occurred during high temperature periods at locations involving unlined metal pipes and high residence times in the distribution systems. The classification of monitoring points for biological assessment in a distribution are specific to a given utility and should be determined after assessing water quality data from distribution system locations determined from existing knowledge and utilization of a computer model to classify residence times.

MONOCHLORAMINES AND FREE CHLORINE

Water Quality

Chlorine Residual

Figure 4.164 shows the comparison of the average change on chlorine residual for free chlorine and chloramine for the different water sources. During the disinfectant comparison study it was found that chloramine decayed faster than free chlorine. The difference in residual consumption between free chlorine and chloramine consumption was 0.2 mg/L as Cl_2 for almost all the waters with the exception of SW which had a difference of 0.9 mg/L as Cl_2. A paired two-sample t-test for means was conducted for each water type and it showed that disinfectant residual consumption in the chloraminated lines was larger than the consumption in the corresponding chlorinated lines under the conditions of the study. The paired t-test showed that the difference was significant at the 95% confidence level.

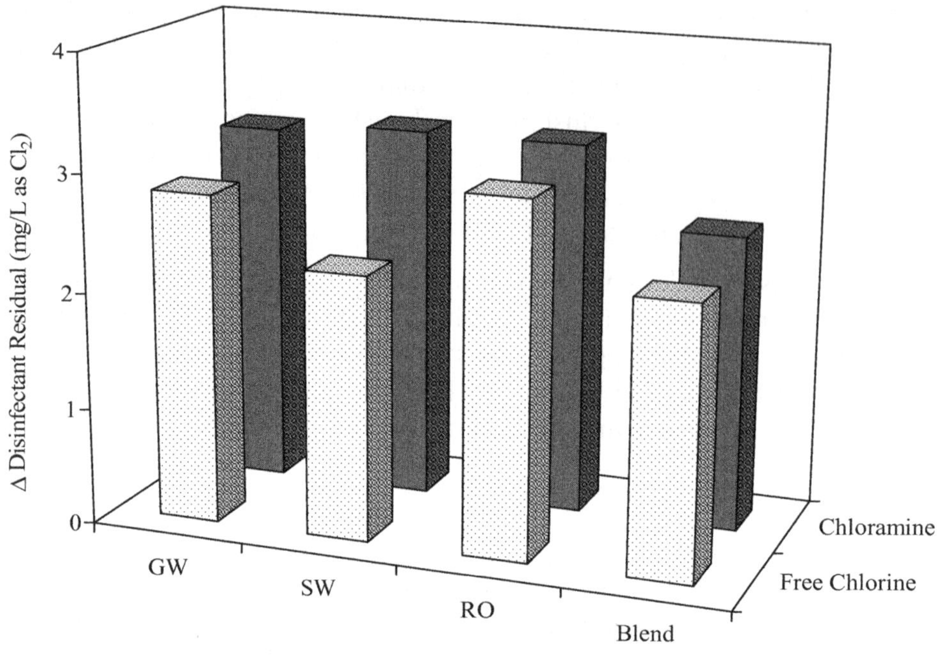

Figure 4.164 Average disinfectant residual change in PDS by water source and disinfectant

Comparison between the source water indicates that between the waters using free chlorine, GW and RO lost the residual faster than SW, and the blend; in the case of waters using chloramine GW, RO and SW lost the residual faster than the blend.

The average influent and effluent free chlorine for PDS 1 to 4 and chloramine for PDS 5 to 8 are shown in Figure 4.165. The average influent concentration was 0.4 mg/L greater for the chlorinated waters than for the chloraminated waters. The average effluent residual for the chlorinated lines was greater (1.3 mg/L as Cl_2) than the effluent residual for the chloraminated lines (0.6 mg/L as Cl_2).

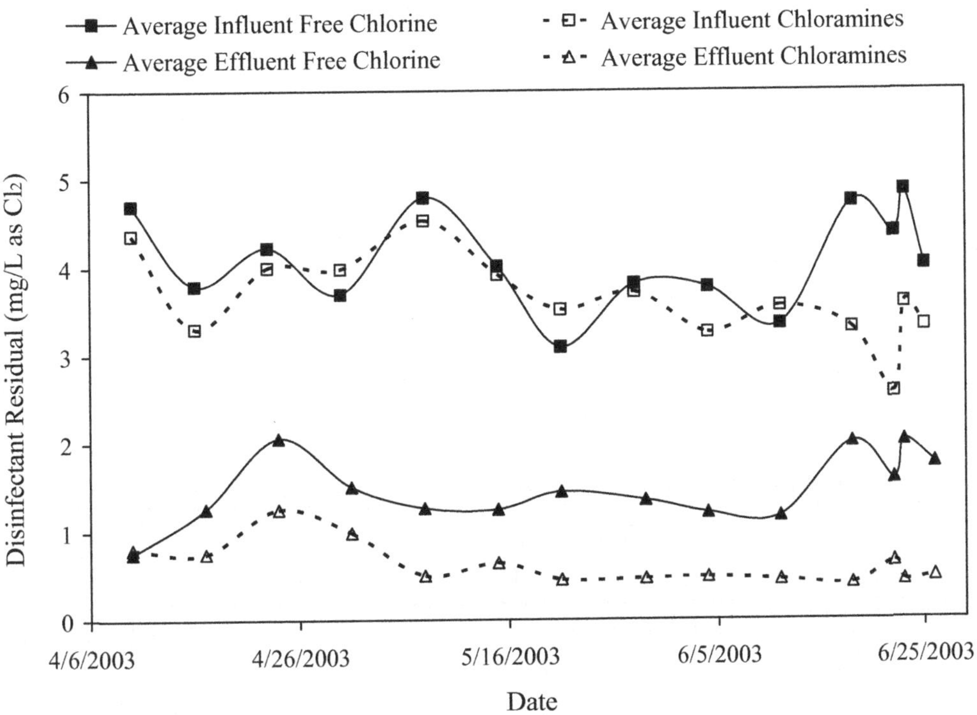

Figure 4.165 Average influent and effluent disinfectant residual for PDS 1 to 8 during disinfectant study

A model was developed for free chlorine and chloramine dissipation by nonlinear regression using the form shown in equation (4.97).

$$Cl_{2(t)} = Cl_{2(0)} \times \exp\left(-\left(K_b \times UV254 + \frac{K_W}{D_p}\right) \times A^{(T-20)} \times t\right) \tag{4.97}$$

where K_b	=	bulk decay constant (cm/hr)
K_W	=	wall decay constant (in/hr)
$Cl_{2(t)}$	=	chlorine concentration at time t (mg/L as Cl_2)
$Cl_{2(0)}$	=	initial chlorine concentration (mg/L as Cl_2)
A	=	temperature correction coefficient
T	=	temperature in degrees C
t	=	time expressed in hours
Dp	=	pipe diameter (inches)

The values obtained for the different variables are presented in Table 4.69 for free chlorine and chloramines. Paired comparison of K_b obtained for each material for free chlorine and chloramine decay shows that the bulk free chlorine dissipation is higher than that of chloramine. As expected, this result indicates that free chlorine is more reactive than chloramine in the bulk liquid.

The values for the wall decay constants were higher for chloramines compared to free chlorine for all the pipe material utilized in the distribution system. The K_W for PVC and LCI pipes are low and similar for free chlorine which indicates that this disinfectant reacts in a similar way with the material of those pipes. In the case of chloramines the value of K_W is also low for PVC and LCI with a slightly larger K_W for LCI. For the metallic pipes, K_W for UCI are larger than the values obtained for G for both free chlorine and chloramines.

The wall decay rate constant is dependent on the diameter of the pipe (ratio volume/wet surface area). The influence of the wall decay decreases with the increase of the pipe diameter. For pipes with a diameter of 6 inches or more, free chlorine residual would dissipate faster than chloramine since free chlorine has a larger rate of decay in the bulk liquid. For small pipe diameters (less than 6 inches), chloramine should be expected to decay faster than free chlorine.

Table 4.69
Free chlorine and chloramine dissipation model coefficients and R^2

Material	Free Chlorine			chloramine		
	Parameter	Value	R^2	Parameter	Value	R^2
PVC	K_b	0.048	0.97	K_b	0.020	0.90
	K_W	0.007		K_W	0.007	
	A	1.15		A	1.16	
LCI	K_b	0.084	0.93	K_b	0.017	0.90
	K_W	0.007		K_W	0.011	
	A	1.18		A	1.16	
UCI	K_b	0.172	0.97	K_b	0.147	0.87
	K_W	0.063		K_W	0.084	
	A	1.13		A	1.11	
G	K_b	0.614	0.93	K_b	0.319	0.87
	K_W	0.054		K_W	0.073	
	A	1.04		A	1.12	

Figure 4.166 and Figure 4.167 illustrate predictions of free chlorine and chloramine residual for PVC and UCI pipes of different diameters. The predictions were developed for water with an initial chlorine residual of 4 mg/L, 2 mg/L of TOC, and a temperature of 25°C. Pipe diameters of 6 inches and 48 inches are presented. From Figure 4.166 it can be seen that in PVC, chloramine lasts longer and maintains a higher concentration than free chlorine for all pipe diameters. Figure 4.167 shows that for unlined cast-iron pipes, chloramines are predicted to be as stable as free chlorine in pipes of 6 inches in diameter(the symbols overlap). For larger UCI pipe diameters, chloramines provides better residual stability than free chlorine.

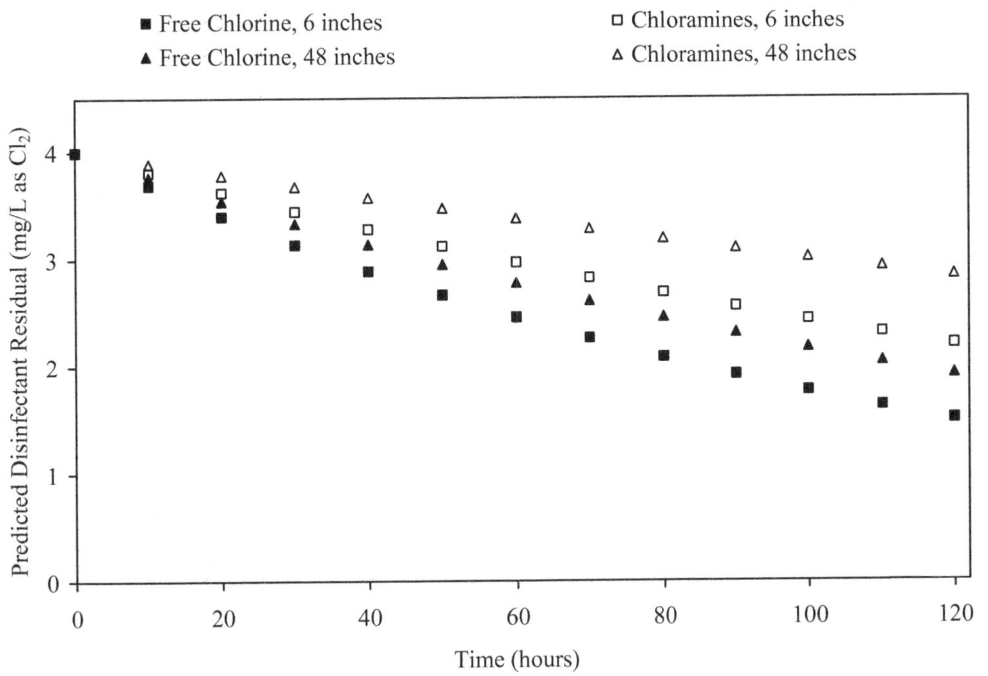

Figure 4.166 Free chlorine and chloramine model predictions for 6 inches and 48 inches PVC pipes

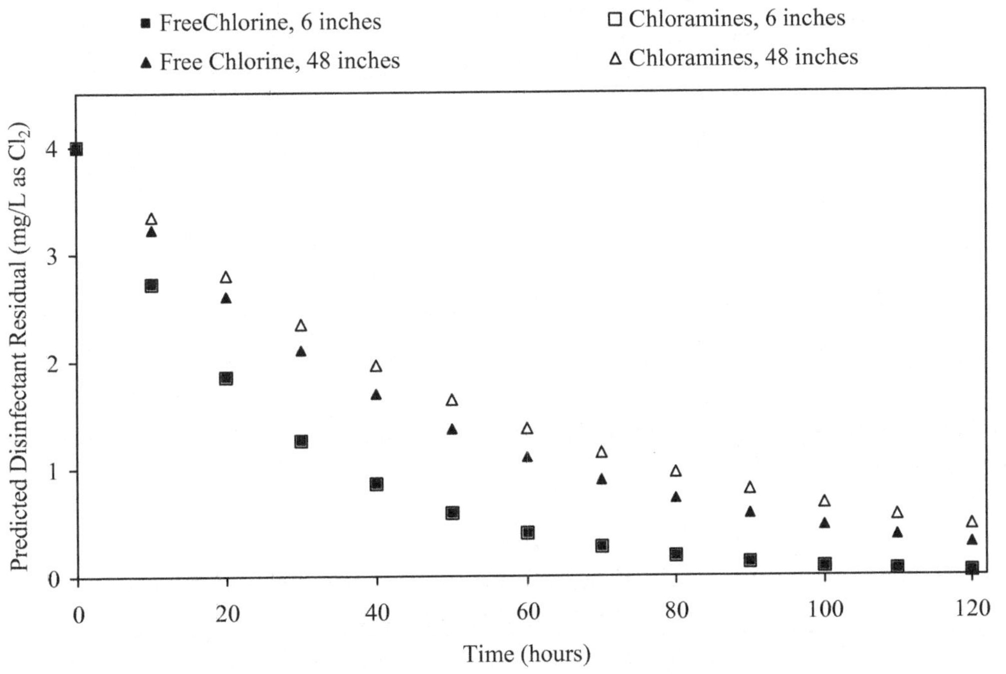

Figure 4.167 Free chlorine and chloramine model predictions for 6 inches and 48 inches UCI pipes

Comparison of actual free chlorine and chloramine decay. The average measured free chlorine and chloramine residual concentrations at each sample port in the PDSs are shown in Figure 4.168. The average rates of decay of free chlorine and chloramines in the different types of pipes are presented in Table 4.70. The rate of chloramine residual loss is subtracted from the rate of free chlorine residual loss in the last column of Table 4.70. A negative number for this calculation means the rate of chloramines loss is less than the rate of free chlorine loss, a positive number means the reverse. There is almost no difference for PVC, LCI, and UCI. However, the rate of chloramine decay in G pipes was 0.06 mg/L-h faster than for free chlorine.

These results show that chloramines were as stable as free chlorine in 6-inch PVC, LCI and UCI pipes and was less stable only in 2-inch galvanized pipes. Since all the lines used during the study were hybrid lines which had 40 ft of G at the end (corresponding to an HRT of 6 hours), that would explain the difference in effluent residual concentration shown in Figure 4.165.

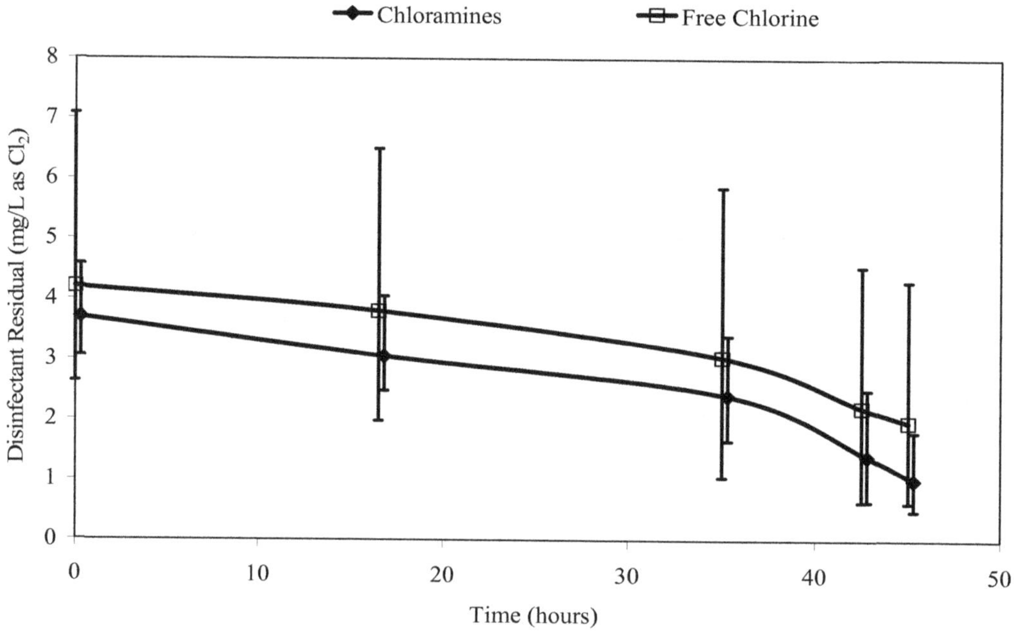

Note: The vertical lines represent the maximum and minimum values for each data point

Figure 4.168 Actual average free chlorine and chloramine residual versus time during comparison of Cl_2 and NH_2Cl

Table 4.70
Assessment of average linear rate of residual loss by material during comparison study of free chlorine and chloramines

Pipe Material	Chloramines Average Decay Rate mg/L-h as NH_2Cl	Free Chlorine Average Decay Rate	Difference NH_2Cl - Cl_2
PVC	0.04	0.03	0.01
LCI	0.03	0.04	-0.01
UCI	0.13	0.11	0.02
G	0.16	0.10	0.06

Temperature

Figure 4.169 presents the temperature trend for the free chlorine and chloramine comparison study. There was no significant difference in temperature among the 8 PDSs. There was also no significant variation in temperature between the influent and effluent of each PDS. The study was started on 04/05/2003 and ended on 07/04/2003. During this period, the temperature increased gradually from 18°C to 32°C and remained above 25°C during the last two months of the study.

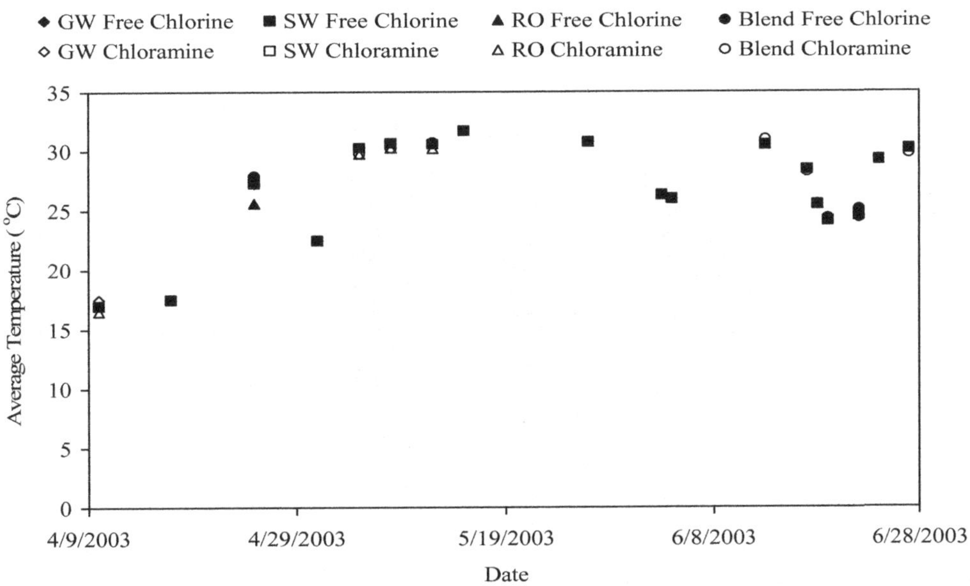

Figure 4.169 Average temperature by source water as a function of time during disinfectant study

DO

The average consumption of DO is presented in Figure 4.170. Greater consumption of DO was observed in the chloraminated PDS, compared to the chlorinated ones. The waters using free chlorine exhibited an average DO reduction of around 1 mg/L while the DO depletion for waters using chloramine averaged between 1.8 and 4 mg/L. The consumption of DO was especially high for the chloraminated RO and SW sources. Paired t-test conducted at the 95% confidence level showed that DO. consumption was larger for total chlorinated waters than for their free chlorinated counterparts.

The feed water sources were not oxygen limited since the average DO in the influent feed was about 7.9 mg/L as O_2. Figure 4.171 illustrates the change in DO consumption with time for each PDS during the disinfectant comparison study. For surface and RO waters using chloramine there is an ncrease in the consumption of DO ranging from 2 mg/L at the beginning of the study to around 4 mg/L at the end. This was coincident with the increasing trend of temperature. GW and the blend suffered a smaller increase going from 2 to 3 mg/L for GW and from 1 to 2 mg/l for the blend. The data points for RO and SW using free chlorine are also included as an example; the consumption of DO for both waters remains constant around 1 mg/L during the three months of duration of the study. It was observed that individual observations of

DO consumption of chloraminated PDSs was 0.5 to 3.5 mg O_2/L greater than their chlorinated counterparts, depending on the water source.

The consumption of DO may be explained by biological activity and iron release, as greater DO consumption was coincident with higher effluent HPCs, turbidity, apparent color and total iron release.

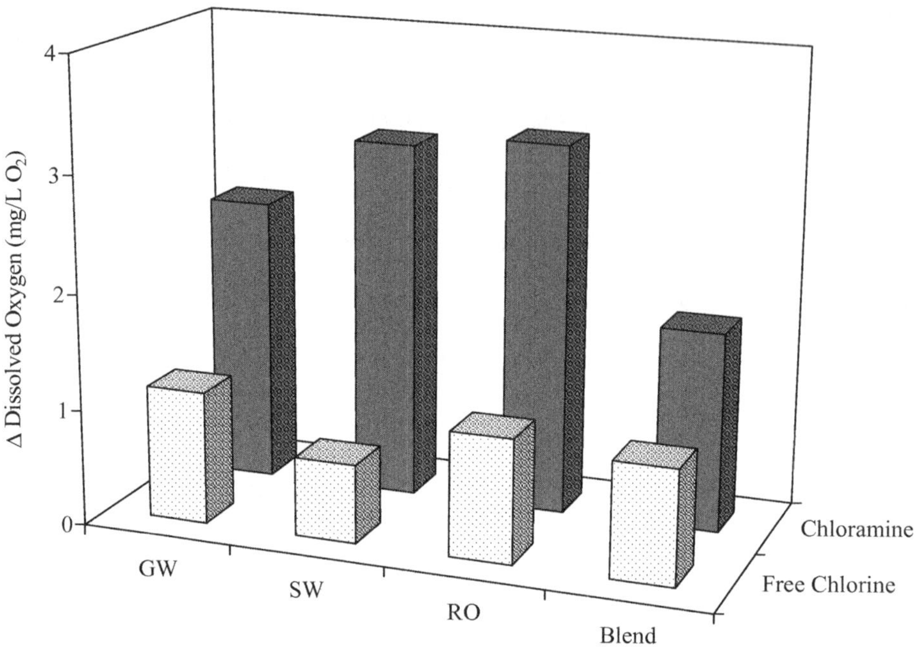

Figure 4.170 Average DO change by water source and residual during disinfectant study

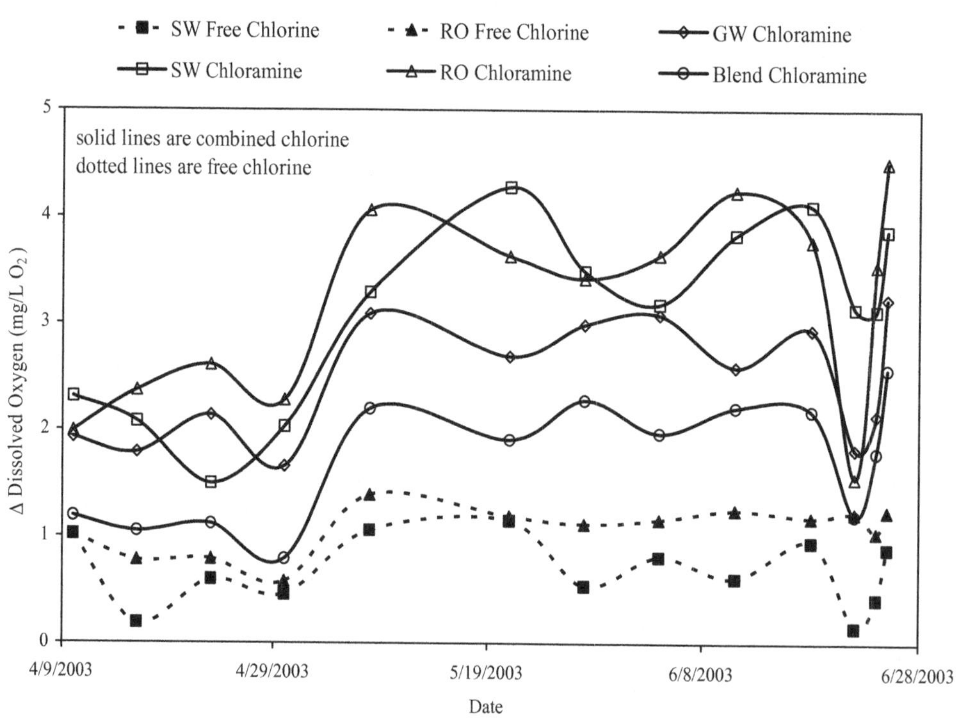

Figure 4.171 DO demand by disinfectant type and water source during disinfectant study

Ammonia

There was no ammonia present in the free chlorinated waters. However, in chloraminated waters ammonia is released due to the dissipation of chloramines. Free ammonia can be utilized by AOBs as food source facilitating the occurrence of nitrification. Figure 4.172 presents the average delta ammonia concentration for each PDS.

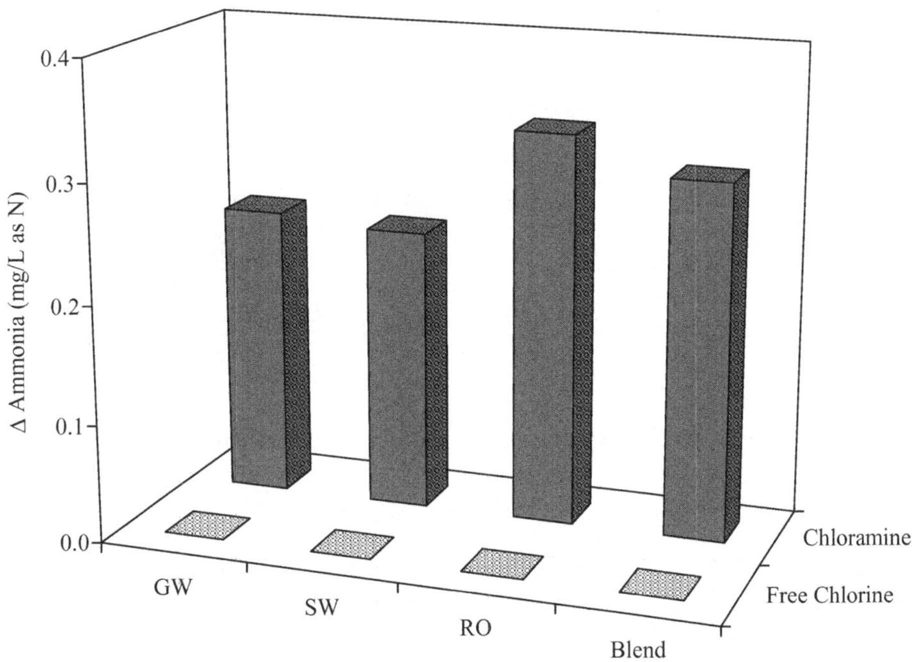

Figure 4.172 Average ammonia change by water source and residual type during disinfectant study

Turbidity

There was a noticeable increase of turbidity between influent and effluent for the PDSs fed with chloraminated water relative to chlorinated PDSs. Figure 4.173 shows the average of delta turbidity (difference between effluent and influent turbidity), sorted by water source and type of disinfectant. The highest increase in average turbidity was observed for SW and RO ($\approx$ 4 NTU) using monochloramines.

As shown in Figure 4.174 almost no changes in delta turbidity occurred during flow through the PDSs except in the chloraminated PDS transporting 100 % RO and the PDS transporting 100 % SW. Delta turbidity continually increased from slightly more than 1 NTU to almost 6 NTU during the three-month study for the twelve weekly observations of delta turbidity. Similar results were obtained for apparent color, which was correlated with turbidity as shown in Figure 4.7.

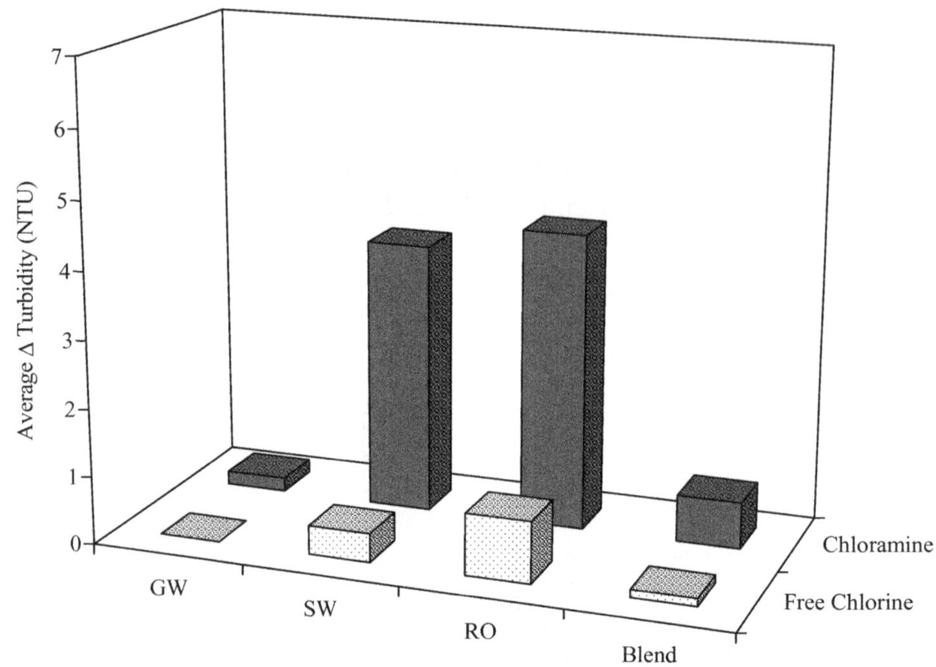

Figure 4.173 Average delta turbidity by water source and residual during disinfectant study

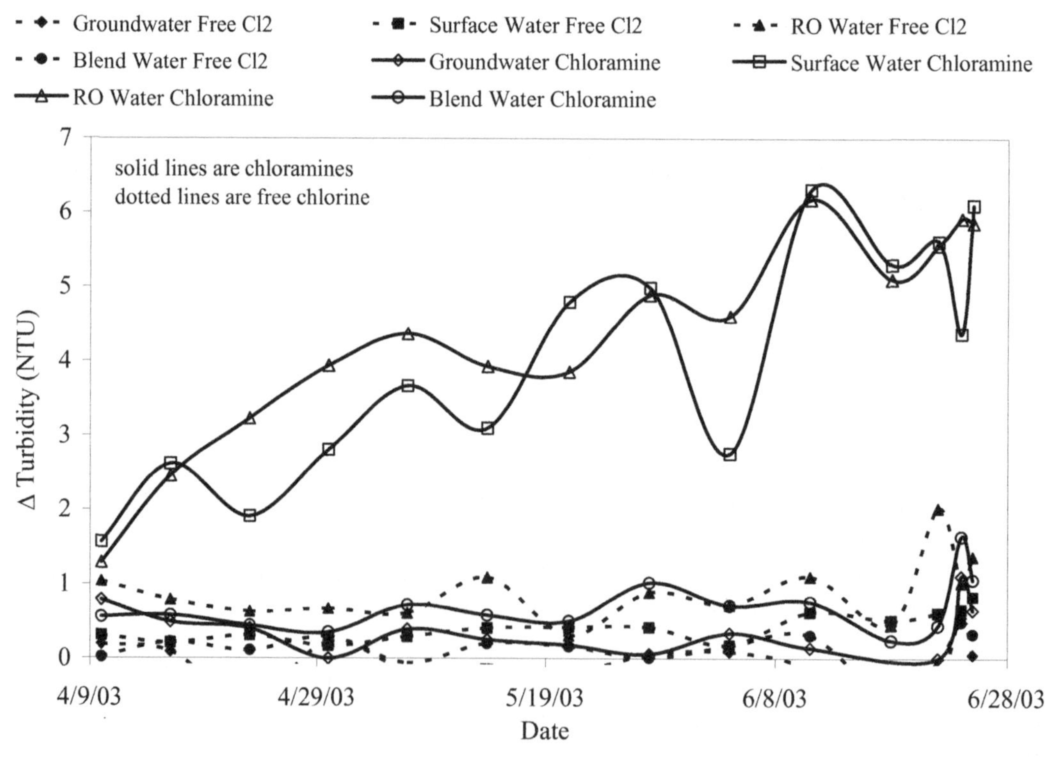

Figure 4.174 Delta turbidity by disinfectant and water source during disinfectant study

The higher turbidity in the effluent from RO and SW sources could be due to release of biofilm or chemical scale on the wall of UCI and G portions of the PDSs since RO and SW had low alkalinity ($\approx$ 50 mg/L as $CaCO_3$).

Total Iron

Similar to the trend of the turbidity and apparent color release, the results for total iron showed that total iron concentrations were higher for the chloraminated PDS than for the chlorinated ones. Chlorinated and chloraminated PDS fed with the most corrosive waters, RO and SW, had higher average total iron release than GW and the blend as presented in Figure 4.175.

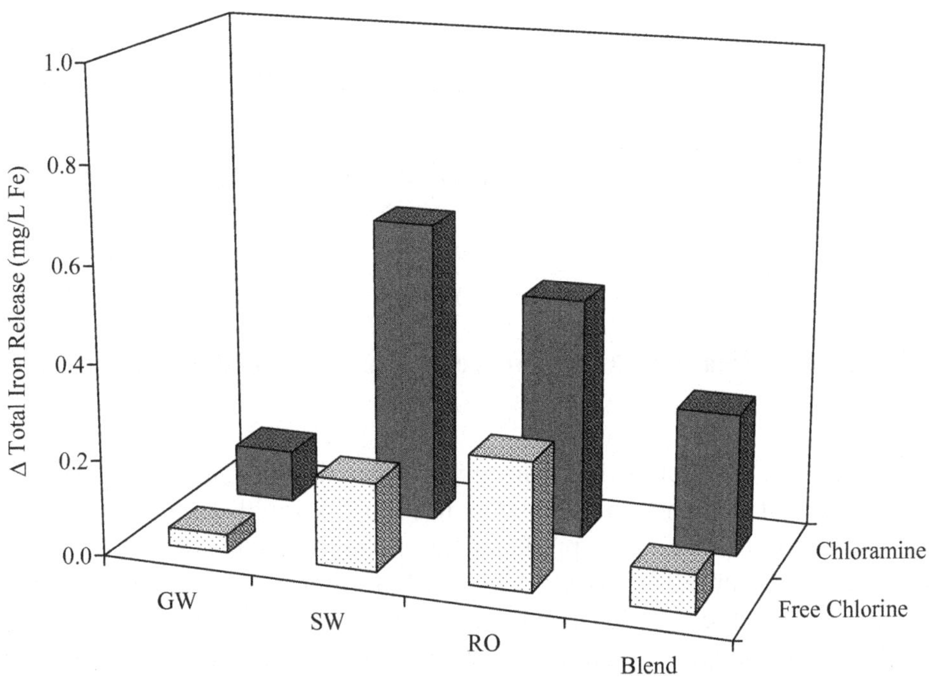

Figure 4.175 Average total iron release by water source and residual type during disinfectant study

SW and RO waters using chloramine had total iron concentration around 0.6 mg/L. in contrast, the same water when using free chlorine had total iron concentration of around 0.25 mg/L The difference in total iron release between chlorinated and chloraminated waters was verified using statistical analysis. A paired t-test for two samples means was conducted at the 95% confidence interval. The t-test found that the average total iron release was higher for the chloraminated waters than for the chlorinated waters.

Figure 4.176 shows delta total iron during the duration of the disinfectant comparison study for the eight PDS. Total iron release was stable and less than 0.2 mg/L for all PDSs using a free chlorine residual as shown by the dotted lines in Figure 4.176. Total iron release gradually increased to 0.70 mg/L in PDS 6 and 7, which received chloraminated RO and SW. The same increase was not observed in PDS 5 and 8, which were a chloraminated GW and Blend.

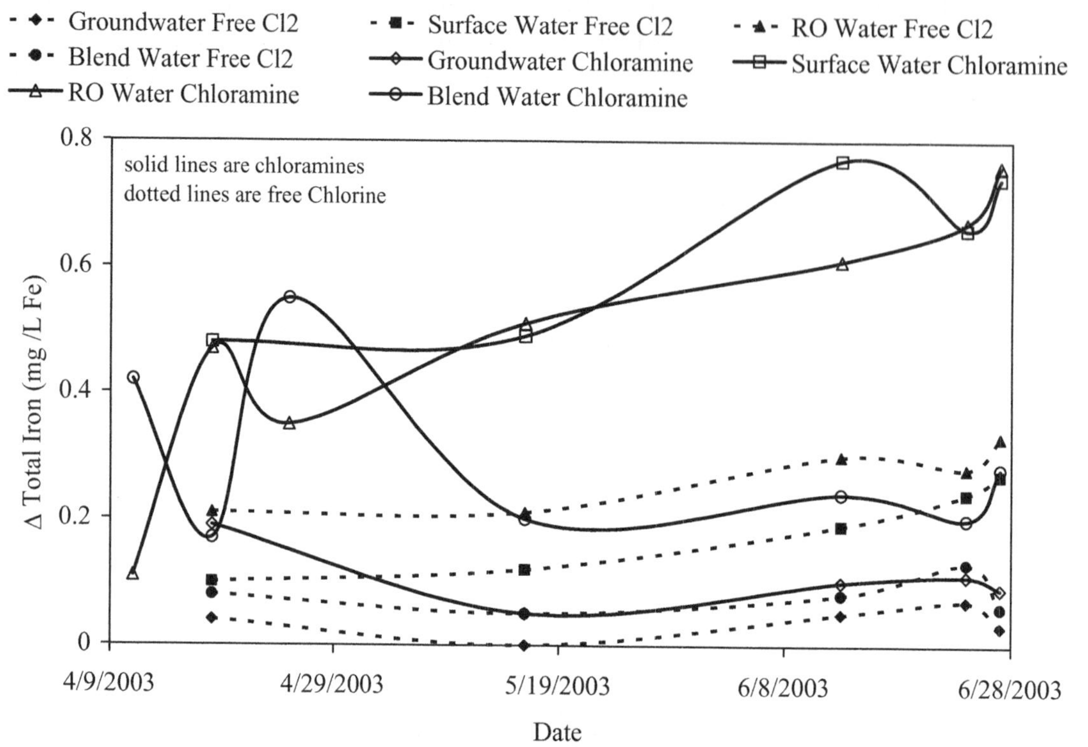

Figure 4.176 Total iron release by water source and residual type during disinfectant study

Iron release was predicted for varying source water quality by incorporating the relationship between total iron and apparent color in Figure 4.8 with the model for apparent color release shown in equation (4.3). The predicted order and concentrations of total iron release were GW<blend<RO<SW and 0.03, 0.09, 0.25 and 0.38 mg/L as Fe, respectively. The model predictions were similar to the average iron release for chloraminated GW (0.1 mg Fe/L) and the blend (0.2 mg Fe/L) but under predicted iron release for RO (0.5 mg Fe/L) and SW (0.63 mg Fe/L). There was no model development for iron release in a free chlorine environment. However, total iron release in the chlorinated lines was lower than the predicted total iron release in the chloraminated PDSs. The release of total iron came primarily from the UCI pipe, which had less residual and more ammonia relative to the UCI pipe in the free chlorine environment. Chloramine may increase iron release in the iron-based pipes (UCI and G) due to formation of iron complexes and direct reaction with the pipe wall, or associated biological growth or reactions. Previous studies (Vikesland and Valentine 2002b) had shown that chloramine reacts with iron at the pipe wall to form insoluble Fe (III), ammonia and chlorides as shown in equation (4.98).

$$2Fe(II)_{surf} + NH_2Cl + H^+ \rightarrow 2Fe(III) + NH_3 + Cl^- \qquad (4.98)$$

Copper

Total copper release was assessed for chloramine and free chlorine corrosion loop effluents. Chloraminated source waters released more total copper than did chlorinated source waters from the corrosion loops. Figure 4.177 shows average copper release by source and disinfectant. The difference in total copper release from identical chloraminated sources relative to chlorinated sources was statistically significant for GW (+0.3 mg/L), SW (+0.1 mg/L) and the blend (+0.1 mg/L), but was not statistically significant for RO finished water (+0.02 mg/L) as shown by a paired t-test for two means at the 95% confidence level.

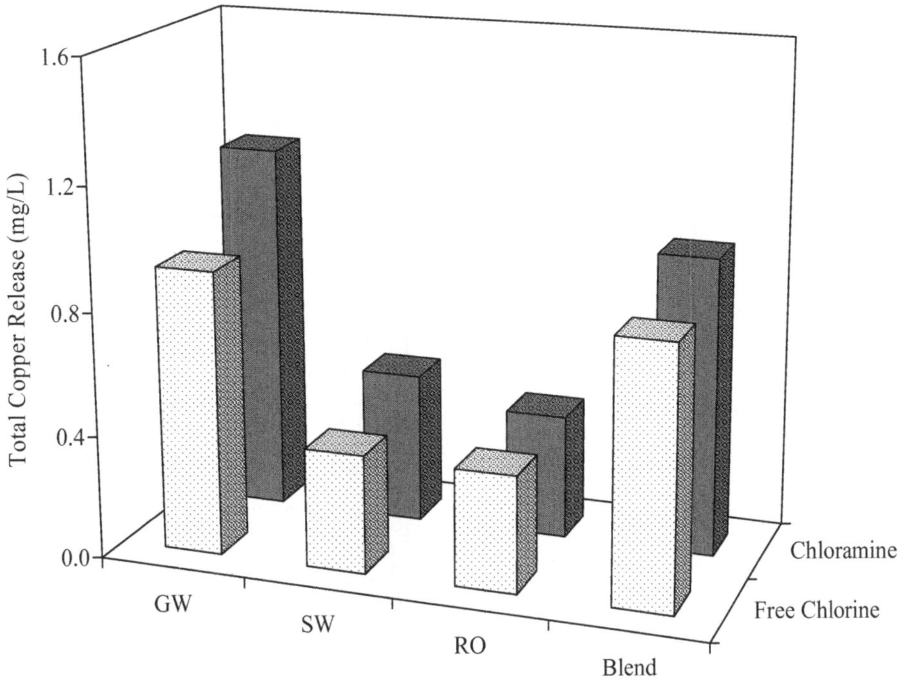

Figure 4.177 Average copper release by water source and type of disinfectant during disinfectant study

The copper release model as shown in equation (4.46) was used to predict copper concentrations for the PDSs used in the comparison study. Predicted total copper release for the chloraminated PDSs was 0.13 mg/L for RO, 0.18 mg/L for SW, 0.27 for GW and 0.27 mg/L for the blend as compared with actual total copper released of 0.40 mg/L, 0.49 mg/L, 1.2 mg/L and 0.96 mg/L respectively. The model under predicted total copper released for all sources. Ammonia-copper complexes do exist in chloraminated waters and increase dissolved copper release relative to chlorinated waters that have no ammonia. However, their contribution to dissolved copper release is less than 1 % compared to free chlorinated waters, and does not account for the differences observed here. This observation suggests that a biological mechanism may be responsible for the increased total copper released when chloraminated sources were used.

Lead

The effluent from the corrosion loops for PDSs 1 to 8 were analyzed to determine the effect on lead release of using either free chlorine or chloramine as disinfectant. Lead release was found to be higher for GW, RO water and the blend of GW/SW/RO for water using chloramines when compared with chlorinated waters. Chloraminated RO and blend water had average concentration of 127 μg/L and 8 μg/L respectively, compared with 9 μg/L and 2 μg/L for the chlorinated versions. There was not significant difference for GW and SW. Figure 4.178 shows the average lead release sorted by type of water and disinfectant utilized.

Total lead release to chloraminated finished waters in distribution systems in the District of Columbia have exceeded the lead action level (Leonnig and Cohn 2004) for the last two years. The elevated lead levels have been attributed to changes in water chemistry and interference with corrosion control methods caused by monochloramines following the change from a free chlorine to chloramine residual.

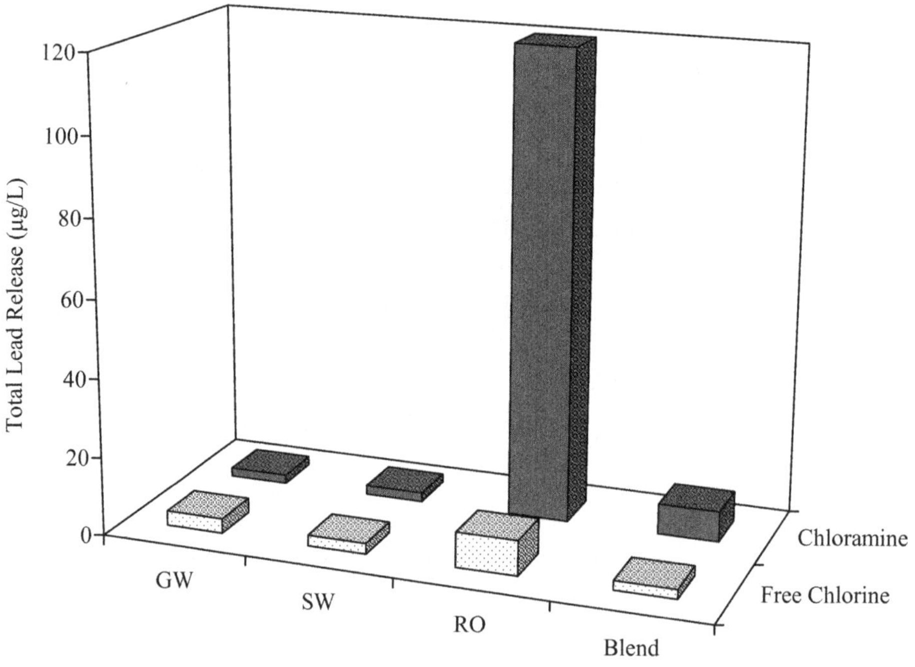

Figure 4.178 Average lead release by water source and type of disinfectant during disinfectant study

In this work, a paired t-test of means was used to determine the significance of total lead release from chlorinated and chloraminated waters. The t-test indicated that lead release was higher in chloraminated than in chlorinated RO and blended water at the 95% confidence level. There was no statistical difference between chloraminated and chlorinated total lead release from the corrosion loops receiving finished GW and SW. The lead release model shown in equation (4.52) was used to predict total lead release. The predicted and actual total lead released were statistically equal save the chloraminated RO source, which had an average and predicted total lead release of 118 μg/L and 24 μg/L respectively.

PDS Biological Activity Comparison by Disinfectant

Several parameters were monitored during the disinfectant comparison study to determine the effect of each disinfectant on nitrification and biological activity in the PDSs with respect to disinfectant by source and material. These parameters were nitrites, the sum of nitrite and nitrate (NO_x), bulk-HPCs, bulk-AOBs, AOC and HPCs, AOBs and PEPA in the biofilms.

Nitrites

Figure 4.179 presents the average increased nitrites through the investigation of free chlorine and chloramines. There was no ammonia added to the free chlorine waters, hence no significant nitrite or nitrate was produced in those PDSs. Nitrites increased in the chloraminated PDSs, indicating that nitrification occurred in the chloraminated waters in the presence of a chloramine residual. However, the production of nitrites in the chloraminated PDS was very limited with averages at or below 0.1 mg/L NO_2-N. The MCL for nitrite (1 mg/L NO_2-N) was not exceeded as the maximum nitrite produced was 0.2 mg/L as NO_2-N. Nitrite production was higher in chloraminated RO and SW sources as shown in Figure 4.180.

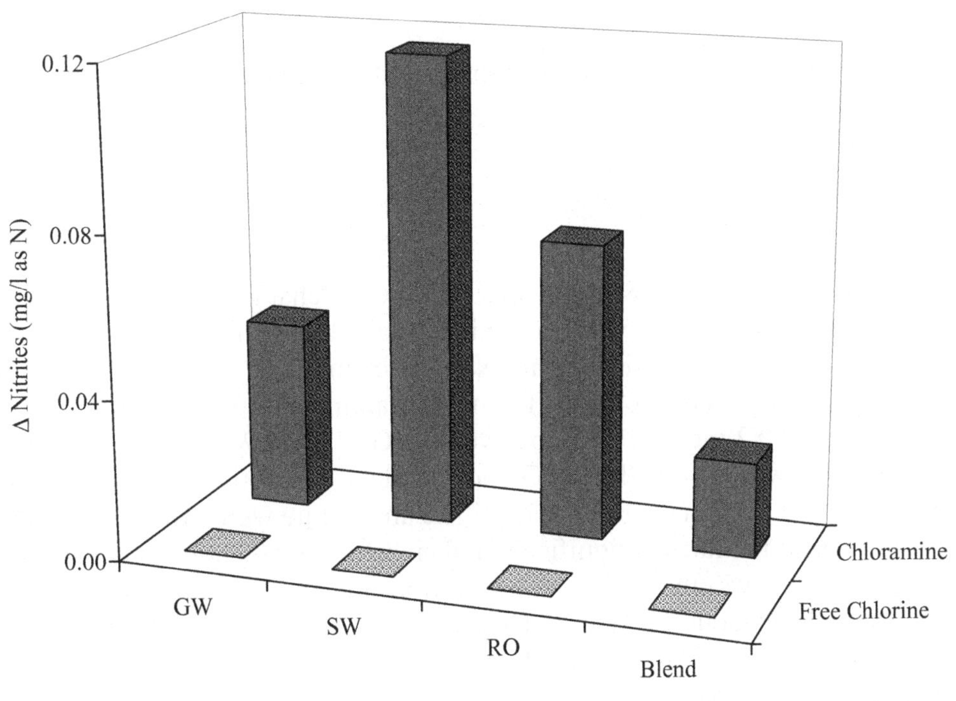

Figure 4.179 Average delta nitrites by water source and residual during disinfectant study

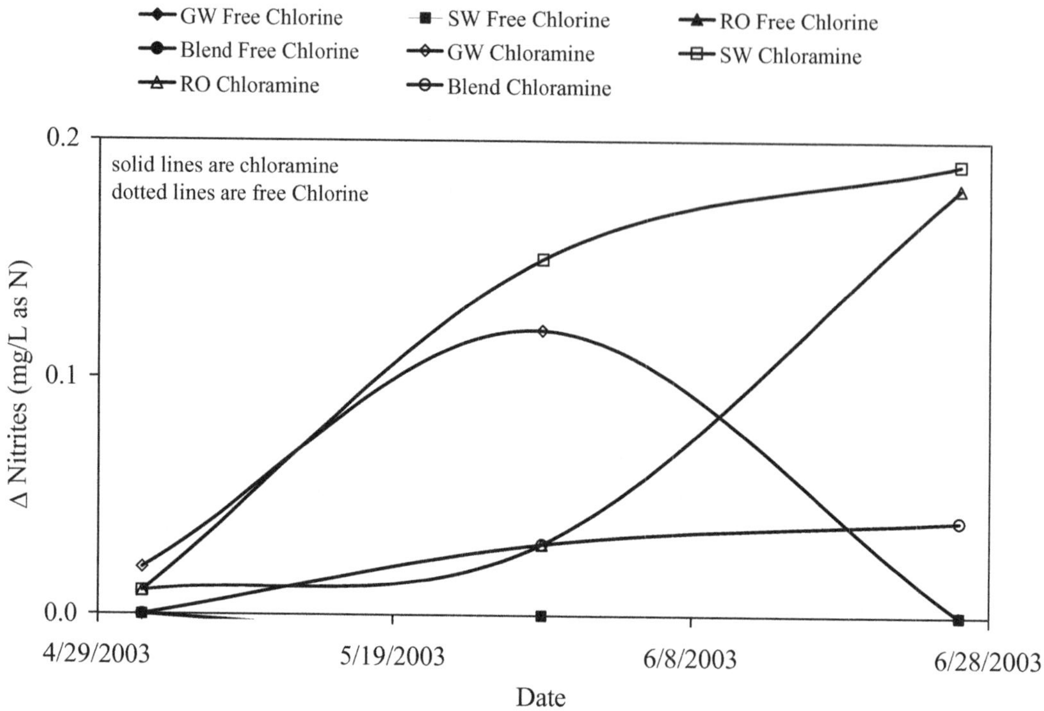

Figure 4.180 Nitrites production by water source and residual type during disinfectant study

NOx-N (sum of nitrite and nitrate)

Figure 4.181 and Figure 4.182 show that there was a change in the concentration of NOx-N in the chloraminated PDSs. Since no ammonia was available for nitrifying bacteria in the free chlorinated water, there was no significant NOx-N change for those waters. Figure 4.181 presents the average NOx-N change for different chloraminated water sources and disinfectants. An average ΔNO_x-N of 0.2 mg N/L was observed for GW and SW and an average Δ NOx-N of 0.12 and 0.10 mg N/L was observed for RO and the blend respectively. The data indicates that nitrification was present and small amounts of nitrite and nitrate were produced. Although small, the present of nitrites and nitrates is significant and signifies nitirification is present.

Figure 4.182 shows that NO_x-N generation in the chloraminated PDSs generally increased with time throughout the study. The relative production of nitrites and nitrates (NO_x-N) is relatively small and did not approach the MCL for nitrite nor nitrate. No change was observed in the free chlorinated waters

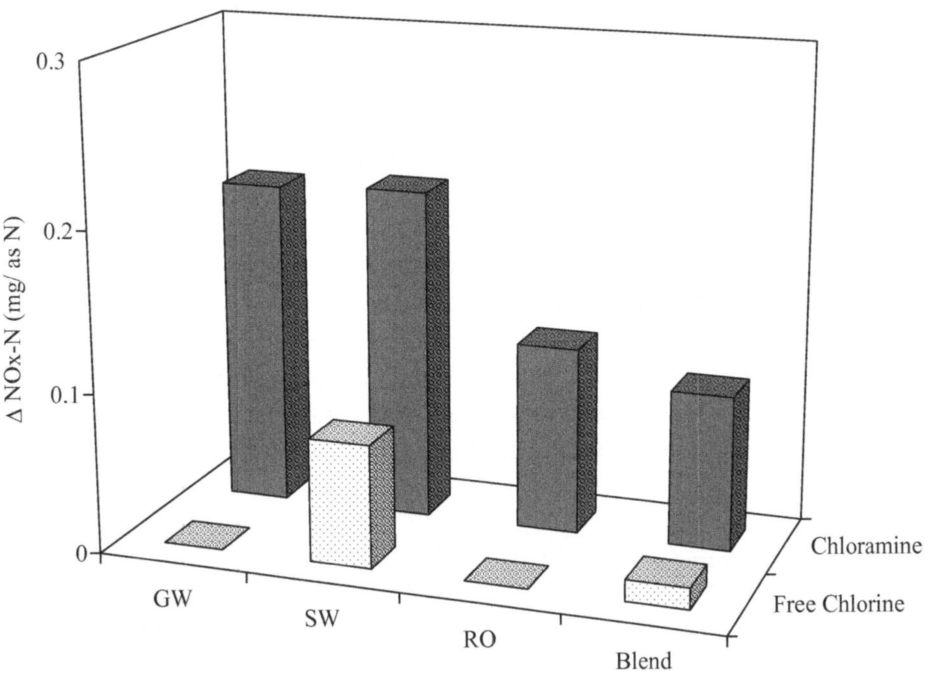

Figure 4.181 Average change on NO_X-N by water source and residual type during disinfectant study

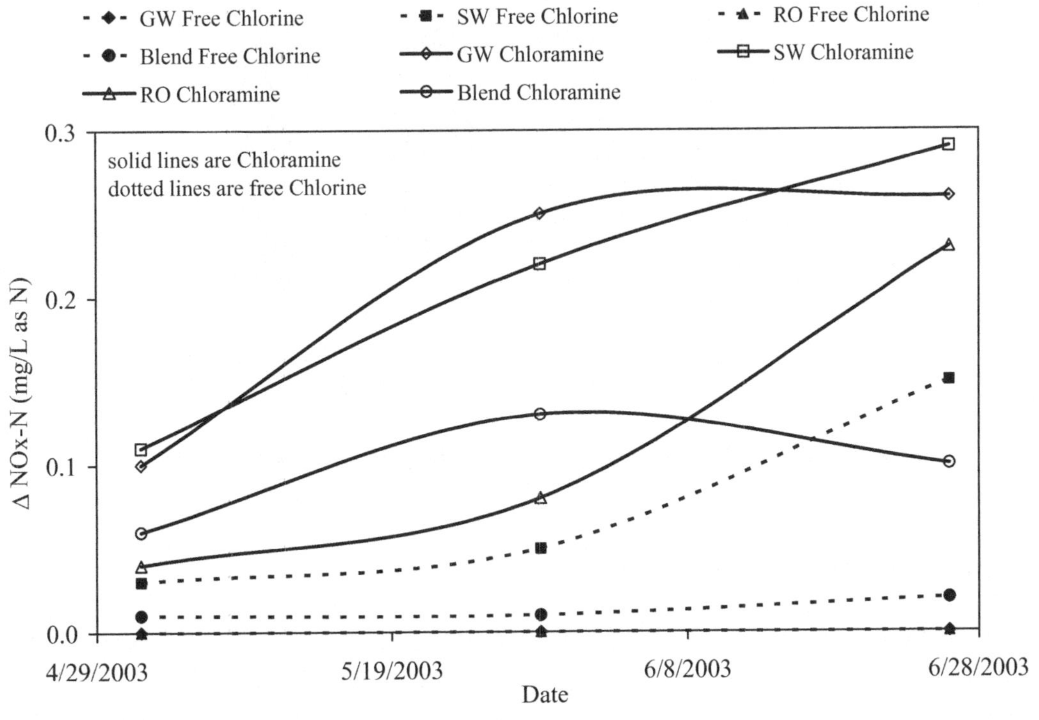

Figure 4.182 NO_X-N generation with time by source and residual during disinfectant study

FISH AOB

In order to quantify microbial changes associated with the type of disinfectant, AOB were measured in the PDS using Fluorescent In-Situ Hybridization (FISH). Only one sampling event was done for PDS 1-8, at the end of disinfectant comparison study. The data in Figure 4.183 shows AOBs were present in the bulk in all PDSs regardless of disinfectant. FISH AOB results showed that AOB in PDS effluent was greater for chloraminated PDS than for chlorinated PDS as shown in Figure 4.183. The difference might be due to the higher residual in the free chlorine lines and not to the type of disinfectant residual in the water.

Chloraminated PDS fed by surface hosted more AOB and showed a greater potential for nitrification than the same type of chlorinated waters. Chloraminated PDS fed by GW, RO and the blend presented more AOB relative to their free chlorinated counterparts, however the difference was less than one order of magnitude and are considered not significant.

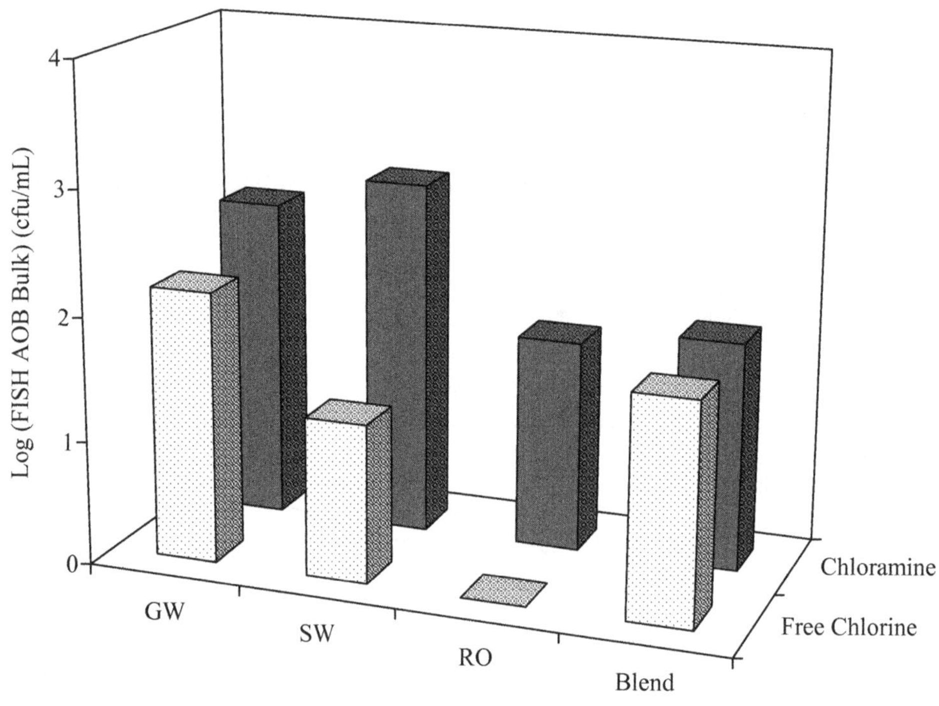

Figure 4.183 Bulk FISH AOB by water source during disinfectant study

Heterotrophic Plate Count (HPC)

The effluent HPC for the PDS was analyzed weekly to identify trends in bacterial inactivation of the waters based on the type of disinfectant utilized. Figure 4.184 and Figure 4.185 show the effluent HPCs for the lines using free chlorine and monochloramines respectively.

Figure 4.184 shows an initial increase in effluent HPC for all the chlorinated lines during the first two weeks of the study, followed by a steady decline towards the end of the phase.

After the change from chloramine to free chlorine, HPC for all the lines increased from an initial 10^3 cfu/mL measured on 4/10/2003 to a maximum of 10^6 cfu/mL measured on 5/01/2003. After this date, effluent HPC decreased with time for all the chlorinated PDSs reaching HPC of 10^2 to 10^3 cfu/mL at the end of the study (90 days of free chlorine use).

The HPC elevation at the beginning of the phase may have been caused by sloughing of biofilm material due to switch from chloraminated to chlorinated environment, disturbing the steady-state equilibrium between bulk liquid and biofilm bacterial populations. As both populations (bulk liquid and biofilm) acclimated to this new environment, effluent HPC gradually decreased until the end of the phase as shown in Figure 4.169.

Effluent HPC for the chloraminated PDS remained between 10^3 and 10^4 cfu/mL during the entire study as shown in Figure 4.185. Influent free chlorine residual was not controlled as well as influent chloramine residual and varied from approximately from 7 to 2.5 mg/L as Cl_2, which exceeded the high and low influent monochloramine concentration of 4.6 to 3.1 mg/L as Cl_2. The average free chlorine influent concentration was 0.4 mg/L higher than the average monochloramine influent concentration. The effluent residual in the chloraminated PDS was consistently lower than for the chlorinated PDS as shown in Figure 4.165. The fast decline of the combined residual within the PDS lines, compounded with the weaker oxidant characteristics of monochloramines relative to free chlorine translated in better inactivation of HPC by free chlorine compared to chloramines. At the end of phase VI, the average HPCs in the free chlorine PDS effluents were 1.3 order of magnitude less than in the chloraminated PDS effluents.

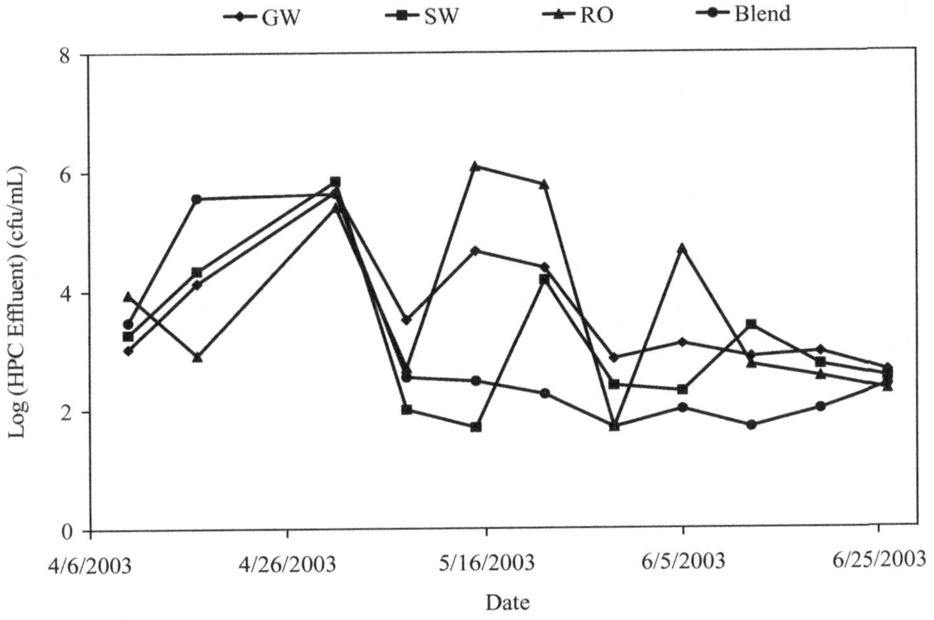

Figure 4.184 Effluent HPC with time by water source for free chlorine lines during disinfectant study

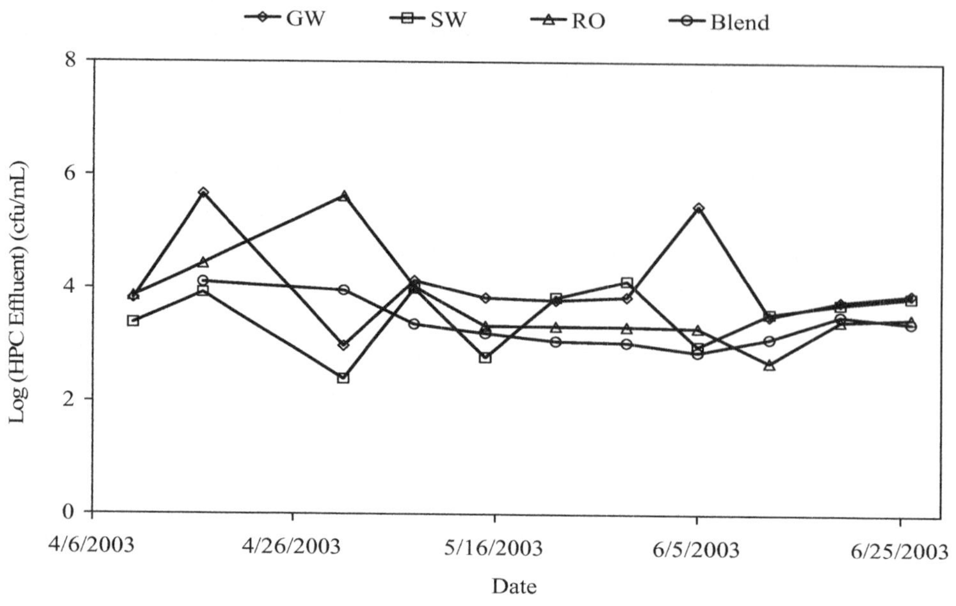

Figure 4.185 Effluent HPC with time by water source for chloraminated lines during disinfectant study

Assimilable Organic Carbon (AOC)

There was one sampling event for AOC on at the end of the study. In order to assess biostability the absolute value of ΔAOC (difference between influent and effluent values) has been calculated and these results are shown in Figure 4.186. Chlorinated GW, RO and the blended waters were more stable than their chloraminated counterparts. However, given the precision of the assay (10μg C/L), the difference in biostability between free chlorination and chloramination for a given water source based on ΔAOC was not significant. Comparing the different water sources, the blended water and SW were more biologically stable than GW and RO in all likelihood.

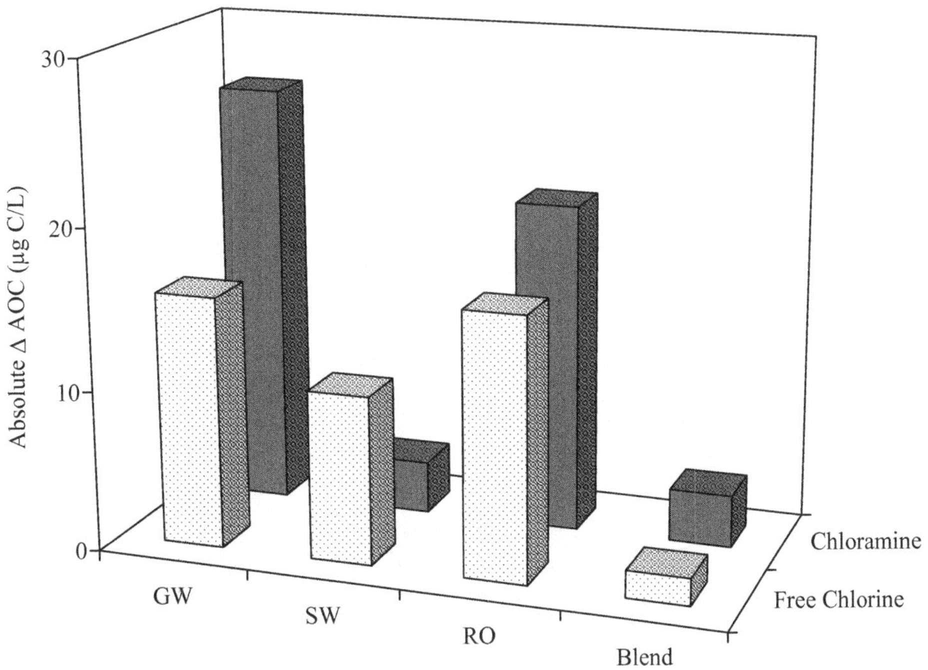

Figure 4.186 Absolute delta AOC by water source and residual type during disinfectant study

Biofilm Biological Activity by Disinfectant

Biofilm Quantification Using the Potential of Exo-Proteolytic Activity (PEPA) Assay

Comparison of PEPA for the different water sources showed that chloraminated PDSs had larger inventory of cells than free chlorinated lines as shown in Figure 4.187. A paired t-test for two sample means was performed for each water source at the 95% confidence level. The results of the test indicate that PDSs transmitting the chloraminated GW and blended sources had greater PEPA than PDSs transmitting the chlorinated GW and blended sources. PEPA in the PDSs transmitting the chloraminated SW or RO finished waters were greater than PEPA in the PDSs transmitting the chlorinated SW or RO but not statistically different. The highest PEPA was observed in the PDS transmitting chlorinated or chloraminated GW.

Average PEPA is shown by disinfectant and material in Figure 4.188. PEPAs were greater in the presence of chloramines than in the presence of free chlorine; however the difference was less than one order of magnitude for the three month investigation. PEPAs for all supporting materials were nearly equally in the presence of chloramines, and varied slightly more in the presence of free chlorine. Cell density as shown by PEPA does not appear to be affected by the supporting material. Each column in Figure 4.187 represents 8 coupons/data points (2 of each material).

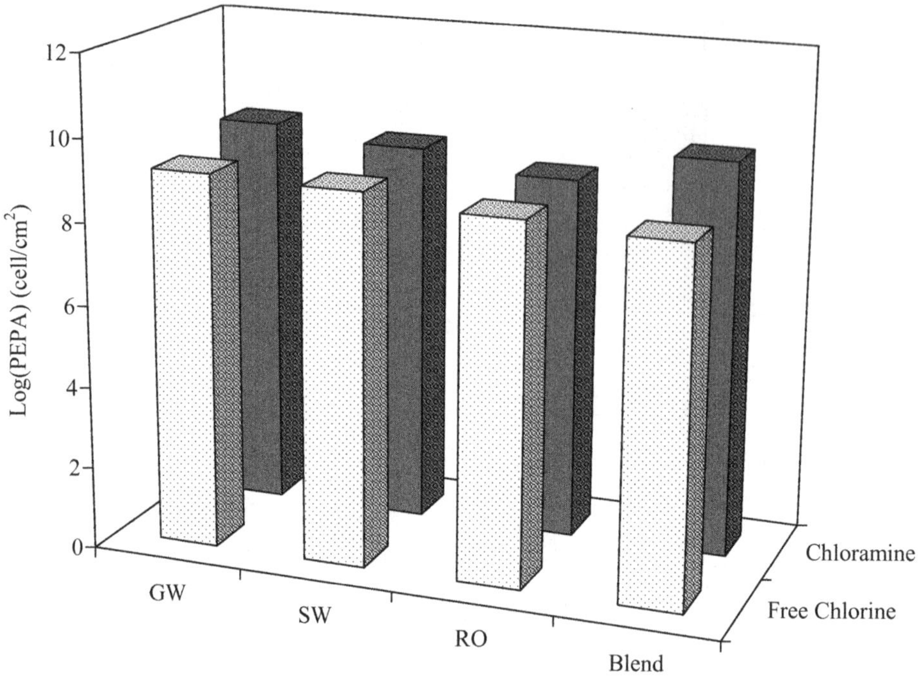

Figure 4.187 PEPA cell count by water source and disinfectant type during disinfectant study

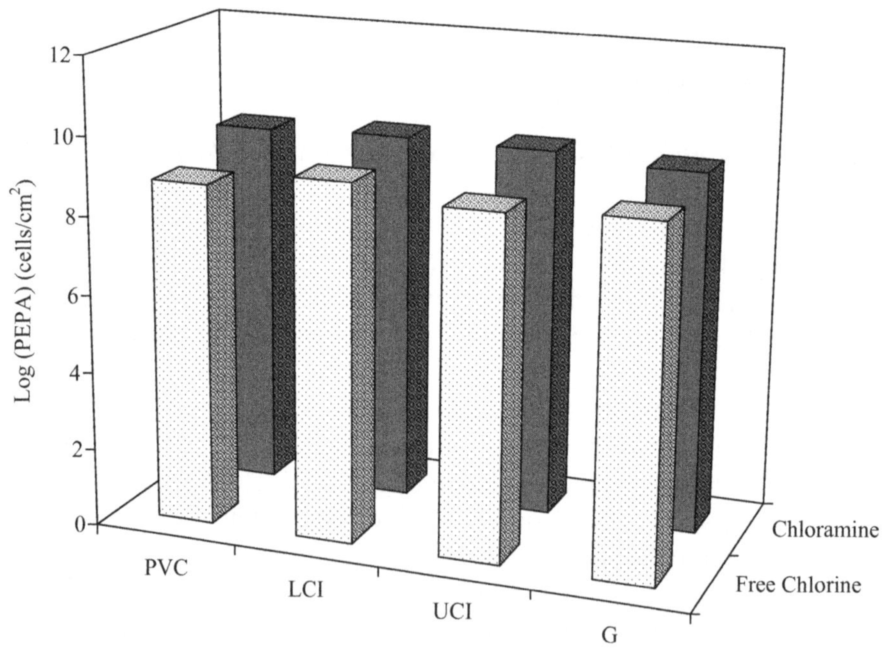

Figure 4.188 PEPA by coupon material and disinfectant type during disinfectant study

Biofilm Heterotrophic Plate Count (BFHPC)

Figure 4.189 shows the biofilm heterotrophic plate count on the coupons by water source and disinfectant. Biofilm HPCs were found to be approximately an order of magnitude higher in the presence of the GW and blend sources in the chloraminated PDS than in the chlorinated PDS. Although there was no significant difference in biofilm HPCs for the RO and SW sources, SW BFHPC was greater in the presence of free chlorine than in the presence of chloramines.

Biofilm HPCs are shown in Figure 4.190 for the four coupon materials. BFHPC were about 3 orders of magnitude lower than PEPA because BFHPC only measured the culturable (on R2A agar medium) fraction of the biofilm. Similar to PEPA, biofilm HPC were greater for chloraminated waters (PDS 5 to 8) than for chlorinated waters (PDS 1 to 4). A paired t-test performed between the corresponding coupons in each PDS showed that BFHPC was lower for chlorinated waters than for the chloraminated waters and the difference was statistically significant at the 95% confidence level. However, the differences are too small to make any statement related to disinfectant effect on biofilm proliferation, especially since the average influent free chlorine concentration was 0.4 greater than the average influent chloramine concentration.

The highest biofilm HPC regardless of type of disinfectant were found on the unlined metallic coupons (UCI and G), which have a greater surface roughness than PVC and LCI. For each coupon material with the exception of PVC, biofilm HPC counts were higher in the PDS receiving the chloraminated water sources than in the PDS receiving the chlorinated water sources. A paired t-test performed between the different coupons inside each cradle showed that biofilm HPC on the PVC coupons were lower than for the other materials and the difference was significant at the 95% confidence level.

The general order of biofilm HPCs by material is the same as the order of apparent surface roughness by material. UCI and G coupons have higher surface roughness than PVC and LCI. PVC has the least surface roughness and the least biofilm HPCs. The G biofilm HPCs are only slightly more than the LCI biofilm HPCS, which is the only deviation from this observation. All other variables being equal, it is rational that biofilms would increase as surface area increases. Consequently, more biofilm mass would be expected on coupons or in pipes that had higher surface roughness since surface area increases with surface roughness.

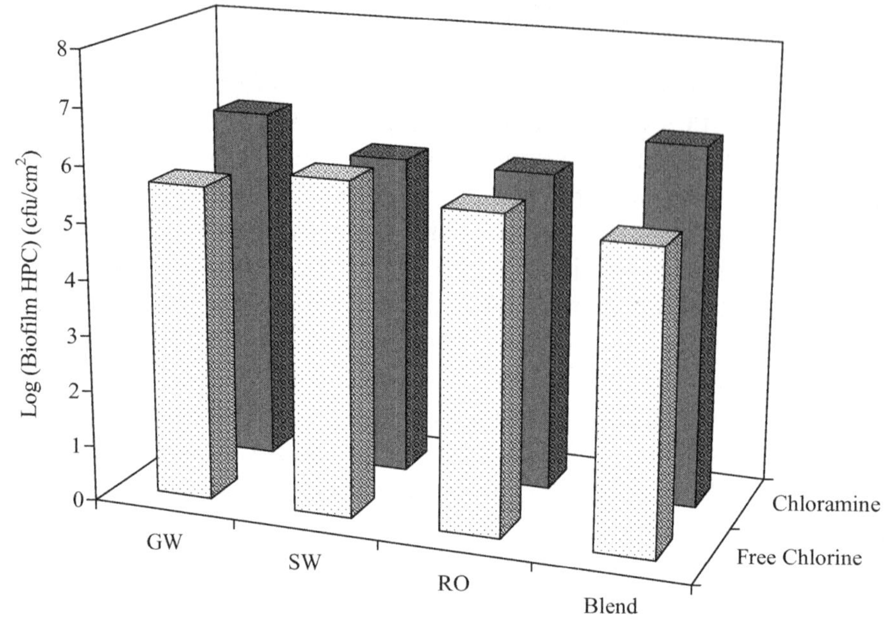

Figure 4.189 Biofilm HPC by source water and disinfectant during disinfectant study

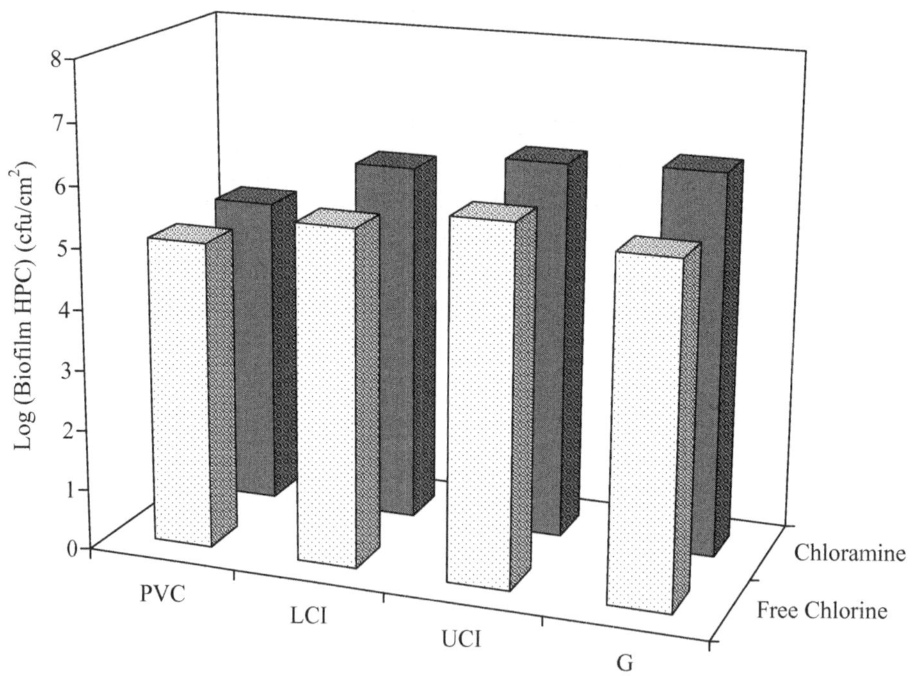

Figure 4.190 Biofilm HPC by coupon material and disinfectant during disinfectant study

Ammonia Oxidizing Bacteria Enumerated by Fluorescent In-Situ Hybridization

The biofilm ammonia oxidizing bacteria were enumerated by the FISH technique on the biological coupons and the results are shown in Figure 4.191 by source and residual and in Figure 4.192 by coupon and residual. Generally, the AOB density varied from 10^3 cfu/cm^2 to 10^5 cfu/cm^2. As previously noted, AOBs were present in PDSs 1-4 after three months of free chlorine application.

As shown in Figure 4.191, biofilm AOBs were higher in chloraminated GW and in the blend than in the chlorinated GW and blend; however chlorinated SW and RO produced higher biofilm AOBs than chloraminated SW and RO. No trend for biofilm AOBs is apparent by source water, which was expected since nitrification was previously found to be independent of source. Hence, AOBs in the biofilms would likewise be expected to be independent of source.

Biofilm AOB populations enumerated by FISH technique and separated by coupon material and residual are presented in Figure 4.192. In general, the biofilms on coupons receiving chloraminated waters had higher AOB populations than biofilms on coupons receiving chlorinated waters. With respect to materials, the order of AOB populations in biofilms were G>UCI>LCI>PVC, which is the same observation that was made for HPCs and correlated to surface roughness. PVC biofilm AOBs were at least one order of magnitude lower than G and UCI biofilm AOBs. A paired t-test indicated that the difference was significant at the 95% confidence level. The lower biofilm AOB on the PVC coupons can be explained by the lower biofilm inventory supported by the smoother surface of this material.

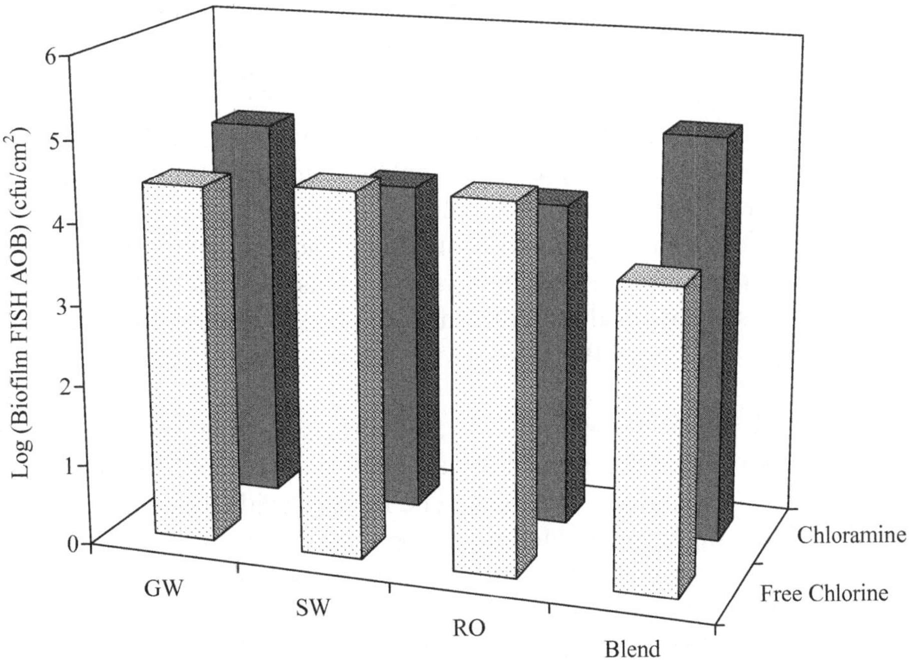

Figure 4.191 Biofilm FISH AOB by water source and residual during disinfectant study

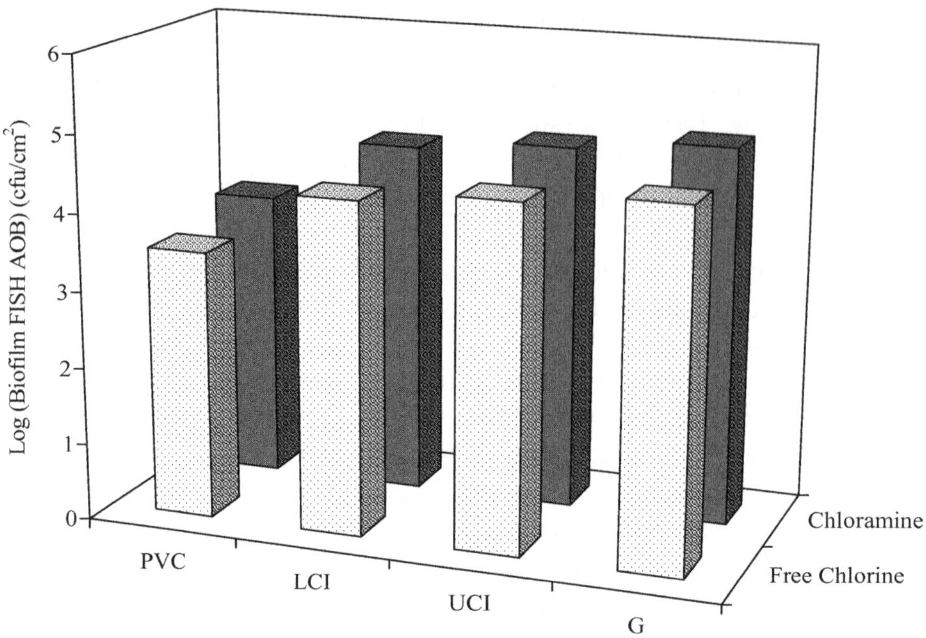

Figure 4.192 Biofilm FISH AOB by coupon material and residual during disinfectant study

Summary

Water Quality

The average effluent chloramine residual was 0.7 mg/L as Cl_2 lower than the average effluent free chlorine residual during the comparison study in the pilot distribution system. The difference was in part due to the influent free chlorine being in average 0.4 mg/L as Cl_2 greater than the influent chloramine residual. RO, GW and the blended finished source waters averaged 0.2 mg/L as Cl_2 less chlorine demand when free chlorine was used for residual maintenance than when chloramine were used for residual maintenance. The average difference in SW chlorine demand over the 2 day HRT was 0.9 mg/L with chloramine still having the most residual demand.

Calculation of the linear average rate of residual loss by pipe material showed that there was practically no difference in the rate of decay between free chlorine and chloramines for PVC, LCI and UCI under the study conditions. The rate of residual loss in the galvanized pipe was calculated to be 0.07 mg/L*hr greater for chloramines than for free chlorine. The water running through the hybrid lines used for the comparison study expended 5.5 hours flowing through 2-inches G pipes which translated to an extra loss of 0.4 mg/L as Cl_2 for the chloramines lines compared to the free chlorine lines.

Dissipation models developed for the decay of free chlorine and chloramine show that free chlorine dissipates more rapidly in the bulk liquid and less rapidly on the pipe wall. The higher rate of dissipation in the bulk is due to the more rapid reaction of free chlorine with dissolved organic compounds as measured by UV-254 absorbance compared to chloramine reactivity with the same. However, the reactivity of free chlorine with the pipe wall is less than the chloramine reactivity with the pipe wall for UCI and G pipes. The residual models incorporated temperature, UV-254, time and pipe material and clearly showed that free chlorine residual will be greater than chloramine residual for equal temperature and UV-254 in small diameter G pipe (1-ft and less) and UCI pipes (less than 6 inches). Chloramine residuals would exceed free chlorine residuals for equal temperatures and UV-254 for large sized pipes. There is not much difference in the wall reaction constants for PCV and LCI pipes; hence, chloramine residuals would exceed free chlorine residuals in all sizes of PVC and LCI pipes.

Simultaneous observations during the comparison of free chlorine to chloramine showed that the release of total iron, turbidity, apparent color, total copper and total lead were greater in the chloraminated PDSs than in chlorinated PDSs for identical source waters, PDSs composition, temperatures and residence times. The reason(s) for the higher concentrations to total released iron, copper, and lead, turbidity and apparent color are not clear. Chloramines have been shown to chemically participate in reactions with iron and increase iron release (equation (4.98)). Equilibrium analysis of ammonia-copper complexes indicated ammonia was responsible for less than 1 % of the dissolved copper species, and not responsible chemically for the increased release of total copper in the presence of chloramine. Thermodynamic constants for ammonia-lead complexes were not available. Biological mechanism(s) may also be associated with increased released of iron, copper, and lead, turbidity and apparent color in the presence of chloramine relative to free chlorine.

Biostability

In general, the difference in biological indicators was not high enough (greater than one order of magnitude) to make conclusive statements concerning the relative distribution system biostability in the presence of free chlorine or chloramines. The data show that there was a higher average residual, less ammonia, less nitrite and less nitrate in the free chlorine PDSs than in the chloramine PDSs. However, there was very little nitrification occurring in the chloramine PDSs. AOBs were present in the bulk in all PDSs regardless of disinfectant. There were more biofilm HPCs and AOBs in 7 of 8 observations in the coupons taken from the chloramine lines relative to the free chlorine coupons; however, these differences are not striking.

Bulk and biofilm HPCs were lower in the presence of free chlorine than chloramine. Free chlorine did not appreciably inhibit AOB growth, the concentrations of AOB within the biofilm were similar for chlorinated and chloraminated waters. The lower values of absolute delta AOC indicated that chlorinated waters had higher biostability than chloraminated waters.

Biofilm densities as indicated by PEPA, biofilm HPCs and biofilm FISH AOBs were higher for the chloraminated GW and blended sources than PEPA, biofilm HPC and biofilm FISH AOB for the chlorinated GW and blend. There was not a significant difference among biofilm densities associated with finished RO and SW. PEPA, biofilm HPC, and biofilm AOB were lower for PVC coupons than for UCI, LCI, and G coupons regardless of water source and disinfectant. The highest PEPA, biofilm HPC and biofilm AOB were found on UCI and G coupons.

The AOB biofilm data shows that nitrification was independent of source and that AOB biofilm density increased as the surface roughness of the supporting material increased. It is significant that AOBs are present in the free chlorine biofilms after three months of exposure to free chlorine. Hence, a three month free chlorine burn did not destroy all of the AOBs in the biofilm, and only reduced the AOBs in the biofilm by no more than one order of magnitude on average.

CHAPTER 5
CONCLUSIONS

APPLICATION

Central to application of the findings in this work is recognition that the results presented herein can not be directly transferred to actual distribution systems without consideration of the pipe material, pipe geometry, pipe hydraulics and water chemistry of the finished water in the distribution system. The equivalent pipe lengths in this work for an actual distribution system with an average velocity of 0.33 ft/sec were 10.8 and 27 miles for a two and 5 day HRT. Such conditions are not commonly seen in actual distribution systems for reaches of galvanized or unlined cast iron pipe materials. However, the models developed from the results of this work allow prediction of water quality if they are properly utilized. Hence, determination of equivalent pipe lengths or direct application of the mass transfer models for iron release, turbidity release, apparent color release and residual dissipation, and consideration residual dissipation on, nitrification and biostability must be done to accurately predict distribution system water quality at any location. Prediction of total copper and lead concentrations can be done directly from the steady state water quality models as those parameters are sampled from a water column that has been stagnant for at least six or more hours, and were not affected by pipe geometry, material and hydraulics in this work.

IRON RELEASE

Release of iron corrosion products was investigated using pilot distribution systems (PDS) constructed of aged pipe materials: PVC, lined cast iron, unlined cast iron, and galvanized steel. Iron concentrations were highly correlated with apparent color determinations. Water quality effects were examined by varying input water quality using blends of seven different finished waters. Steady-state water quality models were developed based on long-term (three months) blend durations. Dynamic water quality models were also developed using short-term data associated with transition periods at the time of blend changes. A practical model was developed for prediction of water quality impairment (apparent color) that incorporated pipe material, pipe geometry, hydraulic conditions, and bulk water quality. Surface characterization of pipe coupons was completed using XRD and XPS techniques. Thermodynamic models were considered based on the controlling pipe scales.

Surface Characterization

XRD and XPS surface characterization techniques successfully identified the chemical composition of corrosion scales in drinking water distribution pipes. Clearly, the nature and quantities of iron corrosion scales are different for different pipe materials, with unlined cast iron having predominantly iron-based products, while galvanized iron pipes have iron-based products as well as zinc-based corrosion scales. The similarities in some of the corrosion scales identified (α-FeOOH, γ-FeOOH, Fe_2O_3, Fe_3O_4, and $FeCO_3$) in the unlined and galvanized iron pipes reflect the base material (iron) in both pipes.

- Significant positive correlation was observed between Fe_2O_3 peak area percentages of cast iron pipes and total and dissolved iron levels observed in the pilot study. However, Fe_2O_3 solubility is very low and actual dissolved iron concentration is much higher than Fe_2O_3 solubility. Since $FeCO_3$ could decompose easily and transform to Fe_2O_3 if exposed to air, a large portion of Fe_2O_3 peak area observed in the XPS study may come from the transformation of $FeCO_3$.
- Lined cast iron pipes and PVC pipes do not generate corrosion scales of their own; rather, small quantities of ferric oxide and calcium carbonates are deposited due to calcium over-saturation and ferric hydroxide precipitation. No iron release problems related to pipe corrosion are expected in either the lined or PVC pipes.
- In this study, despite the age of the pipes, the zinc coating in the galvanized iron pipes was still active, hence the observed formation of ZnO and $Zn_5(CO_3)_2(OH)_6$. There was less total iron release from galvanized pipes, compared to unlined cast iron as zinc acts as a sacrificial anode to protect iron; however the total iron release was significant and would cause consumer complaints if received in drinking water.
- The presence of iron corrosion products in galvanized pipes indicated that the zinc coating was not uniform or had corroded over time. Galvanized pipes will behave like unlined cast iron pipes to the extent that the zinc coating has been removed or lost. This conclusion is supported by similar dependence of total iron release on alkalinity, sulfates and temperature in the hybrid and galvanized empirical models.
- Increased quantities of amorphous corrosion products were observed by switching from 100% conventionally treated ground water to any blend of conventionally treated ground water/enhanced coagulated surface water/RO water indicating an increase in iron release.
- The order of source waters with regard to Fe_2O_3 percent area, which was a direct indication of $FeCO_3$ percent area, was enhanced coagulated surface water > RO > blended conventionally treated ground water/enhanced coagulated surface water/RO > conventionally treated ground water and was the also the order of sources waters with regard to total iron release: This observation supports siderite as the controlling solid phase.

Thermodynamic Modeling

Thermodynamic equilibrium models were developed with assumed controlling solid phases as $Fe(OH)_3$, $Fe(OH)_2$ and $FeCO_3$. The predicted values of the $Fe(OH)_2$ model are much higher (1000 times) than actual field total iron data. The $Fe(OH)_3$ model grossly under-predicts iron release and the predicted values are 200,000 times lower than actual field dissolved iron data. The siderite ($FeCO_3$) thermodynamic model was found to be a closest match to field observations. The predicted iron release by the siderite model was less than the field observations of total iron release; it was however, more than the field observations of the soluble iron release. The major feature of the siderite model is decreased iron release with increased alkalinity and pH.

- As identified in the surface characterization, siderite is easily oxidized in an aerobic environment. It has been postulated to exist in the intermediate zone between the iron metal and the overlying oxidized corrosion scales. Therefore, the release of ferrous iron into the bulk and the availability of alkalinity may be diffusion-limited in this region; thereby requiring increased time to achieve steady-state following changes in alkalinity or other water quality parameters.
- Thermodynamic considerations of ferrous/ferric salts of chlorides and sulfates do not have a significant contribution to iron release. However, these anions increase the conductivity of the bulk water facilitating electron transfer and subsequent corrosion or corrosion-scale deterioration.
- Increasing temperature would increase the rate of diffusion of ferrous iron into the bulk, which would be seen as increased iron release.
- Though important mechanistic inferences can be made based on the thermodynamic siderite model, this model has limited use in actual field conditions due to several complicating factors like geometry and flow conditions.

Response Characterization

For characterization of the release of iron corrosion products, apparent color (or turbidity) were determined to be acceptable surrogates. This finding is consistent with a determination that the majority of iron (and color) is associated with particulates. Apparent color was selected as a response variable for modeling efforts due to ease of field measurement.

Empirical Models for Prediction of Elevation of Apparent Color

Release of iron corrosion products occurred within unlined cast iron and galvanized pipe materials. The other pipe materials (PVC and lined iron pipe) did not exhibit release of corrosion products.

Hybrid PDS

- An empirical model relating total iron release to water quality was developed using 12 months of operating data for the hybrid PDS. Three months of additional verification data were obtained that included manipulation of alkalinity, chloride, and sulfate concentrations. Alkalinity, chloride, sulfate, sodium, dissolved oxygen, temperature, and hydraulic residence time were identified by statistical procedures as having a significant effect on release of corrosion products. Confounding in the experimental design complicated absolute resolution of the singular effect of each variable.
- Alkalinity exhibited a significant negative correlation with color release. Increasing alkalinity is beneficial in reducing the iron release from the intermediate corrosion scales due to the thermodynamic considerations of siderite solubility.
- Chloride, sulfate, sodium, dissolved oxygen, temperature, and hydraulic residence time exhibited a significant positive correlation with color release. Chloride, sulfate

and sodium are hypothesized to increase iron release due to increased conductivity and subsequently increased corrosion of the base iron metal. Dissolved oxygen is the primary electron acceptor for any corrosion process to occur.

- Calcium and pH were not demonstrated to have a significant effect on color release for this dataset. This observation was attributed to stabilization of the source waters (with respect to $CaCO_3$ solubility). Stabilization of waters is an established practice in the industry and therefore was not a variable under the conditions of the study. The results of the empirical model should be applicable to stabilized waters and not otherwise. The empirical model numerical prediction of color release was specific to the geometry and hydraulic conditions of the hybrid pilot distribution systems.
- The relative flux of corrosion products was assessed for different sources (or blends) based on their individual water chemistry. Both the enhanced coagulated surface and finished water RO sources presented a potential concern for release of corrosion products. This adverse effect is attributed to altered water chemistry in comparison to the conventionally treated ground water source. Specific differences included a reduction in alkalinity (both enhanced coagulated surface and finished water RO sources), an increase in chloride (finished water RO), an increase in sodium (both sources, especially finished water RO), and an increase in sulfate (both sources, especially enhanced coagulated surface water).

Galvanized PDS

A similar empirical model was developed for galvanized steel pipe.

- Elevation of temperature and sulfate increased total iron, apparent color and turbidity release from galvanized steel pipe.
- Elevation of alkalinity and silica reduced total iron, apparent color and turbidity release from galvanized steel pipe.
- The predicted variation of total iron, apparent color and turbidity release with water quality for the empirical galvanized steel model was similar to the predicted variation of total iron, apparent color and turbidity release with water quality for the empirical galvanized steel model (for example, temperature, sulfate, and alkalinity). Because the empirical models are specific for the pipe geometry and hydraulic conditions for the PDS, comparison of actual impairment of water quality for a prototype system must incorporate actual site-specific features of pipe geometry and hydraulics.
- The identified iron species in the scales of galvanized steel and unlined cast iron coupons were similar. Consequently, thermodynamic models based on the solubility of these compounds would yield similar predictions.

Mechanism(s) of Total Iron Release

- The detailed mechanism of iron release in distribution systems is complex and depends on a number of factors such as pipe age, bulk water quality, pipe geometry and flow conditions. Ambiguity exists in the literature about the exact nature of the iron release. One theory is that iron is released in a particulate form from the corrosion scales as a result of physical or chemical detachment. This theory would resolve the iron release based on hydraulic transients in the distribution system. Corrosion theory espouses the effect of electrochemical reactions that would lead to iron release in the ionic form, entrapment in the corrosion scales as a hydroxide or salt complex and later released as soluble iron based on the solubility of the controlling solid phase.

- Based on thermodynamic considerations, the solid controlling phase was best described by siderite solubility. The thermodynamic siderite model over-predicts soluble iron but under-predicts total iron release in the PDS. Since the predominant form of iron released is in the particulate form, applicability of the thermodynamic model is limited to evaluating the qualitative effects of water chemistry, especially alkalinity and pH.
- Quantitative assessment of iron release with water chemistry required empirical modeling approach. The empirical approach was more robust than the thermodynamic approach in predicting iron release as a function of a wider range of water quality parameters than were used in the thermodynamic model. However, the mechanistic description of iron release from corrosion scales is provided by the thermodynamic siderite model and reinforced by the surface characterization studies.
- The iron flux model was developed to describe the effect of hydraulic transients on the release of iron in a distribution system. Iron flux was first related to Reynolds number for a constant water quality and then the empirical model was incorporated for varying water quality.
- Based on this study, it was possible to relate the two mechanistic theories of iron release in distribution systems into a unified quantitative and descriptive model that incorporated the effect of both water quality and hydraulic conditions.

Iron Flux Model Development

Examination of operating data at different hydraulic residence time (laminar flow conditions) revealed a constant iron flux (K_m). Development of a mass balance for prediction of iron concentrations was pursued using a constant flux term. For these conditions:

- The increase in iron concentration was observed to follow a zero-order model with respect to iron concentration.
- Because the iron is predominantly present in a particulate form, relationships that address only soluble species may not adequately define total iron present in the distribution system. Chemical equilibrium relationships (for example, solubility) would not be expected to reflect total iron concentrations. Physical processes such as diffusion of soluble species may also be inadequate to characterize total iron. The absence of any dependence of iron flux values on the bulk phase total iron concentration is consistent with a dominant iron form that is particulate.
- Using a constant water quality source, experimental determination of the relation between iron flux and hydraulic conditions (expressed as Reynolds Number) for

galvanized and unlined cast iron pipe was accomplished. A constant flux region was established for laminar conditions (Reynolds Numbers < 2000). For turbulent conditions (Reynolds Numbers > 2000), the flux increased in proportion to the Reynolds Number.

- Adjustment of the flux (K_m) to incorporate variation in water chemistry was accomplished using the empirical color prediction model. This approach had the capability to predict water quality changes related to iron release in a distribution system with known pipe material, diameter, length, flow velocity and bulk water quality.
- The elevation of iron concentration was cumulative. Therefore, distribution networks can be analyzed by sequential addition of incremental increases in iron concentration for pipe reaches with variant conditions (material, geometry, and velocity).
- For systems that operate at laminar flow conditions, the iron flux (K_m) was determined to be independent of Reynolds Number. The resulting increase in total iron concentration (or apparent color, or turbidity) would be proportional to the hydraulic residence time or pipe length and inversely proportional to the pipe diameter. This indicated that for extreme conditions (very long HRT), adverse water quality was possible regardless of the water chemistry. To the extent that water chemistry and/or pipe material and condition caused an increase in K_m, the adverse water quality would be experienced at less severe hydraulic residence times.
- For systems that operate at fully turbulent conditions, the iron flux (K_m) was determined to be proportional to the Reynolds Number. For large Reynolds Number, the resulting increase in total iron concentration would be independent of Reynolds Number or pipe diameter, but proportional to pipe length:
- The relative iron release for the different source waters followed the predictions of the empirical model. The conventionally treated ground water source (high alkalinity, low sulfate, low chloride) exhibited minimal release of iron corrosion products. The conventionally treated ground water source matched the chemistry of the historical water exposure that produced the corrosion products on the pipe surfaces. In contrast, a water quality similar to the enhanced coagulated surface water (low alkalinity, high sulfate) is predicted to result in elevated iron concentrations. Finished waters of similar composition to the finished water RO source (low alkalinity, high sodium, high chloride) pose the greatest potential for water quality impairment.

Mitigation Measures – Alkalinity Supplementation

- Control of release of iron corrosion products, and the associated increase in apparent color, may be mitigated by supplementation of alkalinity. Two alternatives are identified for the alkalinity source: sodium bicarbonate ($NaHCO_3$) and lime/carbon dioxide ($Ca(OH)_2 + CO_2$).
- Simulations were completed to investigate the benefits associated with alkalinity supplementation for the finished surface water and desalination product waters. Both alkalinity sources were examined. Calculations were completed for the hybrid PDS

geometry (5 day HRT) and the single material galvanized PDS (5 day HRT) using the previously developed empirical models. These hydraulic conditions would not be typical of the majority of the Member Governments' distribution system, as corresponding equivalent pipe lengths in the full-scale system would be many miles. The hydraulic conditions represented by these simulations for the unlined cast iron and galvanized pipe materials would match stagnant regions (dead zones and/or dead ends) of the distribution system. Quantitative procedures developed in the report can be used to investigate specific hydraulic conditions, pipe geometries, and water quality of interest.

Surface Water Source – Color Mitigation

- For the surface water, supplementation with sodium bicarbonate to elevate the alkalinity achieved only a modest reduction in color release in the hybrid line. The simulated color release without alkalinity supplementation was 11 cpu at an alkalinity of 50 mg/L as $CaCO_3$. Elevation of the alkalinity to 80, 100, and 120 mg/L as $CaCO_3$ resulted in apparent color increases of 10, 10, and 9 cpu, respectively. The lack of sensitivity is attributed to the simultaneous effect of the sodium ions provided with the bicarbonate.
- For the same alteration in alkalinity, the apparent color release for the galvanized PDS was reduced considerably from 41 to 19, 13, and 10 cpu, respectively. It is noted that the empirical model for the galvanized pipe material did not include a dependence on sodium. It would seem probable for any increase in conductivity to adversely affect water quality due to an increase in iron corrosion rates. Thus the simulated results for the galvanized pipe material may overstate the benefits associated with sodium bicarbonate addition. The corresponding values for the galvanized PDS geometry are identical to predictions with sodium bicarbonate because the galvanized empirical equation does not include a dependence on sodium (or calcium) concentration.
- Alkalinity supplementation of the surface water with lime/carbon dioxide was more effective for control of color release. For the same alkalinity targets (80, 100, and 120 mg/L as $CaCO_3$), use of the calcium-based treatment resulted in apparent color releases of 7, 6, and 5 cpu for the hybrid PDS geometry.

Desalinated Source – Color Mitigation

- Similar trends were noted for alkalinity supplementation of the desalinated finished water source. Use of sodium bicarbonate to elevate alkalinity from 50 to 80, 100, and 120 mg/L as $CaCO_3$ mitigated color release in the hybrid PDS geometry from 28 cpu to 22, 20, and 19 cpu, respectively.
- Lime/carbon dioxide was more effective in the hybrid pipe system, yielding color release values of 18, 15, and 13 cpu for the same alkalinity targets.
- For the galvanized PDS geometry, the color release values were reduced from 13 cpu to 6, 4, and 3 cpu for the identical alkalinities.

COPPER RELEASE

Total copper release was investigated using corrosion loops that utilized the same common copper pipe that is used for home plumbing systems. SEM, XPS and XRD were used to characterize and identify chemical composition of solid copper scales on identical copper coupons that were typically exposed for a period of three months in cradles receiving the same finished water as the corrosion loops. The solid copper scales identified by surface characterization were modeled thermodynamically as the controlling solid phase for copper solubility and compared to actual total copper release data from the field. Empirical models for copper release as a function of bulk water quality during transition and at steady-state were developed to predict and proactively manage total copper release.

Surface Characterization and Theoretical Model Development

- SEM, XRD, and XPS characterization of copper surfaces
  - Three months of exposure to drinking water environments were inadequate to form enough divalent corrosion products on copper coupons for XRD detection. Cuprite (Cu_2O) was the only crystalline structure observed in corrosion layer using XRD. Changing water quality has no observed effect on variation of crystalline structure.
  - The major corrosion scale layer on the copper coupons consisted of cuprite (Cu_2O) as no consistent formation of Cu(II) species was observed in the scale.
  - The morphology of the top layer of the surface scale varied with water quality. A high alkalinity and low chloride finished conventionally treated ground water environment produced a more compact top scale layer than a low alkalinity and high chloride finished RO environment.
  - Tenorite (CuO), cuprite (Cu_2O) and cupric hydroxide ($Cu(OH)_2$) were the major cupric components on the very top of the copper layer. Other Cu(II) species could be formed given the right environment. Cupric hydroxide and tenorite are the major divalent copper forms in the scale and cuprite is the major monovalent copper form in the scale. These indicated a continuous disappropriation reaction in the very top layer where monovalent copper is oxidized to divalent copper.
  - Cuprite (Cu_2O) variation in the top layer was dependent on incubation period of incubation. Cu(II) species were not dependent on aging.
- Thermodynamic modeling
  - Thermodynamic modeling found the $Cu(OH)_2$ solid phase most accurately described total copper release
  - Theoretical modeling illustrated that alkalinity and pH were the major factors affecting dissolved copper release.
  - The theoretical approach was useful qualitatively, but is limited quantitatively due to the impact of particulate copper. However the trend of total copper release with pH and alkalinity was consistent with the trend predicted by the $Cu(OH)_2$ solid phase equilibrium model.

Water Quality Effects on Copper Release

- Experimental data indicated that copper corrosion by-product releases depended heavily on water quality, especially on alkalinity and pH, and also to a lesser extent on sulfates, silica and temperature for the conditions of this investigation.
- The release of total copper from home plumbing systems can be adversely affected by blending finished waters from surface, ground and saline sources. Adverse effects on total copper release can be mitigated by controlling water quality of blended finished waters primarily by controlling alkalinity and pH.
- The blending ratios of finished waters produced from combining ground, surface and saline sources can be monitored in drinking water distribution systems using key water quality variables common to each source. In this work, the percent of finished conventionally treated ground water in blended distribution system water could be estimated by the concentration of alkalinity and calcium in the blend. The percent of enhanced coagulated surface water in the blend could be monitored by the percent of sulfates in the blend. The percent of RO water in the blend could be monitored by the percent of silica or chlorides in the blend.
- The release of total copper was primarily dependent on pH and alkalinity for the conditions of this study. Total copper release increased as alkalinity increased and pH decreased.
- Normalization of calcium and silica with respect to alkalinity was necessary to determine the effects of calcium and silica on total copper release. The concentrations of calcium in the blended PDS water did not affect total copper release. Silica did affect total copper release. Total copper release decreased as silica increased indicating naturally occurring silica functioned as a corrosion inhibitor.
- Empirical Models for Prediction of Copper Release
    - Empirical models for the 50th and 90th percentiles were developed using 12 months of operating data for the hybrid and pure material PDS. Six months of additional verification data were obtained that involved manipulation of alkalinity, chloride, and sulfate concentrations.
    - Total copper release increased with increasing alkalinity, temperature, sulfate and decreased with increasing silica.
    - Dissolved oxygen and chloramine residual did not significantly affect total copper release. This observation may be attributable to the solubility controlling mechanisms and confounding effects.

Simulation of Copper Release in Pipe Distribution System and Copper Rule Compliance

- A method of proactively managing distribution system water quality for controlling copper concentrations has been developed and demonstrated using Ternary diagrams to describe the effects of blended finished water quality from surface, ground and saline sources on total copper release using empirically developed models as a function of source finished water quality. This method can be used by the drinking water industry to predict the effect of blending three different finished waters on any water quality parameter given that the models are available.

- The non-linear model for predicting the ninetieth percentile copper concentrations were developed for varying water quality produced by blending finished conventionally treated ground water, enhanced coagulated surface water and RO finished water. Sensitivity analysis indicated the copper action level would be violated for conventionally treated ground water or blends that contained high alkalinity.
- The models predicted violation of the copper action level if the blend contained more than 60% conventionally treated ground water. Although the copper model was developed from copper pipe equilibrium, there could be other factors such a specific source water quality or electrical grounding schemes that would require the copper model to be calibrated for a specific source or location. Such calibration is easily done by incorporating the source or location ninetieth copper percentile as a ratio of the model predicted ninetieth percentile and the location ninetieth percentile using the same water chemistry.

LEAD RELEASE

Total lead release was investigated using corrosion loops that utilized lead coupons that simulated the lead present in for home plumbing systems. SEM, XPS and XRD were used to characterize and identify chemical composition of solid lead scales on identical lead coupons that were typically exposed for a period of three months in cradles receiving the same finished water as the corrosion loops. The solid lead scales identified by surface characterization were modeled thermodynamically as the controlling solid phase for lead solubility and compared to actual total lead release data from the field. Empirical models for lead release as a function of bulk water quality during transition and at steady-state were developed to predict and proactively manage total lead release.

Surface Characterization and Theoretical Model Development

- Surface Characterization
  - Corrosion products formed on the lead coupons consisted of lead dioxide (PbO_2), lead oxide (PbO), cerrusite ($PbCO_3$), and hydrocerussite ($Pb_3(CO_3)_2(OH)_2$). Among these corrosion products, cerrusite and hydrocerussite peaks were more pronounced. An abundance of amorphous phase lead species was also observed.
  - $Pb_3(OH)_2(CO_3)_2$ and $PbCO_3$ peak area percentage obtained from XPS analysis showed a fairly good linear correlation with lead release: that is, higher $Pb_3(OH)_2(CO_3)_2$ area percentage resulted in higher lead release while higher $PbCO_3$ area percentage resulted in lower lead release. This finding suggested that $Pb_3(OH)_2(CO_3)_2$ may be the controlling solid phase which determined lead concentration in drinking water, and that $PbCO_3$ also has an important role in controlling lead release
  - Based on the results of surface characterization, hydrocerussite ($Pb_3(OH)_2(CO_3)_2$) and ($PbCO_3$) models were developed to predict lead concentrations in water on

the basis of the assumption that corrosion scales are in equilibrium with dissolved lead ions. The models showed that lead levels in water are primarily determined by pH and alkalinity. The model also predicts that complexes with sulfate and chloride also contribute to total lead concentration, but to a much lesser degree.

- Thermodynamic Modeling
  - The thermodynamically developed equilibrium model indicated that alkalinity and pH are the major factors affecting dissolved lead release. The theoretical model provides useful qualitative information, but is limited quantitatively due to the impact of particulate lead. However the change in total lead release with pH and alkalinity is consistent with the trend predicted by either the cerrusite ($PbCO_3$) or hydrocerussite ($Pb_3(CO_3)_2(OH)_2$) solid phase equilibrium model.

Water Quality Effect on Lead Release and Empirical Models

- Water Quality Effects on Lead Release
  - Total lead release increased as temperature, chloride, and alkalinity increased.
  - Total lead release decreased as sulfate and pH increased. The effect of sulfate as predicted by the empirical model was contrary to that predicted by the equilibrium model, as that model only considered soluble species and this model considers both particle and soluble lead release.
  - Calcium and dissolved oxygen did not affect total lead release.
  - The actual effects on total lead release and the predicted effects of soluble lead release using an equilibrium model based on cerrusite as the controlling followed the same trend by water quality variable. That is, increasing temperature, chloride and alkalinity corresponded to increasing total lead release in the field and predicted soluble lead release in the equilibrium model; whereas, increasing sulfate and pH corresponded to decreasing total lead release in the field and predicted soluble lead release in the equilibrium model
- Empirical Lead Model
  - An empirical model was developed using ninety percent of fifteen months of operating data for the hybrid and single material PDS. An additional 10 percent of operation data was used for verification and was not included in model development/ The verification data included manipulation of alkalinity, chloride, and sulfate concentrations.
  - The empirical model for total lead release over predicts total lead release as seen by the MGs; it is proposed that the difference in predicted lead release and actual observed lead release in the regional water systems may be due to the area of the lead coupons used in the pilot system to simulate the lead surfaces in the MG consumers home plumbing systems was too large. However the lead model can be calibrated by location and source by reducing the model predicted ninetieth percentile by the ratio of the location ninetieth percentile over the model predicted ninetieth percentile using the location water chemistry.
  - Temperature, alkalinity, pH, chloride, and sulfate were identified by a statistical procedure as having a significant effect on release of lead. Confounding effects in

the experimental design complicated absolute resolution of the singular effect of each variable.

- The empirical model numerical prediction of lead release is specific to water chemistry. Both the conventionally treated ground and finished water RO sources presented a potential concern for release of corrosion products. This adverse effect was attributed to different water chemistry relative to the enhanced coagulated surface water source and its conventional treatment processes with ferric sulfate as a coagulant.
- Sensitivity analysis on empirical model indicated that alkalinity, chloride, and sulfate were important (sensitive) parameters, with chloride being the most sensitive.
- Higher temperature also increases lead release noticeably: Lead release at 35°C is almost double the lead release at 15°C.

Lead Rule Compliance and Simulation of Lead Release

- Application of the empirically developed model for total lead release indicated that violations of the lead action level will occur for some blends of finished waters produced from ground, surface or saline sources. However, violations of the lead action level are historically rare for MGs using ground water. The empirical lead model over predicts total lead release for MGs and needs to be calibrated for MG use by adjusting the predicted ninetieth percentile by a ratio of the actual total lead ninetieth percentile divided by the predicted total lead ninetieth percentile using the site or location water chemistry.
- Violation of the lead action level is not anticipated from the anticipated water quality from blending finished waters produced from surface, ground and saline sources. However, the lead model predicted that a conventionally treated ground water, enhanced coagulated surface water and RO finished water would violate the lead action level if chlorides in those finished waters exceeded 18 mg/L, 65 mg/L and 48 mg/L, respectively. Based on the model preditions, it would be important that the incorporation of any new water source containing appreciable levels of chloride be evaluated more thoroughly with respect to potential adverse effects on lead release.

TRANSITION

The effects of changing water quality due to blending of finished waters produced from surface, ground or saline sources on the transition of solid phase copper and solid phase lead and the transition of bulk phase total iron, total copper, total lead and biological stability as indicated by HPCs was investigated by analyzing lead and copper coupons and by analyzing total iron, total copper, total lead and HPC samples over time of exposure to a changed steady-state water quality. The transition of total iron, total copper, total lead and HPC was mathematically modeled.

Copper Surface Characterization

- There were not enough divalent corrosion products on copper coupons for XRD detection after one, two, three or six months of incubation in conventionally treated, ground water, enhanced coagulated surface water, finished RO water or blended finished water.
- Cuprite (Cu_2O) was consistently observed in both bulk and very top corrosion layer in all samples following one, two, three and six months of incubation; however, tenorite (CuO) and cupric hydroxide were only found in very top layer for the same periods. Based on XPS spectra analysis, solid phase equilibrium was attained after three months of exposure to an aqueous environment, consequently three months of exposure are adequate to achieve solid phase equilibrium for copper. Changing water quality did not affect the time to reach equilibrium

Lead Surface Characterization

- The main crystalline corrosion products formed on the lead coupons consisted of lead dioxide (PbO_2), lead oxide (PbO), cerussite ($PbCO_3$), and hydrocerussite ($Pb_3(CO_3)_2(OH)_2$) as indicated by all samples taken after one, two, three and six months of incubation in an aqueous environment. As indicated by XRD analysis, the thickness of the controlling solid phase continuously increased during incubation. XPS spectra analysis indicated the percentage of hydrocerussite reached equilibrium after three months; hence, a three period is adequate to achieve equilibrium for the controlling solid phase for lead..

Iron Surface Characterization

- The Fe_2O_3 peak area percentage did not vary significantly with incubation time (one, three, or six months), implying Fe_2O_3 had reached equilibrium in one month; hence, the implication that $FeCO_3$ had likewise reached equilibrium in the same period. Therefore, the iron release level will be stable in less than one month after the old cast iron pipe contacts the water. This conclusion supports the calculated time to reach steady-state as 26 days determined for the total iron release transition to bulk phase in unlined cast iron pipes.
- Though the Fe_2O_3 peak area percentage for a particular source had minor variations under various incubation times, significant variability exists between different sources. For instance the Fe_2O_3 peak area percentage for enhanced coagulated surface water was higher than RO, blend of conventionally treated ground water/enhanced coagulated surface water/RO and conventionally treated ground water in that order: this order corresponded to the release of total iron..

Copper Transient Release

- Model Development
  - ➢ Transition of total copper release from one steady-state water quality condition to another steady-state water quality condition was accurately described by a first-order model based on steady-state total copper release before and after a water quality change. The copper transition model allows evaluation of total copper release from typical household plumbing following a change in water quality.
  - ➢ As predicted by the first-order model and observed in actual field data, forty-two days are required to reach steady-state total copper release following a steady-state water quality change.
  - ➢ Consequently, if water quality is constantly changing a steady-state condition will not be reached; however, total copper release will always progress towards steady-state. If the actual copper is greater than that predicted at steady-state based on current water quality, actual copper will decrease to achieve equilibrium. If actual copper is less than that predicted at steady-state based on current water quality, actual copper will increase to achieve equilibrium.
  - ➢ The transition time provides utilities or water supplies time to mitigate undesirable release of total copper by changing blend ratios, chemical addition or other methods that would reduce the release of total copper as predicted by the steady-state water quality model.
  - ➢ The time required before a violation occurs should be determined using the total copper release transition model to determine the time for violation of the copper action level. A 42 day period for steady-state to occur does not imply that the LCR would be violated only after 42 days. For instance the copper action level changing would be violated in ten days when the finished water is changed from a 60/30/10 blend of conventionally treated ground water, enhanced coagulated surface water and RO finished water to a conventionally treated ground water, but would be violated in forty days when changing from RO to conventionally treated ground water. The time for violation difference is due primarily to alkalinity differences, but the transition model is water quality specific and must be used to estimate the exact time of violation as that time varies by the existing copper level and water quality.
  - ➢ Member governments that anticipate a change in water quality that impacts copper release should be aware that though the anticipated impact may take 42 days to stabilize, any remedial action taken by changing water quality will require equal time to reach steady-state. Hence, use of the transition model is required to predict total copper release over time.

Lead Transient Release

- Model Development
  - ➢ Transition of total lead release from one steady-state water quality condition to another steady-state water quality condition was accurately described by a first-order model based on steady-state total lead release before and after a water quality change. The lead transition model allows evaluation of total lead release

from typical household plumbing following a change in water quality. The lead transition model lead will increase with increasing chlorides, alkalinity and temperature, and decrease with increasing sulfates and pH. Total lead release is very sensitive to chloride concentration based on the model. It would be prudent to thoroughly consider chloride concentrations in new water sources as they may impact lead release.

- The empirical water quality model and hence the transition model for total lead release over predicts total lead release as seen in the regional utilities. The transitional model for total lead release can be calibrated exactly like the empirical water quality model for total lead release. This calibration can be general or by location and source, and consist of reducing the model predicted ninetieth percentile by the ratio of the location ninetieth percentile over the model predicted ninetieth percentile using the location water chemistry.
- As predicted by the first-order model and observed in actual field data, thirty-seven days are required to reach steady-state total lead release following a steady-state water quality change.
  - Consequently, if water quality is constantly changing a steady-state condition will not be reached; however, total lead release will always progress towards steady-state. If the actual lead is greater than that predicted at steady-state based on current water quality, actual lead will decrease to achieve equilibrium. If actual lead is less than that predicted at steady-state based on current water quality, actual lead will increase to achieve equilibrium.
  - The transition time provides utilities or water supplies time to mitigate undesirable release of total lead by changing blend ratios, chemical addition or other methods that would reduce the release of total lead as predicted by the steady-state water quality model.
  - The time required before a violation occurs should be determined using the total lead release transition model to determine the time for violation of the lead action level. A 37 day period for steady-state to occur does not imply that the LCR would be violated only after 37 days. For instance the lead action level changing would be violated in ten days when the finished water is changed from a 60/30/10 blend of conventionally treated ground water/enhanced coagulated surface water/RO to a conventionally treated ground water, but would be violated in forty days when changing from RO to conventionally treated ground water. The time for violation difference is due primarily to alkalinity differences, but the transition model is water quality specific and must be used to estimate the exact time of violation as that time varies by the existing lead level and water quality.
  - Member governments that anticipate a change in water quality that impacts lead release should be aware that though the anticipated impact may take 37 days to stabilize, any remedial action taken by changing water quality will require equal time to reach steady-state. Hence, use of the transition model is required to predict total lead release over time.

Iron Transient Release

- A transient iron release model was developed, which was consistent with the observed exponential relationship of observed release following phase changes in both the galvanized and unlined cast iron pipes.
- The rate constant at 20°C ($k_{20^{\circ}C}$) for transient total iron release model was found to be 0.116 day^{-1} and θ was found to be 1.02 for the unlined cast iron pipe. This indicated that at least 26 days are needed for the total iron response in UCI pipes to reach a steady-state following a perturbation in water quality.
- The rate constant at 20°C ($k_{20^{\circ}C}$) for transient total iron release model was found to be 0.068 day^{-1} and θ was found to be 1.07 for the galvanized iron pipe. This indicated that at least 44 days are needed for the total iron response in galvanized pipes to reach a steady-state following a perturbation in water quality.
- A disturbance in water quality with one day duration would cause only a slight change in iron concentration. For the unlined cast iron pipe, with a rate constant of 0.116 day^{-1}, the one-day duration would result in a fractional response of 11% of the steady-state change in iron concentration. For example, if the anticipated steady-state increase in iron concentration associated with a blend change is 1.5 mg/L, but the duration of the blend change is only one day, then the actual experienced increase in iron concentration would be 0.16 mg/L (=0.11 x 1.5 mg/L). For the galvanized pipe (rate constant = 0.068 day^{-1}), the corresponding increase in iron concentration after one day would be only 7% of the steady-state change. Thus these dynamics suggest that unless a change in water quality is sustained for many days, the associated impairment in iron concentration would be much smaller than predicted by the steady-state models discussed previously.
- The effect of water quality on the transient response was consistent with the steady-state predicted driving forces. That is higher iron release at elevated chlorides, sulfates, sodium, DO, temperature and HRT and lower iron release due to elevated alkalinity. Thus conventionally treated ground water and blends rich in conventionally treated ground water (> 60% conventionally treated ground water) would exhibit a comparatively lower iron release compared to finished water RO and surface water sources.
- Simulations on the effect of laminar or turbulent flow conditions indicate higher potential release at laminar conditions.
- The data from the high frequency PDS 14 indicate a relatively constant iron for the high frequency PDS, despite rapid changes in the input blend (blends were changed every 12 hours). The response of the high frequency PDS is centered about an expected mean for a blend of the three principal sources. The transient response of the high blend frequency line PDS 14 was consistent with the low rate constant (k) that would require at least 26 days for steady-state total iron release.
- The data base that was available for analysis of the dynamic response was limited to laminar flow conditions. Application of the conclusions to other conditions is cautioned at this time.

HPC Transition

- HPC populations increased immediately in all PDSs regardless of blend (water quality) using chloramine or chlorine residuals following a change in steady-state water quality. There were two possible causal factors for the observed HPC release, which were a physical disruption of biofilms caused by switching of PDS feed lines to change finished water chemistry or a chemical disruption of biofilm caused by the change in water quality. Both of these mechanisms need to be investigated in the future.
- Modeling HPC transition from the time of steady-state water quality change with a second-order model found that maximum HPCs were achieved approximately five days after the water quality change and that HPCs would return to the HPC concentrations prior to the water quality change in approximately ten days after the change. The time to reach maximum HPCs was related to a mixing model using alkalinity as a surrogate that confirmed alkalinity transition was complete in approximately eight days.
- HPC transition was also modeled as a function of residual, time and temperature by non-linear regression. The actual and predicted HPCs showed that HPC populations decreased as residual increased; decreased as temperature decreased and increased with time. The actual and predicted data for free chlorine and chloramine residuals demonstrated that exponential growth occurred when either residual approached zero. HPC populations associated with chloramines were significantly (1-2 logs) higher than HPC populations associated with free chlorine

CHLORINE RESIDUALS

Free chlorine were monitored during the four month equilibratiion period and during the three month comparison of free chlorine and chloramine. Chloramine residuals were monitored for eighteen months. All residual were monitored with respect to pipe material, geometry and water quality. This information was collected and analyzed to quantitatively describe residuals in drinking water distribution systems over time.

- Models
  - First-order models for free chlorine and chloramine dissipation were developed as a function of pipe material, pipe geometry, initial chlorine dose, water quality (temperature and UV254) and time.
  - The rate of decay of free chlorine and chloramine residual was dependent on pipe material, temperature and water quality. Lower rates of free chlorine and chloramine decay occurred in PVC and lined cast iron pipes while free chlorine and chloramine decayed more rapidly in unlined iron-based pipes (UCI and G).
  - Decline of free chlorine and chloramine residual increased in PDSs irrespective of material as time, UV-254 (organic content) and temperature increased.
  - The models for free chlorine and chloramine predicted disinfectant residuals in the pilot distribution system at the 95% confidence level using independent data

that was not used to develop the models. Hence the models for free chlorine and chloramine residuals were verified.

- Reaction rate constants
  - The overall rate of free chlorine and chloramine decay was expressed using two rate constants: one for the bulk liquid reactions, K_b, and the other for pipe wall reactions, K_W. The bulk rate constant was dependent on UV-254, temperature and time and the wall rate constant is dependent on temperature, pipe diameter and time. Both kinetic constants for free chlorine and chloramine residual were developed by non-linear regression.
  - For both free chlorine and chloramine, the bulk water reaction rates (K_b) for PVC and lined cast iron were lower than the bulk water reaction rates (K_b) for unlined cast iron and galvanized steel, which indicates that the unlined metal materials or particles released from the UCI and G pipe walls influenced the bulk decay rate and that the lined materials from the PVC and LCI pipe walls do not influence the bulk decay rate.
  - The wall reaction rates (K_W) obtained for free chlorine and chloramine residual were low and similar for PVC and lined cast iron pipes, which indicates that both materials are relatively inert and react in a similar way with chlorine residual. Unlined cast iron and galvanized steel pipes have K_W's that are between 7 and 10 times larger than PVC and lined cast iron pipes. This may be explained by the reactivity of free chlorine and chloramine with the ferrous material of the pipe or additional biological activity on the unlined metallic pipes.
- Water Quality
  - Water temperature affects the rate of chlorine dissipation in a distribution system. The rate of free chlorine and chloramine decay increased with an increase in water temperature and was modeled using an expression similar to the Arrhenius equation. The effect of temperature on chlorine dissipation can make it difficult to maintain a residual in the distribution system in summertime conditions.
  - The rate of free chlorine and chloramine decay was also dependent on the organic content of the water. Free chlorine or chloramine residual was maintained longer in waters with low organics than in waters with high concentration of organics. No relation between pH and the rate of chlorine decay was observed, which most likely was due to the small range of values utilized (7.9 to 8.3).
- Geometry and Hydraulic Conditions
  - The pipe diameter affects the rate at which free chlorine and chloramine dissipate in the distribution system because chlorine decay at the pipe wall is dependent on the wetted surface area to volume ratio. Free chlorine and chloramine would decay faster in a small-diameter pipe than in a large-diameter pipe due to the larger surface area to volume ratio. When the diameter of the pipe is increased, that ratio diminishes and the relative effect of the wall decay is reduced. For PVC and lined iron pipes, with little or no chlorine demand at the wall, an increase in pipe diameter has minimal effect on the overall rate of decay. However, in unlined cast iron and galvanized iron pipe, which have a large chlorine demand at the pipe wall, an increase in pipe diameter would have a significant impact on the time the residual could be maintained.

- Investigation into the impact of changing hydraulic conditions (Reynolds Number) on chloramine decay was constrained by the limited length of the pilot distribution system. Under laminar conditions, residual dissipation rate increased with increasing Reynolds Number. Under fully turbulent flow conditions mass transfer limitations are overcome as indicated by a constant overall dissipation rate at increasing Reynolds Numbers.

BIOSTABILITY

Biostability was assessed using bulk measurements of AOC, BDOC, HPCs through the PDSs and biofilm measurements of general cell mass (PEPA) and biofilm HPCs (BFHPC). The measurements were correlated to temperature, residual, DOC, pH, calcium, alkalinity, sulfates, chlorides, silica, DO and time (HRT) by process and distribution system material visually and statistically.

Bulk Liquid

- BOM, i.e. AOC did not directly correlate with heterotrophic proliferation. Previous studies showed correlations between HPC and AOC, however in this study AOC was typically not limiting even at the lowest levels (approximately 50 μg/L in reverse osmosis treated groundwater). The AOC provided by the different finished waters under study never approached the required concentrations for biostability without residual (10 μg/L, van der Kooij, 1992). However some AOC were below the threshold for biostability recommended by LeChevallier *et al.* (1987) in presence of a residual, i.e. 50-100 μg/L, and some were well above 250 μg/L. HPC below 10^4 cfu/mL were much less frequently observed when AOC exceeded 125μg/L. This observation was consistent with LeChevallier assertion that even when residual was initially provided, heterotrophic proliferation and residual consumption can occur in waters with AOC above 100 μg/L.
- No total coliforms were found in more than three hundred influent or effluent PDS samples taken during the course of this investigation. Biological instability was never observed with respect to coliforms, only with respect to HPC proliferation.
- Bulk liquid heterotrophic biostability was defined in this work as conditions where HPC populations of greater than three logs were observed. HPC populations of 10^3 was considered the beginning of the transition zone and greater than four log HPCs as the actual zone of biological instability.
  - Selection of three and four logs HPC as the beginning of the transition and biological instability zones was correlated to a chloramine residual of approximately 0.1 and 0.5 mg/L as Cl_2.
  - As the residuals associated with biological instability were below the regulatory minimum of 0.6 mg/L as Cl_2, biological stability was about residual and was maintained as long as residual regulatory requirements were met.
- An empirical model relating effluent Log HPC to pipe material, AOC and residual was developed. AOC was only significant at the 73 % CL. Residual impacted Log

HPC significantly more than AOC for the anticipated ranges of residual and AOC in drinking water. Predicted Log HPC increased significantly as residual approached zero.

- A second empirical model was developed with only statistically significant terms at the 95 % CL, which related Log HPC to pipe material and temperature. Log HPC increased as temperature increased and residual decreased. The impact of residual was more significant than the impact of temperature on Log HPC as residual approaches zero, but the impact of temperature is most significant at residuals greater than 1 mg/L on Log HPC.
- Given utilities can do nothing to control temperature in a distribution system, residual is the only practically significant parameter for controlling HPCs and maintaining biological stability.

Biofilms

- A non-linear model was developed for biofilm density (PEPA) as a function of pipe material, temperature, total residual, Log HPC and HRT. Temperature impacted PEPA significantly more than HRT, Log HPC_{eff} and residual, all of which had relatively little impact on PEPA. The model predicts biofilms will always be active for environments common to drinking water distribution systems, which was consistent with a hypothesis that biofilms act as reservoirs for microorganisms and are a constant source of microorganisms to the bulk water under normal conditions.
- An empirical model was developed for $|\Delta AOC|$ that was a function of AOC_{inf} and temperature. The model was similar to a first-order model in that the amount of AOC change was dependent on the amount of AOC present, and was typical of models that describe BOD consumption by microorganisms.
- A Biocide:Food ratio ((total residual - mg/L as Cl_2):(AOC - µg/L as C)) was developed and correlated to $|\Delta AOC|$. The graphical correlation showed that when the ratio was high, $|\Delta AOC|$ was very low: and when the ratio was low, $|\Delta AOC|$ was very high. This relationship and observation indicates that biological stability can be achieved by residual maintenance or AOC reduction.
- DO and total residual varied in the same manner such that they were confounding and could not be used together for development of regression models; however, monitoring dissolved oxygen in a drinking water distribution system can be used as an indicator of biological instability. DO monitoring is relatively inexpensive, can be automated and practical for field application. However residual, nitrite, and nitrate are more informative parameters and automated probes are possible for these parameters as well.

Biological Annular Reactors (BARs)

- BAR experiments were conducted that compared biostability of a blended nanofiltered water (G4) with a surface water treated by enhanced coagulation, ozonation and biological activated carbon filtration (S1). The $|\Delta AOC|$ and biofilm inventory in the BAR containing the nanofiltered blend (G4) was greater than the |ΔAOC| and biofilm inventory in the BAR containing the enhanced coagulated surface water (S1). There were no differences observed between BARs with respect to bulk liquid heterotrophic biostability (HPCs), hence there was no difference in biological stability of either source.

- BAR experiments were conducted that compared biostability of a conventionally treated groundwater (G1) with a surface water treated by enhanced coagulation, ozonation and biological activated carbon filtration (S1). No differences were observed between BARs with respect to bulk HPCs or biofilm inventories (BFHPC and PEPA); hence, there was no difference in S1 and G1 biostability.

Process Comparison

- Finished water biostability of conventionally treated ground water (G1), softened ground water (G2), enhanced coagulated, ozonated and BAC filtered surface water (S1), RO membrane filtered ground water (RO), enhanced coagulated-filtered and nanofiltered surface water (S2), softened blend of S1, G1 and RO (G3) and a nanofiltered blend of S1, G1 and RO (G4) was assessed by a visual and statistical comparison of average AOC, BDOC, Log HPC and biofilm (Log PEPA).
  - All bulk water biostability parameters (AOC, BDOC, and Log HPC) were lower in RO finished water and higher in conventionally treated ground water. Average finished water AOC indicated nanofiltration reduced AOC relative to coagulation or softening; however this relationship was not observed for BDOC or Log HPC.
  - Average process biofilm density (PEPA) of treated ground waters with the exception of RO was ordered as G1>G2>G3>G4. The average process biofilm density of treated surface waters was ordered as S1>S2. The order of biofilm densities of treated surface and ground waters excepting RO indicated that biostability increased as the degree of treatment increased with nanofiltered finished waters being the most biostable. Average process biofilm density (PEPA) was the highest in RO finished water, which was seemingly out of sequence.
  - Biofilm cell density (PEPA) varied by material, was greatest in the unlined metal galvanized steel and unlined cast iron coupons and least in the lined PVC and lined cast iron coupons. The order and average cell densities by material were PVC ($10^{8.37}$)< LCI ($10^{8.83}$)< UCI ($10^{9.01}$)< G ($10^{9.16}$).

NITRIFICATION

Nitrification was investigated during an uncontrolled nitrification episode, a controlled nitrification study and during model development. Nitrification will occur to some degree in every distribution system that uses chloramines for residual maintenance. However it is highly unlikely that either nitrites or nitrates will exceed the associated MCLs. However, nitrification is the result of microbiological activity in drinking water distribution systems and may indicate opportunities for contamination by other microorganisms. Hence, nitrification should be controlled and managed proactively in drinking water distribution systems, but it is unrealistic to expect no nitrites or nitrates in drinking water distribution systems using chloramines for residual maintenance, especially in subtropical climates as exist in the southern United States. However, the comparison of free chlorine and chloramines clearly showed that practically no nitrification occurred during the three month comparison study in summertime conditions, which are most prone to produce nitrification; and showed that nitrification could be successfully controlled using chloramines if a strong residual was maintained and dead zones or poor mixing conditions were avoided.

Episode

- Process
  - Biochemical nitrification is a complex microbiological process. The variation of nitrites, nitrates, pH, DO and ammonia-N varied in a systematic manner with residence time in the pilot distribution systems. Nitrification is further complicated by the oxidation of nitrites to nitrates by chloramines.
  - Although ammonia-N could come from any ammonia source, ammonia N to support biochemical nitrification came from free ammonia-N and chloramine-N associated with the production of chloramines.
  - Distribution system residence time is a primary control variable for biochemical nitrification. Reducing distribution system residence time to maintain a strong residual will reduce biochemical nitrification.
- Water Quality
  - Biochemical nitrification was associated with loss of residual, pH and initially increasing ammonia-N, which is most like to occur at temperatures common to spring and summer.
  - Dissolved oxygen can be a limiting factor for nitrifier growth if anaerobic conditions are developed. However utilities should never develop anaerobic conditions in the distribution system and should never try to control nitrification by DO limitation.
  - No significant alkalinity variation was observed in the pilot distribution systems during biochemical nitrification.
- Controlling Nitrification During the Uncontrolled Episode
  - Increasing the chloramine residual from 4 to 4.5 mg/L as Cl_2 was ineffective for reducing nitrification because the increase of 0.5 mg/L as Cl_2 chloramine residual was inadequate to maintain residual throughout the distribution system. In the absence of a residual, nitrification proliferated..
  - Increasing the Cl_2:NH_3 from 4:1 to 5:1 ratio was ineffective for reducing nitrification because increasing the Cl_2:NH_3 ratio changed the available NH_3-N from 1 to 0.8 mg/L once the residual was completely dissipated. The increase ratio did nothing to maintain chloramine residual throughout the distribution system. In the absence of a residual, nitrification proliferated.
  - Reducing the hydraulic detention time from 5 to 2 days was effective for reducing but not stopping nitrification. The reduced HRT restored the chloramine residual, which was responsible for reducing nitrification.
  - Implementing a free chlorine burn at 5 mg/L as Cl_2 for one week stopped nitrification.

Controlled Study

- Water Quality
  - Nitrification could be reliably monitored by monitoring NO_2-N. There was never any incident observed in this work where NO_2-N was not present when NO_x-N was measured. NO_X-N (the sum of NO_2-N and NO_3-N) concentrations were 137% of nitrite concentrations. Nitrate concentrations were about 37% of the nitrite concentrations.
- Operation
  - The results in this work clearly showed nothing is more important in operation than maintaining a strong residual throughout the distribution system.
  - Nitrification was reduced as Cl_2:NH_3 ratio was increased to 5:1 and possibly was reduced for Cl_2:NH_3 ratios slightly greater than 5:1 and less than 6:1. Cl_2:NH_3 ratios greater than 5:1 should be used with caution as taste and odor problems may develop at Cl_2:NH_3 ratios greater than 5:1.
  - Nitrification was reduced as influent ammonia was reduced. Influent ammonia can be controlled by the Cl_2:NH_3 ratio. It is noted that if all the nitrogen in the maximum allowable chloramine residual (4 mg/L as Cl_2) is converted to NO_2-N, the resulting NO_2-N concentration will be 0.8 mg/L and would not violate the 1 mg/L NO_2-N MCL. The maximum nitrate conversion from the maximum allowable chloramine dose is 0.8 NO_3-N, which is 8 % of the 10 mg/L NO_3-N MCL.
  - Flushing is essential to controlling nitrification in chloraminated drinking water distribution systems.
    - Nitrification was reduced by increasing flush frequencies from one to two weeks.
    - Nitrification was reduced by increasing the concentration of disinfectant residual in the water used to flush the system (flush strength). Flush strength was increased from 4 to 10 mg/L chloramines as Cl_2. Flush strength affected nitrification more than flush frequency for the conditions of this investigation.
- Material
  - The nitrification potential was related to pipe material. The order of material reactivity was G > UCI > LCI > PVC. The order for nitrification was UCI > LCI > G> PVC. PVC had the least nitrification. The nitrification in G PDS was limited by DO. Denitrification may occur in the UCI PDS that resulted in lower nitrite level.
  - Biofilms were correlated to pipe material in that unlined metal pipes (UCI and G) had higher PEPA, HPC and AOB cell densities than lined pipes (PVC and LCI), which means that biofilm densities increased as residual demand of the pipe wall or the biofilm increases and that biofilm densities varied with chloramine residual. Replacing distribution system pipes with high chlorine demand with pipe that have lower chlorine demand will reduce biofilms and the potential for nitrifications. This is an acceptable but costly method for improving distribution system water quality that is available to utilities.

- AOBs
  - Nitrite production was correlated to biofilm AOBs and biofilm HPCs were correlated to biofilm AOBs; hence, AOBs were directly related to biological nitrification.
    - General mass balances of AOB in the bulk and biofilm indicated that AOB mass was higher in the bulk than biofilm but that significant nitrification could occur in either. The biofilm was concluded to be a reservoir for microorganisms, which existed in a pseudo equilibrium condition with the bulk. Nitrifiers released from the biofilm to the bulk can only be kept in check by a strong chlorine residual in these waters. If the strong residual is lost, nitrification will evolve to a new steady-state and undesirable conditions will develop in the bulk, which include loss of DO, taste and odors and release of other microorganisms from the biofilm to the bulk.
    - Correlations of biofilm AOB to bulk residual, biofilm HPCs to bulk nitrites, PEPA to AOB biofilm, PEPA to AOB bulk, and PEPA to nitrites were weak. Typically biofilm correlations to bulk nitrification were weak and did not provide any evidence that regular biofilm monitoring would benefit utility maintenance of drinking water distribution systems. Nevertheless, biofilms are the reservoirs for nitrifying bacteria (and opportunistic pathogens) that are available to inoculate the bulk liquid given the right conditions (available DO and ammonia, low residual). Control of biofilms is essential and can only be achieved by maintaining a strong residual in these waters.

Modeling

- A statistical model was developed predicted NO_2-N and NO_X-N formation for a given pipe material, temperature, influent ammonia-N, effluent DO and chloramine residual. NO_2-N and NO_X-N increased with temperature, influent ammonia-N and decreasing effluent DO and residual. No other inorganic, organic or biological parameters were statistically significant to NO_2-N and NO_X-N formation.

Nitrification Management

- The results in this work clearly showed nothing is more important in operation than maintaining a strong residual throughout the distribution system
- Any utility using chloramines must aggressively manage distribution system water quality to control nitrification. High Cl_2:NH_3 ratios (5:1), residual maintenance, distribution system flushing and water quality monitoring are essential to nitrification control. Nitrification will be more difficult to control as water temperature increases in distribution systems using chloramines for residual maintenance.
- Guidelines for monitoring biological stability and nitrification in drinking water distribution systems were developed. The guidelines separated monitoring frequencies by location, temperature and parameters. The most intensely monitored conditions occurred during high temperature periods at locations involving unlined

metal pipes and high residence times in the distribution systems. The classification of monitoring points for biological assessment in distribution systems are specific to a given utility and should be determined after assessing water quality data from distribution system locations determined from existing knowledge and utilization of a computer model to classify residence times..

MONOCHLORAMINES AND FREE CHLORINE

Monochloramines and free chlorine were compared in a three month investigation using identical PDSs and finished waters. Inorganic, organic and microbiological water quality parameters were monitored in the PDSs influent and effluent over a twelve week period.

Rate of Release

- Residual data and dissipation models developed during this investigation showed that free chlorine dissipates more rapidly in the bulk liquid and less rapidly on the pipe wall than chloramine. The higher rate of dissipation in the bulk is due to the more rapid reaction of free chlorine with dissolved organic compounds as measured by UV-254 absorbance compared to chloramine reactivity with the same. However, the reactivity of free chlorine with the pipe wall is less than chloramine reactivity with the pipe wall for UCI and G pipes.
- The free chlorine and chloramine residual model incorporated temperature, UV-254, time and pipe material and clearly showed that free chlorine residual will be greater than chloramine residual for equal temperature and UV-254 in distribution system G pipes of 1' and less and UCI pipes of less than 6". There is not much difference in the wall reaction constants for PCV and LCI pipes; hence, chloramine residuals would exceed free chlorine residuals in all sizes of PVC and LCI pipes. Chloramine residuals would exceed free chlorine residuals for equal temperatures and UV-254 for large sized pipes regardless of pipe material.
- Chlorine demand was 0.2 mg/L as Cl_2 less in PDSs receiving RO, conventionally treated ground water and blended water that used free chlorine than in identical PDSs using the same sources and chloramine. Chlorine demand was 0.9 mg/L as Cl_2 less in a PDS receiving enhanced coagulated surface water that used free chlorine than in identical PDS using chloramine. Analysis of the linear rate of chlorine decay in the PDSs showed that the galvanized section of the PDS lines was responsible for the difference in residual demand between free chlorine and chloramines.

Water Quality

- Metals
  - The release of total iron, turbidity, apparent color, total copper and total lead were greater in the chloraminated PDSs than in chlorinated PDSs for identical source waters, PDSs, temperatures and residence times. Biological mechanism(s) may be associated with increased released of iron, copper, and lead, turbidity and apparent color in the presence of chloramine relative to free chlorine. Metal

release increased with temperature. Copper release was higher for chloraminated waters than for chlorinated waters for conventionally treated ground water, enhanced coagulated surface water and the blend. Lead release was higher for RO and the blend in the presence of chloramines relative to free chlorine.

Biostability and Nitrification

Nitrification was monitored in all PDS during the comparison of Cl_2 and NH_2Cl. Nitrification did not occur in the PDSs using free chlorine for residual maintenance. Trace nitrification did occur in the PDSs using chloramines for residual maintenance but it was practically insignificant. AOBs were present in the bulk water and biofilms in both residual environments. The data collected in this work showed that nitrification was controlled during the worst season (summertime) for nitrification because a strong chloramine residual was maintained and dead zones were avoided.

- Bulk
  - HPC growth in bulk and in the biofilm was greater in the presence of monochloramines relative to free chlorine. Effluent HPC populations were constant in PDSs carrying a chloramine residual. Effluent HPC populations initially increased from 10^4 to 10^6 cfu/ml, and then steadily decreased and stabilized to less than 10^2 cfu/ml in the presence of free chlorine during the last four weeks of the study. The initial HPC increase was attributable to sloughing from the biofilm.
  - AOB growth in the bulk liquid was reduced in enhanced coagulated surface water in the presence of free chlorine relative to the same waters in the presence of chloramine. AOB growth was not significantly different for either form of residual for conventionally treated ground water, blended and RO finished waters.
  - The presence of free ammonia from decomposition of monochloramines facilitated the occurrence of nitrification in the chloraminated waters. However, nitrification was limited as indicated by the observed concentration of nitrite and NO_X-N.
  - DO consumption was higher in the presence of chloramines relative to free chlorine, which was attributed to higher bacterial activity in the presence of chloramines relative to free chlorine.
- Biofilms
  - Biofilm densities (PEPA) were greater in the presence of chloramines than in the presence of free chlorine in conventionally treated ground water and blended water. The difference was less than one order of magnitude after the three month investigation.
  - There were no significant differences in biofilm HPCs between residual types for the RO and enhanced coagulated surface water sources.
  - No trend for biofilm AOBs was apparent by source water because the AOB growth was initiated by the use of chloramines which was common to all source waters. The presence of AOBs was significant in the lines with free chlorine even after three months of operation.

- In general, the biofilms on coupons receiving waters with chloramine had higher AOB populations than biofilms on coupons receiving waters with free chlorine.
- The order of AOB populations in biofilms with respect to materials was G>UCI>LCI>PVC, which is the same apparent order of decreasing surface roughness. PVC biofilm AOBs were at least one order of magnitude lower than G and UCI biofilm AOBs. PVC had lowest biofilm activity indicated by the lowest biofilm HPC and biofilm AOB inventories.

CHAPTER 6
RECOMMENDATIONS

The TBW drinking water distribution facility is unique in that it provides an environment where field scale evaluation of water produced from new and existing processes can be evaluated in actual distribution system pipes. The facility should be continuously used for research, training and education to improve drinking water supply to municipalities. Central to application of the findings in this work is recognition that the results presented herein can not be directly transferred to actual distribution systems without consideration of the pipe material, pipe geometry, pipe hydraulics and water chemistry of the finished water in the distribution system. The equivalent pipe lengths in this work for an actual distribution system with an average velocity of 0.33 ft/sec were 10.8 and 27 miles for a two and 5 day HRT. Such conditions are not commonly seen in actual distribution systems for reaches of unlined cast iron or galvanized pipe materials. However, the models developed from the results of this work allow prediction of water quality if they are properly utilized. Hence, determination of equivalent pipe lengths or direct application of the mass transfer models for iron release, turbidity release, apparent color release and residual dissipation, and consideration residual dissipation on, nitrification and biostability must be done to accurately predict distribution system water quality at any location.

IRON RELEASE

Surface Characterization

- Further studies looking at both iron and zinc in the galvanized iron pipes are required. The use of galvanized iron pipes in which the zinc protection has been exhausted may yield similar results to unlined iron pipes carrying similar blends. Experiments that measure electrochemical noise may significantly resolve issues regarding thermodynamic corrosion and iron release into the bulk as well as give concise comparisons between different pipe materials and water chemistries.

Thermodynamic Modeling

- Despite the limitations of thermodynamic models to predict total iron concentration, incorporating a driving force (concentration gradient) and hydraulic effects into thermodynamic models may enhance their prediction capabilities for dissolved iron concentrations. Increasing alkalinity mitigates iron release, while decreasing pH increases iron release. It is important to further investigate the impacts of other water chemistry parameters on iron release in order to understand the mechanisms of iron release and develop mitigation measures.

Recommendations for Conduct of Distribution System Testing

Based on the field experience, several suggestions are offered for the benefit of subsequent research involving pilot-scale distribution testing:

- For purposes of characterization of release of iron corrosion products, apparent color (or turbidity) is a suitable surrogate. Field measurement of apparent color is more attractive than iron determination.
- Excavation and transport of pipe materials results in a disturbance in the film characteristics. An equilibration period of several months after such disruption is required prior to valid data collection. Stable output apparent color would be evidence that experimentation could commence.
- Collection of all pipe samples for a particular material from the same location would be desirable to minimize pipe variability as an unintended variable in data interpretation.
- Alteration of water chemistry must consider confounding effects associated with salt addition.
- Transport of pipe material to a remote site for conduct of the PDS testing is feasible.
- Transport of raw and/or treated water to a remote site for conduct of PDS testing is feasible.

Additional Study

- The limited length of pipe constrained the pilot distribution system flux experiments. The nominal length of the pipe was only 26.3 m (86.3 ft) for the unlined cast iron pipe and 38.3 m (125.7 ft) for the galvanized iron pipe, making it difficult to distinguish between input and output iron concentrations at high Reynolds Numbers. Completion of complementary studies on a full-scale system with much greater pipe lengths would be valuable to confirm the relationship between flux (K_m) and Reynolds Number. In particular, data obtained from a system with a greater pipe length (and presumably, a greater measured increase in iron concentration) would be desired to estimate the slope term (β). The predicted steady-state impairment in water quality under fully turbulent flow conditions is proportional to β.
- The quantity of water that was available to conduct the flux experiments was limited due to water production and surface water transport constraints. Consequently, it was not possible to extend the duration of each flux experiment beyond 9.5 pipe volumes. Although steady-state flux appears to have been achieved in the current study, extended duration experiments would be desirable to verify steady-state flux. In conjunction with full-scale testing of large pipe reaches, an expanded duration for the flux experiments would facilitate definition of steady-state concentration changes during flux experiments.
- The pipe materials were restricted to a single diameter for PDS construction. Verification of the dependence of iron release on pipe diameter would require experimental data for different diameters.
- The model development relies in part on an empirical analysis of data for pipe materials specific to the Tampa Bay Water service area. Examination of data from other distribution systems would be desirable to expand the potential application of the developed models.

- The use of corrosion inhibitors was not included in the current study. Completion of complementary studies with various inhibitors would be necessary to develop prediction capability. It is noted that the water chemistry impact associated with inhibitors is unlikely to be expressed in the empirical model developed for the PDS database.
- Investigations of the transient response should be conducted under turbulent flow conditions with variable pipe diameter to confirm the time constants determined in this study under laminar flow.

COPPER RELEASE

The release of all metals should be investigated regarding impact on dissolved and particulate forms. Particle characterization and mechanisms of particle release should be investigated in order to minimize metal release in drinking water.

Model Verification and Calibration

- The model development relies in part on an empirical analysis of data for pipe materials specific to the Tampa Bay Way service area. Examination of data from other distribution systems would be desirable to expand the potential application of the developed models to other utilities and to the materials in their distribution systems.

Additional Study

- Extreme water quality should be examined to verify or modify empirical model that could expand the applicability.
- Fundamental investigation of particle release from copper surface would benefit model development and utility control of all released metals in drinking water and should be done.
- The use of corrosion inhibitors was not included in the current study. Completion of complementary studies with various inhibitors would be necessary to develop prediction capability. It is noted that the water chemistry impact associated with inhibitors is unlikely to be expressed in the empirical model developed for the PDS database

LEAD RELEASE

Model Verification and Calibration

- The model development relied in part on an empirical analysis of data for pipe materials specific to the Tampa Bay Way service area. Examination of data from other distribution systems would be desirable to expand the potential application of the developed models to other utilities and their distribution systems.

- In the lead release study, most of lead concentrations were below detection limits. To improve the model predictive precision, one might use a lead material pipe in addition to copper loop.

Additional Study

- Fundamental investigation of particle release from copper surface would benefit model development and utility control of all released metals in drinking water and should be done.
- Extreme water quality should be examined to verify or modify the empirical model in order to expand the applicability of the model.
- The use of corrosion inhibitors was not included in the current study. Completion of complementary studies with various inhibitors would be necessary to develop prediction capability. It is noted that the water chemistry impact associated with inhibitors is unlikely to be expressed in the empirical model developed for the PDS database

TRANSITION

- Additional studies should be done to determine the effects of varying pipe geometries and hydraulic conditions on transition of total iron, total lead, total copper and microorganisms. The scale and biofilms on the inner walls of distribution system piping are reservoirs of undesirable inorganic contaminants, organic contaminants and microorganisms. The impact of releasing these contaminants should be evaluated with respect to material, hydraulic, geometric and water quality effects as the potential of significant adverse consumer health effects is uncomfortably high.
- Verification studies in actual distribution systems on the transition models for the release of total iron, total lead, total copper and microorganisms are highly recommended. Such studies can verify the time for proactive actions by utilities to offset adverse impact of changing water quality. These studies should be conducted at the worst case environments in actual distribution systems and provide a means of selective proactive control of worst case impacts.

Copper Surfaces

- Longer incubation periods are suggested to investigate the aging of corrosion scales formed on copper surface to determine the predominant species controlling the metal release. These studies would allow a more accurate and practical theoretical consideration for use when predicting the release.

Lead Surfaces

- Longer incubation periods are suggested to investigate the aging of corrosion scales formed on lead surface to determine the predominant species controlling the metal release. This will allow for more accurate and practical theoretical consideration when predicting the release.

Transient Metal Release

- Investigations of the transient response should be conducted under turbulent flow conditions with variable pipe diameter to confirm the time constants determined in this study under laminar flow.
- The model development relies in part on an empirical analysis of data for pipe materials specific to the Tampa Bay Water service area. Examination of data from other distribution systems would be desirable to expand the potential application of the developed models.

HPC Transition

- Additional studies should be done to determine to determine the effects of water quality transition on microorganism with respect to hydraulic effects, pipe geometry, pipe material and temperature as the potential adverse health impact is significant.

RESIDUAL

Maintaining Distribution System Residuals

The following steps may be taken to ensure a good control of chlorine residual levels and increase the time chlorine residual is maintained in a distribution system:

- Unlined cast iron and galvanized steel pipes should be replaced as soon as possible because of the high rates of free chlorine and chloramine residual dissipation in these pipes and the associated complete loss of residual in the distribution systems.
- Monitoring of chlorine residuals in the distribution system should reflect increased monitoring based on location, season and pipe material. Dead end or high residence time locations, galvanized steel or unlined cast iron pipe use locations and high temperature seasons warrant increased monitoring and residual maintenance. This is especially important for waters using chloramine, which are prone to nitrification.
- The models for free chlorine and chloramine residual could be used to predict chlorine concentrations at different points within the distribution system and can determine the need for locations where rechlorination, flushing or other means of residual maintenance should be implemented.

Recommendations for Additional Study

- Calibration of the chlorine residual model in actual distribution systems for varying seasons, pipe materials, water quality conditions, hydraulic conditions and pipe geometries is recommended.

- The models for free chlorine and chloramine residual decay were developed under laminar regime conditions which might limit their use in full scale distribution systems. Furthermore, the limited length of pipe utilized in the development of the residual models constrained the velocity of flow that could be achieved while maintaining the required HRT. Free chlorine and chloramine dissipation studies in full-scale distributions systems are required to verify the models especially under turbulent flow regimes.
- The pipe materials were restricted to a single diameter used in the PDS construction. Verification of the effect of pipe diameter on free chlorine and chloramine dissipation rates would require a greater range of pipe diameters for each material.

BIOSTABILITY RECOMMENDATIONS

Monitoring Recommendations for Biostability in Distribution Systems.

- Regular monitoring of temperature, pH, DO, residual, HPCs, nitrites and nitrates are recommended for evaluation of distribution system biostability. The monitoring points should be related to distribution system residence time in order to develop water quality profiles that would allow assessment of biological stability within the distribution with reference to location, season and pipe material.
- As budget permits, unlined metal pipes should be removed from the distribution system and replaced with PVC, lined ductile iron or other pipes with relatively inert inner linings.
- HPC was determined to be a good indicator of biostability. Correlation of existing HPC data and additional HPC monitoring as necessary to residual and parameters that can be used to assess biostability within the distribution system is therefore highly recommended.
- Maintaining disinfectant residual within the distribution system is crucial and thorough monitoring of this parameter is recommended. Development of residual profiles throughout the distribution is recommended to identify locations of low residual by season. Average combined residual in the distribution system should equal or greater than 0.6 mg Cl_2/L at all times. A chloramine residual of 2.0 mg Cl_2/L or more is considered optimum for bacterial regrowth control.
- Monitoring of dissolved oxygen and development of DO profiles are recommended in the distribution system as DO consumption was shown to increase with heterotrophic proliferation and correlate with residual depletion. However, DO should not be considered sole indicator of biostability as its depletion can be caused by several other confounded phenomena. DO depletion in combination with other indicators such as residual loss can provide meaningful information to assess biological stability of the water within the distribution system.
- AOC may provide information on biostability and heterotrophic proliferation, especially when significant residual is not available. However, AOC did not indicate bio-instability as readily as either HPC or residual depletion. Therefore, monitoring of AOC in the finished water or within the distribution system is not recommended on

a regular basis. An AOC baseline can be established and additional AOC measurement may be beneficial when new sources, treatment processes or chemicals are introduced into the water supply system.

- Regular biofilm monitoring is not recommended for the distribution system, but as with AOC; an biofilm baseline can be established and additional biofilm measurement may be beneficial when new sources, treatment processes or chemicals are introduced into the water supply system. No correlation was established between biofilm inventories and heterotrophic proliferation. However, more work is required to evaluate the contribution of biofilm sloughing to bulk liquid HPC. This contribution has not been proven in this study.

Process and Distribution System Operation

- Shorter hydraulic residence times in distribution systems help maintain residual and therefore microbiological quality. Where possible, elimination of dead ends or reduction of residence times in problem locations in the distribution systems is recommended.
- Correlation of flushing schedules to distribution system problems areas as identified by water quality profiles and operator expertise is recommended for improvement of distribution system water quality.
- Free chlorine burns stopped nitrification in the pilot distribution systems and could be used in distribution systems for the control of nitrification and other microorganisms. Periodic free chlorine burns should not be used in lieu of elimination or reduction of dead zones and the maintenance of a strong residual in the distribution system. The frequency and locations for the burns are utility specific. Adequate monitoring must be done before and after the burns to allow adequate evaluation of the burns and determine frequency and location. The monitoring information before and after burns should be correlated to distribution system water quality profiles.

NITRIFICATION

- Nitrification should be controlled and managed proactively in drinking water distribution systems, but it is unrealistic to expect no nitrites or nitrates in drinking water distribution systems using chloramines for residual maintenance, especially in subtropical climates as exist in the southern United States. However, as shown in the comparison of chloramines and free chlorine, nitrification was practically nonexistent when chloramines were used for residual maintenance, which demonstrated nitrification was controlled in the worst (summertime) season because chloramine residuals were maintained and dead zones in the pilot distribution systems were avoided.

Monitoring

- Biochemical nitrification is a complex microbiological process, but the variation of nitrites, nitrates, pH, DO, residual, HPCs with temperature can be tracked through the distribution system with reference to time, distribution system material and locations to identify problem areas and locations for proactive nitrification control. Monitoring of these water quality parameters and development of water quality profiles in the distribution system by time, location and material in the distribution systems is recommended for nitrification control.
- Nitrification can be reliably monitored by monitoring NO_2-N. There was never any incident observed in this work where NO_2-N was not present when NO_x-N was measured. NO_X-N (the sum of NO_2-N and NO_3-N) concentrations were 137% of nitrite concentrations. Nitrate concentrations were about 37% of the nitrite concentrations. However, nitrite monitoring is not recommended to replace monitoring of the previously recommended water quality parameters for profile development
- Guidelines for monitoring biological stability and nitrification in drinking water distribution systems should be developed for specific systems and determine monitoring frequencies by location, temperature and water quality parameters beginning with the recommended parameters in this report. The most intense monitoring should occur during high temperature periods at locations involving unlined metal pipes and high residence times in the distribution systems.

Operation

- Maintaining the strongest permissible residual throughout the distribution is highly recommended for control of nitrification. A strong chloramine residual will not eliminate nitrification but it will control nitrification in non-stagnant conditions.
- A Cl_2:NH_3 ratio of 5:1 is recommended for production of chloramines and possibly to greater than 5:1 and less than 6:1. Cl_2:NH_3 ratios greater than 5:1 should be used with caution as taste and odor problems may develop at Cl_2:NH_3 ratios greater than 5:1.
- Ammonia-N should not be used indiscriminately to boost chloramine residual. Boosting of chloramine residual should be done after ammonia-N is determined such that a Cl_2:NH_3 ratio of 5:1 is always maintained. Complete mix conditions (highly turbulent flow) should be employed when chloramines are formed to avoid breakpoint chlorination due to inadequate mixing. Excess ammonia-N should always be minimized.
- Flushing is essential to controlling nitrification in chloraminated drinking water distribution systems and is recommended to be well thought out before implementation. Flushing frequencies and possibly strength of flushing solutions should be determined following a system review and assessment of distribution system water quality, and should vary by location and season. Extensive analysis of distribution system water quality and residence times is necessary to optimize a

flushing program, and should be done utilizing computer models for water age and water quality if possible. Such records should be compiled and improved over time to optimize distribution system water quality.

- Implementing a free chlorine burn periodically will stop nitrification; however nitrification will return if the same distribution system practices are maintained Maintenance of a strong chloramine residual in the distribution and reduction or elimination of dead zones can control nitrification.

Modification

- As budget permits, replacement of unlined metal pipes such as galvanized steel and unlined iron with lined pipes such as PVC and lined ductile iron in the drinking water distribution system is highly recommended.. Biofilm growth, residual dissipation and hydraulic capacity are adversely impacted with regard to water quality and water transmission by unlined metal pipes.
- Reduction of distribution system residence time in problem areas by rerouting as budget permits is recommended for reducing nitrification and all distribution system water quality problems.

Modeling

- Development and verification of fundamential models that describe nitrification in drinking water distribution systems should be developed and verified such that these models can be utilized by the drinking water community.

COMPARISON OF FREE CHLORINE AND CHLORAMINES

- Careful monitoring of residual concentrations during summertime should be implemented to avoid total loss of disinfectant residual in the distribution system. This is important especially for waters using chloramine that can suffer nitrification episodes.
- The release of metallic corrosion products from the pipe wall was larger for chloraminated waters than for the chlorinated waters; however chloramines can be successfully used for residual maintenance in the TBW MG distribution systems for meeting all drinking water regulations. Ammonia is known to form complexes with the metallic ions increasing the release of corrosion products into the water. The use of corrosion inhibitors to reduce corrosion rates and corrosion products release should be investigated to determine the effects of different inhibitors on all aspects of drinking water quality in distribution systems transmitting changing water quality.

Additional Study

- Completion of complementary studies on a full-scale distribution system with higher velocities and greater range of pipe diameters would be desirable to study further the effect of hydraulic conditions on free chlorine and chloramine dissipation in distribution systems. Studies in a full scale system would also test the applicability of the current model under fully turbulent conditions.

- Higher chlorine demand was predicted for unlined metallic pipe presumably due to iron release from the pipe wall. Completion of complementary studies using corrosion inhibitors would be desirable to determine their effect on chlorine dissipation.
- Based on the comparison of free chlorine versus chloramine, water utilities should consider employing practices to minimize consumption of chloramine at the pipe wall to maintain the chloramine residual longer in the distribution system.
- Control measures for corrosion products release should be implemented to help maintain the chlorine residual. Free chlorine and chloramines concentration decrease due to nitrification and biofilm activity at the wall. Further research on the impact of corrosion inhibitors addition to control corrosion products release, reduce biofilm accumulation, and hence maintain residual is suggested.

REFERENCES

Anon. 1994. Using Silicate to Lower Lead Levels in Drinking Water. *Public Works*, 125(10):83-84.

APHA, AWWA, and WEF (American Public Health Association, American Water Works Association, and Water Environment Federation). 1995. Standard Methods for the Examination of Water and Wastewater. 19th ed. Washington, D.C.: APHA.

Appenzeller, B.M.R., M. Batté, L. Mathieu, J.C. Block, V. Lahoussine, J. Cavard, and D. Gatel. 2001. Effect of Adding Phosphate in Drinking Water on Bacterial Growth in Slightly and Highly Corroded Pipes. *Water Res.*, 35(4):1100-1105.

Ascon-Cabrera, M.A., D. Thomas, and J.M. Lebeault. 1995. Activity of Synchronized Cells of a Steady-state Biofilm Recirculated Reactor During Xenobiotic Biodegradation. *Appl. and Envir. Microbio.*, 61(3):920-925.

AWWA (American Water Works Association). 1996. Water: STATS Survey, Denver, Colo.: AWWA.

AWWA (American Water Works Association). 1999. AWWA Report: $325 billion for Pipes. *AWWA Mainstream,* February 1999.

AwwaRF (Awwa Research Foundation). 1985. Internal Corrosion of Water Distribution Systems Cooperative Research Report, AwwaRF and DVGW-Forschungsstelle.

Bade, K., W. Manz, and U. Szewzyk. 2000. Behavior of Sulfate Reducing Bacteria Under Oligotrophic Conditions and Oxygen Stress in Particle-Free Systems Related to Drinking Water. *FEMS Microbio. Ecol.*, 32(3):215-223.

Beccaria, A.M., E.D. Mor, G. Bruno, and G. Poggi. 1985. Corrosion of Lead in Sea Water. *British Corr. Jour.*, 17(2):87-91.

Beckett, M.A., V.L. Snoeyink, K. Jim, P. Sarin, D. Lytle, W. Driven, and J. Clement. 1998. A Pipe Loop System for Evaluating Iron Uptake in Distribution Systems, In *Proceedings of the AWWA Water Quality Technology Conference.* San Diego, Calif.:AWWA.

Beevers, D.G., E. Erskine, M. Robertson, Beattie A.D., B.C. Campbell, A. Goldberg, M.R. Moore and V.M. Hawthorne. 1976. Blood Lead and Hypertension. *The Lancet,* 2(7975):1-3.

Birden H.H., E.J. Calabrese, and A. Stoddard. 1985. Lead Dissolution from Soldered Joints. *Jour. AWWA*, 77(11):66-70.

Biswas, P., C. Lu, and R.M. Clark. 1993. A Model for Chlorine Concentration Decay in Pipes. *Water Res.*, 27(12):1715-1724.

Block J.C., I. Sibille, D. Gatel, D.J. Reasoner, B. Lykins and R.M. Clark. 1997. Biodiversity in Drinking Water Distribution System. In *The microbiological Quality of Water.* 63-71. London: J.G. Jones and D. Sutcliffe (eds), Royal Soc. Public Hlth Hyg. Publishers.

Block J.C., L. Mathieu, P. Servais, D. Fontvieille, and P. Werner. 1992. Indigenous Bacterial Inocula for Measuring the Biodegradable Dissolved Organic Carbon (BDOC) in Waters. *Water Res.*, 26(4):481-486.

Block, J.C., K. Haudidier, J.L. Paquin, J. Miazga, and Y. Lévi. 1993. Biofilm Accumulation in Drinking Water Distribution Systems. *Biofouling*, 6(4):333-343.

Boffardi, B.P. 1988. Lead in Drinking Water – Causes and Cures. *Public Works*, 119(12):67-70.

Boffardi, B.P. 1990. Minimization of Lead Corrosion in Drinking Water. *Materials Performance*, 29(8):45-49.

Boffardi, B.P. 1992. Corrosion and Deposit Control in Mill Water Supply. In *Proceedings of the 1992 TAPPI Engineering Conference*. Houston, Texas: TAPPI.

Boffardi, B.P. 1995. Lead Corrosion. *Jour. NEWWA*, 109(2):121-131.

Brodtmann, N.V. and P.J. Russo. 1979. Use of Chloramine for Reduction of Trihalomethanes and Disinfection of Drinking Water. *Jour. AWWA*, 71(1):40-42.

Broo, A.E., B. Berghult and T. Hedberg. 1998. Copper Corrosion in Water Distribution Systems – The Influence of NOM on the Solubility of Copper Corrosion Products. *Corr. Sci.*, 40(9):1479-1489.

Bryers, J.D. 1987. Biologically Active Surfaces: Processes Governing the Formation and Persistence of Biofilms. *Biotech. Prog.*, 3(2):57-68.

Buchanan S.D., R. Diseker, J. Daniel, T. Floodman and T. Sinks. 1994. Evaluating Gastrointestinal Irritation among Humans from Copper in Drinking Water. Interim Report- Center of Disease Control and Prevention (CDC), Atlanta and Nebraskan Health Department. Lincoln Neb.: CDC.

Burlingame, G.A., and G.L. Brock. 1985. Water Quality Deterioration in Treated-water Storage Tanks. In *Proceedings of AWWA ACE.* Washington, D.C.: AWWA.

Butler G. and H. Ison. 1966. *Corrosion and its Prevention in Waters*. New York, N.Y.: Reinhold.

Byrd, J.J., H.S. Xu, and R.R. Colwell. 1991. Viable but Nonculturable Bacteria in Drinking Water. *App. and Envir. Microb.,* 57(3):875-878.

Campbell, B.C., A.D. Beattie, M.R. Moore, A. Goldberg and A.G. Reid. 1977. Renal Insufficiency Associated with Excessive Lead Exposure. *British Med. Jour.*, 1(6059):482-485.

Carter, T.J., E.W. Rice, S.G. Buchberger, and Y. Lee. 2000. Relationship Between Levels of Heterotrophic Bacteria and Water Quality Parameters in Drinking Water Distribution System. *Water Res.*, 34(5):1495-1502.

Chalyi, V.P. 1963. Mechanism of Aging of Individual Metal Hydroxides and Hydroxide Systems. Russ. *Jour. of Inorg. Chem.*, 8(2):269-273.

Characklis, W.G., and K.C. Marshall (eds). 1990. *Biofilms*. New York: John Wiley Publishers.

Characklis, W.G., and P.A. Wilderer (eds). 1989. *Structure and Function of Biofilms*. Berlin: Dahlem Workshop. New York: John Wiley Publishers.

Chen, C.A., A.S. McAnally, and S. Kumaraswamy. 1994. Lead and Copper Corrosion Control. *Envir. Sci. and Health.*, Part A. *Envir. Sci. and Eng.*, 29(8):1587-1606.

Chen, M., S. Daroub, L. Ma, W.G. Harris, and X. Cao. 2002. Characterization of Lead in Soils of a Rifle/Pistol Shooting Range in Central Florida. *Soil and Sediment Contamin.,* 11(1):1-17.

Chuttani, H.K., P.S. Gupta, S. Gulati and D.N. Gupta. 1965. Acute Copper Sulfate Poisoning. *Amer. Jour. of Med.*, 39(11):1185-1162.

Clement, J., M. Hayes, P. Sarin, J. Bebee, K. Jim, M. Beckett, W.M. Kriven, and V.L. Snoeyink. 2002. *Development of Red Water Control Strategies*: AWWARF.

Colling, J.H., P.A.E. Whincup, and C.R. Hayes. 1987. Measurement of Plumbosolvency Propensity to Guide the Control of Lead in Tapwaters. *Jour. Institution of Water and Envir.*, 1(3):263-269.

Colwell, R.R. and D.J. Grimes (eds). 2000. *Nonculturable Microorganisms in the Environment.* Washington: ASM Press.

Crissot-Laruade, M., N. Merlet, B. Legube. 1999. Formation de Biofilms Dans les Réseaux de Distribution d'eau à Faible Potentiel Nutritionnel: Exemple d'une eau Thermale. *Jour. Europ. Hydrol.*, 30:201-219.

Croué, J.P., G.V. Korshin and M. Benjamin (eds). 2000. *Characterization of Natural Organic Matter in Drinking Water*. AWWA Research Foundation and AWWA, USA.

Cullen, C.J. 2002. Equilibration of Pilot-Scale Distribution Systems. Master's Thesis. University of Central Florida, Orlando, FL.

Cunliffe, D.A. 1991. Bacterial Nitrification in Chloraminated Water Supplies. *App. and Envir. Microb.*, 57(11):3399–3402.

David, O., J. Clark and K. Voeller. 1972. Lead and Hyperactivity. *The Lancet*, 2(7783):900-903.

De Beer, D., R. Srinivasan, and P. Stewart. 1994. Direct measurement of Chlorine Penetration into Biofilms During Disinfection. *App. and Envir. Microb.*, 60(12):4339-4344.

Deb, A.K., R.A. Sabir, and W. Garlow. 1990. The Study of Lead in The Water Supply System of Washington, D. C. In *Proceedings of the 1990 Specialty Conference: National Conference on Envir. Eng*. Arlington, VA: ASCE.

DeBerard, H.I. 1917. Chloramine at Denver (Colo.) Solves Aftergrowth Problem. *Eng. News-Record*. 79:210.

Devender, K.J. 1995. Microbial Colonization of the Surface of Stainless Steel Coupons in a Deionized Water System. *Water Res.*, 29(8):1869-1876.

Doggett, M.S. 2000. Characterization of Fungal Biofilms within a Municipal Water Distribution System. *App. and Envir. and Microb.*, 66(3):1249-1251.

Duirk S.E., B. Gombert, J. Choi, and R.L. Valentine. 2002. Monochloramine Loss in the Presence of Humic Acid. *Jour. of Envir. Monitor.*, 4(1):85-90.

Dye, J.F. 1964. Review of Anticorrosion Water Treatment. *Jour. AWWA*, 56(4): 457-465.

Edwards, M., and N. Sprague. 2001. Organic Matter and Copper Corrosion By-product: a Mechanistic Study. *Corr. Sci.*, 43(1):1-18.

Edwards, M., M. Schock, and T. E. Meyer. 1996. Alkalinity, pH, and Copper Corrosion By-Product Release. *Jour. AWWA*, 88(3):81-94.

Eilbeck, W.J. 1984. Redox Control in Breakpoint Chlorination of Ammonia and Metal Ammine Complexes. *Water Res.*, 18(1):21-24.

Elzenga, C.H.J., A. Graveland and J.G.M. Smeenk. 1987. Corrosion by Mixing Water of Different Qualities. *Water Supp.*, 5(3/4): 9-12.

Elzenga, C.H.J., A. Graveland. 1980. Proposal of the European Communities for Directive Relating to The Quality of Water for Human Consumption, Actual Metal Levels in Drinking Water and Some Possibilities for Reduction by Central Water Conditioning by Waterworks. Subject 3. *13th Congres et Exposition Internationaux des Distributions d'Eau*. Paris, France: IWSA.

Escobar, I.C. and A.A. Randall. 2000. Sample Storage Impact on the Assimilable Organic Carbon (AOC) Bioassay. *Water Res.*, 34(5):1680-1686.

Evans, U.R. 1960. *The Corrosion and Oxidation of Metals: Scientific Principles and Practical Application*. London: E.Arnold.

Facey, R.M., and D.W. Smith. 1995. Soft, Low-temperature Water-distribution Corrosion: Yellowknife, NWT. *Jour. Cold Regions Eng.*, 9(1):23-40.

Feben, D. 1935. Nitrifying Bacteria in Water Supplies. *Jour. AWWA*, 27:439-445.

Feng, Y., W.K. Teo, K.S. Siow, K.L. Tan and A.K. Hsieh. 1996. The Corrosion Behavior of Copper in Neutral Tap Water. *Corr. Sci.*, 38(3):369-385.

Fransolet, G., A. Depelchin, G. Villers, R. Goossens, and W.J. Masschelein. 1988. The Role of Bicarbonate in Bacterial Growth in Oligotrophic Waters. *Jour. AWWA*, 80(11):57-61.

Frateur I., C. Deslouis, L. Kiéné, Y. Lévi, and B. Tribollet. 1999. Free Chlorine Consumption Induced by Cast Iron Corrosion in Drinking Water Distribution Systems. *Water Res.*, 33(8):1781-1790.

Frey, M.M. 1989. The AWWA Lead Information Survey: A Final Report. *Jour. AWWA*. 81(9):64-68.

Frost, A.A., and R.G. Pearson. 1961. *Kinetics and Mechanisms*. New York: John Wiley and Sons.

Geldreich, E.E. 1996. *Microbial Quality of Water Supply in Distribution Systems*. Boca Raton, Florida: CRC Press (Lewis Publishers).

Gilcreas, F.W. 1941. Chlorine-Ammonia Treatment of Water Committee Report. *Jour. AWWA*. 33(12):2079-2087.

Giljanovic, N.S. 1999. Water Quality Evaluation by Index in Dalmatia. *Water Res.*, 333(16):3423-3440.

Haas, C.N. 1999. Benefits of Using a Disinfectant Residual. *Jour. AWWA*. 91(1):65-69.

Haas, C.N. 2000. Disinfection. *In Water Quality & Treatment: A Handbook of Community Water Supplies*. 5th edition. McGraw-Hill, Inc.: New York.

Hallam, N.B., J.R. West, C.F. Forster, J.C. Powell, and I. Spencer. 2002. The Decay of Chlorine Associated with the Pipe Wall in Water Distribution Systems. *Water Res.*, 36(14):3479-3488.

Hamann, C.H., A. Hammett, and W. Vielstich. 1998. *Electrochemistry*. New York, N.Y.: Wiley-VCH.

Harrington, G.W., D.R. Noguera, A.I. Kandou and D.J. VanHoven. 2002. Pilot-scale evaluation of nitrification control strategies. *Jour. AWWA*, 94(11):78-89.

Hatch, G.B. and O. Rice. 1959. Influence of water composition on the Corrosion of steel. *Jour. AWWA*, 51(6):719-727.

Hedberg, T. and E. Johansson. 1987 Protection of Pipes against Corrosion. *Water Supp.*, 5(3–4): SS20/1-SS20/10.

Hem, L.J., E.A. Vik and A.B. Langen. 2001. Water Treatment to Reduce Internal Corrosion in the Drinking Water Distribution System in Oslo. *Water Sci. & Technol.: Water Supply*, 1(3):91-96.

Heuer, J.K., and J.F. Stubbins. 1999. An XPS Characterization of $FeCO_3$ Films from CO_2 Corrosion. *Corr. Sci.* 41(7):1231-1243.

Holt, J.G., N.R. Krieg, P.H.A. Sneath, J.T. Staley and S.T. Williams. 1994. *Bergey's Manual of Determinative Bacteriology*, 9th ed. Baltimore, Md: Williams & Wilkins.

Hong, P.K.A. and Y.Y. McCauley. 1998. Corrosion and Leaching of Copper Tubing Exposed to Chlorinated Drinking Water. *Water, Air and Soil Pollution*, 108(3-4):457-471.

Hua F., J. R. West, R. A. Barker, and C. F. Forster. 1999. Modeling of Chlorine Decay in Municipal Water Supplies. *Water Res.*, 33(12):2735-2746.

Ike, N.R., R.L. Wolfe, and E.G. Means. 1988. Nitrifying Bacteria in a Chloraminated Drinking Water System. *Water Sci. and Technol.*, 20(11):153-165.

Imran, S.A.V. Effect of Water Quality on Red Water Release in Iron Drinking Water Distribution Systems. PhD Dissertation. University of Central Florida. 2003.

Ives, D.J.G., and A.E. Rawson. 1962. Copper Corrosion. I. Thermodynamic Aspects. *Jour. of the Electrochem. Soc.*, 109:447-466.

Jeong, B.Y., and M.H. Kim. 2001. Corrosion Characteristics of Duplex Surface-treated Spheroidal Graphite Cast Iron. *Surface and Coating Technol.*, 141(2-3):262-268.

Johnson, B., R. Yorton, T. Tran, and J. Kim. 1993. Evaluation of Corrosion Control Alternatives to Meet the Lead and Copper Rule for Eastern Massachusetts. *Jour. NEWWA.*, 107(1):24-45.

Jones, D.A. 1996. *Principles and Prevention of Corrosion*. 2nd Ed, New York: Prentice Hall.

Karalekas, P.C., R.C. Ryan and F.B. Taylor. 1983. Control of Lead, Copper and Iron Pipe Corrosion in Boston. *Jour. AWWA*, 75(2):92-95.

Kashinkunti, R.D., D.H. Metz, J. DeMarco, and D.J. Hartman. 1999. How to Reduce Lead Corrosion without Increasing Iron Release in the Distribution System. In *Proc. of AWWA WQTC*, Tampa, Fla.: AWWA WQTC.

Kerr, C. J., K.S. Osborn, G.D. Robson and P.S. Handley. 1999. The Relationship Between Pipe Material and Biofilm Formation in a Laboratory Model System. *Jour. of Appl. Microbiol. Sym. Supp.*, 85:229S-238S.

Kielemoes, J., F. Hammes, and W. Verstraete. 2000. Measurement of Microbial Colonization of two Types of Stainless Steel. *Envir. Sci, and Technol.*, 2(7):831-843.

Kiéné L., W. Lu, and Y. Lévi. 1998. Relative Importance of the Phenomena Responsible for Chlorine Decay in Drinking Water Distribution Systems. *Water Sci. and Technol.*, 38(6):219-237.

Kuch, A. 1988. Investigations of The Reduction and Re-oxidation Kinetics of Iron (III) Oxide Scales Formed in Waters. *Corrosion Sci.*, 28(3):221-231.

Kuch, A., and I. Wagner. 1983. A Mass Transfer Model to Describe Lead Concentration in Drinking Water. *Water Res.*, 17(10):1303-1307.

Landrigan, P.J., R.H. Whitworth, R.W. Baloh, N.W. Staehling, W.F. Barthel and B.F. Rosenblum. 1975. Neuropsychological Dysfunction in Children with Chronic Level Lead Adsorption. *The Lancet*, 1(7909): 708-712.

Larson, T.E. and R.V. Skold. 1958. Laboratory Studies Relating Mineral Water Quality of Water to Corrosion of Steel and Cast Iron. *Corr.-NACE, 15:285t.*

Larson, T.E., and R.V. Skold. 1957. Corrosion and Tuberculation of Cast Iron. *Jour. AWWA*, 49(10):1294-1302.

Laurent, P. and P. Servais. 1995. Fixed Bacterial Biomass Estimated by Potential Exoproteolytic Activity. *Canad. Jour. of Microbiol.*, 41(8):749-752.

LeChevallier M.W., N.E. Shaw, L.A. Kaplan, and T.L. Bott. 1993. Development of a Rapid Assimilable Organic Carbon Method for Water. *Appl. Environ. Microbiol.*, 59(5):1526-1531.

LeChevallier M.W., N.J. Welch, and D.B. Smith. 1995. Full-scale Studies of Factors Related to Coliform Regrowth in Drinking Water. *Appl. Environ. Microbiol.*, 62(7): 2201-2211.

LeChevallier M.W., T. Babcock, and R.G. Lee. 1987. Examination and Characterization of Distribution System Biofilm. *Appl. Environ. Microbiol.*, 53(12):2714-2724.

LeChevallier M.W., W. Shulz, and R.G. Lee. 1991. Bacterial Nutrients in Drinking Water. *Appl. Environ. Microbiol.*, 57(3):857-862.

Lee, R.G., W.C. Berker and D.W. Collins. 1989. Lead at Tap: Sources and Control. *Jour. AWWA*, 81(7):52-62.

Lee, S.H.; J.T. O'Connor, and S.K. Banerji. 1980. Biologically Mediated Corrosion and its Effects on Water Quality in Distribution Systems. *Jour. AWWA.*, 72(11):636-645.

Légrand L. and P. LeRoy. 1995. *Prévention de la Corrosion et de l'Entarage dans les réseaux de Distribution d'Eau*. Neuilly-sur-Seine, France:CIFEC.

Légrand L. and P. LeRoy. 1990. *Prevention of Corrosion and Scaling in Water Supply Systems*. New York: Ellis Harwood.

Leonnig, C.D., and D. Cohn. 2004. Lead Found in D.C. Water may Mean EPA Tests are Flawed. *The Orlando Sentinel*, March 7, 2004. pp A24.

Levi, Y., G. Randon, C. Denojean, A. Plagiardini, G. Vajente, and H. Champsaur. 1992. Study of a Distribution Network with a Very Low Level of Dissolved Organic Matter: the Network of Nice. *Revue des Sciences de l'Eau*. 5:225-243.

Lieu, N.I., R.L. Wolfe, and E.G. Means. 1993. Optimizing Chloramine Disinfection for the Control of Nitrification. *Jour. AWWA*, 85(2):84-90.

Lin, J., M. Ellaway, and R. Adrien. 2001. Study of Corrosion Material Accumulated on The Inner Wall of Steel Water Pipe. *Corrosion Sci*., 43(11):2065-2081.

Lipponen, M.,T. T. , P. J. Martikainen, K. Regina, J. Kangasjärvi, J. Tuomainen, K. Servomaa, R. Vasara, and M. H. Suutari. 1998. Nitrifying Bacteria and Nitrification in Drinking Water Pipelines. *8th Symposium on Microbial Ecology (ISME-8)*, Halifax, Cd.

Lipponen, M.T.T., M.J. Suutari, and P.T. Martikainen. 2002. Ocurrence of Nitrifying Bacteria and Nitrification in Finnish Drinking Water Distribution Systems. *Water Res*., 36(17):4319-4329.

Lowe, J.P. 1993. *Quantum Chemistry, Second Edition*. New York: Academic Press.

Marangou, V.S., and K. Savvides. 2001. First Desalination Plant in Cyprus – Product Water Aggresivity and Corrosion Control. *Desalination*, 138(1-3):251-258.

Marani, D., G. Macchi, and M. Pagano. 1995. Lead Precipitation in the Presence of Sulfate and Carbonate. Testing of Thermodynamic Predictions. *Water Res*., 29(4):1085-1092.

Marco, J.F., M. Gracia, J.R. Gancedo, M.A. Martin-Luengo, and G. Joseph. 2000. Characterization of the Corrosion Products Formed on Carbon Steel after Exposure to The Open Atmosphere in the Antarctic and Easter Island. *Corr. Sci*., 42(4):753-771.

McCauley, R. F. 1960. Calcite coating protects water pipes. Water and Sewage Works. 107:276-81.

McGuire, M.J. 1997. Complaint Assessment for Tucson Water. Prepared for City of Tucson, Ariz.

McIntyre P. and A.D. Mercer. eds. 1993. Corrosion and Related Aspects of Materials for Potable Water Supplies. London, U.K., Maney Publishing.

McNeill, L. S., and M. Edwards. 2002. The Importance of Temperature in Assessing Iron Pipe Corrosion in Water Distribution Systems. *Envir. Monitor. and Assess*., 77(3):229-242.

McNeill, L.S., and M. Edwards. 2001. Iron Pipe Corrosion in Distribution Systems. *Jour. AWWA*, 93(7):88-100.

Means, E.G., T.S. Tanaka, D.J. Otsuka, and M.J. McGuire. 1986. Effects of Chlorine and Ammonia Application Points on Bactericidal Efficiency. *Jour. AWWA*, 78(1):62-69.

Melo, L.F., and M.J. Vieira. (1999). Physical Stability and Biological Activity of Biofilms under Turbulent Flow and Low Substrate Concentration. *Bioprocess Eng*., 20(4):363-368.

Merkel, T.H., H.J. Gross, W. Werner, T. Dahlke, S. Reicherter, G. Beuchle, and S. H. Eberle. 2002. Copper Corrosion By-product Release in Long-term Stagnation Experiments. *Water Res*., 36(6):1547-1555.

Merrill, D.T. and R.L. Sanks. 1978. Corrosion control by deposition of $CaCO_3$ films: A practical approach for plant operators. *Jour. AWWA*. 70(1):12-18.

Miller J.D.A. and R.A. King. 1975. Biodeterioration of metals. In *Microbial Aspects of the Deterioration of Materials*. R.J. Gilbert and D.W. Lovelock, Editors. London: Academic Press Inc.

Mitcham, R.P., M.W. Shelley, and C.M. Wheadon. 1983. Free Chorine versus Ammonia-chlorine: Disinfection, Trihalomethane Formation, and Zooplankton Removal. *Jour. AWWA*, 75(4):196-198.

Miyata, Y., and S. Asakura. 2002. Oxygen Reduction Reaction at Rust Free Iron Surface in Neutral Unbuffered Chloride Solutions. *Corr. Sci.*, 44(3):589-602.

Morin, P., V. Gauthier, S. Saby, and J. C. Block. 1999. Bacterial Resistance to Chlorine Through Attachment to Particles and Pipes Surfaces in Drinking Water Distribution Systems. 171-190. In *Biofilms in aquatic systems*: Royal Society of Chemistry, Cambridge, UK.

Murrell N. E. 1988. Impact of Lead and Other Metallic Solders on Water Quality. Draft USEPA Rept. Grant CR810958-01-1: USEPA.

Najm I.N., M. LeChevallier, C. Norton, C. Volk, A.A. Randall, I.C. Escobar, L. Kiene, and C. Campos. 2000. *Case Studies of the Impacts of Treatment Changes on Biostability in Full Scale Distribution Systems*. AWWARF and AWWA Report.

Neden, D.G., R.J. Jones, J.R. Smith, G.J. Gregory, and G.W. Foust. 1992. Comparing Chlorination and Chloramination for Controlling Bacterial Regrowth. *Jour. AWWA*, 84(7):80-88.

Negrin, J.M., W.M. Heyer, and S. Cheng. 1990. Evaluation of Nitrification Control Strategies at the Los Angeles Department of Water and Power. In *Proceedings of AWWA*. LA, Calif.: AWWA.

Niquette, P., P. Servais and R. Savoir. 2000. Impacts of Pipe Materials on Densities of Fixed Bacterial Biomass in a Drinking Water Distribution System. *Water Res.*, 34(6):1952-1956.

Norman, T.S., L.L. Harms, and R.W. Looyenga. 1980. The Use of Chloramines to Prevent Trihalomethane Formation. *Jour. AWWA*. 72(3):176-181.

Norton, C.D., and M.W. LeChevallier. 1997. Chloramination: Its Effect on Distribution System Water Quality. *Jour. AWWA.*, 89(7):66-77.

Obrecht, M.F. and L.L. Quill, 1961. How Temperature, Velocity of Potable Water affect Corrosion of Copper and Copper Alloy Piping. *Heating, piping and Air Conditioning*, 33(4): 129-134.

Odell L.H., G.J. Kirmeyer, A. Wilczak, J.G. Jancangelo, J.P. Marcinko, and R.L. Wolfe. 1996. Controlling Nitrification in Chloraminated Systems. *Jour. AWWA.*, 88(7):86-98.

Oh, S.J., D.C. Cook, and H.E. Townsend. 1999. Atmospheric Corrosion of Differ Steels in Marine, Rural and Industrial Environments. *Corr. Sci.*, 41(9):1687-1702.

Oliphant, R.J. 1983. Lead Contamination of Potable Water Arising from Soldered Joint. *Water Supply*, 1(2-3):5-12.

Pankow, J.F. 1991. *Aquatic Chemistry Concept*. Chelsea, Mich.: Lewis Publishers.

Paparazzo, E., and L. Moretto. 1995. X-ray Photoemission Spectroscopy and Scanning Auger Microscopy Studies of a Roman Lead Pipe 'Fistula'. *Surface and Interface Analysis*, 23(2):69-76.

Patterson, J.W., and J.E. O'Brien. 1979. Control of Lead Corrosion. *Jour. AWWA*, 71(5):264-271.

Patterson, J.W., E.B. Richard, and D. Marani. 1991. Alkaline Prediction and Aging of Copper from Dilute Cupric Nitrite. *Envir. Sci. and Technol.*, 25(10):1780-1787.

Pedersen, K. 1990. Biofilm Development on Stainless Steel and PVC Surfaces in Drinking Water. *Water Res.*, 24(2):239-243.

Percival, S.L., J.T. Walker, and P.R. Hunter (eds). 2000. *Microbiological Aspects of Biofilms and Drinking Water*. Washington, D.C. CRC Press.

Pisigan, R.A. and J.E. Singley. 1984. Evaluation of Waters Corrosivity using Langelier Index and Relation to Corrosion Rate Models. *Corrosion*, Paper No. 149.

Pisigan, R.A. and J.E. Singley. 1985a. Effects of Water Quality Parameters on the Corrosion of Galvanized Steel. *Jour. AWWA*, 77(11):76-82.

Pisigan, R.A. and J.E. Singley. 1985b. Technical Note: Experimental Determination of the Calcium Carbonate Saturation States of Water Systems. *Jour. AWWA*, 77(10):76-82.

Pisigan, R.A., and J.E. Singley. 1987. Influence of Buffer Capacity, Chlorine Residual, and Flow Rate on Corrosion of Mild Steel and Copper. *Jour. AWWA*, 79(2): 62-70.

Powell, J.C., N.B. Hallam, J.R. West, C.F. Forster, and J. Simms. 2000. Factors Which Control Bulk Chlorine Decay Rates. *Water Res.*, 34(1):117-126.

Price, S., and F.T. Jefferson. 1997. Corrosion Control Strategies for Changing Water Supplies in Tucson, Arizona. *Jour. NEWWA.*, 111 (3):285-293.

Race, J. 1918. Chlorination and Chloramine. *Jour. AWWA.*, 5(3):63-82.

Race, J. 1931. Discussion of Pre-Ammoniation of Filtered Water. *Jour. AWWA.*, 23(3):411-412.

Refait, Ph., J.B. Memet, C. Bon, R. Sabot, and J.M. R.. Genin. 2003. Formation of the Fe(II)-Fe(III) Hydroxysulfate Green Rust During the Marine Corrosion of Steel. *Corrosion Sci..* 45(4):833-845.

Refait, Ph., M. Abdelmoula, and J.M. R. Genin. 1998. Mechanisms of Formation and Structure of Green Rust one in Aqueous Corrosion of Iron in the Presence of Chloride Ions. *Corrosion Sci.,* 40(9):1547-1560.

Regan, J.M., G.W. Harrington, and D.R. Noguera. 2002. Ammonia- and Nitrite-Oxidizing Bacterial Communities in a Pilot-scale Chloraminated Drinking Water Distribution System. *App. and Envir. Microbiol.*, 68(1):73-81.

Reiber, S.H. 1989. Copper Plumbing Surfaces: An Electrochemical Study. *Jour. AWWA*, 81(7):114-122.

Risser, T.M. 1997. The Use of Potassium Carbonate for Corrosion Control in Small Systems. *Jour. NEWWA.*, 111(3):253-257.

Rittman, B.E. 1989. Detachment from Biofilms. In W.G. Characklis and P.A. Wilderer (eds), *Structure and Function of Biofilms*. New York : John Wiley and Sons.

Rittmann, B. E., and P. L. McCarty. 2001. *Environmental Biotechnology: Principals and Applications*. New York : McGraw-Hill.

Rogers, J., A.B. Dowsett, P.J. Dennis, J.V. Lee, and C.W. Keevil. 1994. Influence of Plumbing Materials on Biofilm Formation and Growth of Legionella Pneumophilla in Potable Water Systems. *Appl. Envir. Microbiol.*, 60(6):1842-1851.

Rossman L.A., R.M. Clark, and W.M. Grayman. 1994. Modeling Chlorine Residual in Drinking Water Distribution Systems. *Jour. of Envir. Eng.*, 120(4):803-820.

Rossman, L.A. 2000. *EPANET User Manual. Risk Reduction Engineering Laboratory.* Cincinnati, Ohio: USEPA.

Rossum, J.R. 1987. Dead Ends, Red Water and Scrap Piles. *Jour. AWWA*, 79(7):113-115.

Rossum, J.R., and D.T. Merrill. 1983. An Evaluation of the Calcium Carbonate Saturation Indexes. *Jour. AWWA.*, 75(2):95-100.

Roszak, D.B., and R.R. Colwell. 1987. Survival Strategies of Bacteria in the Natural Environment. *Microbiol. Rev.*, 51(3):365-379.

Ruiz, N., S. Seal, and D. Reinhart. 2000. Surface Chemical Reactivity in Selected Zero-valent Iron Samples Used in Groundwater Remediation. *Jour. of Hazard. Materials*, 80(1-3):107-117.

Rushing, J.C., L.S. McNeill, M. Edwards. 2002. Some Effects of Aqueous Silica on Corrosion of Iron, *Water Res.*, 37(5):1080-1090.

Saby, S. 1999. Résistance à l'hypochlorite de Sodium des Bactéries dans les Eaux Potables. Mémoire de thèse de doctorat de l'Université Henri Poincaré, Nancy 1. France.

Saby, S., P. Leroy, J. Freney, and J.C. Block. 2001. Les Mécanismes de Défense Bactérienne en Réponse à la Désinfection par le Chlore. *Hygiènes* (sous presse).

Sander, A., B. Berghult, A.E. Broo, E.L. Johansson, and T. Hedberg. 1996. Iron Corrosion in Drinking Water Distribution Systems - The Effect of pH, Calcium and Hydrogen Carbonate. *Corrosion Sci.*, 38(3): 443-455.

Sander, A., B. Berghult, E. Ahlberg, A.E. Broo, E.L. Johansson, and T. Hedberg. 1997. Iron Corrosion in Drinking Water Distribution Systems – Surface Complexation Aspects. *Corrosion Sci.*, 39(1):77-93.

Sarin, P., J. Bebee, M.A. Beckett, K.K. Jim, D.A. Lytle, J.A. Clement, W.M. Kriven, and V.L. Snoeyink. 2000. Mechanisms of Iron Release in Corroded Iron/Steel Pipes in Water Distribution Systems. *Proceedings 2000 AWWA Annual Conference*. Denver, Colo.: AWWA.

Sarin, P., V.L. Snoeyink, J. Bebee, W.M. Kriven, and J.A. Clement. 2001. Physico-chemical Characteristics of Corrosion Scales in Old Iron Pipes. *Water Res.*, 35(12):2961-2969.

Savoye, S., L. Legrand, G. Sagon, S. Lecomte, A. Chausse, R. Messina and Toulhoat. 2001. Experimental Investigation on Iron Corrosion Products Formed in Bicarbonate/carbonate Containing Solutions at 90 °C. *Corr. Sci.*, 43(11):2049–2064.

Sazou, D., M. Pagistas. 2002. Nitrate Ion Effect on the Passive Film Breakdown and Current Oscillations at Iron Surface Polarized in Chloride-Containing Sulfuric Acid Solutions. *Electrochemical Acta*, 47(10):1567–1578.

Schimpff, W.K. and H.E. Pearson. 1980. Water Quality Effects Related to Blending Waters in Distribution Systems, EPA-600/2-80-132. Cincinnati, Ohio: USEPA.

Schock, M. R. 1980. Response of Lead Solubility to Dissolved Carbonate in Drinking Water. *Jour. AWWA.*, 72(12):695-704.

Schock, M.R. 1989. Understanding Corrosion Control Strategies for Lead. *Jour. AWWA.* 81(7):88-100.

Schock, M.R. 1999. *Internal Corrosion and Deposition Control. Water Quality and Treatment. 5th Edition*. AWWA, McGraw Hill Inc. USA.

Schock, M.R., and M.C. Gardels. 1983. Plumbosolvency Reduction by High pH and Low Carbonate-Solubility Relationship. *Jour. AWWA.*, 75(2):87-91.

Schock, M.R., D.A. Lytle, and J.A. Clement. 1995. Effect of pH, DIC, Orthophosphate and Sulfate on Drinking Water Cuprosolvency. EPA Report: EPA/600/R-95/085, Cincinnati, Ohio: USEPA.

Scottish Executive: Environment and Rural Affairs Department. 2002. Drinking Water Quality in Scotland 2001. Available: <http://www.scotland.gov.uk/library5/environment/dwq01-00.asp>. [cited by May 20, 2003].

Seal, S., K. Sapre, A. Kale, V. Desai, M. Gopal and P. Jepson. 2000. Effect of Multiphase Flow on Corrosion of C-steel in Presence of Inhibitor: A Surface Morphological and Chemical Study. *Corrosion Sci.*, 42(9): 1623-1634.

Servais, P., P. Laurent, G. Billen, and D. Gatel. 1995. Development of a Model of BDOC and Bacterial Biomass Fluctuations in Distribution Systems. *Revue des Sciences de l'Eau*, 8(4):427-462.

Servais, P., P. Laurent, G. Billen, and Y. Lévi. 1992. Etude de la colonisation bactérienne des réseaux de distribution. *Techn. Sci. Méthodes L'Eau*, 6: 321-326.

Shams El Din, A.M. 1986. The Problem of "Red Waters": A New Approach to Its Solution. *Desalination*, 60(1):75-88.

Sheiham, I., and P.J. Jackson. 1981. The Scientific Basis for Control of Lead in Drinking Water. *Jour. Inst. Wat. Engr. and Scientists*, 35(6):491-515.

Shull, K.E. 1980. An Experimental Approach to Corrosion Control. *Jour. AWWA*, 72(5):280-285.

Sibille, I., L. Mathieu, J.L. Paquin, D. Gatel and J.C. Block. 1997. Microbial Characteristics of a Distribution System Fed with Nanofiltered Drinking Water. *Water Res.*, 31(9):2318-2326.

Simard, S., H. Menard, and L. Brossard. 1998. Localized Corrosion of 1024 Mild Steel in Slightly Alkaline Solution with Cl^- Ions. *Jour. of App. Electrochem.*, 28(2):151-160.

Singley, J. E., B. A. Beaudet, P. H. Markey, D. W. DeBerry, J. R. Kidwell, and D. A. Malish. 1985. Corrosion Prevention and Control in Water Treatment and Supply Systems. *Pollution Technology Review*. New Jersey: Noyes Publications. 122: 12-17.

Singley, J.E. 1981. The Search for a Corrosion Index. *Jour. AWWA*, 73(11):578-582.

Skadsen, J. 1993. Nitrification in a Distribution System. *Jour. AWWA.*, 85(7):95–103.

Smith, S.E., T. Ta, D.M. Holt, A. Delangoue, J.S. Colbourne, A.H.L Chamberlain, and B.J. Lloyd. 1999. A Pipeline Testing Facility for the Examination of Pipe-Wall Deposits and Red-Water Events in Drinking Water. *Jour. of the Chartered Institution of Water and Envir. Management*, 13(1):7–15.

Srinath, E.G., R.C. Loehr, and T.B.S. Prakasam. 1976. Nitrifying Organism Concentration and Activity. *Jour. of Envir. Eng.*, 102(2):449-463.

Stewart, P. S. and J. B. Raquepas. 1995. Implications of Reaction-diffusion Theory for the Disinfection Theory for the Disinfection of Microbial Biofilms by Reactive Antimicrobial Agents. *Chem. Eng. Sci.,* 50(19):3099-3104.

Stone, A., D. Spyridakis, M. Benjamin, J. Ferguson, and S. Osterbus. 1987. The Effect of Short-term Changes in Water Quality on Copper and Zinc Corrosion rate. *Jour. AWWA.*, 79(2):75-82.

Stoodley, P., D. de Beer, and Z. Lewandowski. 1994. Liquid Flow in Biofilm Systems. *Appl. Envir. Microbiol.*, 60(8):2711-2716.

Storz, G., and M. Zheng. 2000. *Oxidative stress*. In Storz G. and Hengge-Aronis R. (eds), Bacterial Stress Responses. Washington, D.C.: ASM Press.

Streicher, L. 1956. Effects of Water Quality on Various Metals. *Jour. AWWA*, 48(3):219-238.

Stumm, W. 1990. *Aquatic Chemical Kinetics: Reaction Rate if Processes in Natural Waters.* New York: John Wiely & Sons, Inc.

Sutherland, I.W. 1999. *Biofilm Exopolysaccharides in Microbial Extracellular Polymeric Substances*. Wingender, J., Neu, T.R., Flemming, H.-C. (Eds.): New York, New York: Springer.

Taylor, J.S., J.D. Dietz, S.K. Hong, L.A. Mulford, and A.A. Randall. 2001. Required Treatment and Water Quality Criteria for Distribution System Blending of Treated Surface, Ground and Saline Sources: Periodic Report 2. University of Central Florida, Orlando.

Townsend, H.E., T.C. Simpson, and G.L. Johnson. 1995. Structure of Rust on Weathering Steel in Rural and Industry Environments. *Corrosion*, 50(7):546-554.

Treweek, G.P., J. Glicker., B. Chow, and M. Sprinker. 1985. Pilot Plant Simulation of Corrosion in Domestic Pipe Materials. *Jour. AWWA*, 77(10):74-82.

Trussell, R.R. and J.F. Thomas. 1971. Discussion of the Chemical Character of Water Mixtures. *Jour. AWWA*, 63(1):49-51.

Tuovinen, O.H., K.S. Button, A. Vuorinen, L. Carison, D.M. Mair, and L.A. Yut. 1980. Bacterial, Chemical, and Mineralogical Characteristics of Tubercles in Distribution Pipelines. *Jour. AWWA.*, 72(11):626-635.

Tuthill, A.H., S. Lamb, G. Kobrin, and R.E. Avery. 1998. Effects of Chlorine on Common Materials in Fresh Water. Presented in *Corrosion* in San Diego, Calif.:NACE.

U.S. Environmental Protection Agency. 1990. Guidance Manual for Compliance with the Filtration and Disinfection Requirements for Public Water Systems Using Surface Water Sources. By Malcolm Pirnie Inc. and HDR Engineering Inc.: USEPA.

U.S. Environmental Protection Agency. 1998. National Primary Drinking Water Regulations: Disinfection and Disinfection By-Products: Final Rule. Federal Register. 63: 241:69389-69476.

Ugorets, M.Z., E.A. Buketov, and K.M. Akhmetov. 1968. Thermographic Study of the Dehydration of Copper Hydroxide in Alkaline Solutions. *Russ. Jour. of Inorg. Chem.*, 13(6):1525-1529.

United States Army Environmental Hygiene Agency. 2003. Lead in Potable Water Supplies. [online]. Available: <http://chppm-www.apgea.army.mil/dehe/pgm31/IPs/IP43.pdf>. [cited June 10, 2003].

Valentine, R.L. 1984. *Disappearance of Monochloramine in the Presence of Nitrite. Water Chlorination: Chemistry, Environmental Impact and Health Effects.* Lewis Publ.: Chelsea, Mich.

Van der Kooij, D. 1990. Assimilable Organic Carbon (AOC) in Drinking Water. In *Drinking Water Microbiology: Progress and Recent Development.* Edited by G.A. McFeters. New York: Springer-Verlag.

Van der Kooij, D., and H.R. Veenendaal. 1992. Assessment of the Biofilm Formation Characteristics of Drinking Water. In *AWWA WQTC Proceedings*, 1099-1110. Toronto.

Van der Kooij, D., H.R. Veenendaal, C. Baars-Lorist, D.W. Van der Klift and Y.C. Drost. 1995. Biofilm Formation on Surfaces of Glass and Teflon Exposed to Treated Water. *Water Res.*, 29(7):1655-1662.

Van der Wende, E., W.G. Characklis, and D.B. Smith. 1989. Biofilms and Bacterial Drinking Water Quality. *Water Res.*, 23(10):1313-1322.

Van Rijsbergen, D., P.G.G. Slaats, D. Van der Kooij and J.H.G. Vreeburg. 1998. Water Quality in Distribution System Operation, Storage, Cross-section. *Water Supp.*, 18(1):226-227.

Vasconcelos J.J., L.A. Rossman, W. M. Grayman, P. F. Boulos and R. M. Clark. 1997. Kinetics of Chlorine Decay. *Jour. AWWA.*, 89(7):54-65.

Vatankhah, G., M. Drogowska, and H. Menard. 1998. Electrodissolution of Iron in Sodium Sulfate and Sodium Bicarbonate Solutions at pH 8. *Jour. of App. Electrochem.*, 28(2):173-183.

Vayenas, D.V., S. Pavlou, and G. Lyberatos. 1997. Development of a Dynamic Model Describing Nitritification and Nitratification in Trickling Filters. *Water Research*, 31(5):1135-1147.

Verner S.S. 1990. *Handbook for Preparing Quality Assurance Project Plans for Environmental Measurements: A Guide for Implementing Quality Assurance in the Research Laboratory*. Rockville, Md.: Technical Resources Inc.

Vikesland, P.J., and R.L. Valentine. 2000. Reaction Pathways Involved in the Reduction of Monochloramine by Ferrous Iron. *Envir. Sci. and Technol.*, 34(1):83-90.

Vikesland, P.J., and R.L. Valentine. 2002a. Iron Oxide Surface-catalyzed Oxidation of Ferrous Iron by Monochloramine: Implications of Oxide Type and Carbonate on Reactivity. *Envir. Sci. and Technol.*, 36(3):512-519.

Vikesland, P.J., and R.L. Valentine. 2002b. Modeling the Kinetics of Ferrous Iron Oxidation by Monochloramine. *Envir. Sci. and Technol.*, 36(4):662-668.

Vikesland, P.J., K. Ozekin, and R.L. Valentine. 2001. Monochloramine Decay in Model and Distribution System Waters. *Water Res.*, 35(7): 1766-1776.

Vinci, A. 1991. Reducing Lead Contamination in Potable Water Using Sodium Bicarbonate and Sodium Hydroxide. In *Proceedings – AWWA Annual Conference*. Denver, Colo.: AWWA.

Volk, C., E. Dundore, E. Schiermann, and J. Lechevallier. 2000. Practical Evaluation of Iron Corrosion Control in a Drinking Water Distribution System. *Water Res.*, 34(6):1967-1974.

Walski T., D.V. Chase, and D. Savic. 2001. *Water Distribution Modeling*. 1st Ed. Haestad Press.

Watson, S.W., E. Bock, H. Harms, H.P. Koops, and A.B. Hooper, eds. 1989. Nitrifying Bacteria. *Bergey's Manual of Systematic Bacteriology*. Volume 3. Baltimore, Md.: Lippincott, Williams & Wilkins.

White, G.C. 1999. *Handbook of Chlorination and Alternative Disinfectants*. 4th edition. New York, N.Y.: John Wiley & Sons, Inc.

Wieckowski, A. 1999. *Interfacial Electrochemistry: Theory, Experiment, and Applications*. New York, N.Y.: Marcel Dekker.

Wilczak, A., J.G. Jacangelo, J.P. Marcinko, L.H. Odell, G.J. Kirmeyer, and R.L. Wolfe. 1996. Occurrence of Nitrification in Chloraminated Distribution Systems. *Jour. AWWA.*, 88(7):74-85.

Wolfe, R.L., E.G. Means, M.K. Davis, and S.E. Barrett. 1988. Biological Nitrification in Covered Reservoirs Containing Chloraminated water. *Jour. AWWA*, 80(9):109-114.

Wolfe, R.L., N.I. Lieu, G. Izaguirre, and E.G. Means. 1990. Ammonia-oxidizing Bacteria in a Chloraminated Distribution System: Seasonal Occurrence, Distribution, and Disinfection Resistance. *App. and Envir. Microbio.*, 56(2):451-462.

Wood, A.P., Aurikko, J.P., Kelly, D.P. 2004. A challenge for 21st Century Molecular Biology and Biochemistry: what are the Causes of Obligate Autotrophy and Methanotrophy?. *FEMS Microbiological Review.*, 28(3):335-352.

Wong, C.S. and P. Berrang. 1976. Contamination of Tap water by Lead pipe and Solder. *Environ. Contam. and Toxic.*, 15(5):530-534.

Wong-Chong, G.M., and R.C. Loehr. 1975. Kinetics of Microbial Nitrification. *Water Res.*, 9(12):1099-1106.

Young, V., and P.C. McCaslin. 1985. Angular Distribution X-ray Photoelectron Spectroscopy Studies on Compacted Lead Ion Selective Membrane Powders. *Analy. Chem.*, 57(4):880-886.

Zacheus, O.M., E.I. Ilvanainen, T.K. Nissinen, M.J. Lethola, and P.J. Martikainen. 2000. Bacterial Biofilm Formation on Polyvinyl Chloride, Polyethylene and Stainless Steel Exposed to Ozonated Water. *Water Res*, 34(1):63-70.

ABBREVIATIONS

AAS	Atomic Absorption Spectrophotometer
AOB	Ammonia Oxidizing Bacteria
AOC	Assimilable Organic Carbon
$\|\Delta AOC\|$	Absolute change of AOC between influent and effluent
BAR	Biological Annular Reactor
BDOC	Biological Dissolved Organic Carbon
BFHPC	Biofilm Heterotrophic Plate Count
^{o}C	Degrees Celsius
cfu	Coliform forming units
CL	Confidence Level
cpu	Cobalt-platinum color units
D	Diameter
D/DBP	Disinfection/Disinfection By-Products
deg C	Degrees Celsius
DO	Dissolved Oxygen
DPB	Disinfection By-Products
DPD	N,N-dietyl-p-phenylenediamine
ΔC	Change (delta) in Concentration
E.Coli	Escherichia coli
EPA	Environmental Protection Agency
FISH	Fluorescent In-Situ Hybridization
Flame AE	Flame Atomic Emission
G	Galvanized Steel
G1	Conventionally treated Groundwater
G2	Softened Groundwater
G3	Softened Blend
G4	Nanofiltered Blend
GFAAS	Graphite Furnace Atomic Absorption Spectrophotometer
GW	Groundwater
HPC	Heterotrophic Plate Counts
HRT	Hydraulic Residence Time
IC	Ion Chromatograph
IMS	Integrated membrane Sytem
K_B	Bulk Liquid Decay constat
K_F	mass transfer Coeffcient

Km	Flux (M/L2.T)
K_W	Wall Decay Constant
L	Length
LCI	Lined Cast Iron
mg/L	Milligrams per Liter
MPN	Most Probable Number
µg/L	Micrograms per Liter
NA	Not Applicable
NH_3-N	Ammonia-Nitrogen
NIST	National Institute of Science and Technology
NOM	Natural Organic Matter
NO_2-N	Nitrite-Nitrogen
NOx	Nitrate + Nitrite
NO_X-N	(Nitrite+Nitrate)-Nitrogen
NPDOC	Non-Particulate Dissolved Organic Carbon
NSO	FISH probe label
NTU	Nephelometric Turbidity Unit
PDS	Pilot Distribution System
PEPA	Potential of ExoProteolytic Activity
pH	Potential Hydrogen
ppb	Parts per billion
ppm	Parts per million
PVC	PolyVinyl Chloride
QA/QC	Quality Assurance/Quality Control
RO	Reverse Osmosis
RPD	Relative Percent Deviation
RSD	Relative Standard Deviation
S1	Ozonated treated Surface Water
S2	Nanofiltered surface water (IMS)
SM	Standard Methods
SW	Surface Water
TBW	Tampa Bay Water
TC	Total Coliform
TDS	Total Dissolved Solids
THM	TriHaloMethane
TKN	Total Kjeldahl Nitrogen
TOC	Total Organic Carbon

U	Velocity
UCF	University of Central Florida
UCI	Unlined Cast Iron
US	United States
USEPA	United States Environmental Protection Agency
UV	Ultra Violet
UV-254	Ultra-Violet absorption at 254nm (standard wavelength)
Vis spec	Visible Spectrum
WL	Warning Limit

IWA Publishing's authorised EU representative for General Product Safety Regulations is Diane D'Arras, 15 rue Duret, 75116 Paris, France, e-mail: safety@iwap.co.uk.

Printed and bound by CPI Group (UK) Ltd, Croydon, CR0 4YY
17/09/2026
02218810-0006